Enhanced Oil Recovery Series

CHEMICAL METHODS

Series Editor

ABDOLHOSSEIN HEMMATI-SARAPARDEH

University of Kerman, Kerman, Iran

CHEMICAL METHODS

Edited by

ABDOLHOSSEIN HEMMATI-SARAPARDEH

University of Kerman, Kerman, Iran

MAHIN SCHAFFIE

Shahid Bahonar University of Kerman, Kerman, Iran

MOHAMMAD RANJBAR

Shahid Bahonar University of Kerman, Kerman, Iran

MINGZHE DONG

University of Calgary, Calgary, Canada

ZHAOMIN LI

China University of Petroleum (East China), Beijing, China

ELSEVIER

Gulf Professional Publishing
An imprint of Elsevier

Gulf Professional Publishing is an imprint of Elsevier
50 Hampshire Street, 5th Floor, Cambridge, MA 02139, United States
The Boulevard, Langford Lane, Kidlington, Oxford, OX5 1GB, United Kingdom

Library of Congress Cataloging-in-Publication Data
A catalog record for this book is available from the Library of Congress

British Library Cataloguing-in-Publication Data
A catalogue record for this book is available from the British Library

ISBN: 978-0-12-821931-7

For information on all Gulf Professional publications
visit our website at https://www.elsevier.com/books-and-journals

Publisher: Charlotte Cockle
Senior Acquisitions Editor: Katie Hammon
Editorial Project Manager: Timothy Bennett
Production Project Manager: Kamesh Ramajogi
Cover Designer: Christian J. Bilbow

Typeset by STRAIVE, India

Contents

Contributors

Arastoo Abdi
Enhanced Oil Recovery (EOR) Research Centre, IOR/EOR Research Institute, Shiraz University, Shiraz, Iran

Yaser Ahmadi
Petroleum and Chemical Engineering Department, Ilam University, Ilam, Iran

Forough Ameli
School of Chemical Engineering, Petroleum and Gas, Iran University of Science and Technology, Tehran, Iran

Morteza Asemani
Department of Geology, Faculty of Earth Sciences, Kharazmi University, Tehran, Iran

Sefatallah Ashoorian
Institute of Petroleum Engineering, School of Chemical Engineering, College of Engineering, University of Tehran, Tehran, Iran

Marjan Ashrafizadeh
Department of Chemical Engineering, College of Engineering, University of Tehran, Tehran, Iran

Saman Bagherpour
Separation Processes & Nanotechnology Lab, Faculty of Caspian, College of Engineering, University of Tehran, Tehran, Iran

Zahra Bahmani
Enhanced Oil Recovery (EOR) Research Centre, IOR/EOR Research Institute, Shiraz University, Shiraz, Iran

Ali Mirzaalian Dastjerdi
Chemical and Petroleum Engineering Department, Sharif University of Technology, Tehran, Iran

Hassan Divandari
Department of Petroleum Engineering, Shahid Bahonar University of Kerman, Kerman, Iran

Morteza Jami
School of Petroleum and Chemical Engineering, Universiti Teknologi Brunei, Bandar Seri Begawan, Brunei Darussalam

Sabber Khandoozi
Enhanced Oil Recovery (EOR) Research Centre, IOR/EOR Research Institute; Petroleum Engineering Department, School of Chemical and Petroleum Engineering, Shiraz University, Shiraz, Iran

Hossein Mehrjoo
Department of Petroleum Engineering, Shahid Bahonar University of Kerman, Kerman, Iran

Saber Moghadam
School of Chemical Engineering, Petroleum and Gas, Iran University of Science and Technology, Tehran, Iran

Majid Mohammadi
Department of Energy Engineering, Qom University of Technology, Qom, Iran

Saeid Norouzi-Apourvari
Department of Petroleum Engineering, Shahid Bahonar University of Kerman, Kerman, Iran

Meysam Nourani
Ugelstad Laboratory, Department of Chemical Engineering, Norwegian University of Science and Technology (NTNU), Trondheim, Norway

Ali Rahimi
Department of Petroleum Engineering, Marvdasht Branch, Islamic Azad University, Marvdasht, Iran

Behnam Ranjbar
Enhanced Oil Recovery (EOR) Research Centre, IOR/EOR Research Institute, Shiraz University, Shiraz, Iran

Amin Rezaei
Department of Petroleum Engineering, School of Chemical and Petroleum Engineering, Shiraz University, Shiraz, Iran

Masoud Riazi
Enhanced Oil Recovery (EOR) Research Centre, IOR/EOR Research Institute; Petroleum Engineering Department, School of Chemical and Petroleum Engineering, Shiraz University, Shiraz, Iran

Mohsen Riazi
Department of Petroleum Engineering, Shahid Bahonar University of Kerman, Kerman; Enhanced Oil Recovery (EOR) Research Center, IOR/EOR Research Institute, Shiraz University, Shiraz, Iran

Saeid Sadeghnejad
Department of Petroleum Engineering, Faculty of Chemical Engineering, Tarbiat Modares University, Tehran, Iran

Mehdi Safari
Department of Petroleum Engineering, Curtin University, Miri, Malaysia

Mehdi Sedighi
Department of Chemical Engineering, University of Qom, Qom, Iran

Samiye Shahmarvand
School of Chemical Engineering, Petroleum and Gas, Iran University of Science and Technology, Tehran, Iran

Amin Sharifi
Enhanced Oil Recovery (EOR) Research Centre, IOR/EOR Research Institute; Petroleum Engineering Department, School of Chemical and Petroleum Engineering, Shiraz University, Shiraz, Iran

Preface

Oil production is generally divided into three stages, namely primary, secondary, and tertiary [enhanced oil recovery (EOR)]. After the primary and secondary stages, only part of the oil is produced and the remaining hydrocarbons should be produced by EOR. Generally, EOR can produce up to 75% oil. EOR techniques are also applicable to unconventional reservoirs such as tight gas, oil sands, coal bed methane, and shale oil reservoirs. EOR methods are generally classified into thermal recovery, gas injection, chemical injection, and new emerging technologies such as microbial, electrical, mechanical, and chemical leaching.

Chemical Methods, the first book in the Enhanced Oil Recovery Series, helps engineers to focus on the latest developments in one of the fast-growing areas of EOR. The book presents the latest research and practical applications specific to chemical EOR methods. Chapter 1 focuses on the fundamentals and applications of chemical EOR methods. In Chapters 2–5, the main chemical methods, namely polymer flooding, use of surfactants, alkaline flooding, and alkaline surfactant polymers, are described. Chapter 6 presents the technique of improved oil recovery by gel technology for water shutoff and conformance control. Chapters 7–9 describe new emerging chemical methods such as smart water flooding, natural surfactant flooding, and hybrid chemical methods. Chemical EOR methods in unconventional reservoirs are covered in Chapter 10. The last chapter, Chapter 11, explains formation damage during chemical flooding.

All the chapters have been peer reviewed by the editors and have been accepted after considering the editors' comments and suggestions. The book will be helpful for all scientists/researchers involved in petroleum engineering, EOR, and reservoir engineering. It will also help graduates, postgraduates, and PhD candidates to understand the fundamentals and applications of chemical methods.

Abdolhossein Hemmati-Sarapardeh
Editor of Enhanced Oil Recovery Series

Acknowledgments

I express my sincere thanks to all the editors and authors for their timely contributions. The successful completion of this book is the result of the cooperation of many individuals, and I express my gratitude to all of them, especially the editors of this book—Prof. Mahin Schaffie, Prof. Mohammad Ranjbar, Prof. Mingzhe Dong, and Zhaomin Li.

I also acknowledge the support provided by Timothy Bennett, Ali Afzal-Khan, Katie Hammon, and Kamesh Ramajogi from Elsevier in completing the publication process. I express my profound gratitude and thanks to Amin Rezaei from Shiraz University for designing the cover image for this book.

Abdolhossein Hemmati-Sarapardeh
Editor of Enhanced Oil Recovery Series

CHAPTER 1

Introduction to chemical enhanced oil recovery

Yaser Ahmadi[a], Majid Mohammadi[b], and Mehdi Sedighi[c]
[a]Petroleum and Chemical Engineering Department, Ilam University, Ilam, Iran
[b]Department of Energy Engineering, Qom University of Technology, Qom, Iran
[c]Department of Chemical Engineering, University of Qom, Qom, Iran

1.1 Introduction

With the rapid industrialization of economies, the necessity for natural resources has grown rapidly. The easy-to-develop oilfields have long been identified. Finding new economically viable fields, however, is becoming increasingly difficult [1]. Therefore, the essential goal of oil industries is to improve the recovery factor of mature oilfields by using EOR technologies [2, 3]. Fig. 1.1 shows that the yearly oil and gas demand is on the rise; therefore, petroleum-related countries should boost oil exploitation. Fig. 1.2 depicts three phases of oil recovery. Almost two-thirds of the original oil in place (OOIP) remains before the EOR phase [6]. There are several reasons for the remaining oil, including heterogeneities, capillary forces, and unfavorable mobility ratios [7]. Table 1.1 shows the substantial amounts of oil that have been recovered after two stages of recovery.

To improve overall displacement efficiency (macroscopic and microscopic) and recover remaining oil in a reservoir, two main EOR techniques need to be applied, which can be either thermal or nonthermal [16–19]. It has been established that nonthermal methods are more effective than thermal methods to recover trapped oil. Figs. 1.3 and 1.4 illustrate the classification of EOR methods with additional details.

In comparison to all nonthermal methods, chemical methods have been shown to have higher efficiency and lower cost. This is likely a consequence of lowering the IFT, altering wettability, and enhancing viscosity by the best possible method [20]. Fig. 1.5 shows common chemical EOR methods, along with a combination of those methods that will be discussed further. In the severe situations, such as elevated pressure and temperature and high salinity, the alkaline-surfactant-polymer combination has more efficacy [22].

The primary goal of chemical EOR is to act on one or more of the following aspects: mobility (application of water-polymer solutions to increase viscosity), wettability of the rock, and interfacial tension. Consequently, factors that affect recovery should be

Chemical Methods
https://doi.org/10.1016/B978-0-12-821931-7.00002-X

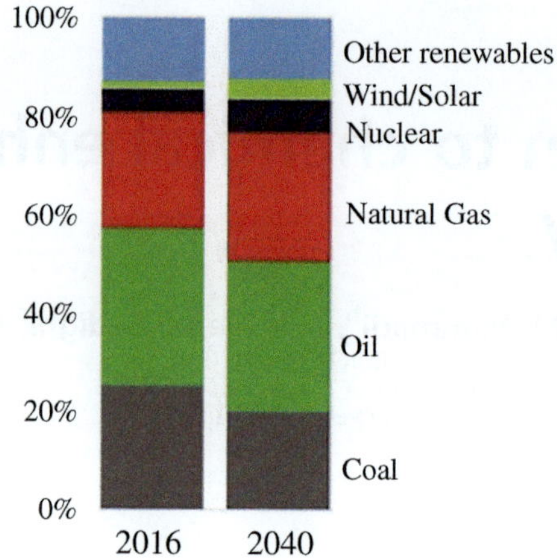

Fig. 1.1 World crude oil demand [4].

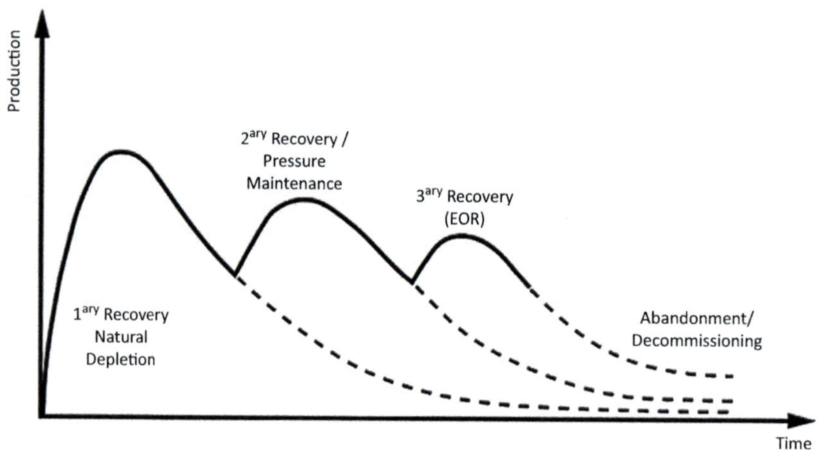

Fig. 1.2 Production framework for the different recovery stages [5].

Table 1.1 Recovery factors after primary and secondary stages.

Oil recovery factor (%)	Reference
30–40	Sengupta et al. [8]
40–50	Mujahidin et al. [9]
~30	Kumar and Mandal [10]
<50	Ahmadi et al. [11]
20–30	Maurya et al. [12]
30–50	Riazi and Golkari [13]
30–50	Kok [14]
~50	Ali et al. [15]

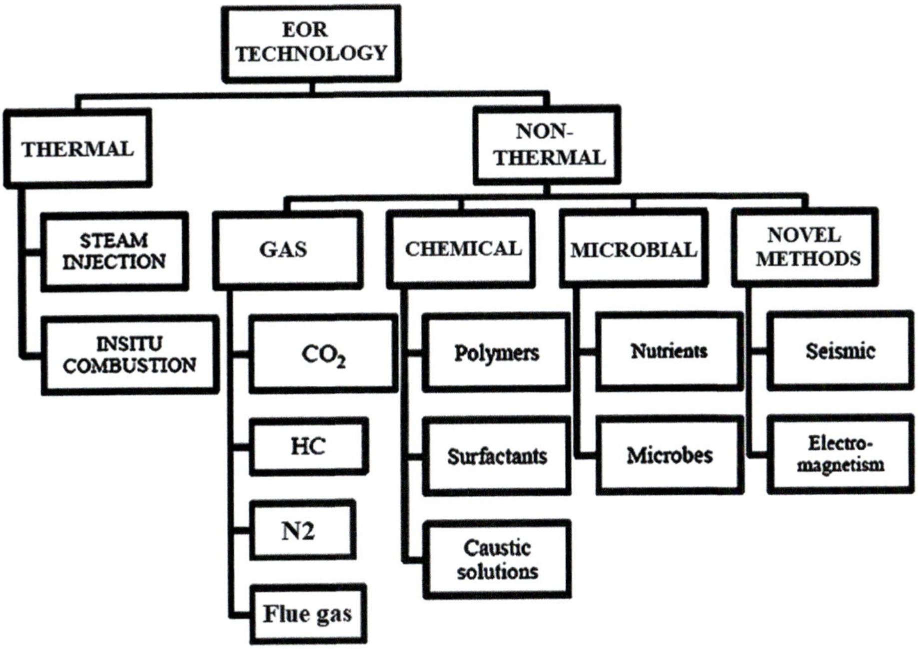

Fig. 1.3 Classification of EOR techniques [6].

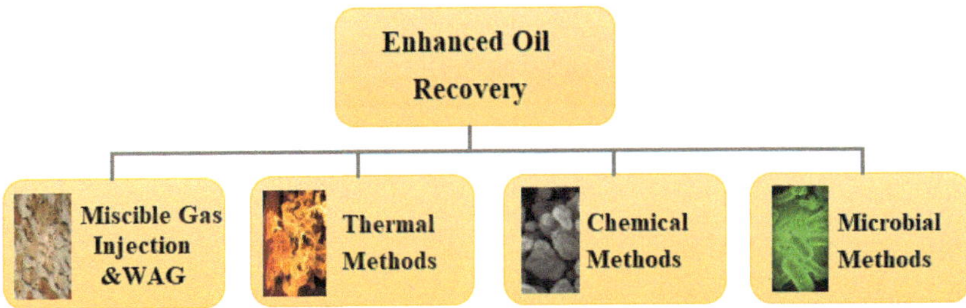

Fig. 1.4 EOR techniques.

thoroughly studied before developing novel chemicals for EOR [5]. To improve recovery, it is necessary to understand both the dynamic and the static aspect of the whole system at diverse scales (Fig. 1.6).

1.2 Chemical EOR methods

1.2.1 Polymer flooding

Polymer flooding improves the water flooding process by delaying water breakthrough time and viscous fingering problems. Moreover, in polymer flooding, the viscosity of the

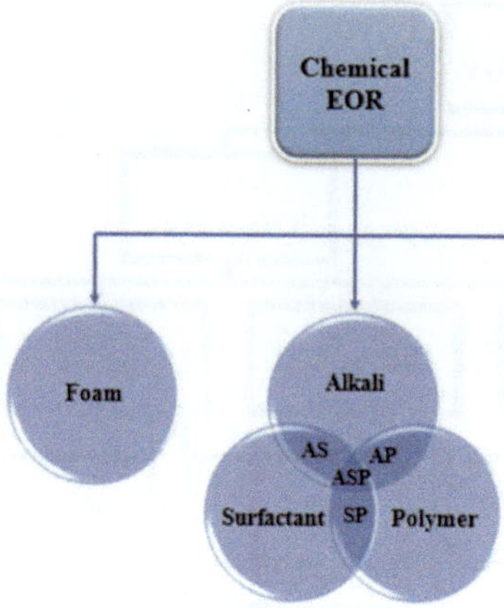

Fig. 1.5 Common chemical EOR methods [21].

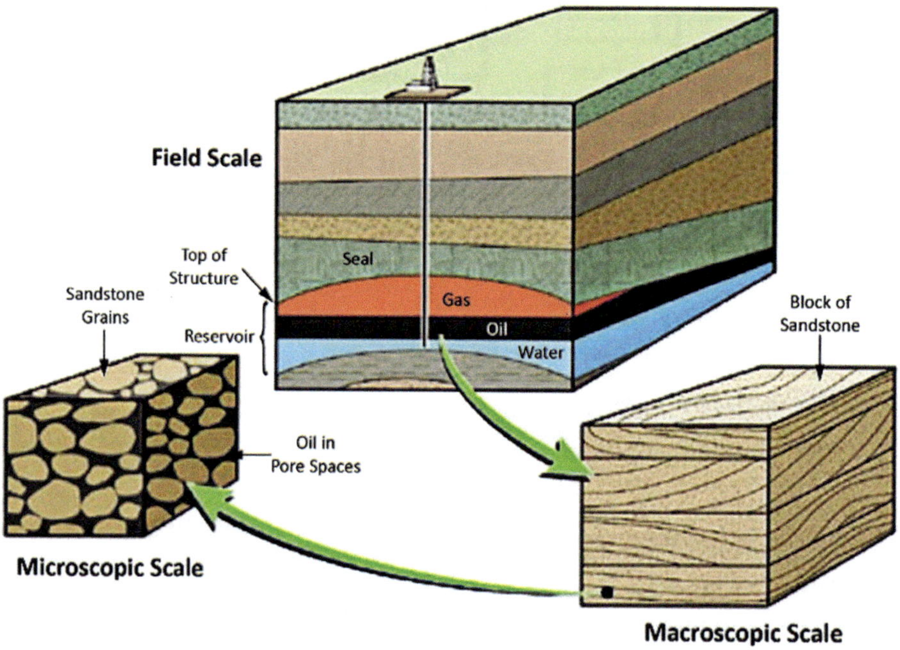

Fig. 1.6 Different scales involved in EOR [5].

water is enhanced with high-molecular-weight polymers [23]. Pye and Sandiford, in 1964, confirmed that small injections of water-soluble polymer could decrease water mobility. Hence, the latter parameter is reduced compared to oil [24, 25]. Fig. 1.7 shows how polymer flooding affects mobility ratio compared with water flooding. The present graph indicates that the mobility ratio is enhanced in the presence of polymer. The fact

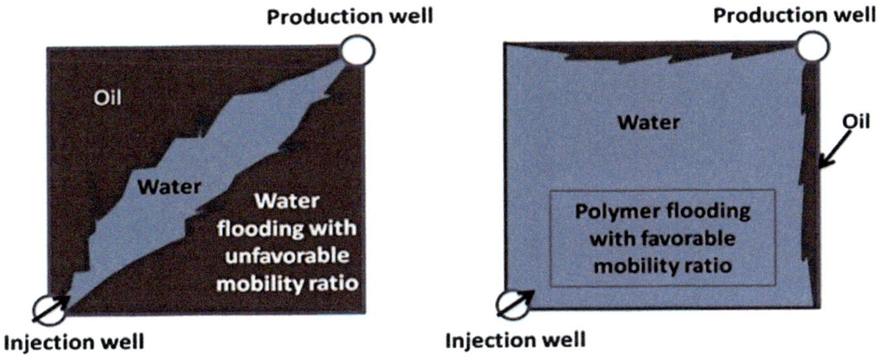

Fig. 1.7 Effects of polymer flooding on the mobility ratio [26].

that polymer flooding is a valuable method has been proven in Oman, Chinese, and Canada countries [27].

1.2.1.1 Mechanisms of polymer flooding

The main mechanisms for polymer flooding are as follows:

1. mobility control
2. reduction of permeability
3. polymers nature

In the next section, these three mechanisms are discussed in detail.

Mobility ratio

Polymer flooding enhances oil recovery through an overall process of disproportionate permeability reduction, mobility control, and the viscoelastic properties of the polymers. The Buckly-Leverett formula describes the mobility ratio in terms of

$$M = {}^{K_w \mu_0}\big/_{K_0 \mu_w} \tag{1.1}$$

where M: mobility ratio, μ_0: viscosity of oil (cP), μ_w: viscosity of water (cP), K_0: oil permeability (mD),and K_w: water permeability (mD).

Two different ranges are used:

1. $M > 1$: water mobility is greater than oil mobility, which is not suitable for us (Fig. 1.8A).
2. $M \leq 1$: water mobility is smaller than oil mobility, which is favorable in the displacement process (Fig. 1.8B).

Reduction of permeability

Adsorption on the pore surface or inside confined channels can occur for polymer molecules [7]. Polymer retention is the fundamental interaction between the porous medium

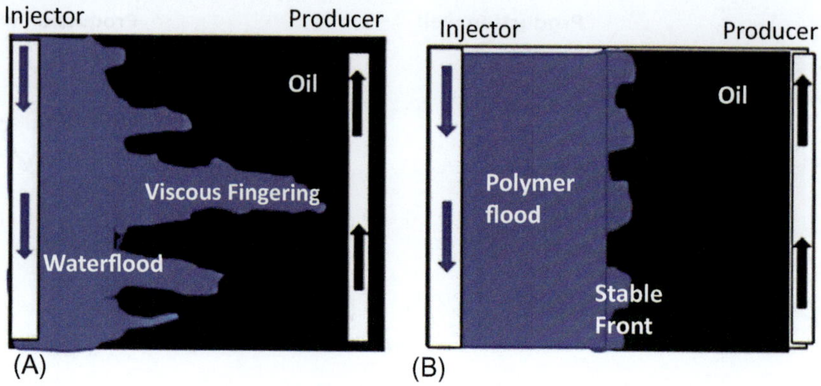

Fig. 1.8 Typical mobility ratio of (A) water flooding ($M > 1.0$) and (B) polymer flooding ($M \leq 1.0$) process [21].

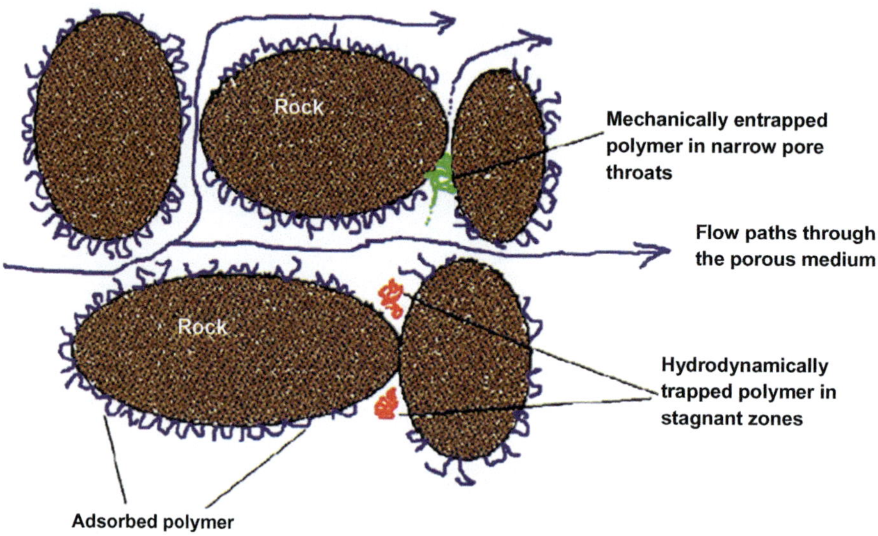

Fig. 1.9 Retention mechanisms in porous media [21].

and polymer molecules [28]. Hence, the permeability of the rock could be decreased through polymer flooding [29]. Fig. 1.9 indicates the interaction between rock grain and polymer molecules. There are two criteria in this regard: resistant factor (*RF*) and residual resistant factor (*RRF*). Eq. (1.2) shows that the *RF* is a term commonly applied to describe the resistance to flow through a polymer solution relative to the water flow:

$$RF = \frac{\left(\frac{K}{\mu}\right)\text{water flooding}}{\left(\frac{K}{\mu}\right)\text{polymer flooding}} = \frac{\Delta P\,\text{polymer}}{\Delta P\,\text{water}} \tag{1.2}$$

The *RRF* indicates how much the permeability is reduced in response to the polymer flooding into the rock due to the adsorption of the polymer molecules [30]:

$$RRF = \frac{\left(\frac{K}{\mu}\right)\text{before polymer flooding}}{\left(\frac{K}{\mu}\right)\text{after polymer flooding}} = \frac{\Delta P\,\text{water after polymer flooding}}{\Delta P\,\text{water before polymer flooding}} \tag{1.3}$$

Polymer nature

Another mechanism is defined as the elasticity of the polymer. As the polymer's elasticity increases, it is expected that the resistance to flow through porous media and the displacement efficiency will increase. Accordingly, lower residual oil saturation is expected [31]. According to this mechanism, polymer flows better in porous media. Consequently, higher macroscopic and microscopic efficiency is expected in the presence of polymers [31].

1.2.1.2 Polymer type

There are two types of polymers:

1. Biopolymers such as xanthan gum.
2. Synthetic polymers such as hydrolyzed polyacrylamide (HPAM).

Fig. 1.10 shows the xanthan gum polymer structure. A substantial increase in mobility ratio was observed with this polymer [33]. On the other hand, xanthan gum is a very useful polymer in high salinity [34]. Fig. 1.11 shows the HPAM structure. HPAM is antibacterial, thus making it a good polymer to use in field applications [35].

Since HPAM and other anionic polymers bind very strongly to the surface of carbonate rocks, most polymer floodings have been used in sandstone. Polymer flooding, however, is not limited to carbonate rocks of this type [36]. Polymer flooding has its own set of requirements [37]. Table 1.2 lists the specifications for polymer flooding.

Polymer flooding studied, in the beginning, was limited to experiments only. However, many researchers have documented some applications in the field [39]. In Southern Oman, HPAM was utilized in the Marmul field in a sandstone reservoir. The water cut was reduced from 50% to 20% after the polymer flooding [39]. Yabin et al. have shown the usefulness of polymer's in the Daqing oilfield. In this area, polymer flooding began in 1996. The additional recovery factor has been documented to be 12% higher than the water flooding [40]. Since 2000, polymer flooding was restricted to high-salinity

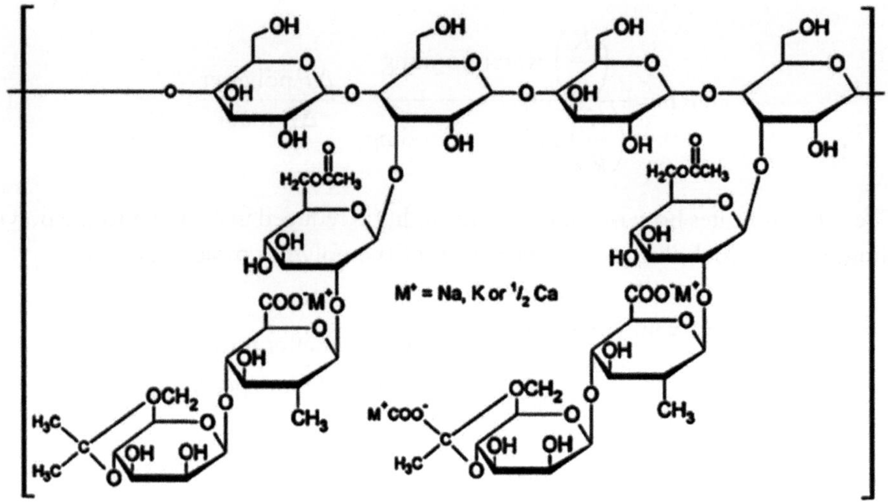

Fig. 1.10 Xanthan gum structure [32].

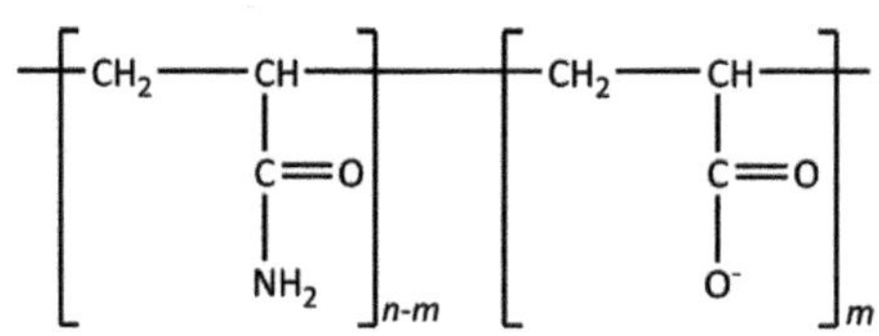

Fig. 1.11 HPAM structure [21].

Table 1.2 Screening requirements of polymer flooding [38].

Factor	
Permeability, mD	>10
Porosity, %	>10
Thickness, ft	>10
Oil viscosity, cp	<10,000
Temperature, F	<250
Formation salinity, ppm	<250,000
Residual oil saturation, %	>30

reservoirs as well as elevated temperature; therefore, new HPAM polymers have been synthesized. An investigation of polymer flooding in the Sanand oilfield, in the United States, was reported by Tiwari et al.. A water-soluble polymer, HPAM, was used in this study, and the reservoir contained 20 cp oil at 85°C [41]. Only a few documents refer to xanthan polymer compared to HPAM [42]. Xanthan gum was first used at Vorhop-Knesebeck in Norway and Eddesse-Nord in Germany [43].

1.2.2 Surfactant flooding

Surfactant flooding is a method of enhanced oil recovery (EOR) that increases efficiency by reducing IFT and altering wettability [44]. Fig. 1.12 indicates the schematic of surfactant flooding in EOR.

Fig. 1.13 shows the surfactant's head and tail [19]. It is clear from the graph that the polar head is hydrophilic and soluble in water, while the tail part is hydrophobic and soluble in oil.

Surfactants belong to four types: cationic, anionic, nonionic, and zwitterionic. Their tails are the same, but their head parts are different. The main points about these four categories are as follows:

1. Two examples of anionic surfactants are sulfate and sulfonate. The latter is stable at higher temperatures and salinity [45]. These surfactants are usually used in sandstone reservoirs.

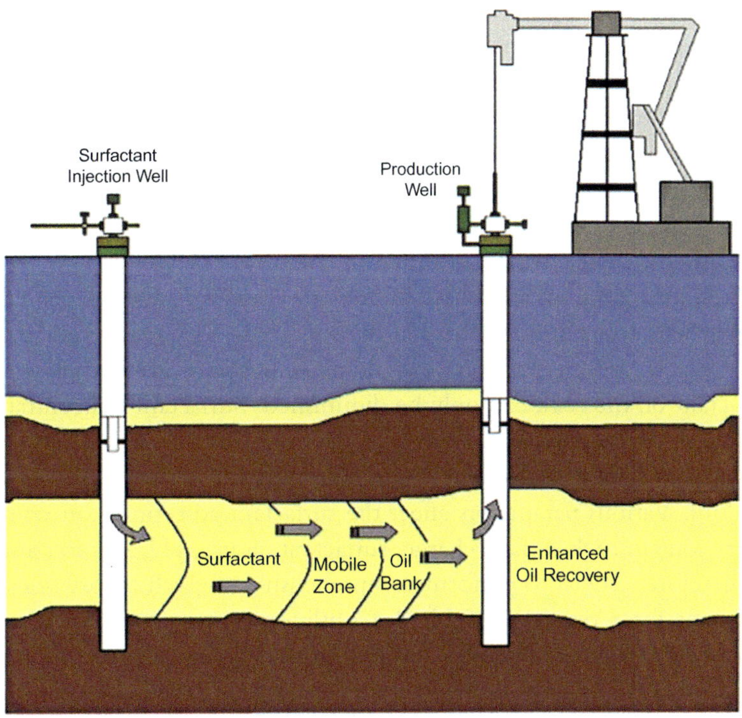

Fig. 1.12 Schematics representation of surfactant flooding in enhanced oil recovery [44].

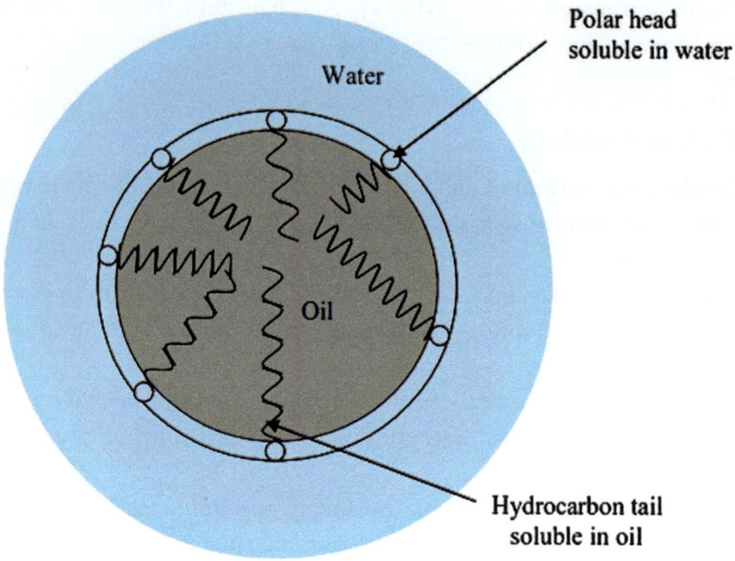

Fig. 1.13 Surfactant's tail and head [19].

2. In contrast to anionic surfactants, the cationic ones are more costly, and are usually useful in carbonate reservoirs and applications where alternating wettability is required [46].
3. The nonionic surfactants are more efficient in decreasing IFT than the previously mentioned surfactants, but their stability is less [47].
4. Zwitterionic compounds contain both cationic and anionic groups and are more stable at elevated temperatures and salinities [48].

Table 1.3 shows the typical surfactants used in chemical EOR.

Adsorption, precipitation, and phase transfer are the three main mechanisms used in the flooding process to retain surfactants in porous media. Surfactants resistant to salinity and temperature can prevent precipitation and phase trapping. Nevertheless, the adsorption of surfactant on the rock can only be diminished. Surfactant concentrations in the injected solution are reduced as a result of adsorption, thereby lowering the oil-water IFT. This, in turn, affects the efficiency of the surfactant-based EOR in mobilizing trapped oil [50]. Various parameters affect the surfactant adsorption on reservoir rock. These factors contain rock surface charge, surfactant nature, pH, salinity, and temperature. It is, therefore, crucial to perform comprehensive research before selecting a particular surfactant for a reservoir. Table 1.4 outlines numerous reports that highlight the influence of temperature, salinity, and pH on the adsorption of surfactants.

Studies showing the effects of surfactant usage in the field are few in comparison with experimental tests. According to Table 1.5, a variety of surfactants are used in EOR techniques involving chemicals.

Table 1.3 Typical example of surfactants with relevant application [5, 49].

Type of surfactant	Advantages	Dominant Mechanisms	Schematic
Cationic	Stable solution in brine, efficient in carbonate EOR	IFT reduction, Wettability alteration	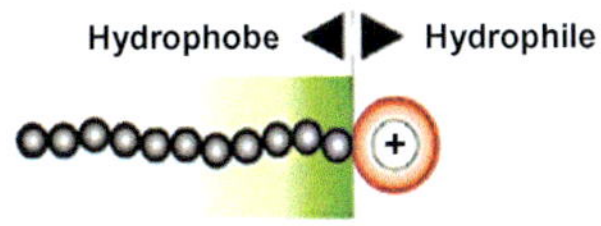
Anionic	Efficient in sandstone	IFT reduction	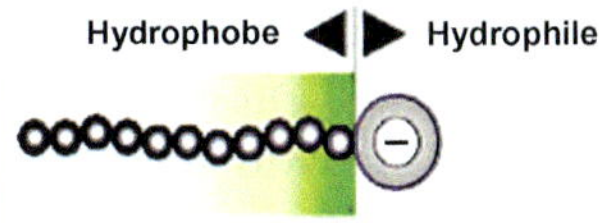
Nonionic	Efficient in harsh and high salinity water reservoir	IFT reduction	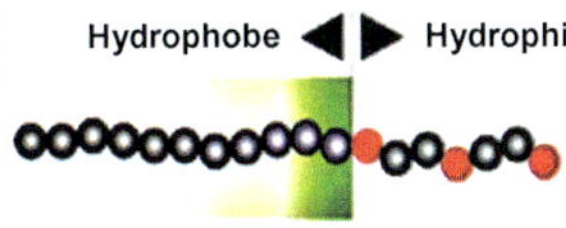
Zwitterionic	Compatible in high temperature	IFT reduction, Wettability alteration	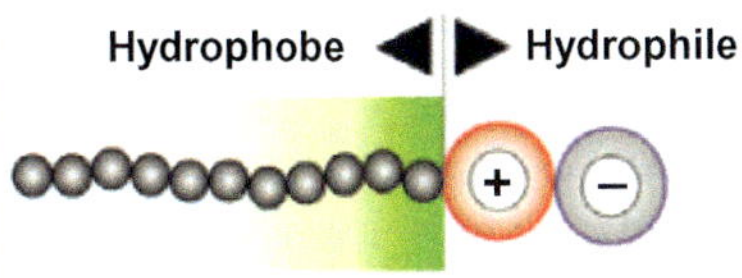

Table 1.4 Effect of temperature, salinity as well as pH on the adsorption of surfactant.

Surfactant	Solid surface	Range	Results	Reference
Temperature				
Synthesized surfactant contains primarily of C18 carbon chain and a sulfonate headgroup, anionic	Sandstone	25°C and 70°C	Surfactant adsorption at 25°C was 0.96 mg/g, and was reduced to 0.75 mg/g at 70°C	[51]
SDS, anionic	Kaolinite	25.85°C, 49.85°C, and 79.85°C	In the presence and the absence of electrolytes, surfactant adsorption reduced as the temperature raised	[52]
APG, nonionic	Sandstone	20°C, 50°C, 80°C, and 115°C	The alteration in adsorption was slight due to the stable surfactant hydrophilicity despite temperature adjustment	[53]

Continued

Table 1.4 Effect of temperature, salinity as well as pH on the adsorption of surfactant—cont'd

Surfactant	Solid surface	Range	Results	Reference
Saponin, nonionic	Shale-sandstone surface	28°C, 40°C, 60°C,and 75°C	At different surfactant concentrations, surfactant adsorption reduces as the temperature raises	[54]
Salinity				
Aerosol-OT, anionic	Kaolinite, Quartz, Illite, Montmorillonite	0, 20,000, 35,000, and 60,000 ppm	Adsorption rates were low and high in quartz and kaolinite, respectively	[55]
CTAB, cationic TX-100, nonionic	Shale	0, 10,000, and 30,000 ppm	A high quartz content in the shale contributed significantly to the adsorption of CTAB	[56]
DTAC, cationic TTAC, cationic OTAC, cationic	Carbonate	13,987.7, 27,975.4, 55,950.7, 111,901.4, and 223,802.8 ppm	Adsorption on limestone surfaces raised as the salinity intensified	[57]
C16DmCB, zwitterionic	Carbonate Sandstone	0, 15,000, 30,000, and 45,000 ppm	Adsorption was higher in carbonate rock compared to sandstone rock	[58]
pH				
TDA, anionic IOS, anionic GAC, anionic	Indiana limestone	7.5–12	The adsorption reduced linearly as the pH exceeded an amount of 9	[59]
SDS, anionic	Porous SiO_2 powder gel	3–7	SDS adsorption on SiO_2 surface reduced as the pH raised	[60]
Alcohol alkoxy sulfate, anionic	Clay	6–11	The adsorption of alcohol alkoxysulfate decreases when pH is raised in the presence of sodium ions. However, the opposite is observed when pH is raised in the presence of calcium ions	[61]

Table 1.4 Effect of temperature, salinity as well as pH on the adsorption of surfactant—cont'd

Surfactant	Solid surface	Range	Results	Reference
CAPB, zwitterionic	Dolomite	7.3–11.2	Na_2CO_3 raised the solution pH from 7.3 to 11.2 and decreased CAPB adsorption from 0.79 to 0.32 mg/g-rock	[62]

Table 1.5 Surfactant flooding-field cases.

Country	Field	Formation type	Surfactant	Results	Reference
China	Daqing	Sandstone	Nonionic alkanol acid amide Surfactant	Recovery = 30%	[63]
Indonesia	Tanjung	Sandstone	Anionic surfactant A127	Surfactant flooding significantly increased oil recovery	[64]
Indonesia	Semoga	Carbonate	Nonionic surfactants	The water cut was 8% lower and the 3-month oil recovery was 5800 bbls higher	[65, 66]
USA	Cottonwood Creek	Carbonate	Polyoxyethylene alcohol	70% of the 23 treated wells with surfactant-soak failed	[67, 68]
USA	Yates	Carbonate	Nonionic ethoxy alcohol	The average oil-production rate for one well raised from 35 barrels per day to 67 barrels	[69]

1.2.3 Alkali flooding

As one of the successful chemical oil recovery techniques, the alkali flooding method relies on the reaction between caustic alkali and organic acid (forming soap) [70] (Eq. 1.4).

$$HA + \mathrm{OH}^- \rightarrow A^- + \mathrm{H_2O} \quad (1.4)$$

where HA is an acidic component and A^- is the soap one. Fig. 1.14 demonstrates the chemical pattern of the alkali-oil chemistry in the reservoir rock. The organic acid is

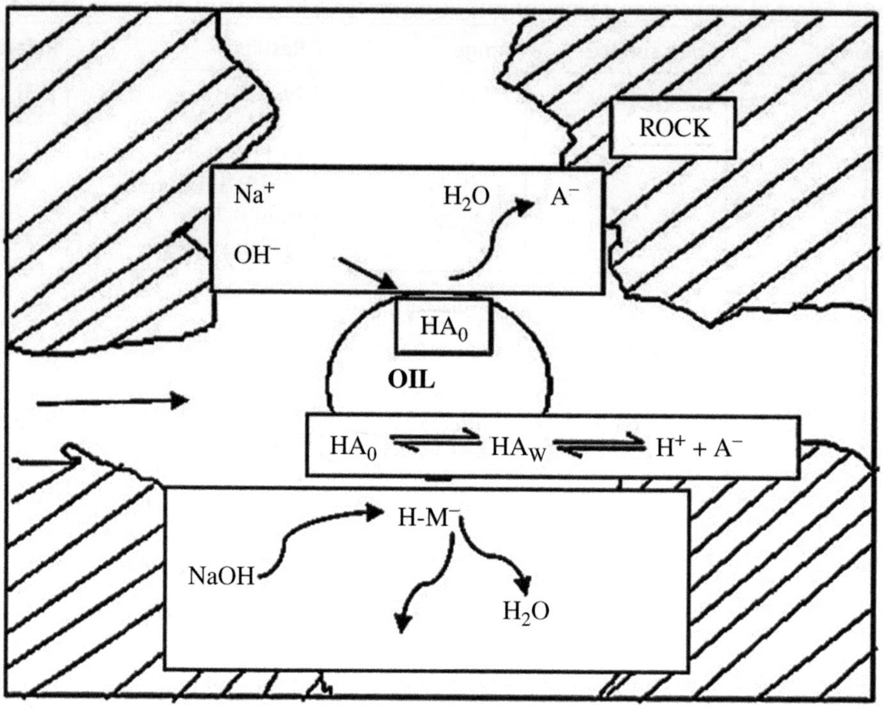

Fig. 1.14 Model of alkaline reactions in porous media [21].

achieved from the acidic part of the crude oil. By reducing IFT and emulsifying crude oil with the soap produced, higher recovery can be achieved [71].

There are many mechanisms for increasing recovery in alkali flooding, such as lowering IFT, alternating wettability, and oil swelling [72]. The following represent four types of alkaline flooding:

1. sodium hydroxide (NaOH)
2. sodium metaborate ($NaBO_2$)
3. sodium bicarbonate ($NaHCO_3$)
4. sodium carbonate (Na_2CO_3)

Alkali selection is influenced by the formation type. Sodium hydroxide, for instance, is known to react strongly with sandstone at elevated temperatures, thus increasing porosity and, subsequently, sandstone weight loss [73]. The most preferred alkali is sodium carbonate because it is a low-cost alkali and exhibits better transport properties in porous media. Sodium bicarbonate is usually preferred in the formation of clay-type material [74]. Fig. 1.15 shows the effective mechanisms in alkali flooding.

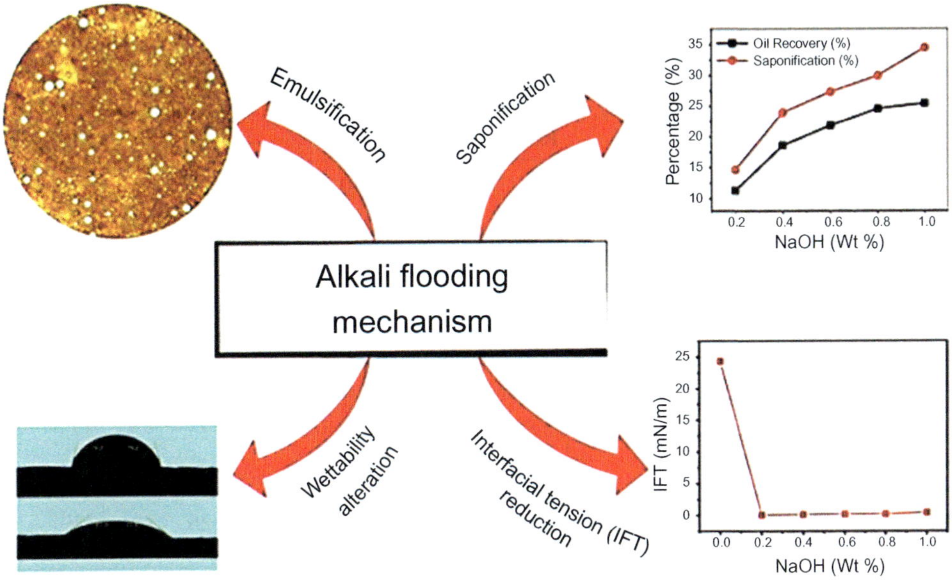

Fig. 1.15 The major mechanisms in flooding of alkali [75].

As shown in Table 1.6, light-moderate crude oils from fields in Texas-Frlo as well as Louisiana showed the most significant variation in oil recovery because of the higher amounts of naphthenic acid [75].

1.2.4 Foam flooding

Foam is the dispersion of gas in a liquid [80], which is used to increase gas mobility and compliance control within porous media. By reducing gas mobility, foam further concentrates gas into unoccupied parts of the reservoir, helping gas injections recover more oil. Foam flooding is more operative than typical water flooding, gas flooding, or even water alternating gas methods because foam flooding increases the mobility ratio as injectant viscosity increases [81]. Moreover, foam flooding is used in thief zone with different ranges of permeability. Recent field experiences [82–85] have shown that the foam enhances the sweep efficiency and can divert the injected gas. By reducing the gas/oil ratio in wells suffering from high gas production, foam may also extend the oil plateau period at mature oilfields. A representative foam usage for EOR is shown in Fig. 1.16.

1.2.5 Chemical EOR combined techniques

In the previous sections, the fundamental techniques for EOR were provided and studied. The properties of polymers, surfactants, and alkalis were also discussed. However,

Table 1.6 Comparison of various crude oil recovery in alkali flooding.

Crude oil	Rock	Viscosity, (mPa s)	Acid value (mg of KOH/gm)	Alkaline	Recovery %	Reference
Heavy oil (Alberta)	Sandpack	1360	1.07	NaOH(0.3 wt%)/ Na_2CO_3(0.3 wt%)	15	[76]
Heavy oil (Binnam)	Sandpack	2000	2.69	NaOH (1 wt%)	20	[71]
Heavy oil (Xia-8)	Sandpack	3950	4.66	$NaBO_2$ (1 wt%)	27	[77]
Heavy oil (Xiaba)	Sandpack	3950	4.66	NaOH (1 wt%)	30	[78]
Light to moderate (Assam)	Sandpack	33.4	2.87	NaOH (1 wt%)	25.5	[75]
Light to moderate (Texas-frlo)	Sandstone	44.2	1.39	NaOH (0.1 wt%)	26	[79]
Light to moderate (Louisiana)	Sandstone	19.2	0.61	NaOH (0.1 wt%)	21	[79]
Light (Louisiana)	Sandstone	29.1	1.02	NaOH (0.1 wt%)	28	[79]

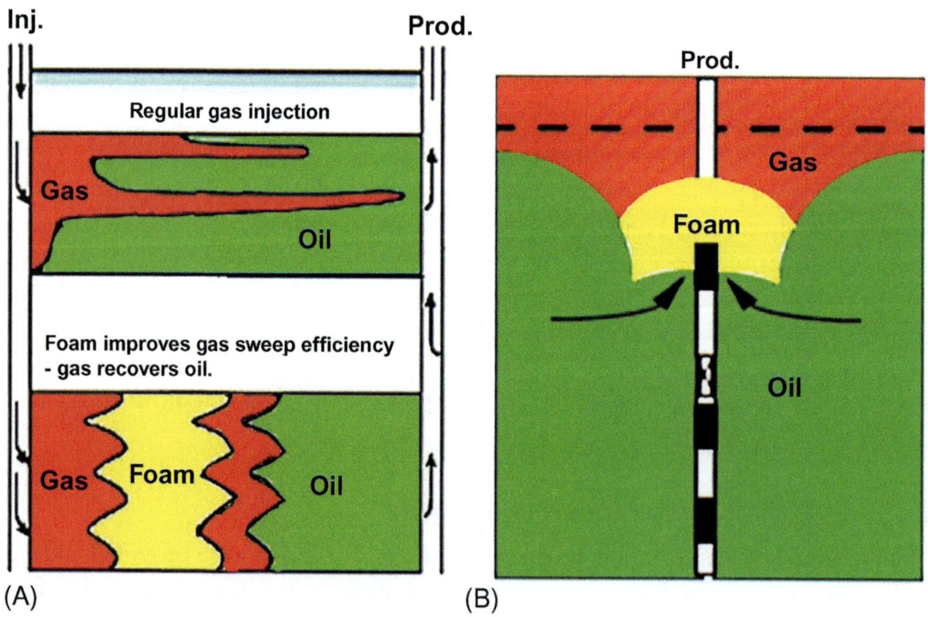

Fig. 1.16 Using foam flooding in thief zone with different ranges of permeability [86].

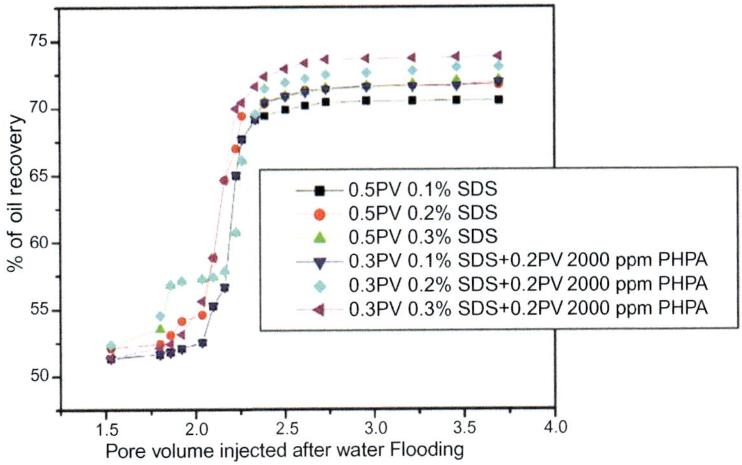

Fig. 1.17 Oil recovery comparison between standard surfactant EOR and a SP flooding [94].

more complicated processes have emerged that combine one or more traditional systems, leading to a whole new set of procedures [87]. Thus, the following processes have been established: alkaline surfactant (AS), alkaline polymer (AP), surfactant polymer (SP) and alkaline surfactant polymer (ASP) [72, 88–93]. It has been observed, however, that these combined methods have higher recovery rates not only due to the linear combination of the benefits of the injected EOR agents, but due to their interaction with a porous media, thus improving the sweeping efficiency (Fig. 1.17).

During surfactant polymer (SP) flooding, depending on the goal, either surfactant or polymer can be injected. The main aims are to avoid fingering problems and lowering adsorption [94]. In alkaline surfactant (AS), alkaline is injected at first followed by surfactant, and the efficiency is higher than alkali or surfactant flooding. In this method, alkali reacts with the naphthenic part of crude oil, and surfactant lowers IFT. There is a well-known fact that alkaline prevents surfactant adsorption on clay; therefore, more surfactants are released in the reservoir, yielding improved oil recovery [95].

The presence of alkali in the AP method causes less polymer to be adsorbed on the rock, and high mobility control is expected in this method (Fig. 1.18) [89]. In this method, the optimum concentration of alkali polymer should be obtained to prevent any possible effects of alkali on the polymer structure [96]. Table 1.7 shows field applications in SP and AP floodings.

1.2.5.1 Alkali-surfactant-polymer (ASP) flooding

This process involves injecting alkaline solutions into the reservoir prior to injecting surfactant and polymer slugs. Flooding with alkaline surfactant polymer aims to improve the

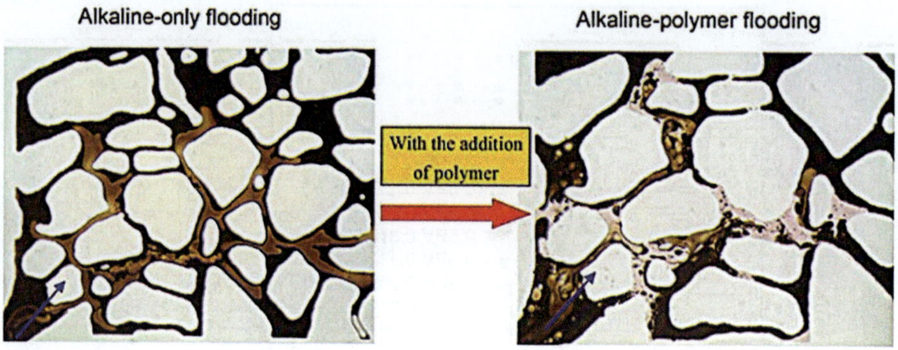

Fig. 1.18 Micromodel images of AP flooding (NaOH alkaline+HPAM polymer) [78].

Table 1.7 SP flooding-field cases.

Country	Field	Formation type	Surfactant	Results	Reference
USA	Reservoir SF	Sandstone	SAT/SST/SC01 cosurfactants, CPXG biopolymer	Recovery: 19%	[97]
China	Dagang	Sandstone	Petroleum sulfonate/AP-P7 polymer	Oil production raised from 27.7 to 51.7 m^3/d	[98]
Germany	Bramberge	Sandstone	Olefin sulfonate	High oil recovery	[99]
China	Xing long tai	Sandstone	Na_2CO_3 alkaline/HPAM polymer	Recovery: 18.5%	[100]

SST, sodium alkyl ethoxylate sulfate; *SAT*, sodium sulfosuccinate; *SC01*, branched sodium diphenyl oxide disulfonate.

efficiency of microscopic as well as macroscopic displacement. Fig. 1.19 shows the schematics of alkali-surfactant-polymer flooding, which is the most stimulating chemical EOR technique [87]. Using this method, both volumetric and pore-scale displacement efficiency can be increased, which can increase oil recovery. A big problem with this method is the surfactant precipitation when it comes in contact with alkali and brine [102].

A common ASP-EOR injection process is illustrated in Fig. 1.20. The first slug consists of an alkali and a surfactant that mobilizes the residual oil trapped in the pores. Polymer slug is then injected to enhance the mobility ratio and the volumetric sweeping performance [103]. Eventually, water slug and freshwater are injected to improve the chemical method.

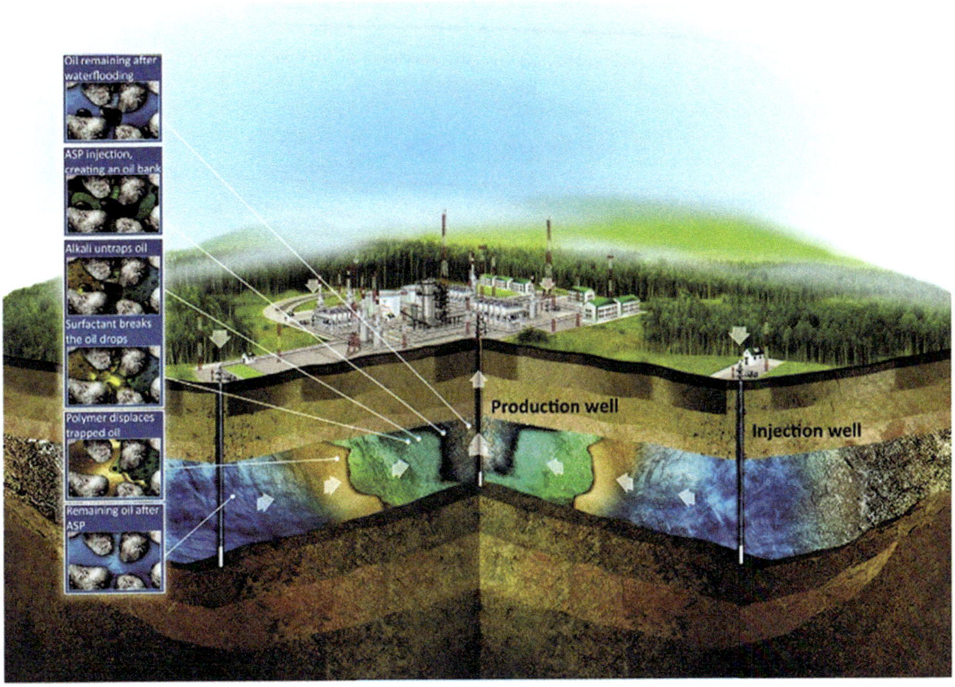

Fig. 1.19 Schematic representation of alkali-surfactant-polymer flooding [101].

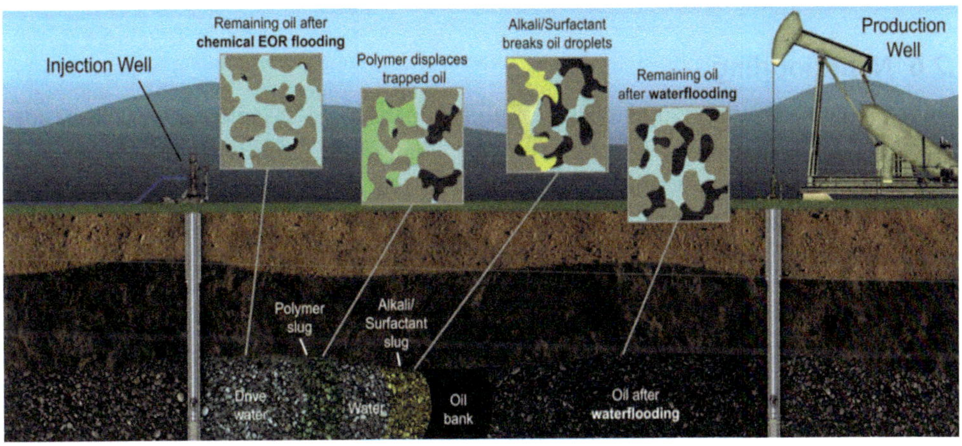

Fig. 1.20 The process of ASP flooding [5].

In recent years, ASP plans have been extensively developed via both experimental and pilot tests to demonstrate their capability [3, 92, 104–114]. Compared to secondary EOR, polymer or even other chemical combination procedures have presented enhanced recovery potential (Fig. 1.21). The ASP flooding is the preferred chemical

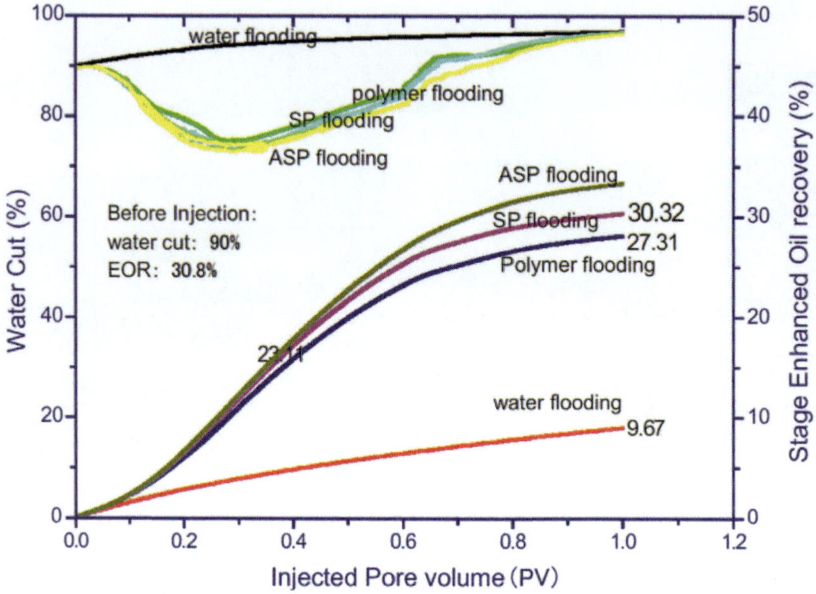

Fig. 1.21 The oil recovery and the water cut in oilfield of Palouge, South Sudan [115].

flooding test in many oilfields (Table 1.8). From 2016 to 2018, the annual production of ASP flooding oil in Daqing oilfield was more than 4 million tons [117].

1.2.6 Nanoflooding

Nanofluids are a relatively new class of liquids containing a base liquid with suspended particles of nanometer scale (1–100 nm) [121]. Nanofluid flooding (also recognized as nanoflooding) is a new chemical EOR approach that involves injecting nanomaterials or nanocomposite fluids into oil reservoirs to improve oil displacement [1, 122]. Fig. 1.22 shows different mechanisms in the presence of nanoparticles, including wettability alteration, IFT reduction, and disjoining pressure.

It has been found that nanoparticles such as polysilicon alter the wettability from oil wet to water wet [124]. Hydrophilic polysilicon NPs adsorb on the pore walls and change rock wettability. Fig. 1.23 shows the distribution of the liquids during the flooding test. The oil is left on the pore wall, and thus the water is trapped in the pores. On the other hand, Fig. 1.24 outlines the picture of the liquid's microscopic distribution that occurs when SiO_2 NPs are flooded in various concentrations. SiO_2 NPs were found to play a functional role in modifying the wettability of the porous surfaces, particularly at high SiO_2 NP concentrations. Thus, the flow of SiO_2 into the medium decreased oil on the pore walls, thereby improving recovery [125].

Table 1.8 ASP flooding-field cases.

Country	Field	Formation type	Material	Recovery %	Reference
China	Gudong	Sandstone	Petroleum sulfonate/ Na_2CO_3/3530S polymer	13.4	[107]
China	Karamay	Sandstone	Na_2CO_3/KPS-1/3530S	18–25	[93, 116]
China	Daqing	Sandstone	ABS/HPAM polymer/ NaOH	30	[117]
China	Northern part of Daqing	Sandstone	ABS/RH/HPAM polymer/NaOH	16.6	[113]
USA	Cambridge Minnelusa	Sandstone/ dolomite	Petrostep B-100 surfactant, Na_2CO_3, Alcoflood 1275A polymer	26.8	[118]
Russia	West Salym	Sandstone	Olefin sulfonate/ FLOPAAM/Na_2CO_3/ $C_4H_{10}O$	16	[119]
India	Viraj	Sandstone	Sodium carbonate/ petroleum sulfonate/ PHPA	32	[120]

RH, rhamnolipid-fermentation-liquor biosurfactant.

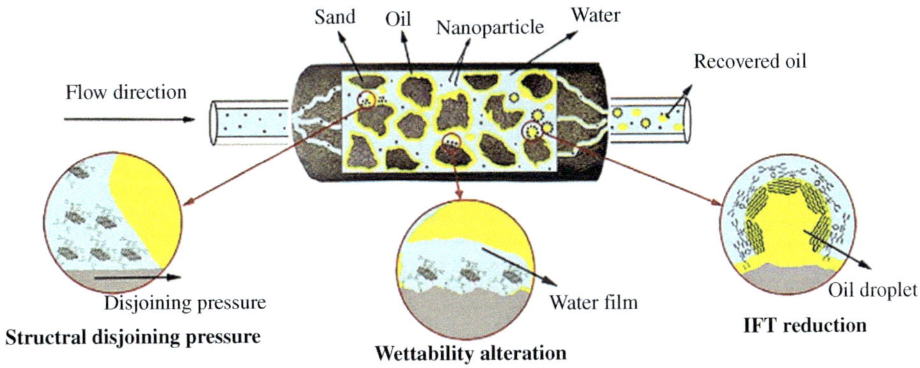

Fig. 1.22 Different mechanisms of EOR in the presence of nanoparticles [123].

Additionally, nanofluid can decrease the IFT between the oil and water phases by improving the capillary number [126]. Fig. 1.25 indicates the droplets used to measure the IFT between crude oil and nanofluids at a concentration of 500 ppm. The plot of the IFT as a function of concentration reveals that at higher concentrations, the IFT is reduced more slowly (Fig. 1.25). When the oil and the nanofluids are close together, the nanoparticles will tend to move toward the oil/water interface.

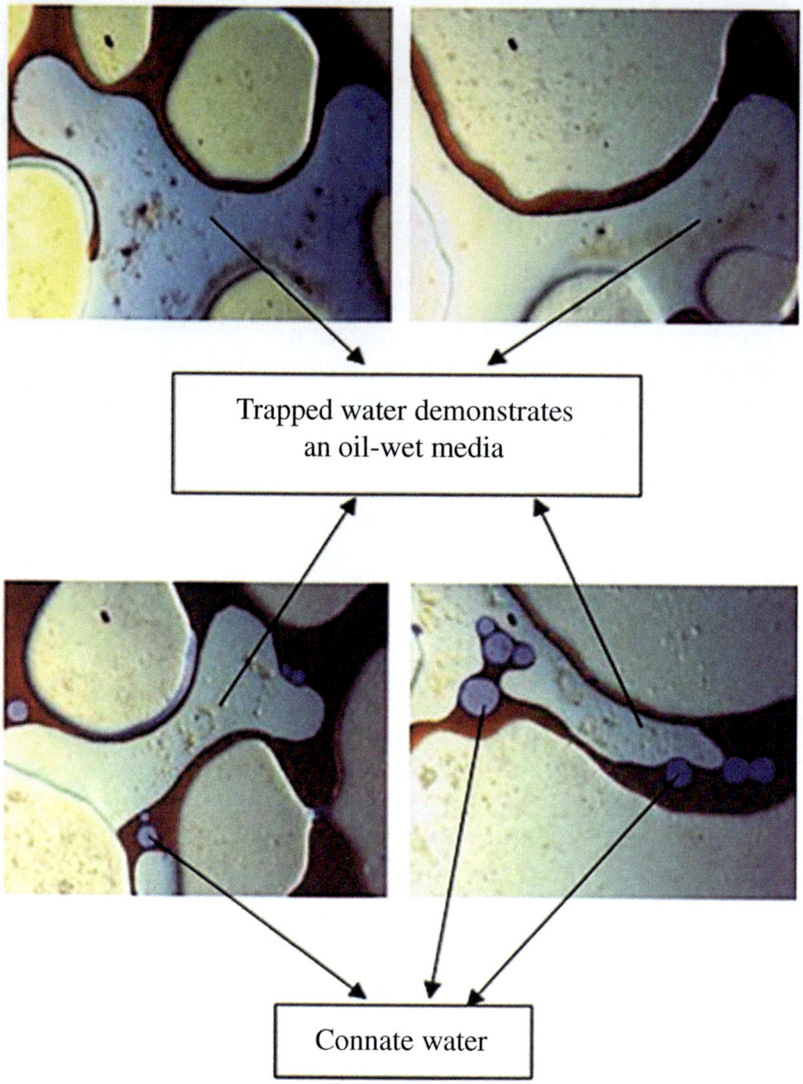

Fig. 1.23 Pore-scale study of water injection [125].

Nanoparticles typically have a size of 1–100 nm, and are employed in water flooding, gas flooding, and chemical EOR to enhance oil recovery [128]. A common problem during chemically enhanced oil recovery is pore plugging and permeability reduction [129]. Nanoparticles such as silica, alumina, and titanium oxide are commonly used to solve these relevant problems due to their small size [130]. These nanoparticles can reach regions that are inaccessible and increase microscopic efficiency. Furthermore,

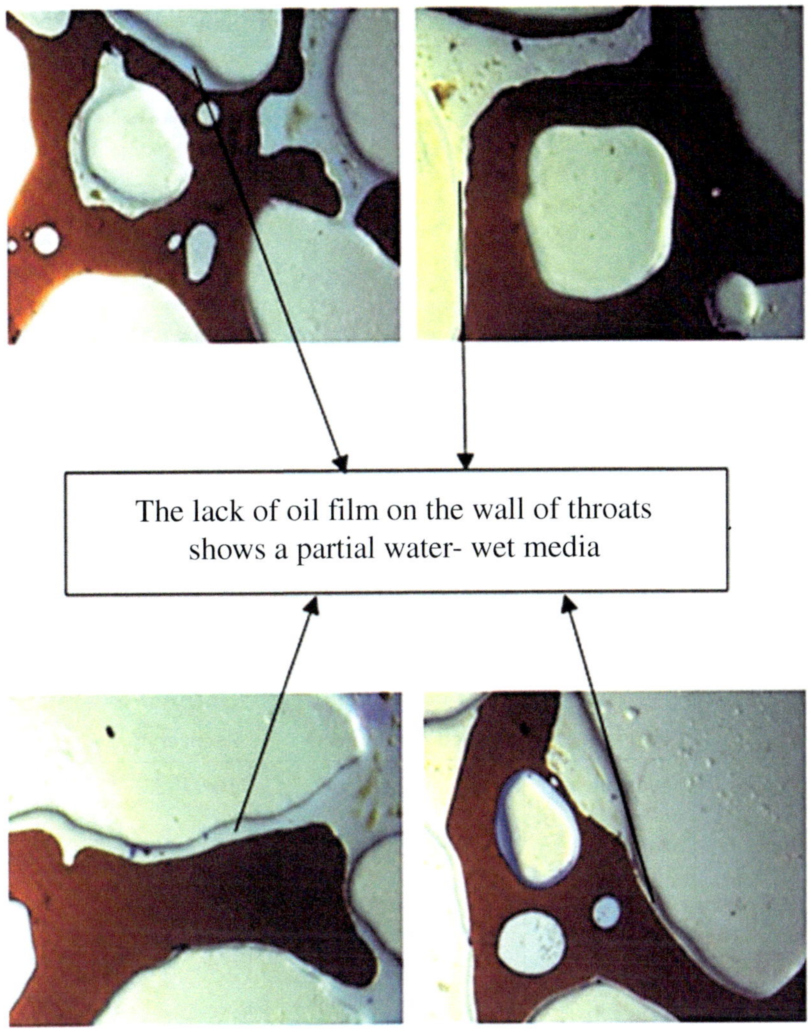

Fig. 1.24 Pore-scale study of NP injection [125].

nanoparticles are adapted to the environment as well, for example, silica is a part of the sandstone reservoir [131]. Table 1.9 shows common nanoparticles that are used in EOR and different relevant parameters.

1.3 Conclusions

The chapter provides a detailed explanation of chemical EOR and its long-term potential for energy future. Different chemical EOR techniques such as polymer flooding,

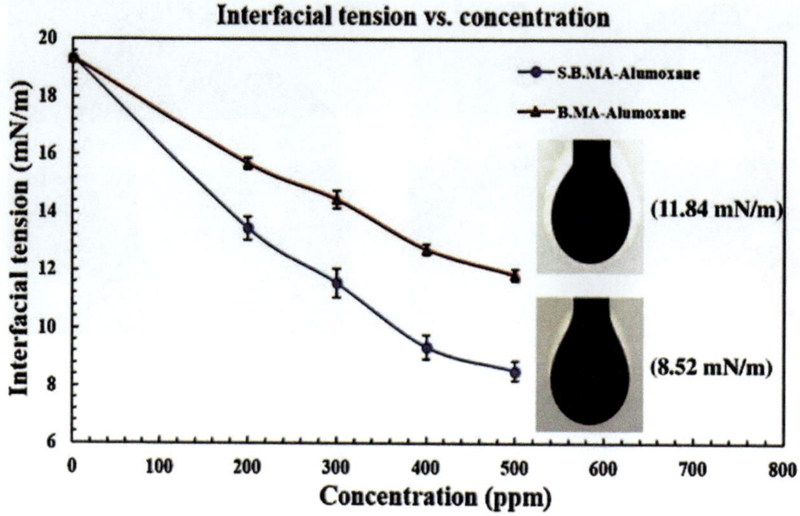

Fig. 1.25 IFT reduction by Carboxylate-Alumoxane nanoparticles [127].

Table 1.9 Common nanoparticles used in EOR and different relevant parameters.

Nanoparticles type	Cores type	Studied parameters	Recovery%	Reference
Alumina-silica-NiO-TiO_2	Sandstone	Nanoparticles type	16.94–23.72	[131, 132]
Silica-alumina	Sandstone	Nanoparticles type	Silica (8.74–13.88), Alumina (4.65–8.12)	[133]
SiO_2	Sandstone	Permeability-pore volume-nanoparticles concentration	5.93–14.29	[124]
SiO_2	Sandstone	Injection time	Secondary(8), Tertiary(2)	[134]
SiO_2	Micromodel	Nanoparticles adsorption	–	[135]
SiO_2	Sandstone	Nanoparticles concentration	4.26–5.32	[135]
SiO_2	Sandstone	Nanoparticles concentration	5–15	[136]
SiO_2	Carbonate	Aging time	9–12, 16–17 (24 h)	[137]
SiO_2	Micromodel	Nanoparticles concentration	28–40	[138]

Table 1.9 Common nanoparticles used in EOR and different relevant parameters—cont'd

Nanoparticles type	Cores type	Studied parameters	Recovery%	Reference
SiO_2	Micromodel	Nanoparticles concentration	8.7–26	[125]
Silica-alumina-TiO_2	Sandstone	Nanoparticles type	7–11	[139]
SiO_2	Sandstone	Nanoparticles size, permeability, temperature, wettability	0–8.41	[140]
SiO_2	Sandstone	Permeability-nanoparticles concentration	0–9.9	[140]
Silica-NiO-Fe_2O_3	Micromodel	Type and Size of nanoparticles	Silica (22.6), NiO (14.6), Fe_2O_3 (8.1)	[141]
Silica	Sandstone	Type and size of nanoparticles	5–10	[142]
Silica	Sandstone	Type and size of nanoparticles	9–19	[143]
Silica	Sandstone	Type and size of nanoparticles	13.3–24.1	[144]
TiO_2	Sandstone	Nanoparticles concentration	10–14	[139]
CuO-NiO-Fe_2O_3	Carbonate	Nanoparticles type	CuO (14.07), NiO (7.59), Fe_2O_3 (14.07)	[145]
TiO_2-SiO_2-PVP	Sandstone	Nanoparticles type, salinity, pH	15	[146, 147]
Polymer-coated ZnO/SiO_2	Carbonate	Nanoparticles concentration, time, and salinity	19.28	[148]
Carboxylate-alumoxane	Sandstone	Nanoparticles type and concentration	17–26	[127]
Fe_3O_4/eggshell-CTAB	Carbonate	Nanoparticles concentration	8.16	[149]

surfactant-polymer, surfactant flooding, alkali flooding, foam flooding, alkali-surfactant flooding, alkali-polymer flooding, alkali-surfactant-polymer, and nanoflooding are discussed. The main results and key points are listed below.

- In chemical EOR, wettability alteration, IFT reduction, and viscosity improvement are the main mechanisms.
- The main mechanisms for polymer flooding are: mobility control, reduction of permeability, and polymers nature.

- Xanthan gum is a very useful polymer in high salinity.
- The same steps as IFT reduction and wettability alteration are used with surfactants. A cationic surfactant is usually used for carbonates, while an anionic surfactant is used for sandstone reservoirs.
- In alkali flooding, by reducing IFT and emulsifying crude oil with the soap produced, higher recovery can be achieved.
- Foam flooding is used in thief zone with different ranges of permeability.
- Volumetric and pore-scale efficiency increase in alkali-surfactant-polymer (ASP) Flooding.
- In severe states like high pressure and temperature, and high salinity, alkali-surfactant-polymer has greater efficacy.
- Nanoparticles typically range between 1 and 100 nm size, offering better performance in water flooding, gas flooding, and chemical EOR process.
- Nanoparticles can reach inaccessible regions and provide microscopic efficiency.

References

[1] Mohammadi M, Sedighi M, Hemmati-Sarapardeh A. Application of nanofluids in enhanced oil recovery: a systematic literature review and organizing framework. In: Nanofluids and their engineering applications; 2019. p. 433–51.

[2] Vishnyakov V, Suleimanov B, Salmanov A, Zeynalov E. Chemical EOR. In: Primer on enhanced oil recovery. Elsevier; 2020. p. 141–59.

[3] Sedaghat M, Mohammadzadeh O, Kord S, Chatzis I. Heavy oil recovery using ASP flooding: a pore-level experimental study in fractured five-spot micromodels. Can J Chem Eng 2016;94(4):779–91.

[4] Pudasainee D, Kurian V, Gupta R. Coal: past, present, and future sustainable use. Future Energy 2020;21–48.

[5] Druetta P, Raffa P, Picchioni F. Chemical enhanced oil recovery and the role of chemical product design. Appl Energy 2019;252:113480.

[6] Gbadamosi AO, Kiwalabye J, Junin R, Augustine A. A review of gas enhanced oil recovery schemes used in the North Sea. J Pet Explor Prod Technol 2018;8(4):1373–87.

[7] Sorbie KS. Polymer retention in porous media. In: Polymer-improved oil recovery. Dordrecht: Springer Netherlands; 1991. p. 126–64.

[8] Sengupta B, Sharma V, Udayabhanu G. A study of the effect of the concentration of constituents on the characteristics of a cross-linked polyacrylamide gel. Pet Sci Technol 2012;30(18):1865–81.

[9] Mujahidin D, Permana Y, Radiman CL. Copolymerization of acrylamide with 9-and 10-acrylamidodecanoic acids. In: Macromolecular Symposia, vol. 353. Wiley Online Library; 2015. p. 198–204.

[10] Kumar S, Mandal A. A comprehensive review on chemically enhanced water alternating gas/CO2 (CEWAG) injection for enhanced oil recovery. J Petrol Sci Eng 2017;157:696–715.

[11] Ahmadi M-A, Ahmad Z, Phung LTK, Kashiwao T, Bahadori A. Evaluation of the ability of the hydrophobic nanoparticles of SiO2 in the EOR process through carbonate rock samples. Pet Sci Technol 2016;34(11 − 12):1048–54.

[12] Maurya NK, Kushwaha P, Mandal A. Studies on interfacial and rheological properties of water soluble polymer grafted nanoparticle for application in enhanced oil recovery. J Taiwan Inst Chem Eng 2017;70:319–30.

[13] Riazi M, Golkari A. The influence of spreading coefficient on carbonated water alternating gas injection in a heavy crude oil. Fuel 2016;178:1–9.

[14] Kok M. Progress and recent utilization trends in petroleum recovery: steam injection and in-situ combustion. Energy Sources, Part A 2012;34(24):2253–9.

[15] Ali H, Soleimani H, Yahya N, Khodapanah L, Sabet M, Demiral BM, et al. Enhanced oil recovery by using electromagnetic-assisted nanofluids: a review. J Mol Liq 2020;113095.
[16] Gbadamosi AO, Junin R, Manan MA, Yekeen N, Augustine A. Hybrid suspension of polymer and nanoparticles for enhanced oil recovery. Polym Bull 2019;76(12):6193–230.
[17] Kamal MS, Hussein IA, Sultan AS. Review on surfactant flooding: phase behavior, retention, IFT, and field applications. Energy Fuel 2017;31(8):7701–20.
[18] Levitt D, Pope GA. Selection and screening of polymers for enhanced-oil recovery. In: SPE symposium on improved oil recovery. Tulsa, Oklahoma, USA: Society of Petroleum Engineers; 2008.
[19] Olajire AA. Review of ASP EOR (alkaline surfactant polymer enhanced oil recovery) technology in the petroleum industry: prospects and challenges. Energy 2014;77:963–82.
[20] Gbadamosi AO, Junin R, Manan MA, Yekeen N, Agi A, Oseh JO. Recent advances and prospects in polymeric nanofluids application for enhanced oil recovery. J Ind Eng Chem 2018;66:1–19.
[21] Gbadamosi AO, Junin R, Manan MA, Agi A, Yusuff AS. An overview of chemical enhanced oil recovery: recent advances and prospects. Int Nano Lett 2019;9(3):171–202.
[22] Mandal A. Chemical flood enhanced oil recovery: a review. Int J Oil, Gas Coal Technol 2015;9 (3):241–64.
[23] Wei B. Advances in polymer flooding. In: Viscoelastic and Viscoplastic Materials. InTechOpen; 2016. p. 1–20.
[24] Pye DJ. Improved secondary recovery by control of water mobility. J Petrol Tech 1964;16(08):911–6.
[25] Sandiford BB. Laboratory and field studies of water floods using polymer solutions to increase oil recoveries. J Petrol Tech 1964;16(08):917–22.
[26] Zerkalov G. Polymer flooding for enhanced oil recovery. Stanford University; 2015.
[27] Sydansk R, Romero-Zeron L. Reservoir conformance improvement, society of petroleum engineers. TX, USA: Richardson; 2011.
[28] Vela S, Peaceman D, Sandvik EI. Evaluation of polymer flooding in a layered reservoir with crossflow, retention, and degradation. Soc Pet Eng J 1976;16(02):82–96.
[29] Zaitoun A, Kohler N. Two-phase flow through porous media: effect of an adsorbed polymer layer. In: SPE annual technical conference and exhibition. Society of Petroleum Engineers; 1988.
[30] Cannella W, Huh C, Seright R. Prediction of xanthan rheology in porous media. In: SPE annual technical conference and exhibition. Society of Petroleum Engineers; 1988.
[31] Urbissinova TS, Trivedi J, Kuru E. Effect of elasticity during viscoelastic polymer flooding: a possible mechanism of increasing the sweep efficiency. J Can Pet Technol 2010;49(12):49–56.
[32] Wever D, Picchioni F, Broekhuis A. Polymers for enhanced oil recovery: a paradigm for structure–property relationship in aqueous solution. Prog Polym Sci 2011;36(11):1558–628.
[33] Sheng JJ. Polymer flooding. Modern chemical enhanced oil recovery: theory and practice. Gulf Professional Publishing; 2011. p. 101–206.
[34] Choppe E, Puaud F, Nicolai T, Benyahia L. Rheology of xanthan solutions as a function of temperature, concentration and ionic strength. Carbohydr Polym 2010;82(4):1228–35.
[35] Pu W, Shen C, Wei B, Yang Y, Li Y. A comprehensive review of polysaccharide biopolymers for enhanced oil recovery (EOR) from flask to field. J Ind Eng Chem 2018;61:1–11.
[36] Jiang Z, Zhu J. Cationic polyacrylamide: synthesis and application in sludge dewatering treatment: a review. Asian J Chem 2014;26(3):629–33.
[37] Chang HL. Polymer flooding technology yesterday, today, and tomorrow. J Petrol Tech 1978;30 (08):1113–28.
[38] Firozjaii AM, Saghafi HR. Review on chemical enhanced oil recovery using polymer flooding: fundamentals, experimental and numerical simulation. Petroleum 2020;6(2):115–22.
[39] Koning E, Mentzer E, Heemskerk J. Evaluation of a pilot polymer flood in the Marmul field, Oman. In: SPE annual technical conference and exhibition. Society of Petroleum Engineers; 1988.
[40] Jiran L. Experimental investigation and simulation of polymer flooding in high temperature high salinity carbonate reservoirs. United Arab Emirates: The Petroleum Institute; 2015.
[41] Marathe RV, Patel NK, Ramachandran K, Maurya CR, Tewari P. Performance of polymer flood in sanand field, India-A case study. In: SPE Asia Pacific oil and gas conference and exhibition. Society of Petroleum Engineers; 2008.

[42] Huh C, Pope GA. Residual oil saturation from polymer floods: laboratory measurements and theoretical interpretation. In: SPE symposium on improved oil recovery. Society of Petroleum Engineers; 2008.
[43] Littmann W, Kleinitz W, Christensen B, Stokke B, Haugvallstad T. Late results of a polymer pilot test: performance, simulation adsorption, and xanthan stability in the reservoir. In: SPE/DOE enhanced oil recovery symposium. Society of Petroleum Engineers; 1992.
[44] Gurgel A, Moura M, Dantas T, Neto EB, Neto AD. A review on chemical flooding methods applied in enhanced oil recovery. Braz J Pet Gas 2008;2(2).
[45] Pal S, Mushtaq M, Banat F, Al Sumaiti AM. Review of surfactant-assisted chemical enhanced oil recovery for carbonate reservoirs: challenges and future perspectives. Pet Sci 2018;15(1):77–102.
[46] Kumar S, Panigrahi P, Saw RK, Mandal A. Interfacial interaction of cationic surfactants and its effect on wettability alteration of oil-wet carbonate rock. Energy Fuel 2016;30(4):2846–57.
[47] Ahmadi MA, Zendehboudi S, Shafiei A, James L. Nonionic surfactant for enhanced oil recovery from carbonates: adsorption kinetics and equilibrium. Ind Eng Chem Res 2012;51(29):9894–905.
[48] Kumar A, Mandal A. Synthesis and physiochemical characterization of zwitterionic surfactant for application in enhanced oil recovery. J Mol Liq 2017;243:61–71.
[49] Negin C, Ali S, Xie Q. Most common surfactants employed in chemical enhanced oil recovery. Petroleum 2017;3(2):197–211.
[50] Belhaj AF, Elraies KA, Mahmood SM, Zulkifli NN, Akbari S, Hussien OS. The effect of surfactant concentration, salinity, temperature, and pH on surfactant adsorption for chemical enhanced oil recovery: a review. J Pet Explor Prod Technol 2020;10(1):125–37.
[51] Azam MR, Tan IM, Ismail L, Mushtaq M, Nadeem M, Sagir M. Kinetics and equilibria of synthesized anionic surfactant onto berea sandstone. J Dispers Sci Technol 2014;35(2):223–30.
[52] Yekeen N, Manan MA, Idris AK, Samin AM. Influence of surfactant and electrolyte concentrations on surfactant adsorption and foaming characteristics. J Petrol Sci Eng 2017;149:612–22.
[53] Wei P, Li J, Xie Y, Huang X, Sun L. Alkyl polyglucosides for potential application in oil recovery process: adsorption behavior in sandstones under high temperature and salinity. J Petrol Sci Eng 2020;189:107057.
[54] Ahmadi MA, Shadizadeh SR, Chen Z. Thermodynamic analysis of adsorption of a naturally derived surfactant onto shale sandstone reservoirs. Eur Phys J Plus 2018;133(10):1–11.
[55] Abbas AH, Moslemizadeh A, Sulaiman WRW, Jaafar MZ, Agi A. An insight into a di-chain surfactant adsorption onto sandstone minerals under different salinity-temperature conditions: chemical EOR applications. Chem Eng Res Des 2020;153:657–65.
[56] Yekeen N, Padmanabhan E, Idris AK, Ibad SM. Surfactant adsorption behaviors onto shale from Malaysian formations: influence of silicon dioxide nanoparticles, surfactant type, temperature, salinity and shale lithology. J Petrol Sci Eng 2019;179:841–54.
[57] Chen L, Zhang G, Wang L, Wu W, Ge J. Zeta potential of limestone in a large range of salinity. Colloids Surf A Physicochem Eng Asp 2014;450:1–8.
[58] Kumar A, Mandal A. Critical investigation of zwitterionic surfactant for enhanced oil recovery from both sandstone and carbonate reservoirs: adsorption, wettability alteration and imbibition studies. Chem Eng Sci 2019;209:115222.
[59] Tagavifar M, Jang S, Sharma H, Wang D, Chang L, Mohanty K, et al. Effect of pH on adsorption of anionic surfactants on limestone: experimental study and surface complexation modeling. Colloids Surf A Physicochem Eng Asp 2018;538:549–58.
[60] Li P, Ishiguro M. Adsorption of anionic surfactant (sodium dodecyl sulfate) on silica. Soil Sci Plant Nutr 2016;62(3):223–9.
[61] Liu Z, Ghatkesar MK, Sudhölter EJ, Singh B, Kumar N. Understanding the cation-dependent surfactant adsorption on clay minerals in oil recovery. Energy Fuel 2019;33(12):12319–29.
[62] Rezaei A, Riazi M, Escrochi M, Elhaei R. Integrating surfactant, alkali and nano-fluid flooding for enhanced oil recovery: a mechanistic experimental study of novel chemical combinations. J Mol Liq 2020;308:113106.
[63] Yin D, Pu H. A numerical simulation study on surfactant flooding and it's field application in Daqing oilfield. In: Europec/EAGE conference and exhibition. Rome, Italy: Society of Petroleum Engineers; 2008.

[64] Aslam B, Ulitha D, Swadesi B, Fauzi I, Marhaendrajana T, Purba F, et al. History match to support interpretation of surfactant flooding pilot test in tanjung field. In: SPE/IATMI Asia Pacific oil & gas conference and exhibition. Jakarta, Indonesia: Society of Petroleum Engineers; 2017.
[65] Rilian NA, Sumestry M, Wahyuningsih. Surfactant stimulation to increase reserves in carbonate reservoir: "a case study in semoga field". In: SPE EUROPEC/EAGE annual conference and exhibition. Spain: Barcelona; 2010.
[66] Syaifudin A. Reservoir simulation to increase recovery in kaji semoga field. In: 28th annual convention proceedings, vol. 2; 2002. p. 1–17.
[67] Weiss WW, Xie X, Weiss J, Subramanian V, Taylor AR, Edens FJ. Artificial intelligence used to evaluate 23 single-well surfactant-soak treatments. SPE Reservs Eval Eng 2006;9(03):209–16.
[68] Xie X, Weiss WW, Tong ZJ, Morrow NR. Improved oil recovery from carbonate reservoirs by chemical stimulation. SPE J 2005;10(03):276–85.
[69] Chen HL, Lucas LR, Nogaret LAD, Yang HD, Kenyon DE. Laboratory monitoring of surfactant imbibition with computerized tomography. SPE Reserv Eval Eng 2001;4(01):16–25.
[70] Gong H, Li Y, Dong M, Ma S, Liu W. Effect of wettability alteration on enhanced heavy oil recovery by alkaline flooding. Colloids Surf A Physicochem Eng Asp 2016;488:28–35.
[71] Pei H, Zhang G, Ge J, Jin L, Ma C. Potential of alkaline flooding to enhance heavy oil recovery through water-in-oil emulsification. Fuel 2013;104:284–93.
[72] Sheng JJ. A comprehensive review of alkaline–surfactant–polymer (ASP) flooding. Asia Pac J Chem Eng 2014;4(9):471–89.
[73] Sydansk RD. Elevated-temperature caustic/sandstone interaction: implications for improving oil recovery (includes associated papers 11348 and 11548). Soc Pet Eng J 1982;22(04):453–62.
[74] Zhang J, Nguyen QP, Flaaten A, Pope GA. Mechanisms of enhanced natural imbibition with novel chemicals. SPE Reserv Eval Eng 2009;12(06):912–20.
[75] Saha R, Uppaluri RV, Tiwari P. Influence of emulsification, interfacial tension, wettability alteration and saponification on residual oil recovery by alkali flooding. J Ind Eng Chem 2018;59:286–96.
[76] Wang J, Dong M, Arhuoma M. Experimental and numerical study of improving heavy oil recovery by alkaline flooding in sandpacks. J Can Pet Technol 2010;49(03):51–7.
[77] Tang M, Zhang G, Ge J, Jiang P, Liu Q, Pei H, et al. Investigation into the mechanisms of heavy oil recovery by novel alkaline flooding. Colloids Surf A Physicochem Eng Asp 2013;421:91–100.
[78] Pei H, Zhang G, Ge J, Zhang L, Wang H. Effect of polymer on the interaction of alkali with heavy oil and its use in improving oil recovery. Colloids Surf A Physicochem Eng Asp 2014;446:57–64.
[79] Ehrlich R, Wygal Jr RJ. Interrelation of crude oil and rock properties with the recovery of oil by caustic Waterflooding. Soc Pet Eng J 1977;17(04):263–70.
[80] Hiraski GJ. The steam-foam process. J Petrol Tech 1989;41(05):449–56.
[81] Yekeen N, Idris AK, Manan MA, Samin AM, Risal AR, Kun TX. Bulk and bubble-scale experimental studies of influence of nanoparticles on foam stability. Chin J Chem Eng 2017;25(3):347–57.
[82] Blaker T, Aarra MG, Skauge A, Rasmussen L, Celius HK, Martinsen HA, et al. Foam for gas mobility control in the snorre field: the FAWAG project. SPE Reserv Eval Eng 2002;5(04):317–23.
[83] Skoreyko FA, Villavicencio AP, Rodriguez Prada H, Nguyen QP. Development of a new foam eor model from laboratory and field data of the naturally fractured cantarell field. In: SPE Reservoir Characterisation and Simulation Conference and Exhibition. Society of Petroleum Engineers; 2011.
[84] Ocampo-Florez A, Restrepo A, Rendon N, Coronado J, Correa JA, Ramirez DA, et al. Foams prove effectiveness for gas injection conformance and sweep efficiency improvement in a low porosity fractured reservoir – field pilots. In: International petroleum technology conference. Kuala Lumpur, Malaysia; 2014.
[85] Ocampo A, Restrepo A, Lopera SH, Mejia JM. Creation of Insitu EOR foams by the injection of surfactant in gas dispersions - lab confirmation and field application. In: SPE improved oil recovery conference. Tulsa, Oklahoma, USA; 2018.
[86] Skauge A, Solbakken J, Ormehaug PA, Aarra MG. Foam generation, propagation and stability in porous medium. Transp Porous Media 2020;131(1):5–21.
[87] Pogaku R, Fuat NHM, Sakar S, Cha ZW, Musa N, Tajudin DNAA, et al. Polymer flooding and its combinations with other chemical injection methods in enhanced oil recovery. Polym Bull 2018;75 (4):1753–74.

[88] Sheng JJ. Polymer flooding—fundamentals and field cases. In: Enhanced oil recovery field case studies. Elsevier; 2013. p. 63–82.
[89] Sheng JJ. Critical review of alkaline-polymer flooding. J Pet Explor Prod Technol 2017;7(1):147–53.
[90] Kamal MS, Hussain SS, Sultan AS. Development of novel amidosulfobetaine surfactant–polymer systems for EOR applications. J Surfactant Deterg 2016;19(5):989–97.
[91] Lai N, Zhang Y, Wu T, Zhou N, Liu Y, Ye Z. Effect of sodium dodecyl benzene sulfonate to the displacement performance of hyperbranched polymer. Russ J Appl Chem 2016;89(1):70–9.
[92] Bataweel MA, Nasr-El-Din HA. ASP vs. SP flooding in high salinity/hardness and temperature in sandstone cores. In: SPE EOR Conference at Oil and Gas West Asia. Society of Petroleum Engineers; 2012.
[93] Qi Q, Gu H, Li D, Dong L. The pilot test of ASP combination flooding in Karamay oil field. In: International oil and gas conference and exhibition in China. Beijing, China; 2000.
[94] Samanta A, Ojha K, Sarkar A, Mandal A. Surfactant and surfactant-polymer flooding for enhanced oil recovery. Adv Pet Explor Dev 2011;2(1):13–8.
[95] Hirasaki GJ, Miller CA, Puerto M. Recent advances in surfactant EOR. SPE J 2011;16(04):889–907.
[96] Jung JC, Zhang K, Chon BH, Choi HJ. Rheology and polymer flooding characteristics of partially hydrolyzed polyacrylamide for enhanced heavy oil recovery. J Appl Polym Sci 2013;127(6):4833–9.
[97] Hsu T-P, Lohateeraparp P, Roberts BL, Wan W, Lin Z, Wang X, et al. Improved oil recovery by chemical flood from high salinity reservoirs-single-well surfactant injection test. In: SPE EOR conference at oil and gas West Asia. Muscat, Oman: Society of Petroleum Engineers; 2012.
[98] Guo Y, Zhang J, Zhang X, Hu J, Wang W, Liang Y. Investigation and application of an associative polymer-surfactant binary system for a successful flooding pilot in a high-temperature, high-salinity, ordinary heavy oil reservoir. In: SPE EOR conference at oil and gas West Asia. Muscat, Oman; 2018.
[99] Tabary R, Douarche F, Bazin B, Lemouzy PM, Moreau P, Morvan M. Design of a Surfactant/polymer process in a hard brine context: a case study applied to Bramberge reservoir. In: SPE EOR conference at oil and gas West Asia. Muscat, Oman; 2012.
[100] Zhang J, Wang K, He F, Zhang F. Ultimate evaluation of the alkali/polymer combination flooding pilot test in XingLongTai oil field. In: SPE Asia Pacific improved oil recovery conference. Kuala Lumpur, Malaysia; 1999.
[101] Salym Petroleum. Enhanced oil recovery [Available from:] https://salympetroleum.com/technologies/newoil/asp/; 2013.
[102] Qing J, Zhou B, Zhang R, Chen Z, Zhou Y. Development and application of a silicate scale inhibitor for ASP flooding production scale. In: International symposium on oilfield scale. Aberdeen, United Kingdom: Society of Petroleum Engineers; 2002.
[103] Dang C, Nghiem L, Nguyen N, Yang C, Chen Z, Bae W. Modeling and optimization of alkaline-surfactant-polymer flooding and hybrid enhanced oil recovery processes. J Petrol Sci Eng 2018;169:578–601.
[104] Atthawutthisin N. Numerical simulation of low salinity water flooding assisted with chemical flooding for enhanced oil recovery. Institutt for petroleumsteknologi og anvendt geofysikk; 2012.
[105] Hou J, Liu Z, Zhang S, Yang J. The role of viscoelasticity of alkali/surfactant/polymer solutions in enhanced oil recovery. J Petrol Sci Eng 2005;47(3–4):219–35.
[106] Jain AK, Dhawan AK, Misra TR. ASP flood pilot in Jhalora (K-IV)-a case study. In: SPE oil and gas India conference and exhibition. OnePetro; 2012.
[107] Zhijian Q, Yigen Z, Xiansong Z, Jialin D. A successful ASP flooding pilot in Gudong oil field. In: SPE/DOE improved oil recovery symposium. All Days. Tulsa, Oklahoma; 1998.
[108] Pandey A, Beliveau D, Corbishley DW, Suresh KM. Design of an ASP pilot for the Mangala field: Laboratory evaluations and simulation studies. In: SPE indian oil and gas technical conference and exhibition. Society of Petroleum Engineers; 2008.
[109] Kumar R, Mohanty KK. ASP flooding of viscous oils. In: SPE annual technical conference and exhibition. Society of Petroleum Engineers; 2010.
[110] Samanta A, Bera A, Ojha K, Mandal A. Comparative studies on enhanced oil recovery by alkali–surfactant and polymer flooding. J Pet Explor and Prod Technol 2012;2(2):67–74.

[111] Rai SK, Bera A, Mandal A. Modeling of surfactant and surfactant–polymer flooding for enhanced oil recovery using STARS (CMG) software. J Pet Explor and Prod Technol 2015;5(1):1–11.
[112] Bataweel MA, Nasr-El-Din HA. Rheological study for surfactant-polymer and novel alkali-surfactant-polymer solutions. In: North Africa technical conference and exhibition. OnePetro; 2012.
[113] Daoshan L, Shouliang L, Yi L, Demin W. The effect of biosurfactant on the interfacial tension and adsorption loss of surfactant in ASP flooding. Colloids Surf A Physicochem Eng Asp 2004;244 (1–3):53–60.
[114] Panthi K, Sharma H, Mohanty KK. ASP flood of a viscous oil in a carbonate rock. Fuel 2016; 164:18–27.
[115] Chen X, Feng Q, Wu X, Zhao G. A pilot numerical simulation case study for chemical EOR feasibility evaluation. J Pet Explor Prod Technol 2016;6(2):297–307.
[116] Gu H, Yang R, Guo S, Guan W, Yue X, Pan Q. Study on reservoir engineering: ASP flooding pilot test in Karamay oilfield. In: SPE international oil and gas conference and exhibition in China. Beijing, China: Society of Petroleum Engineers; 1998.
[117] Sun C, Guo H, Li Y, Jiang G, Ma R. Alkali effect on alkali-surfactant-polymer (ASP) flooding enhanced oil recovery performance: two large-scale field tests' evidence. J Chem 2020;2020:2829565.
[118] Vargo J, Turner J, Bob V, Pitts MJ, Wyatt K, Surkalo H, et al. Alkaline-surfactant-polymer flooding of the Cambridge Minnelusa field. SPE Reserv Eval Eng 2000;3(06):552–8.
[119] Volokitin Y, Shuster M, Karpan V, Koltsov I, Mikhaylenko E, Bondar M, et al. Results of alkaline-surfactant-polymer flooding pilot at west Salym field. In: SPE EOR conference at oil and gas West Asia. Muscat, Oman; 2018.
[120] Pratap M, Gauma MS. Field implementation of alkaline-surfactant-polymer (ASP) flooding: a maiden effort in India. In: SPE Asia Pacific oil and gas conference and exhibition. Perth, Australia; 2004.
[121] Saidur R, Leong K, Mohammed HA. A review on applications and challenges of nanofluids. Renew Sustain Energy Rev 2011;15(3):1646–68.
[122] El-Diasty AI, Aly AM. Understanding the mechanism of nanoparticles applications in enhanced oil recovery. In: SPE North Africa technical conference and exhibition. Cairo, Egypt: Society of Petroleum Engineers; 2015.
[123] Radnia H, Rashidi A, Nazar ARS, Eskandari MM, Jalilian M. A novel nanofluid based on sulfonated graphene for enhanced oil recovery. J Mol Liq 2018;271:795–806.
[124] Hendraningrat L, Li S, Torsæter O. A coreflood investigation of nanofluid enhanced oil recovery. J Petrol Sci Eng 2013;111:128–38.
[125] Maghzi A, Mohammadi S, Ghazanfari MH, Kharrat R, Masihi M. Monitoring wettability alteration by silica nanoparticles during water flooding to heavy oils in five-spot systems: a pore-level investigation. Exp Thermal Fluid Sci 2012;40:168–76.
[126] Li S, Torsæter O. An experimental investigation of EOR mechanisms for nanoparticles fluid in glass micromodel. In: Paper SCA2014–022 was prepared for presentation at the International Symposium of the Society of Core Analysts held in Avignon, France; 2014. p. 8–11.
[127] Bagherpour S, Rashidi A, Mousavi SH, Izadi N, Hamidpour E. Experimental investigation of carboxylate-alumoxane nanoparticles for the enhanced oil recovery performance. Colloids Surf A Physicochem Eng Asp 2019;563:37–49.
[128] Krishnamoorti R. Technology tomorrow: extracting the benefits of nanotechnology for the oil industry. J Petrol Tech 2006;58(11):24–6.
[129] Ahmadi M, Habibi A, Pourafshari P, Ayatollahi S. Zeta potential investigation and mathematical modeling of nanoparticles deposited on the rock surface to reduce fine migration. In: SPE Middle East oil and gas show and conference. Manama, Bahrain: Society of Petroleum Engineers; 2011.
[130] Alomair OA, Matar KM, Alsaeed YH. Nanofluids application for heavy oil recovery. In: SPE Asia Pacific oil & gas conference and exhibition. Adelaide, Australia: Society of Petroleum Engineers; 2014.
[131] Zhang LD, Liu Y. Preparation and application technology for ultrafine powder. J North China Inst Technol 2001;22:32–43.
[132] Sun Q, Li Z, Li S, Jiang L, Wang J, Wang P. Utilization of surfactant-stabilized foam for enhanced oil recovery by adding nanoparticles. Energy Fuel 2014;28(4):2384–94.

[133] Salem Ragab AM, Hannora AE. A comparative investigation of nano particle effects for improved oil recovery–experimental work. In: SPE Kuwait oil and gas show and conference. Mishref, Kuwait: Society of Petroleum Engineers; 2015.
[134] Torsater O, Engeset B, Hendraningrat L, Suwarno S. Improved oil recovery by nanofluids flooding: An experimental study. In: SPE Kuwait international petroleum conference and exhibition. Kuwait City, Kuwait: Society of Petroleum Engineers; 2012.
[135] Li S, Hendraningrat L, Torsaeter O. Improved oil recovery by hydrophilic silica nanoparticles suspension: 2-phase flow experimental studies. In: International petroleum technology conference. All days. Beijing, China: European Association of Geoscientists & Engineers; 2013.
[136] Li S, Genys M, Wang K, Torsæter O. Experimental study of wettability alteration during Nanofluid enhanced oil recovery process and its effect on oil recovery. In: SPE reservoir characterisation and simulation conference and exhibition. Day 3 wed, Abu Dhabi, UAE; September 16, 2015. p. 2015.
[137] Roustaei A, Bagherzadeh H. Experimental investigation of SiO2 nanoparticles on enhanced oil recovery of carbonate reservoirs. J Pet Explor Prod Technol 2015;5(1):27–33.
[138] Mohebbifar M, Ghazanfari MH, Vossoughi M. Experimental investigation of nano-biomaterial applications for heavy oil recovery in shaly porous models: a pore-level study. J Energy Resour Technol 2014;137(1).
[139] Hendraningrat L, Torsaeter O. Unlocking the potential of metal oxides nanoparticles to enhance the oil recovery. In: Offshore technology conference-Asia. Kuala Lumpur, Malaysia: Society of Petroleum Engineers; 2014.
[140] Hendraningrat L, Li S, Torsæter O. Effect of some parameters influencing enhanced oil recovery process using silica nanoparticles: an experimental investigation. In: SPE reservoir characterization and simulation conference and exhibition. Abu Dhabi, UAE: Society of Petroleum Engineers; 2013.
[141] Kazemzadeh Y, Eshraghi SE, Kazemi K, Sourani S, Mehrabi M, Ahmadi Y. Behavior of Asphaltene adsorption onto the metal oxide nanoparticle surface and its effect on heavy oil recovery. Ind Eng Chem Res 2015;54(1):233–9.
[142] Salem Ragab AM, Hannora AE. An experimental investigation of silica Nano particles for enhanced oil recovery applications. In: SPE North Africa technical conference and exhibition. Cairo, Egypt: Society of Petroleum Engineers; 2015.
[143] El-Diasty AI. The potential of nanoparticles to improve oil recovery in Bahariya formation, Egypt: an experimental study. In: SPE Asia Pacific enhanced oil recovery conference. Kuala Lumpur, Malaysia: Society of Petroleum Engineers; 2015.
[144] Ogolo NA, Olafuyi OA, Onyekonwu MO. Enhanced oil recovery using nanoparticles. In: SPE Saudi Arabia section technical symposium and exhibition. Al-Khobar, Saudi Arabia: Society of Petroleum Engineers; 2012.
[145] Tarek M. Investigating Nano-fluid mixture effects to enhance oil recovery. In: SPE annual technical conference and exhibition. Houston, Texas, USA: Society of Petroleum Engineers; 2015.
[146] Mohammadi M, Dadvar M, Dabir B. TiO2/SiO2 nanofluids as novel inhibitors for the stability of asphaltene particles in crude oil: mechanistic understanding, screening, modeling, and optimization. J Mol Liq 2017;238:326–40.
[147] Mohammadi M, Dadvar M, Dabir B. Application of response surface methodology for optimization of the stability of asphaltene particles in crude oil by TiO2/SiO2 nanofluids under static and dynamic conditions. J Dispers Sci Technol 2018;39:431–42.
[148] Ali JA, Kolo K, Khaksar Manshad A, Stephen KD. Potential application of low-salinity polymeric-nanofluid in carbonate oil reservoirs: IFT reduction, wettability alteration, rheology and emulsification characteristics. J Mol Liq 2019;284:735–47.
[149] Omidi A, Khaksar Manshad A, Moradi S, Ali JA, Sajadi SM, Keshavarz A. Smart- and nano-hybrid chemical EOR flooding using Fe3O4/eggshell nanocomposites. J Mol Liq 2020;316:113880.

CHAPTER 2

Polymer flooding

Forough Ameli, Saber Moghadam, and Samiye Shahmarvand
School of Chemical Engineering, Petroleum and Gas, Iran University of Science and Technology, Tehran, Iran

2.1 Introduction

Among various chemical enhanced oil recovery (cEOR) techniques, polymer flooding is a widely used method for improving the oil recovery factor with various successful field cases. This process is in the category of low-risk enhanced oil recovery processes and its application has greatly increased. By reviewing various studies, it is observed that polymer flooding is among the prospering methods for increasing oil production. There are a significant number of successful pilot plants for polymer flooding, but the number of field cases is limited. This is due to the operating problems including the reduction in the polymer injectivity, late injection of polymer solution in mature oil fields, low pore volume of the injected polymer, and early polymer breakthrough [1]. This process involves the injection of polymer solution to the reservoir to enhance the sweeping efficiency. This increases the mobility ratio between oil and the injected water in porous media [2, 3].

The first case for field study was performed in the late 1950s. For a usual and economically feasible project, polymer concentration of 500–3000 ppm is applied to enhance the production of original oil in place (OOIP) from 5% to 20%. There are various studies indicating the priorities of polymer flooding projects in field studies. This process is widely applied worldwide with various successful onshore cases [4]. In most of the reservoirs, after the water flooding, large amounts of oil are not recovered since the mobility ratio is unfavorable. By forming a proper path between injection and production wells, the injected water flows directly to the production well without pushing the oil in place. This leads to low sweeping efficiency of the produced oil. In this case, polymer solution is injected into the reservoir to enhance the mobility ratio between oil and water by increasing the water viscosity. In a recent study in Daqing oil field, the incremental oil production reached to 300,000 bpd [5]. In this process, the polymer solution was prepared and injected into the reservoir and adequate time was given to reach the favorable pore volume of the reservoir [6]. An efficient polymer flooding process leads to increased injection pressure and resistance factor, enhanced injection profile, and reduced water-intake index [7]. This technique is widely applied in low salinity and temperate reservoirs with low hardness.

Chemical Methods
https://doi.org/10.1016/B978-0-12-821931-7.00003-1

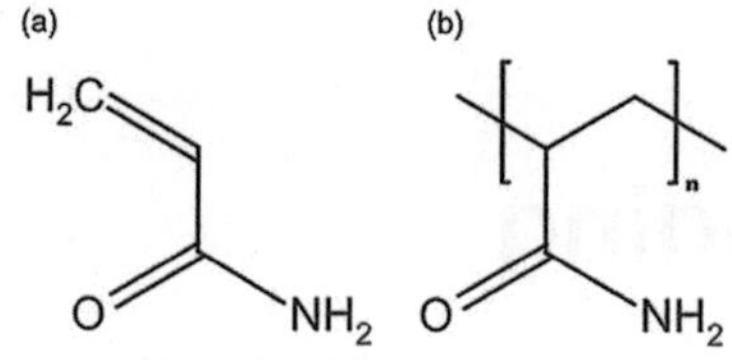

Fig. 2.1 (A) Acrylamide monomer and (B) PAM chain after polymerization [8].

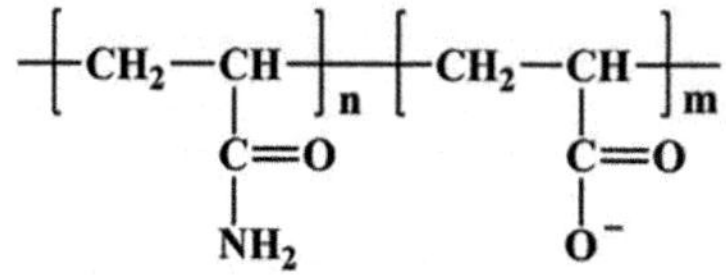

Fig. 2.2 Structure of HPAM [9].

2.2 Classification of EOR polymers

Polymers applied in the polymer flooding process are divided into two main groups, namely, synthetic polymers and biopolymers. Biopolymers are mainly polysaccharides produced through biological reactions. The first one contains a wide range of chemicals [8]. Synthetic polymers have low cost and are more available. Polyacrylamide (PAM) is the simplest and highly applicable polymer. It consists of a high number of acrylamide monomers as illustrated in Fig. 2.1. As presented in this figure, it contains a hydrocarbon chain with amino groups and oxygen atoms. As the monomers undergo a polymerization reaction, viscous polyacrylamide is produced. In this process, acrylamide monomers are applied as initiator, and ammonium persulfate is used as the catalyst [8].

In linear PAM, the mechanical structure of the polymer is easily degraded while it experiences shear stress. So, it is essential that linear PAMs undergo cross-linking to increase the mechanical resistance while flowing in the reservoir rock. To do so, methylenebisacrylamide is applied for polymerization [8].

Hydrolyzed polyacrylamide (HPAM) is the most important polymer synthesized for the polymer flooding process (Fig. 2.2). This polymer is synthesized by partially hydrolyzing polyacrylamide or by copolymerization of acrylic acid and acrylamide. It increases mobility ratio, is soluble in water, and is not expensive. Moreover, the polyelectrolyte behavior is enhanced by using anions on carboxyl group. This leads to increase in viscosity and reduction in adsorption phenomena. Molecular weight and degree of hydrolysis are important parameters in determining oil recovery factor. Other synthetic polymers that have been applied in the polymer flooding process include acrylamide-based copolymers with 2-acrylamido-2-methylpropane sulfonic acid, *N,N*-dimethyl acrylamide, *N*-vinyl pyrrolidone, and branched polyacrylamide. Hydrophobically modified polyacrylamide (HMPAM) is another widely applied polymer in this process.

The main components of this polymer are polyacrylamide and hydrophobic groups. Hydrophobic monomers with low concentration enhance the shear resistance viscosity, temperature, and salinity resistance in comparison to HPAM [10].

Biopolymers are obtained during fermentation. Since these polymers are nonionic, they are not sensitive to salinity and have a semirigid structure. Xanthan gum is a biopolymer with a molecular weight of 2–50 million Da [6]. The priority of a biopolymer to a synthetic polymer is its resistance to salinity. However, it is expensive and susceptible to degradation. Moreover, it has low injectivity. HPAM is the most applied polymer in the Middle East. The salinity and temperature of the reservoirs in this region are up to 180,000 ppm and 120°C, respectively [11]. Xanthan gum is a useful biopolymer applied in enhanced oil recovery processes and is more resistant to high salinity and shear rate, in comparison to HPAM. However, the viscosity of the solution is not increased as that of HPAM. Moreover, high cost and formation damage in the near-wellbore region are other limitations of this polymer. Wellbore damage may occur due to the polymer cross-linking or debris [12]. Using freshwater and prefiltration of the polymer solution might hinder this problem.

2.2.1 Chemistry aspects

Polymers that are applied in polymer flooding projects are produced in liquid or powder form. Both of these techniques have special limitations and advantages. It is more convenient to handle the polymer in liquid form that is more efficient to be used in offshore cases. However, the active amount of the polymer in powder form is more than that of liquid (90% vs 50%). This leads to the requirement of more storage equipment to supply the polymer solution. The preparation equipment for polymer solution is easier in offshore cases and just high-pressure pumps and filtration facilities are required. For polymers in powder form, it is necessary to transmit the powder pneumatically or mechanically, and then dissolve it and prepare it for delusion and aging [4]. The polymer structural analysis leads to describing the polymer, based on the constituent monomers, molecular weight, and microstructure. Polymerization kinetics influences the stated properties and accordingly, polymer applications are specified. For instance, the concentration of the monomer affects the viscosity modification, and molecular weight of the polymer. As a result, defining the optimum properties of the polymer and its characterization lead to an optimum polymer structure for EOR processes [10]. Fig. 2.3 presents the molecular structures of two functional polymers.

It is more beneficial to use fewer polymeric materials in terms of economic and environmental issues. For high-molecular-weight polymers, polymer injection tests become more complicated. Another important parameter is mechanical degradation. Long-molecular chains are more sensitive to high shear rates. The high viscosity of polymer solutions might also lead to low injectivity problems and a reduced increase in oil

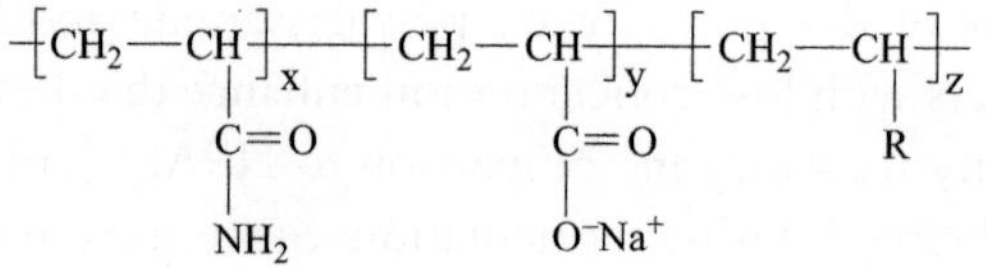

(a) RH-4 (R is carboxybetaine)

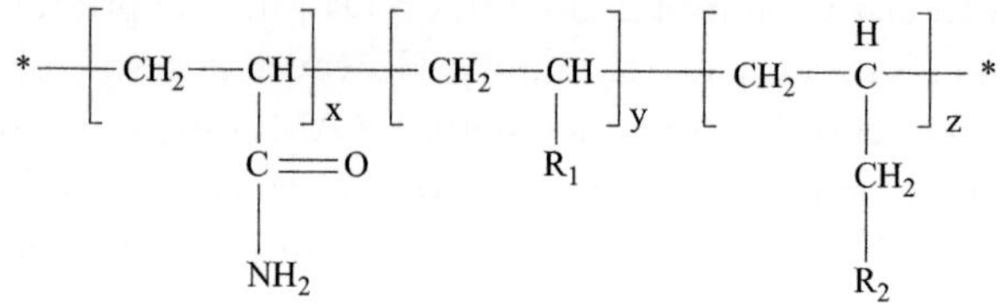

(b) KLP-6 (R1 is -RSO3Na, R2 is -(OCH2CH2) pOCnH2n+1 (n=8-12)

Fig. 2.3 The molecular structures of two functional polymers [13].

recovery factor [10]. Viscosity average molecular weight is determined using the Mark-Houwink equation:

$$[\eta] = KM^{\alpha} \tag{2.1}$$

$[\eta]$ denotes the intrinsic viscosity, K and α are empirical parameters, and M shows the polymer molar mass.

The mechanical and chemical degradation leads to reduction of polymer solution viscosity that reduces the functionality of the injected polymer. SiO_2 nanoparticles have been applied to improve the functionality of the synthesized polymers. Effects of these particles on carboxylic acids and silanes are confirmed by FTIR tests. The addition of SiO_2 nanoparticles leads to increase in the viscosity of xanthan gum. This is due to the formation of 3D structures between polymer chains and nanoparticles. The addition of the modified SiO_2 to 3-(methacryloyloxy)propyl] trimethoxy silane (MPS), octyl triethoxy silane (OTES), and oleic acid-method A (OAA) to synthesize HPAM enhances its thickening effect. On the other hand, the addition of the modified nanoparticles to HPAM and xanthan gum enhances its resistance to temperature and salinity, more than that of unmodified SiO_2. There are various numerical methods for predicting the viscosity of these polymers using a multilayer perceptron (MLP) neural network. This technique could be modified with Levenberg-Marquardt (LM), Bayesian regularization (BR), scaled conjugate gradient (SCG) algorithms, among which BR-MLP model has led to the best results [12].

2.3 Polymers features and screening criteria

Various polymers are categorized in one of the two groups of water-based and oil-based. They could be produced synthetically or naturally. The properties of the synthesized

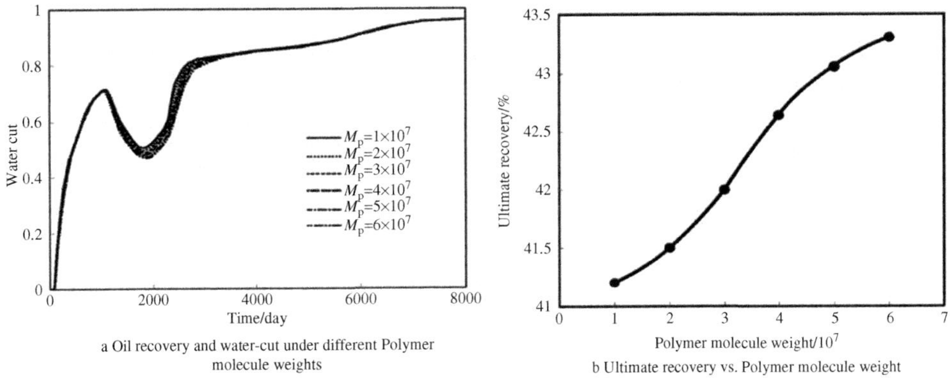

Fig. 2.4 (A) Oil recovery and water cut under different polymer molecule weights and (B) ultimate recovery vs. polymer molecule weight [15].

polymer are a function of temperature, viscosity, pH, shear stress, metal ions, and salinity. Screening of the polymers is performed based on the reservoir and polymer characteristics [14]. Fig. 2.4 represents the functionality of the water cut and oil recovery factor with molecular weight. More recovery factors and polymer relaxation time are obtained by increasing the molecular weight. Moreover, increasing the first normal stress difference and higher effective viscosity, and also lower residual oil are other results [15].

Thermal stability tests give important data on designing a polymer injection test since the synthesized polymer needs to be stable during the polymer flooding test under reservoir conditions. Otherwise, the viscosity of the polymer would be reduced. Precipitation of the polymer might also occur at high salinity conditions in the reservoir. A proper polymer solution actually has less than 50% reduction of viscosity after 6 months of injection to the reservoir. The selected polymer should have good adsorption, thermal, and rheological properties in addition to consistency with the salinity of the reservoir, permeability, temperature, biological, and shear stability, low cost, and high injectivity. High permeability leads to better injectivity of the polymer solution in comparison to low permeability reservoirs [16]. The polymer should have high resistance to degradation at high salinity and temperature environments. The chemical and thermal stability should also be considered. Since sulfonic acid groups enhance the thermal and mechanical stability, 2-acrylamido-2 methylpropane sulfonic acid (AMPS) is added to the polymer structure [17]. The optimum value of ionic strength is 0.9 M. With increase in the amount of AMPS, the ionic strength would increase and low ionic strength would be more appropriate for the synthesis of AMPS/AAm/AAc polymers. Polymers of high molecular weight are more desired for polymer injection tests. Increasing the concentration of the added monomer would lead to increase in the polymer molecular weight [17].

Debris present in the solution and polymer hydration might lead to reducing the injectivity of the prepared polymer and plugging the near-wellbore region. Moreover, while dissolving the polymers in water, they might swell, if the external surface of the particles becomes wet. This might produce a barrier for other particles to dissolve. This process forms gel ball or fish eye, which could be prevented by high mixing rate and slower addition of the polymers. Two new treatment techniques before polymer injection tests include diatomaceous earth (DE) and enzyme clarification [14].

2.3.1 Fingering effects

The fingering effect is referred to an unstable flow of a fluid with another fluid of lower viscosity. These fluids may be immiscible or miscible. It is said that displacing fluid is finger to the other fluid. Defining the stability limits of fingering phenomena is very important. For miscible fluids, the fingering is enhanced by gravity and transverse dispersion. Moreover, for immiscible fluids, fingering is a function of gravity, condensation, and capillary effects. The stability of the fingering can be predicted based on the flowing conditions [18]. The fingering effect is more important at a high injection rate at the same viscosity ratio, which leads to reducing the oil recovery factor. For high permeability rocks, oil recovery factor would be higher at the same operating conditions [19]. Increasing the polymer viscosity would result in reducing the number of fingers. Thus, the sweeping area would increase.

The water mobility in comparison to that of viscous oil could be very high while injected into the reservoir. This situation is very unstable. Since the mobility of water is higher, viscous fingering is generated in the homogeneous reservoir and water is reached to the production well by passing the oil [20]. Viscous fingering occurs when the mobility ratio of the displaced or displacing fluid is more than one. This is an important parameter in nonthermal recovery processes. The generated patterns of viscous fingering have been shown in various experiments conducted in micromodel or porous media. This phenomenon is grown rapidly while the flow rate and viscosity ratio are increased in water-wet porous media. Thus, the breakthrough time and subsequently, the recovery factor are reduced. In a porous media with a viscous fingering effect, the flow is separated in various regions in the porous media as a result of different mechanisms including, shielding, bypassing of the stagnant oil, and finger splitting. Different parameters influence the viscous fingering phenomena including IFT, viscosity ratio, wettability, and fluid velocity. Moreover, the channeling phenomena in porous media would be strengthened during viscous fingering [21]. This occurs while there are high permeability areas in the reservoir. Polymer flooding is a promising technique to treat the viscosity of water.

2.3.2 Screening criteria

Polymer flooding processes have been applied in a wide range of operating conditions including temperature, permeability, oil viscosity, pay zone thickness, salinity, and total dissolved solids. At the beginning of polymer injection, the original oil in place ranged from 36% to 97.1%, and water to oil ratio, varied from 0 to 100. The operating condition ranges for polymer flooding projects are reported in Table 2.1. The applicability of this process in sandstone or carbonate reservoirs was four times that of carbonates. The main economic limitation involves the cost-effectiveness of the polymers. For instance, for the lower cost of HPAM in comparison to xanthan, higher sizes for polymer bank and polymer concentrations could be supplied that leads to higher oil recovery factor. A majority of the cases for polymer flooding are applied in reservoirs with temperatures less than

Table 2.1 Polymer flooding [22].

Technical screening guides'

Crude oil	Wide-range recommendation	Range of current field projects
Gravity (°API)	>15	14–43
Viscosity (cP)	<150 (preferably <100 and >10	1–80
Composition reservoir	Not critical	
Oil saturation (% pv)	>50	50–92
Type of formation	Sandstones preferred but can be used in carbonates	
Net thickness	Not critical	
Average permeability (mD)	>10	10–15,000
Depth (ft)	<9000 (see temperature)	1300–9600
Temperature (°F)	<200 to minimize degradation	80–185

Properties of polymer flood field projects

Property	1980′ median (171 projects)	Marmul	Oerrel	Courtenay	Daqing
Oil/water viscosity ratio at reservoir temperature	9.4	114	39	50	15
Reservoir temperature (°F)	120	115	136	86	113
Permeability (mD)	75	15,000	2000	2000	870
% OOIP present at startup	75	≈92	81.5	78	71
WOR at startup	3	1	4	8	10
HPAM concentration (ppm)	460	1000	1500	900	1000
lbm polymer/acre-ft	25	373	162	520	271
Projected IOR (%OOIP)	4.9	25	−13	30	11
Projected bbl oil/lbm polymer	1.1	1.2	≈1.4	0.96	0.57
Projected bbl oil/acre-ft	27	461	≈230	499	155

200°F. Stability of these polymers at temperatures higher than 175°F has been studied by many researchers. Water salinity has been defined as an important parameter for polymer flooding. HPAM viscosity reduces with increase in the salinity for TDS up to 1%. Diverse opinions exist for injection of HPAM of low salinity to the reservoirs containing high salinity water. Most of the polymer flooding cases include low concentration and bank size of polymer. Moreover, low enhancement in IOR values is observed, which leads to a small change in WOR and oil production decline curves. Several successful polymer flooding projects with their operating parameters are reported in Table 2.1. The common features of these successful cases are useful to define the screening criteria of this process. Polymer flooding was conducted in low temperature (86–136°F) and high permeability (>0.87 Darcy) reservoirs. The oil in place and oil to water viscosity were in the range of 71%–92% and 15%–114%, respectively. The concentration of the injected polymer solutions was in the range of 900–1500 ppm. The polymer solution in the range of 162–520 lbm polymer/acre-ft was injected into low salinity water. The oil recovery factor increased from 11% to 30% at the end of this process [23].

More precise data on screening the suitable polymers in this process are presented in Table 2.2.

Table 2.2 Screening criteria for selection of polymer for EOR projects [14].

No.	Polymer and reservoir properties	Important characteristics/standard range
1.	Cost[a]	HIPAM: 2 Copolymer (PAM): 2 Xanthan gum: 3 HEC: 1 CMC: 1
2.	Filtration properties	Should be without fish eye formation and no significant plugging Filterability through a 1.2-μ filter with filter ratio below 1.5, provided more time for hydration of high mol. wt. polymers New filter test: Millipore API0 filter pad upstream of a 10-μm polycarbonate (Sterlitech Track Etch) membrane filter (both 13 mm in diameter) was used for filtration which matched well to that of berea core experiments, besides being economic and less time-consuming process
3.	Oil viscosity (cP)	4000[b]–0.4 cP existing ranges of oil viscosity reported oil viscosity of cP have proved to be a success widely (10–150 mPa s range)
4.	Water-oil mobility ratio	>1 Published successful range between 0.1 and 42

Table 2.2 Screening criteria for selection of polymer for EOR projects—cont'd

No.	Polymer and reservoir properties	Important characteristics/standard range
5.	Mobile oil saturation (% PV)	>10
6.	Reservoir temperature (°F)	<200
7.	Rock type	Sandstone or carbonate
8.	Reservoir permeability	20–2300 mD Permeability variation (DPV) lies in the range of 0.28–0.80
9.	Salinity and calcium tolerance	<100.000 ppm of salinity Formation chlorides should be less than 20,000 ppm and divalent ions (Ca^{++} and Mg^{++}) should be less than 500 ppm High-pH (>11) sodium silicate and chelating agent-Na_3NTA (trisodium nitrilotriacetate) and Na_4EDTA (letrasodium ethylenediaaminetetraacetate) solutions are an effective preflush for micellar/polymer floods in sandstones which contain high-salinity brine and divalent cations (Ca^{++} and Mg^{++})
10.	Surfactant compatibility	Phase behavior screening be based on: – Absence of macroemulsions or gels – High solubilization ratio – Aqueous stability of surfactant polymer based on injection requirements Structures with branched hydrophobes are more efficient in EOR ORS-41HF, ORS-46 HF, ORS-62, OP10, and CY 1 (local Chinese products), BES (an anionic surfactant), lignosulfonate, KPS-I (anionic surfactant), dodecyl sulfobetaine, YPS-3A (petroleum sulfonate), Petrostep B-100, TDA-13PO-SO4, and C20-24 IOS. Cosurfactants isobutanol, n-butanol, and isopropyl alcohol have been reported in many flooding projects worldwide Petrostep B-100 has being reported a success in the oil Cambridge Minnclusa Field flooding project
11.	Thermal stability	At temp >60°C. polymers tend to hydrolysis as pH decreases further neutrality Cross-linker as flow-diverting agents and increasing stability of polymers has been reported recently in many papers: (Polyacrylamides, high M_w—terephthalaldehyde dihydroxy-naphthalene, HMTA cross-linkers) (acrylamide-based stable polymers—phenol and formaldehyde cross-linker) (scleroglucan and lignosulfate—Cr(III) cross-linker)

Continued

Table 2.2 Screening criteria for selection of polymer for EOR projects—cont'd

No.	Polymer and reservoir properties	Important characteristics/standard range
12.	Mechanical stability	High M_w polymers are more prone to shear degradation. Chemical structure of polymer determines its resistance to flow, and thus, the introduction of monomeric groups such as acrylate, ATBS/sulfonate, and nVP/vinyl-pyrrolidone enhances shear stability
13.	Good injectivity	Reservoirs do not plug till $1000\,cm^3/cm^2$ flowrate of polymer solution and do not induce fractures upon injectivity
14.	Low retention	Desired level less than approximately $20\,\mu g/cm^3$. Values greater than $200\,\mu g/cm^3$ can have serious impact on economies of oil recovery projects
15.	Resistance to microbial degradation	Microbial degradation is aggravated at warn temperatures and under aerobic conditions. Addition of biocides and lower amounts of formaldehyde or sodium hydrosulfite can lower the adverse effects
16.	Oxygen, hydrogen sulfide, and pH sensitivity	Residual impurities cause M_w degradation in the presence of oxygen

[a]The order of cost per pound of polymer, one being the cheapest and four being the costliest. The cost varies with the supplier and amount of order.
[b]Maximum viscosity recorded at Pelican lake, Canada.
DPV, Dykstra-Parsons V-factor; M_w, molecular weight.

To define the proper characteristics of the reservoir for polymer flooding process, screening tools must be applied. Commercial analytical approaches are available in selecting the proper polymer. Moreover, artificial intelligence techniques could also help in screening the suitable polymers to be applied in polymer flooding tests. Another applicable technique is an experimental design in which maximum information is obtained at the lowest cost. Sensitivity analysis is performed on effective parameters and their effects on oil recovery factors are studied. In this technique, the effects of different input parameters are combined to reach the best information for the flooding process. Experimental design analysis (EDA) has been widely applied in various cases including history matching, surrogate models, and screening applications [24].

Fig. 2.5 represents the functionality of average permeability with average oil viscosity in approximately 400 polymer flooding tests. As it is obvious, most of the successful projects have been implemented in reservoirs with viscosity and average permeability of less than 100 cP and 1 Darcy, respectively. On the other hand, in three large-scale projects implemented in Pelican Lake, Bohai Bai "A," and Patos-Marinza, viscosity values are more than those values presented in dashed rectangle region. It is concluded that oil viscosity is not the only parameter that determines the applicability of the polymer flooding in various oil fields [1].

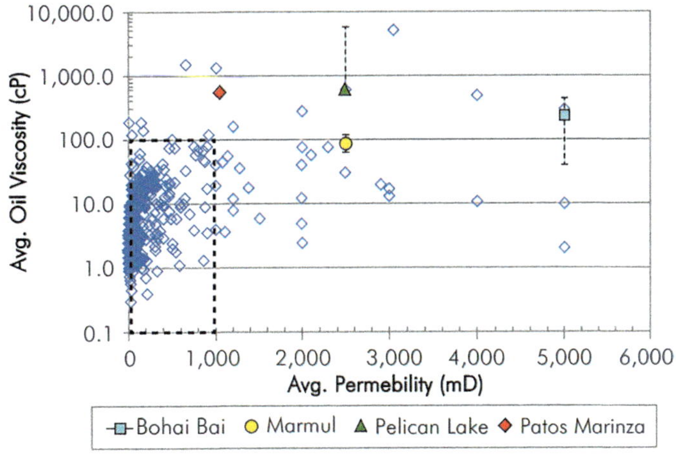

Fig. 2.5 Average viscosity-permeability plot of polymer floods implemented at a pilot or large/commercial scale [1].

2.4 Polymer rheology

Studying the rheology of the polymer at reservoir conditions is an important parameter for designing a polymer flooding process. The polymers are selected based on rheological properties. An EOR process would be effective for mobility ratios of less than 1. In polymer injection process, the mobility ratio is achieved by increasing the water viscosity. As it is required to keep the polymer viscosity at high values, it is important to study on functionality of polymer solution viscosity with salinity and temperature. Based on the studies performed in various oil fields, the preferred viscosity of oil should be less than 150 cP. To attain a proper mobility ratio, polymer viscosity should be optimized as high values of the viscosity are not desired in all processes. Polymer solutions of high viscosity is required, which leads to concentrated polymer solutions. With increase in the viscosity of the solution, the injectivity is reduced. Polymers with high viscosity do not diffuse to low permeability areas. The optimized value of viscosity leads to a mobility ratio of less than 1. The concentration of the polymer solution with economic considerations should lie in the range of 1000–3000 ppm. Two rheological properties should be evaluated for chemical EOR processes. Experiments of steady shear are performed at a shear rate of $7.3\,s^{-1}$. The shear rate selected for the experiments should coincide with that of an actual case in pipelines and reservoir rock, especially at perforated areas and the near-wellbore region. Previously, dynamic rheology test was not performed regularly, but now it is found that elasticity tests are required to define the amount of residual trapped oil. The elastic modulus and complex viscosity are obtained using dynamic tests [16]. The structural properties of polymers make them suitable for polymer flooding tests. The shear thinning behavior of polymer solutions leads to better alignment of polymer chains at high shear rates [10].

$$\tau = K\gamma^n \text{ with } n < 1 \text{ and } \mu_{effective} = K\gamma^n \quad (2.2)$$

where γ represents the shear rate, τ denotes shear stress, $\mu_{effective}$ shows the effective viscosity, K and n are the consistency coefficient and liquidity index, respectively [15].

2.4.1 Polymer sensitivity to mechanical shearing

Most of the polymer solutions are sensitive to mechanical degradation. While the flow pressure is reduced, the shear which is incurred to the polymer chain makes it shorter. Since the flow rate of the system is controlled by choke valves, the mechanical shear leads to the degradation of the polymer structure during flow pressure drop [4].

There is a category of polymers named "shear resistant polymers." A group of these polymers contain delayed viscosity polymers (DVPs) and 2-acrylamide-2-methylpropanesulfonic acid (ATBS) or AMPS monomer. For high shear rates, the chemical composition of the solution must be set to reduce mechanical degradation. To improve the mechanical stability of the polymer solution, ATBS monomer is added to the polymer structure [4]. Another strategy is to encapsulate the polymer at high shear rates. These polymers are called DVP. They are synthesized using inversed emulsification technique and might contain ATBS. Due to the structure of the protective layer, encapsulated polymers are preserved for shear. Based on the pressure and temperature of the reservoir, the protected polymers are set to release from the shell. As the shell is completely destroyed, the polymer viscosity is kept constant [4].

2.4.2 Viscoelastic behavior of the polymer

Viscoelastic behavior is of paramount importance in defining the efficiency of the polymer flooding process. Thus, studying rheological properties of a polymer provides enough information on polymer injection process [10]. The rheological behavior of the polymer is characterized by experimental studies using a rotational rheometer. In this approach, the viscoelastic behavior of the polymer is characterized by various environmental conditions related to porous media. In these experiments, the properties of small amplitude oscillatory shear (SAOS, stress relaxation), steady shear viscosity, extensional thickening, and N1 (normal stress difference) are evaluated [8]. Some aspects of rheology are determined using in situ rheology in which the polymer solution flows into the porous media. Based on the structure and composition of polymers, they are able to store energy (elastic behavior), dissipate it (viscous behavior), or both. These behaviors are defined as shear-thinning and shear-thickening behaviors, respectively. As stated before, polymers might behave differently. Some of them might show high resistance to stretching, compared to other polymers that are stretched under low shear. Most of the polymers applied in EOR processes are of high molecular weight and concentration. Consequently, polymers exhibit viscoelastic behavior. Viscoelastic behavior of polymer solutions is of paramount importance, since the speed of the fluid is changed while flowing in the porous media. This leads to generation of extensional forces while crossing from the region near the walls. These properties are determined using various devices which relate the stress to strain rate in different operating conditions. Elastic turbulence occurs at low Reynolds numbers, which is an additional term in equations of continuum mechanics. The stability of laminar flow is changed for polymers of high elasticity which leads to elastic instability. Elastic turbulence mostly occurs at the pore level in which the stability of the laminar flow is altered with polymers of high elasticity [8].

To study the effect of viscoelasticity, a standard equation is represented for isotropic Newtonian flow for an incompressible fluid. The shear viscosity is a function of shear stress based on the following equation:

$$\eta = \frac{\tau}{\gamma} \tag{2.3}$$

where η denotes the shear viscosity, γ represents the shear rate, and τ shows the shear stress. This equation is linear for Newtonian fluids and is nonlinear for viscoelastic fluids such as HPAM solution. Moreover, for a completely viscous fluid deformation, the stress becomes zero at constant strain. On the other hand, in a completely elastic solid, there is no relaxation. In Fig. 2.6 the behaviors of a Newtonian fluid (instantaneous relaxation) and a viscoelastic fluid (with elastic and viscous behavior) are presented [8].

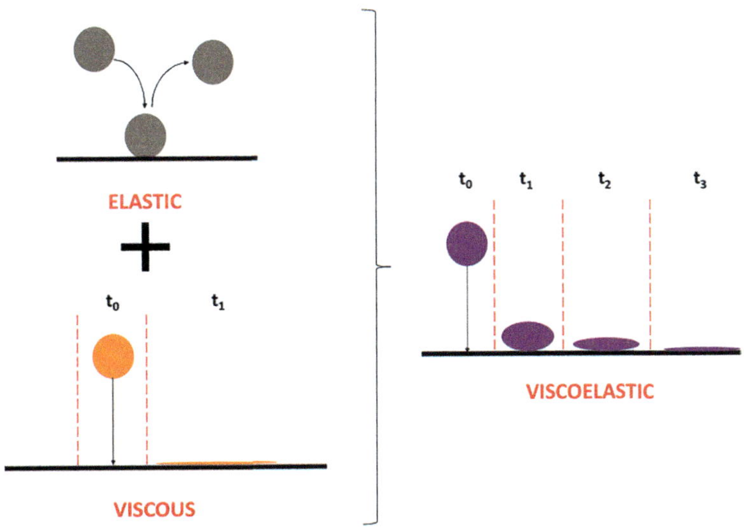

Fig. 2.6 Illustration of elastic, viscous, and viscoelastic behavior [8].

The relaxation module represents the relaxation behavior of the solution.

$$G(t) = \frac{\tau(t)}{\gamma} \tag{2.4}$$

where G denotes the relaxation modulus. The functionality of strain and stress relaxation is linear for $\gamma < 0.5$. This is known as linear-viscoelastic behavior or plateau modulus (Ge). For $\gamma > 1$, this relation is nonlinear [8].

The equation of linear viscoelasticity is described by the following equation:

$$\tau = \int_{-\infty}^{t} \sum\nolimits_{1}^{N} G_k e^{\frac{t-t'}{\lambda_k}} \gamma(t') dt' \tag{2.5}$$

where t states the time and t' is the past time since the infinite past $-\infty$. Since this equation does not show a 3D viscoelastic behavior, it is not possible to predict the normal stresses to 2D shear plane. This means that at low shear rates there is no functionality between shear rate and shear viscosity.

Nonlinear viscoelasticity model is applicable for $\gamma > 1$.

$$G(\gamma, t) = \frac{\tau(t, \gamma)}{\gamma} \tag{2.6}$$

The flow behavior of the polymer solution is represented by shear flow and strain flow. The solution effective viscosity (η_e) is formulated as follows [25].

$$\eta_e = \eta_s + \eta_{st} \tag{2.7}$$

where η_s states the shear viscosity and η_{st} denotes the intrinsic viscosity. Since the polymer molecules rotate while flowing in porous media, effective viscosity is only a function of shear flow. For long enough rotation less flow under strain, polymer deformation is possible in a specified pore geometry. This deformation results in an increase in strain viscosity and consequently, the effective viscosity increases [25].

2.4.3 Theoretical model

Viscoelastic studies are characterized based on the Deborah number (De), which is defined as the product of effective strain rate and relaxation time. In this model, spherical packings are assumed with uniform grain diameters that do not exist in reality. Due to this limitation, a theoretical model proposed by Maxwell is applied to describe the viscoelastic behavior. In this model, it is assumed that the fluid has both viscous and elastic behaviors. For the pure elasticity, Hooke-Model is applied. Viscosity is represented by a dashpot element. The combination of viscous and elastic forces is defined as follows:

$$\varepsilon = \varepsilon_s + \varepsilon_{st} \tag{2.8}$$

Total stress:

$$\sigma = \sigma_s + \sigma_{st} \tag{2.9}$$

Based on the Newtonian law of flow and Hooke's law of elasticity, these equations are expressed as follows:

$$\varepsilon_s = \frac{\sigma_s}{G} \tag{2.10}$$

$$\varepsilon'_{st} = \frac{\sigma_{st}}{\eta} \tag{2.11}$$

The first derivatives of Eqs. (2.8) and (2.10) gives.

$$\varepsilon' = \varepsilon'_s + \varepsilon'_{st} \tag{2.12}$$

$$\varepsilon'_s = \frac{\sigma'_s}{G} \tag{2.13}$$

by combining Eqs. (2.11)–(2.20):

$$\varepsilon' = \frac{\sigma'_s}{G} + \frac{\sigma_{st}}{\eta} \tag{2.14}$$

The relaxation time is defined as the proportion of elasticity to viscosity and is represented as follows:

$$T = \frac{\eta}{G} \tag{2.15}$$

The differential equation of the Maxwell-Model is defined as follows:

$$\sigma' + \frac{\sigma}{T} = G \cdot \varepsilon \tag{2.16}$$

For the creeping flow experiments which are performed at constant shear conditions, the following expression is the solution of this equation:

$$\eta'(t) = \eta \cdot \left(1 - e^{-\frac{t}{T}}\right) \tag{2.17}$$

For a nonstationary shear flow, the viscosity is reduced to.

$$\eta'(t) = \eta \cdot e^{-\frac{t}{T}} \tag{2.18}$$

This model is modified for the characterization of viscoelastic flow behavior. In this model which is based on the experimental studies, by reducing the injection rate or pressure, they follow a time-dependent exponential function. The validity of Darcy's law is lost. Q is the injection rate, ε denotes the necessary strain for the polymer solution. ε' is the strain while the flow is reduced by Q'' to Q'. The following equations are valid.

$$Q = Q' + Q'' \tag{2.19}$$

$$\varepsilon = \varepsilon' + \varepsilon'' \tag{2.20}$$

This relation is applied for the deformation of the molecules:

$$\varepsilon''_t = \varepsilon'' \cdot e^{-\frac{t}{T}} \tag{2.21}$$

with $t_0 \leq t \leq t_1$

The absolute deformation at time t is defined as.

$$\varepsilon_t = \varepsilon' + \varepsilon'' \cdot e^{-\frac{t}{T}} \tag{2.22}$$

The pressure difference and flow are reduced exponentially with time. Pressure variations are represented as follows:

$$\Delta P_t = P' + P'' \cdot e^{-\frac{t}{E'}} \tag{2.23}$$

where E' is an index which is a function of relaxation time and the ratio of effective pore radius. For the exact calculation of this parameter, variations in pressure and flow rate are recorded. Similarly, the flow variation is represented as follows [25]:

$$q_t = q' + q'' \cdot e^{-\frac{t}{E'}} \tag{2.24}$$

2.4.3.1 Normal stress difference, N$_1$

The fluid viscoelastic behavior might be defined by the difference between normal stress values, namely, N_1 and N_2. The fluid elasticity while flowing is defined by N_1. For a viscoelastic fluid, a 3D deformation occurs while flowing, which is expressed by a 3×3 tensor which has six parameters for shear stress and three normal stress parameters (σ_{xx}, σ_{yy}, σ_{zz}). The first normal stress difference based on the introduced deformation tensor is as follows:

$$N_1 = \sigma_{xx} - \sigma_{yy} \tag{2.25}$$

Increasing the oil recovery factor leads to increase in the value of N_1. It determines whether the increased oil recovery factor is a result of flow instability or the increased value of N_1. By increasing N_1, elastic turbulence is improved which leads to excess oil recovery factor [8].

2.4.3.2 Weissenberg number

Weissenberg number is defined by the following equation.

$$W_i = \lambda \dot{\gamma} \tag{2.26}$$

This number states the nonlinear behavior and elasticity of the fluid as a result of the nonisotropic arrangement of the molecular structures of the polymer under shear stress. For viscous fluids $W_i = 0$ and viscoelastic behavior of the fluid is designated for $W_i > 0$. For high shear rates at injection and production wells, W_i is more than that for the fluid flow between wells. Moreover, this parameter is higher in injection wells in comparison to the production wells, since the solution is degraded while flowing in the reservoir and the relaxation time of the polymer would reduce in comparison to the injection well [8].

2.4.3.3 Deborah number

Deborah number is another dimensionless parameter, which describes the viscoelasticity of the polymer solution, defined as.

$$De = \lambda \frac{v_s}{L} \tag{2.27}$$

where L states the fluid flow length and v_s denotes the fluid superficial velocity. This parameter defines the elastic strain of the solution. The fluid is viscoelastic for $De > 1$, and for that of less than 1, it is viscous. This parameter is variable throughout the injection process and it is hard to predict it [8].

2.4.4 Influence of polymer viscoelasticity on its injectivity

The shear-thickening behavior of the polymers has a negative effect on the injectivity of the fluid. Since the flow rate is high near the injection and production wells, the injection pressure might become more than that of formation parting pressure. This leads to fracturing the reservoir rock. The generated fractures of high permeability lead to the explosion of the fluid to a high surface. This leads to a reduction in the shear rate. The shear thickening and viscoelastic behaviors of the fluid reduce and as a result, S_{or} value is almost constant around the wellbore. In some cases, the injection rate should be reduced to avoid fracture generation. This leads to reducing the injectivity of such polymers [26].

2.4.5 Viscoelastic influence of the polymer on residual resistance factor

The capillary number is defined as the ratio of viscous to interfacial forces and is used to study the microscopic displacement of the polymer. This parameter is defined as.

$$N_c = \frac{\mu \times v}{\sigma} \tag{2.28}$$

where σ represents the IFT, μ denotes viscosity, and v states velocity. The characterization of a polymer flooding process is performed by defining the two parameters of resistance factor (RF) and residual resistance factor (RRF). RF states the mobility control as the ratio of the water mobility to that of polymer, using the following equation [27]:

$$RF = \frac{\lambda_w}{\lambda_p} = \frac{K_w/\mu_w}{K_p/\mu_p} \tag{2.29}$$

RRF denotes the reduction in permeability as a result of polymer flooding. This parameter is defined as water permeability before and after polymer flooding, by the following relation [27]:

$$RRF = \frac{K_{wb}}{K_{wa}} \tag{2.30}$$

2.5 Polymer as fracturing fluid in oil reservoir

Hydraulic fracturing is a well-defined technology to increase the oil recovery factor in tight reservoirs. Studying the fracturing fluid is very important in developing this technology. Fracturing fluids are divided into two categories. One of them is guar gum. It is one of the early used fracturing fluids with considerable viscoelastic properties. However, this material might be easily decomposed by bacteria or is susceptible to mechanical degradation resulting from friction. So, new fracturing fluids were then developed namely, VES (viscoelastic surfactant). Surfactants are added to this material to generate micellar structure to show better viscoelastic properties. This material is not resistant to temperature which leads to a higher injected volume of VES and increasing the operating costs. Moreover, the viscosity of VES is reduced at high shear rates. Another fracturing fluid is a polymer. This topic has been extensively studied in recent years for increasing the oil recovery factor from tight reservoirs. Polymers have many advantages in comparison to previously applied materials. It has lower friction at low concentrations, due to the linear structure of the polymer. Traditionally applied polymers had some disadvantages including the poor ability for carrying sand. So a fracturing fluid with high diversity should be used [28].

Water-based fracturing polymers are of paramount importance in hydraulic fracturing. Guar gum is one of the very first materials that was developed as the fracturing fluid. This material in the cross-linked and linear forms is still applied in the industry. By injecting the fracturing fluid into the reservoir at high pressure, some cracks are generated and fractures are propagated in the reservoir. Using viscous fracturing fluids including polymer, guar, and gels is so advantageous. Under the pressure, a layer of the filter cake is generated on the wall of fractures to prevent leaking the fracturing fluid to surrounding rocks. However, for high-molecular-weight polymers, the wall-building capability is low and polymers might invade to formations of high permeability. As a result, an internal filter cake is produced leading to formation damage in porous media [29].

Since formation damage in a tight reservoir arising from the hydraulic fracturing fluids reduced its productivity, another category of materials was generated. It is called "slick water" and is composed of a very low concentration of a linear polymer in water. Low concentration polymer acts as a friction reducer which lowers the friction while the fluid is pumped to the reservoir. No breaker is required in this material, however, the capability of carrying the proppants at low viscosity fluids should be studied more deeply [29].

2.6 Polymer adsorption

Retention of the injected polymer as a result of adsorption or mechanical trapping leads to the reduction of sweeping efficiency. The amount of adsorbed polymer should maintain at the minimum value. High values of polymer retention lead to the reduction of polymer concentration in the solution, which conducts to increasing the required amount of polymer to attain the desired mobility ratio and recovery factor. This might not be economically viable. There are static and dynamic approaches for evaluating polymer retention behavior. Many parameters are effective on the total amount of polymer retention, including temperature, salinity, polymer chemistry, and composition of the reservoir rock. In isoelectric point, there is no net charge on that surface at a specified pH. Above this point, the net charge of the surface would be negative, and below this point, it would be positive. The chance for the adsorption of the polymer is for concentrations lower than 1 mg/g-rock [16]. While polymer is adsorbed on the surface to make an impermeable layer, the effective cross section in the pore throat is reduced. The flow of polymer solution becomes difficult since the pore size is close to the length of macromolecule. Thus, for low rates of injection, the polymer solution does not move with an equal rate of the brine. The main effects of polymer adsorption on pore walls are as follows:

(1) The penetration of polymer solution containing brine is difficult since the steric effect of the adsorbed polymer becomes considerable.
(2) Generation of lubricating effect on the surface which makes the oil easier to flow.
(3) Enhancement of oil mobility by improving rock wettability [14].

In a static approach to determine adsorption, the adsorbent is transferred by diffusion. In this approach, the polymer solution is transferred to the core sample and is kept to reach equilibrium. Then the liquid phase is separated to determine its polymer concentration. Adsorption is calculated with a decrease in polymer concentration in comparison to the initial value [30]:

$$G = \frac{(C_0 - C) \times V}{m} \tag{2.31}$$

where C_0 and C indicate the initial and equilibrium concentrations of polymer, respectively. m represents the mass of solid particles and V denotes the solution volume. In the dynamic approach of defining adsorption, an analogy is described between polymer solutions with chemical flooding. The prepared solution of polymer is pumped to the core. As the initial concentration of the polymer is known and the outlet fluid is analyzed regularly, the current polymer concentration is calculated until it reaches the initial value. The total adsorption is defined by analyzing all the samples [30]. Langmuir isotherm is used for modeling adsorption. The functionality of water salinity and polymer concentration with adsorption could be considered linear or nonlinear.

$$C_f = \frac{a_4(C_{41})}{1 + b_4 C_{41}} \tag{2.32}$$

$$a_4 = \left(a_{41} + a_{42} C_{spe}\right) \tag{2.33}$$

where C_{41} denotes the concentration of the polymer solution and the parameters a_{41}, a_{42} and b_4 are the inputs of the model.

$$ad = \frac{\left(tad1 + tad2 \times x_{nacl}\right) \times ca}{1 + tad3 \times ca} \tag{2.34}$$

x_{nacl} denotes the salinity, ca stands for polymer solution molar fraction, and $tad1$, $tad2$, and $tad3$ are model inputs.

$$C_{ads} = \frac{ac^m}{1 + \mathrm{b}C'} \tag{2.35}$$

$$a = (a1 + a2C\,se)\left(\frac{k_{ref}}{k}\right)^n \tag{2.36}$$

In most of the current simulators, an uneven front for adsorption distribution is not easy to produce. This could arise from the uncertainties in reservoir rock properties or degradation of the polymer in the channels. These problems are quantified while analyzing the samples using a microscope but are not quantitatively defined [30].

2.7 Displacement mechanisms in polymer flooding

It has been proved in previous studies that in the polymer flooding process, the microscopic efficiency is enhanced while residual oil saturation is reduced. This is due to the elasticity of the polymer solution. In this process, the efficiency of the polymer flooding is evaluated by sweeping the "residual oil" after water flooding in different parts of the porous media, including in dead ends, on rock, in pore throats, and unswept residual oil in the heterogeneous part of the rock. In all the above-stated conditions, the residual oil was reduced [5].

Various mechanisms resulting from the viscoelastic behavior of polymer solutions include (a) pulling (b) oil-thread flow (c) stripping, and (d) shear thickening. In the

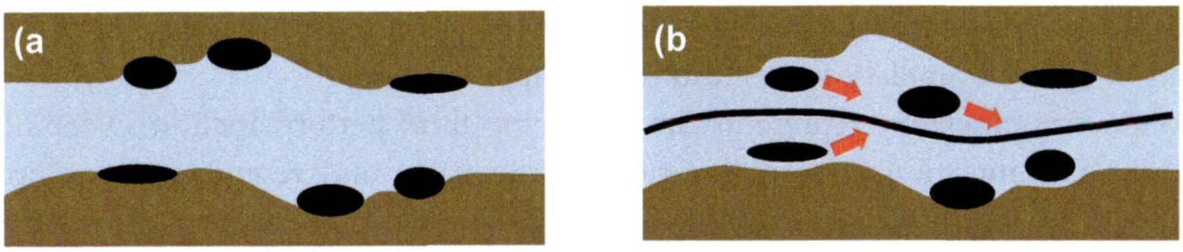

Fig. 2.7 (A) Situation after conventional water flooding and (B) viscoelastic polymer solution is injected [8].

pulling mechanism, the trapped oil is pulled out from the dead-end pores. In the second mechanism, the flow becomes important while the residual oil converts to droplets. Since the polymers have viscoelastic behavior, oil droplets are pulled to oil threads. Threads make the oil flow toward the residual oil in downstream which finally makes an oil bank that would sweep more simply. After water flooding, the oil is isolated and left behind (Fig. 2.7A). By injection of polymer solution, an oil thread is formed as a result of stretching and thinning of the drops (Fig. 2.7B). As the thread forms, the polymer droplets generate an oil bank in downstream that is easy to recover in comparison to oil drops inside the pore structure [8]. The third mechanism namely stripping indicates the stripping of the oil on pore walls. This occurs when the velocity of viscoelastic polymer solution is higher at the wall [8].

In the last mechanism namely shear thickening, the viscosity of the polymer solution is increased as a result of viscoelastic behavior. In this mechanism, the viscosity is not actually increased, but due to elastic turbulence, an additional pressure drop is generated [31].

2.7.1 Evaluation of volumetric sweep efficiency

Oil recovery factor is evaluated using volumetric sweep efficiency, which is defined by the following relation:

$$E_r = E_d \times E_v \tag{2.37}$$

where E_r denotes the recovery efficiency, which states the fraction of original oil in place (OOIP). E_d represents the microscopic oil displacement efficiency, which is the total amount of oil that is swept from a unit rock segment, and E_v shows the volumetric sweep efficiency, which is the portion of the reservoir that has been in contact with the polymer solution. Low sweep efficiency values lead to the reduction of the produced oil. One of the reasons for poor sweeping efficiency is the mobility difference between the sweeping fluid and oil. Water moves rapidly with an irregular front toward the production well. Oil and brine move at the front in contact with the polymer solution, which enhances the oil recovery factor [14].

Reducing the interfacial tension and capillary force leads to the enhancement of microscopic efficiency. Another technique is reducing the oil viscosity. The capillary number is an indicator of displacement efficiency in which viscous forces lead to mobilization of oil, and capillary forces lead to its trapping.

Volumetric efficiency represents the displacing fluid performance in sweeping the reservoir. The injection pattern, location of the wells and reservoir characteristics, gas-oil contacts, mobility ratio, and flow rate are important parameters to increase the efficiency of a polymer flooding process. Sweep efficiency might be defined as the product of vertical sweep efficiency and areal sweep efficiency. In areal sweep efficiency, a fraction of the formation area that is swept by polymer is considered. Vertical sweep efficiency states the total formation volume in the vertical direction, which is swept by the polymer. Microscopic and macroscopic sweep efficiencies are illustrated in Fig. 2.8. As presented before, the oil recovery factor could be increased by increasing the capillary number or the mobility ratio [32].

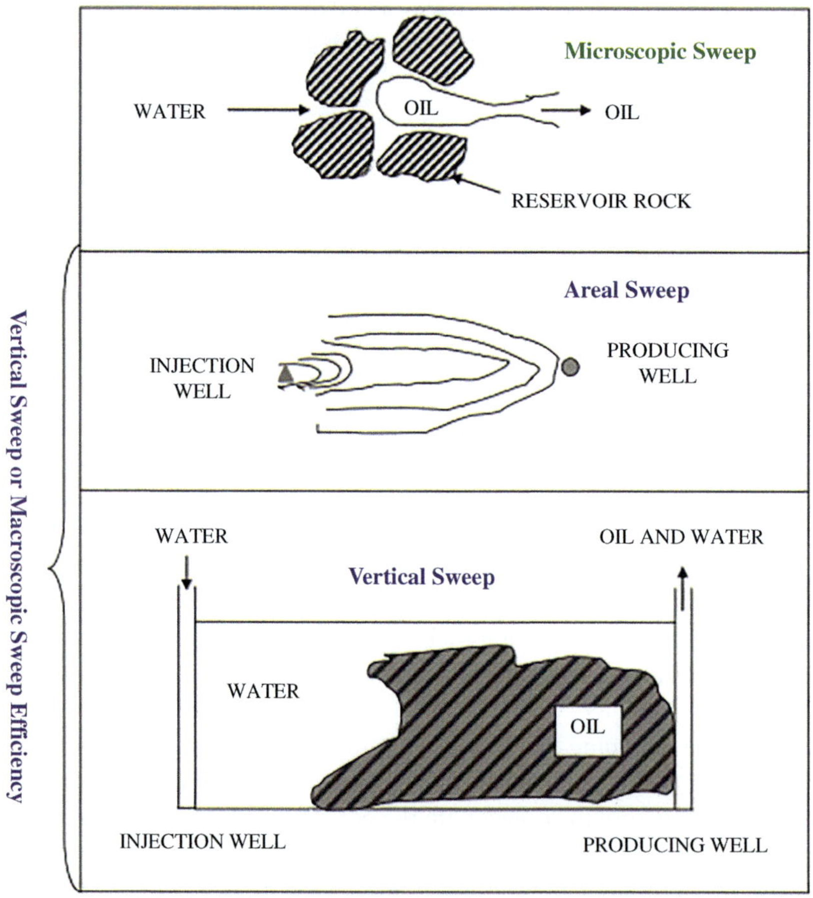

Fig. 2.8 Schematic representation of microscopic and macroscopic sweep efficiencies [32].

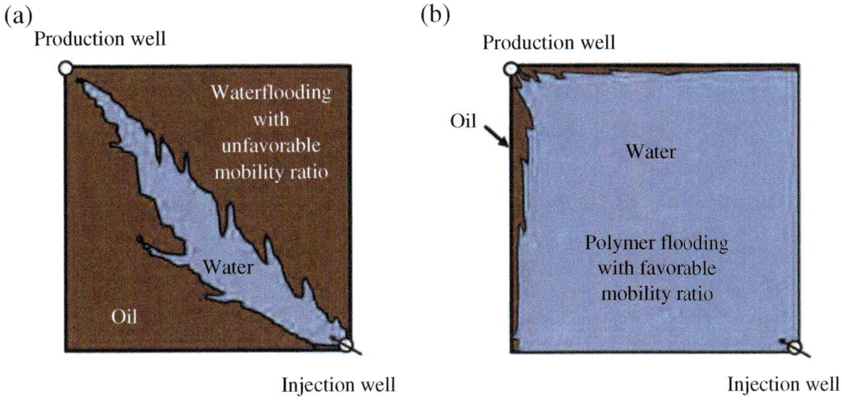

Fig. 2.9 (A) Water flooding ($M > 1$) and (B) polymer augmented water flooding ($M \leq 1$) [32].

2.7.2 Mobility control improvement

In the polymer flooding process, the mobility of the aqueous phase is controlled. The mobility of the polymer is reduced by dissolving in water, the viscosity of the aqueous phase is reduced and the fingering effect diminishes. In this process, the mobility ratio should be reduced to less than 1 to create favorable conditions for polymer flooding. As presented in Fig. 2.9, for mobility ratios more than 1, the viscosity difference between polymer and solution is increased which results in the fingering effect and reduction in oil recovery factor [32].

The mechanical entrapment of polymers in porous media is a reversible process and by a high displacement pressure, the polymers are mobilized again. Polymer retention reduces the viscosity of the polymer solution. However, mobility is improved by decreasing the viscosity of the displacing fluid. As a result, the mobility control is also enhanced by increasing the RRF and adsorption retention in porous media [27].

2.7.3 Influencing on relative permeability

In the polymer flooding process, the viscosity of water is increased by adding the polymer to the injected water. This process also reduces the relative permeability to water [33]. The main purpose of this process is to reduce water mobility by increasing its viscosity and also enhance relative permeability of water. This would increase sweeping efficiency on macroscopic scale [12, 19]. The addition of polymers leads to decreasing the relative permeability of water more than that of oil [27]. Since the adsorbed molecules oppose the flowing of displacing phase, the relative permeability of water is reduced. To study this effect, the permeability reduction factor is introduced [15]. Oil relative permeability is a function of the residual oil saturation, which resists against the polymer molecules elasticity to change the relative permeability. The relative permeability of each phase is defined using the following relation:

$$K_{rl} = K_{rl}^{0}\left(\frac{S_l - S_{rl}}{1 - S_{r0} - S_{wc}}\right)^{\beta_l} \tag{2.38}$$

where K_{rl}^{0} is the endpoint of permeability of phase 1. β_l represents the exponent of relative permeability for this phase [15].

2.8 Fractional flow curve analysis

Using fractional flow curves leads to analyzing the flow behavior in different EOR processes in 1D flow. Neglecting the buoyancy and capillary forces, the fractional flow equation would be.

$$f_w = \frac{1}{1 + \frac{\lambda_o}{\lambda_w}} = \frac{1}{1 + \left(\frac{k_{rw}}{k_{ro}}\right) \times \left(\frac{\mu_o}{\mu_w}\right)} \tag{2.39}$$

where $\lambda_\alpha = k_{r\alpha}/\mu_\alpha$ represents the relative mobility of phase α [34]. The injected polymer solution is analyzed by an additional function in which the input parameters are updated to original water-oil f_w function. In this process, viscosity of the displacing fluid is increased, which should be implemented in equations. An example of f_w vs S_w and variations of water saturation is presented in Fig. 2.10. Analyzing this process designates four steps: (1) Designation of injection point and initial conditions of polymer flooding process on the curve. (2) Point A is obtained by drawing a tangent line from point (D_s,0) on f_w curve. The specific velocity of the polymer front is obtained using the slope. (3) Point B is defined at the intersection of the tangent line and f_w. (4) Line B defines the velocity of the oil front. Chemical adsorption would lead to delay in front. The retardation factor is calculated as follows:

$$D_s = \frac{1 - \varphi}{\varphi} \frac{\rho_s}{\rho_w} \frac{\Gamma_s}{C_{inj}} \tag{2.40}$$

where ρ_s states the rock density, Γ_s is the adsorbed chemical, φ denotes the porosity, C_{inj} indicates the concentration of the injected chemical, and ρ_w shows the density of the polymer solution.

Considering the chemical adsorption leads to changing the places of points A and B on the curve. The results are specified by the decrease in the front propagation [34]. Fig. 2.10 designates two fronts in this process. First, water pushes oil and then the injected polymer displaces the oil toward the production well. In the first front, values of the relative permeability and reservoir conditions, designate the stability of the front. This is not modified easily, since it depends on the inherent properties of the reservoir. In

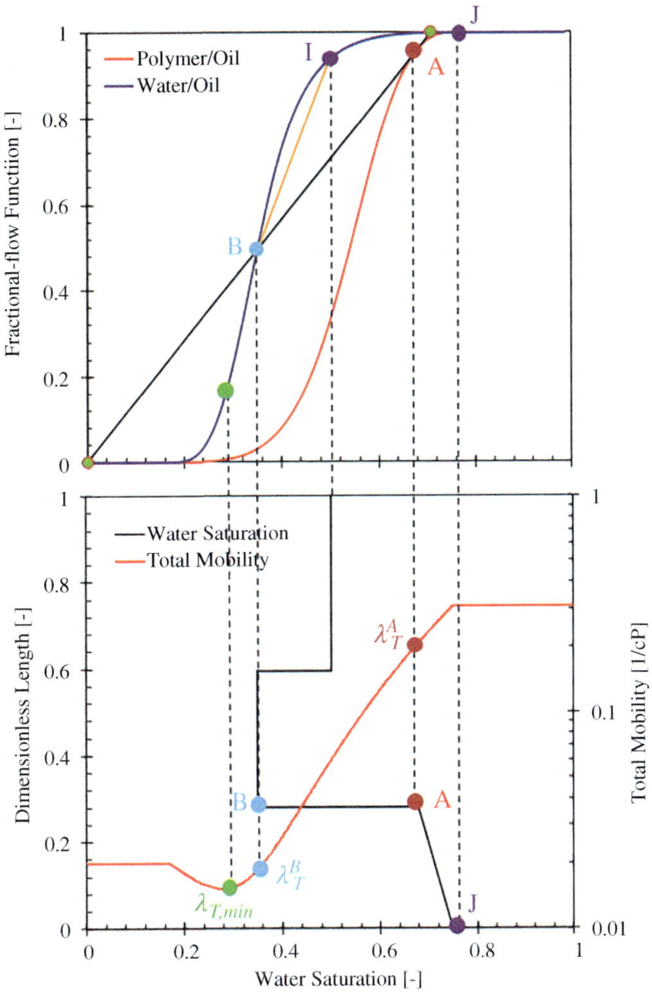

Fig. 2.10 The fractional-flow function vs water saturation (top) the constructed water-saturation profile and total mobility as a function of water saturation (bottom) [34].

the second front, the viscosity ratios between oil and injected polymer, adsorption and the relative permeabilities, define the stability of the front [34].

2.8.1 Continuous polymer injection without adsorption

In Fig. 2.10, the fractional flow curves for polymer and water flooding are presented. A shock and saturation profile is illustrated in the front at 0.2 PV in Fig. 2.11A and B. In this case, using the previously described procedure, the two shocks are reported. The first of which is attributed to the oil front in which water is pushing the oil. The second

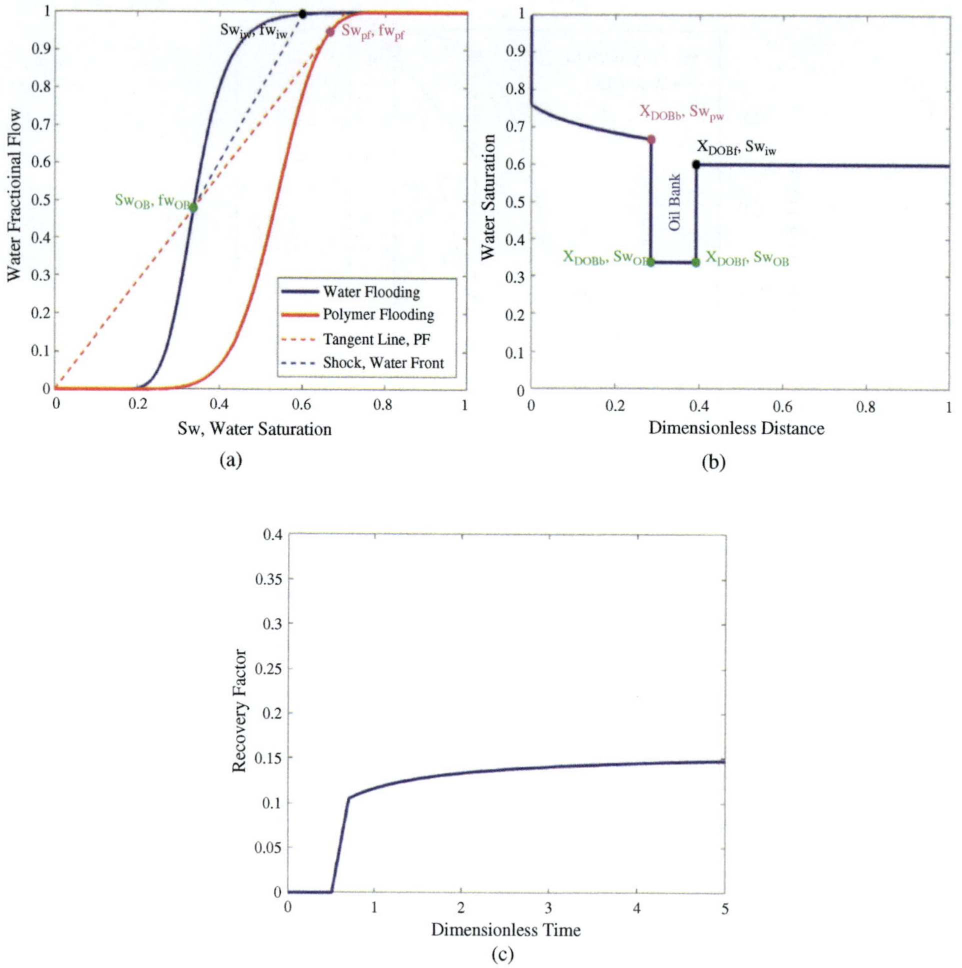

Fig. 2.11 (A) Construction of shock front, (B) water saturation profile at $t_D = 0.20$ PV, and (C) cumulative oil recovery for tertiary polymer flooding, continuous injection, no polymer retention [35].

one is the chemical front in which the injected polymer displaces the oil to the outlet point. A region with constant polymer concentration is developed behind the surfactant shock, in which a clear bank is formed while the viscosity of the polymer solution is increased [35].

2.8.2 Influence of initial oil saturation

Various studies show that at the early stages of oil production with low water cut, polymer flooding would result in reducing the residual oil saturation. For high values of water saturation, the dispersion coefficient in water is lower for higher values of water saturation, which leads to a decrease in the sweeping efficiency of the injected polymer.

A polymer injection test was implemented with no modification on saturation of the residual oil. Results indicated that water saturation at the water flooding shock front was higher than that in the oil bank. Otherwise, a spreading wave was forms between the oil bank and the polymer shock [35].

Cumulative oil recovery for polymer flooding is presented in Fig. 2.11C. Using the proposed method, two shock waves are generated. In this case, the oil saturation was 0.66. Comparison of shock front for water flooding and polymer flooding is presented in Fig. 2.12A and construction of shock front polymer flooding is shown in

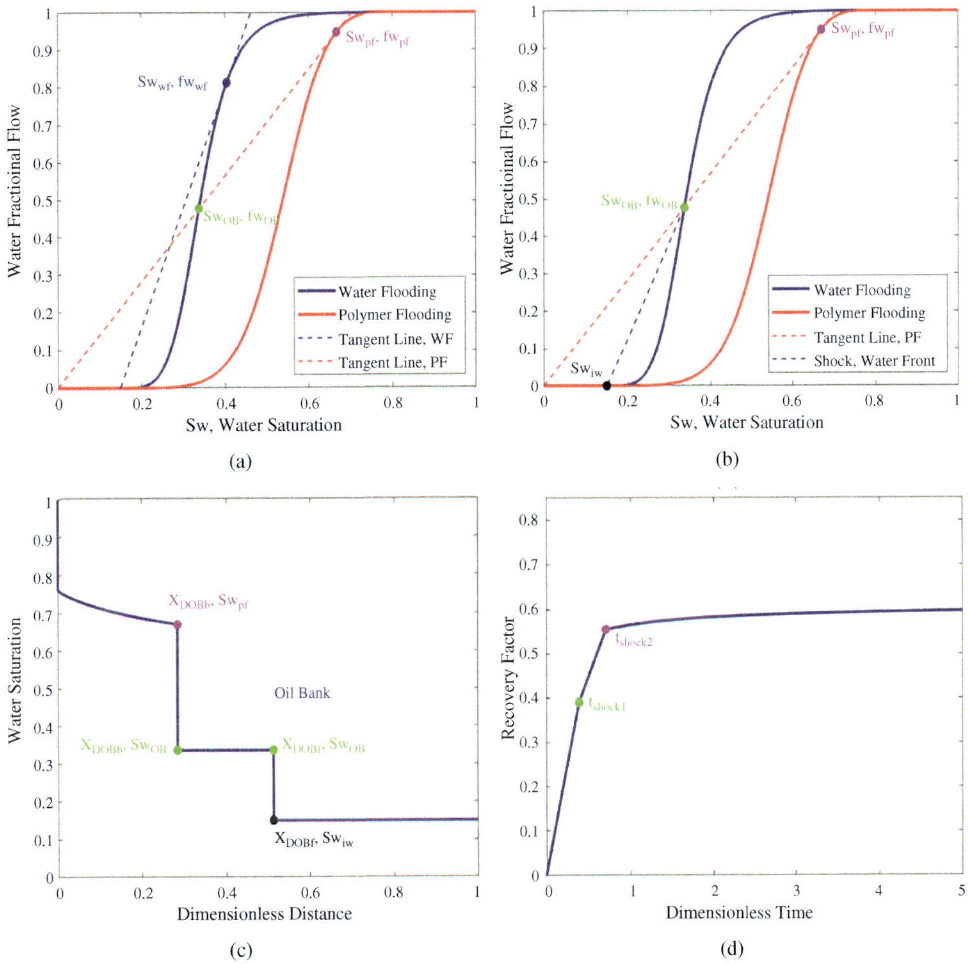

Fig. 2.12 (A) Comparison of shock front for water flooding and polymer flooding, (B) construction of shock front during secondary recovery by polymer flooding, (C) water saturation profile at $t_D = 0.2$ PV during continuous polymer injection, and (D) cumulative oil recovery history during secondary recovery by polymer flooding [35].

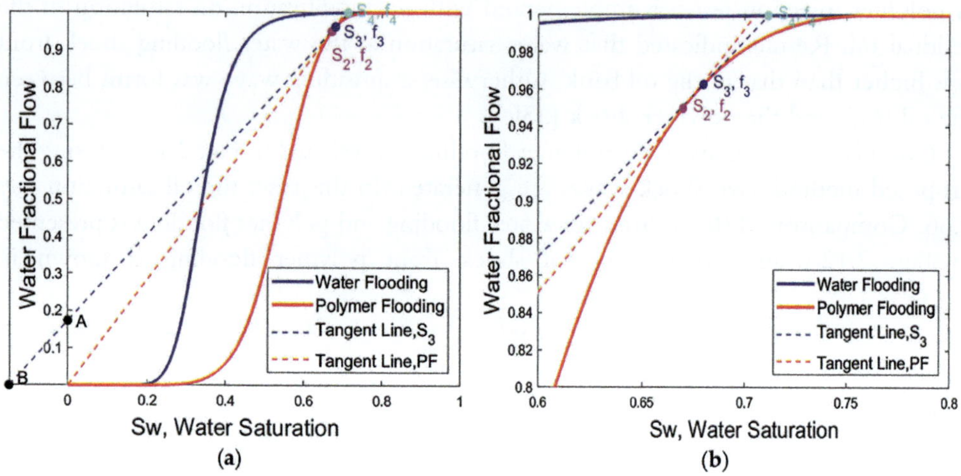

Fig. 2.13 (A) Determination of the shock front during polymer slug injection and (B) magnification of figure (A) for determining shock front location [35].

Fig. 2.12B. Moreover, initial oil saturation has no effect on the surfactant front, but influences oil production (Fig. 2.12C and D). After the breakthrough time (0.39 PV), the oil saturation is reduced to 0.52 after the first shock. After the second one, the oil saturation reduced sharply to 0.05. The remaining oil cut was close to the residual oil saturation after 0.71 PV injection [35].

2.8.3 Influence of polymer slug size

The construction of the shock front in the polymer injection process is illustrated in Fig. 2.13A and is magnified in Fig. 2.13B. The surfactant is injected at 0.30 PV ($t_{d,slug}$) and water injection is performed at 1.44 PV. After polymer injection at 0.3 PV, water slug is injected at 0.13 PV. Fig. 2.14A and B presents the water saturation and shock front at $t_{d,total}$ = 0.43 PV [35].

2.9 Polymer flooding performance

Microscopic and macroscopic efficiencies are important in the polymer flooding process. The residual oil saturation decreases after polymer flooding, which leads to increase in the oil recovery factor during microscopic displacement. The injectivity of polymer solution was studied using various brine salinities, with the two parameters of RF and RRF. These parameters were calculated using the pressure drop values in single-phase flooding tests [36]. RRF is defined by the measurement of permeability reduction and is represented as the ratio of water permeability before and after polymer

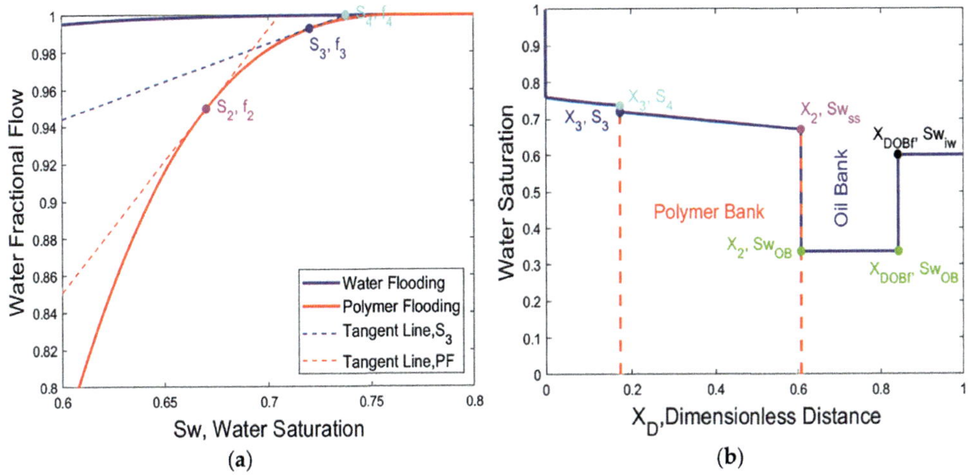

Fig. 2.14 (A) Determination of the shock front during polymer slug injection and (B) profile of water saturation and oil bank at $t_{d,slug}=0.30$ PV, $t_{d,total}=0.43$ PV [35].

injection. For constant injection rates during water flooding process, the RRF is determined by the following equation:

$$RRF = \frac{\Delta P_{wf2}}{\Delta P_{wf1}} \tag{2.41}$$

where P_{wf2} denotes stabilized pressure after polymer injection. To define the actual value of polymer viscosity (in situ viscosity) flowing through the porous media, the following equation is applied:

$$\mu_{in\ situ} = \frac{RF}{RRF} \tag{2.42}$$

Degradation of polymer in the flowing solution is measured from outlet stream viscosity η_e.The change in shear viscosity is attributed to the shear rate of $10\,s^{-1}$ [25, 37].

$$Deg\,(\%) = \frac{\eta_i - \eta_e}{\eta_e - \eta_w} \times 100 \tag{2.43}$$

where η_w and η_i show the brine and injected polymer viscosities, respectively [37]. Various factors affect the polymer flooding efficiency. Increase in the polymer viscosity leads to decrease in the number of fingers. The recovery factor would slightly increase by injection of a low concentration polymer. Viscoelasticity of the polymer would reduce the oil saturation of viscous fluids. Low viscoelasticity effect would not lead to improved oil recovery factor in viscous oils. Wettability is also an important parameter in polymer flooding processes. The remaining oil saturation after the polymer flooding process is a function of wettability. Phase repartitioning is an important factor to control the oil sweeping efficiency in the water flooding process [19].

2.9.1 Polymer flooding in high-temperature reservoir

As previous studies show, the temperature is an effective parameter on the rheology of the polymer solution. By increasing the temperature, the dilatant behavior and critical shear rate of the polymer could be delayed. This behavior is analyzed as follows: as temperature is increased, the solvent viscosity and polymer relaxation time are reduced. Moreover, solvent quality is reduced by increasing the temperature. As solvent quality is reduced, the coil size is decreased. Thus, a high shear rate is needed for the elongation of the polymer. So, extensional viscosity happens at large shear rates [38].

For temperature reservoirs of more than 85 °C, special structures of polymers should be applied to ensure their stability during polymer injection process. Suitable polymers for this purpose contain, *N*-vinyl pyrrolidone (NVP) and acrylamido-tert-butyl-sulfonate (ATBS), which are more expensive than HPAM. Since their molecular weights are low, they should be used at high concentrations to meet the requirements. In some cases, the project might not be economic. On the other hand, in some projects in South America, low concentration brines (<3000 ppm) are applied in polymer flooding process. Application of these brines is very advantageous in polymer flooding process in which less robust polymers could be used and the required viscosity is obtained with lower content of polymer [39].

For reservoir temperatures of more than 200°F, most of the studies are focused on laboratory tests and simulation studies. However, there are some commercial polymer flooding tests in these reservoirs. The main challenge of polymer flooding in these reservoirs is high well spacing. Large volumes of polymer with high thermal stability and thick net pays are required. The residence time would be long, which affects the polymer stability. Another important issue is that these reservoirs are associated with deep formations and infill drilling for polymer injection would be costly [1].

2.10 Polymer flooding in heavy oil recovery

In heavy oil reservoirs, the primary recovery leads to less than 10% increase in production, and several barrels are remained. Polymer flooding in combination with drilling horizontal wells was first applied for Pelican Lake reservoir with high viscosity oil. An unsuccessful pilot was implemented in 1997. Lessons of this project were applied in another study and a successful pilot plant was done in 2006. The oil production rate was increased from 43 to 700 bpd for the next 6 years. Moreover, the water cut remained less than 60%. The viscosity range of oil in candidate reservoirs for polymer flooding was generally less than 150 cP. This pilot plant was the first successful study with viscosity in the range of 1000–2500 cP [40]. This study was then implemented on a field scale.

Important points from this study are summarized as follows: (1) In heavy oil reservoirs, the mobility ratio of 1 does not necessarily lead to the best recovery factor. There exists a balance between mobility ratio and injectivity for the economic issues. (2) Water quality is another important parameter in polymer flooding tests. Other technologies in combination with polymer flooding are also studied for increasing the oil recovery factor in heavy oil reservoirs. These might include polymer injection in hot water, or using surfactant in the tests [40].

There exists an optimum value for the polymer solution viscosity. Out of this range, a sharp decrease in the oil recovery factor is obvious with increase in the oil viscosity. An example of optimum effective viscosity of a polymer solution, as a function of oil viscosity, is reported in Fig. 2.15. Another study was conducted for the minimum value of viscosity. For less viscosity values, no increase in oil recovery factor was obtained [41].

In heavy oils, there is a specified mobility ratio range in which the polymer flooding process would be beneficial. The lower and upper limits of this range are 0.18 and 0.3, respectively, as illustrated in Figs. 2.16 and 2.17. For mobility ratios within this range, the oil recovery factor would increase with increase in the effective viscosity of the polymer. When water is injected into a water-wet reservoir that is saturated with heavy oil, since the viscosity ratio of oil to water is large, fingering phenomena occurs and water channels are created. After the second water injection, the water flows to low resistivity channels. In the polymer flooding process, the mobility of the polymer solution is effective on the capacity of blocking the water channels and diverting the flow to low permeability areas. Moreover, the flow pressure drop for displacing the oil from unswept areas is inversely

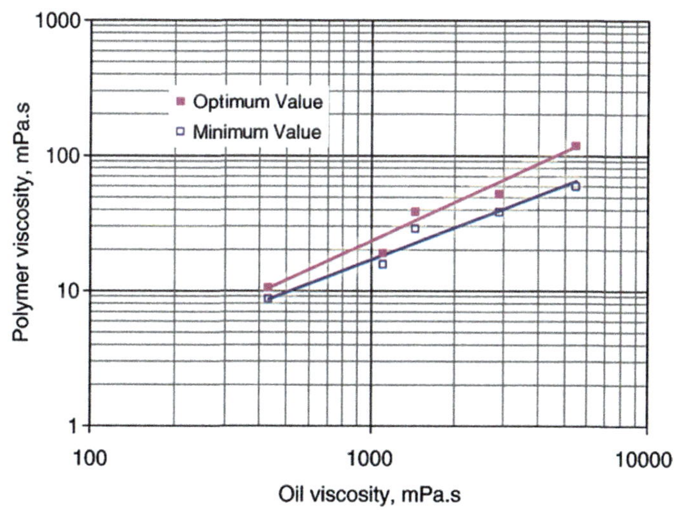

Fig. 2.15 Change in optimum value and minimum value of the effective viscosity of polymer solution with oil viscosity [41].

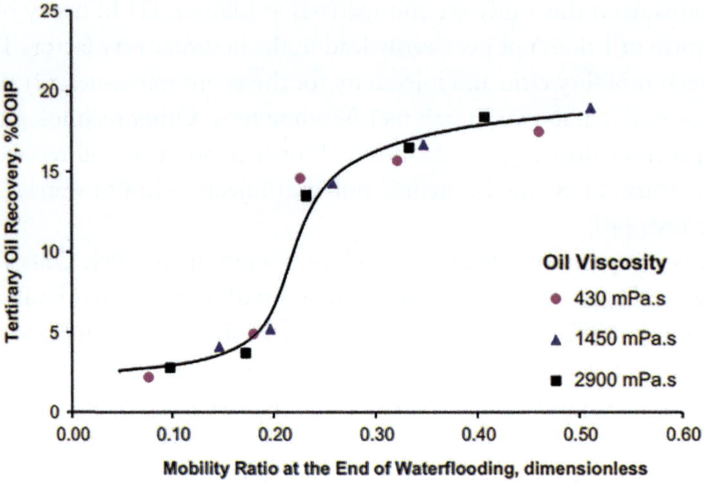

Fig. 2.16 Normalized relationship between tertiary oil recoveries of three oil samples and oil-water mobility ratio at the end of water flooding [42].

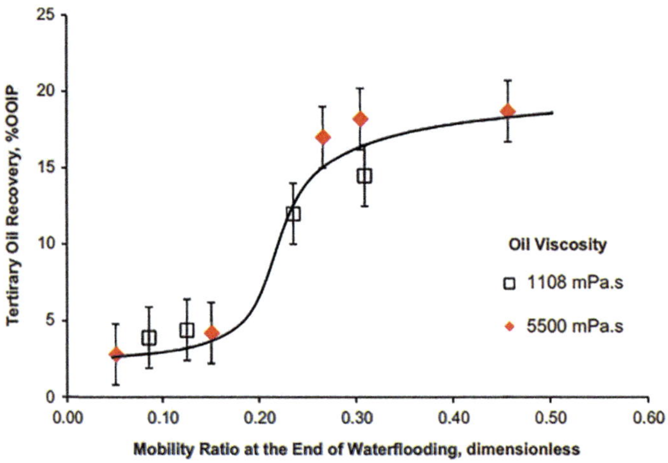

Fig. 2.17 Normalized relationship between tertiary oil recoveries of oil samples (1108 and 5500 mPa s) and oil-water mobility ratio at the end of waterflooding. The error bars represent 72% error in tertiary oil recovery [42].

proportional to oil mobility. As a result, beyond the low limit of the oil-water mobility ratio, the recovery factor would increase, significantly. The same explanation exists for the upper limit. The heavy oil recovery factor is obtained by multiplying the displacement efficiency and volumetric sweeping efficiency. The maximum recovery factor is obtained as follows:

$$R_{max} = (1 - S_{wi} - S_{or})/(1 - S_{wi}) \tag{2.44}$$

where S_{or} represents the residual oil saturation. In most of the studies, the initial water saturation and the residual oil saturation are set at 0.1 and 0.35, respectively. As a result, the maximum oil recovery is almost 60%, which is the upper limit of the stated parameter [42].

This range for polymer viscosity is different for various heavy oils, which limits their application. If the normalization factor is defined as oil-water mobility ratio at the end of initial water flooding, the S-shaped curves fall into a single curve. Using this normalized curve, the polymer flooding process for EOR is easily evaluated. The presented approach is only applicable for heavy oil reservoirs that are highly water-wet. All the tests were performed at an injection rate of $10\,cm^3/h$ and frontal viscosity of 0.5 m/d [42].

In another study, the displacement efficiency of welan gum and xanthan gum was compared, and the results showed higher displacement efficiency of welan gum in comparison to xanthan gum (>7.0% at the normal permeability). Rheological studies revealed that at the same temperature, concentration, and salinity, the loss modulus and storage modulus of the welan gum were higher than xanthan gum. Since more aggregates are formed using welan gum, the displacement efficiency would be higher, although its molecular weight is lower. There are intermolecular structures in welan gum, which leads to drawing and dragging the residual oil. These structures are stable at high salinity and temperature conditions. Flooding experiments were conducted in high and low permeability conditions under the same salinity and temperature, to compare the functionality of the two polymers in terms of displacement efficiency. The results indicated more sweeping efficiency and reduced mobility ratio for welan gum which is due to the higher viscoelasticity of this polymer [43].

2.11 Polymer flooding design and offshore experiences

In polymer injection processes, increasing the injection rate would lead to increased oil recovery factor, and might lead to some problems in terms of equipment capacity. For example, in an offshore oil field, the space required for the platform would be $60–70\,m^3$. For the flow rate of $90\,m^3/h$, the required flow time would be 40 min. This would be beneficial for the dissolution of the polymer in addition to the injection requirements. On the other hand, as the injection rate is increased, there would not be enough time for the dissolution of the polymer in the form of powder. As a result, the flocculated polymer would appear. These are produced due to an increase in the local concentration of the polymer powders and not enough mixing, which leads to the adhesion of particles. Therefore, it is necessary to study on the dispersion of the polymer around the mixing device and also around the dissolution device of the offshore platform to minimize the generated flocs. This would help lay down guidelines for developing the field cases of polymer flooding [44].

The designing constraints are actually different for onshore and offshore polymer flooding processes. Equipment footprint is very important for offshore plants. It is crucial to determine the load and space limitations of floating production storage offloading (FPSO). The form of the polymer to be applied in the field is a function of the local weather and logistics of that area. For example, it is very easy to displace the inverse emulsion polymer in offshore plants, even in unfavorable weather. This approach is very simple in which inversion is done on the fly using enough energy. The solution is then injected into the reservoir after delusion. On the other hand, for powder form of the polymer, the displacement would be easier with less logistics and operating cost [2]. The most important issues to be considered in polymer flooding projects are as follows:

(1) Designing the amount of loads and footprint limitations.
(2) Limiting the installation activities on the platform or FPSO. Special guidelines and constraints should be defined for lifting/handling devices such as crane.
(3) Special guidelines for the loads and their interactions on structures.
(4) There should be a specified classification of the designed equipment in terms of hazards. This should be performed for utilities, motor, and control rooms.
(5) The effect of environmental issues on equipment and the loads on structures.
(6) Equipment should be protected from corrosion (using special paints, stainless steel, grease).
(7) Handle the produced wastes and water. Using specified guidelines for cleaning the fluids.
(8) Using nonshearing injection pumps for polymer injection. As it is a shear-sensitive solution, it should have minimum viscosity degradation.
(9) Since a large quantity of polymer is needed for this process, the supply chain should be confident [2].

Various off-shore projects have been conducted in recent years but due to various constraints, they have not grown significantly. Different studies have been conducted to enhance the concentration of the mother solution to reduce the maturation time of polymer powder and also for inversion of the solution [4]. Moreover, in offshore cases, there exist some limitations in comparison to onshore cases. For instance, due to large well spacing, there might encounter chemical and mechanical degradation, actually at high temperatures. On the other hand, the tack ability of the polymer would reduce at a high salinity environment [15]. These issues make it difficult to use experiences of offshore polymer flooding projects in onshore cases. In a study containing 455 enhanced oil recovery projects, there were only 18 offshore cases. Among which only five cases were conducted using polymer flooding. Just one case was conducted in the depth of 4300 ft. Others were implemented in the depth of 27.9–346 ft. The minimum porosity and permeability of the under test reservoir were 30% and 1000 mD, respectively [7].

In a study conducted in the Dalia Field, high permeability, low temperature, and medium viscosity were the desirable conditions for conducting a successful polymer flooding test. In a polymer flooding project conducted in the Caspian field, the reservoir conditions were low temperature (87°F), high permeability (5000 mD), moderate salinity (15,500 mg/L), and high acidity. In another study, the screening criteria for selecting the polymer flooding project were as follows: (1) reservoir temperature <199°F, (2) oil viscosity <200 cP, (3) salinity <100,000 mg/L, and (4) permeability >10 mD [7].

2.11.1 Salinity

In most of the polymer flooding projects, seawater of high salinity (9000 mg/L) is injected into the reservoir. Due to the shielding effect of cations present in the polymer structure, the polymer molecule would roll up and effective viscosity would reduce significantly. The effect of salinity on the polymer flooding process was studied. The results of the study in terms of water cut and ultimate oil recovery factor are presented in Fig. 2.18. The selected salinities were 0, 3000, 6000, 9000, 12,000, 15,000, and 18,000 mg/L. For lower salinity values, RF would reduce significantly. For high salinity, the oil recovery factor would attain a constant value of 42% [15].

2.11.2 Well space

Well spacing is an important parameter for offshore oil fields. For large well space values, there would be much exploitation time and seepage. The results of a model with well spacing of 200, 300, 400, 500, and 600 m and an injection rate of 135 m^3/day are presented in Fig. 2.19. With increase in the well spacing, the oil recovery factor would be reduced. For smaller well spacing the RF would decrease faster [15].

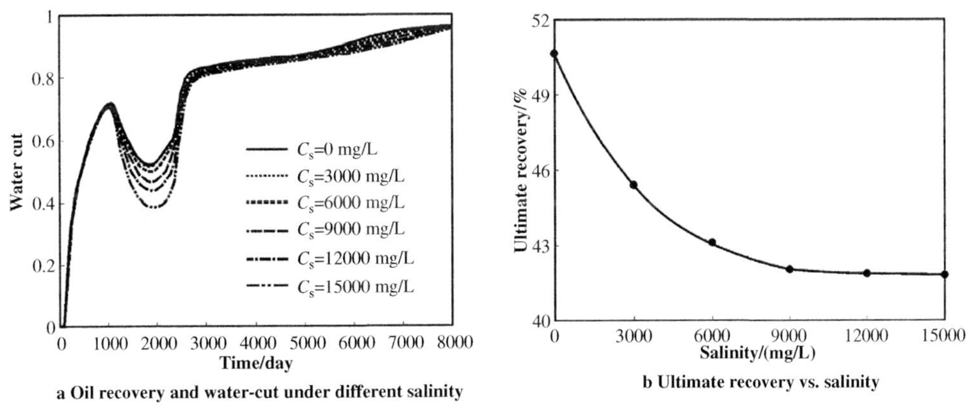

Fig. 2.18 (A) Oil recovery and water cut under different salinity and (B) ultimate recovery vs salinity [15].

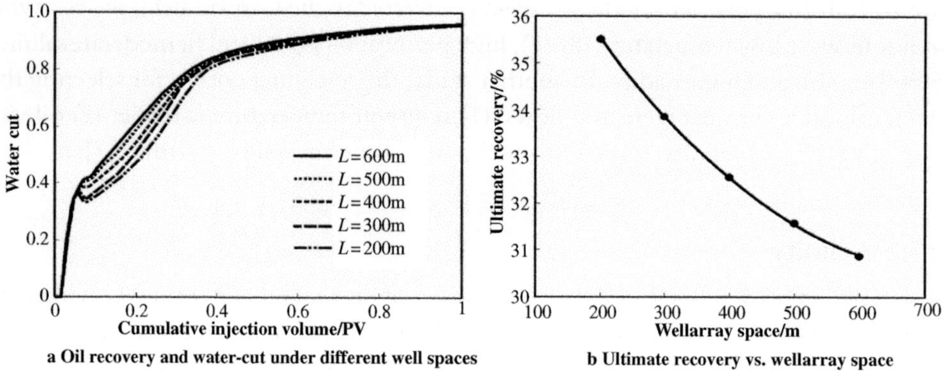

Fig. 2.19 (A) Oil recovery and water cut under different well spaces and (B) ultimate recovery vs well array space [15].

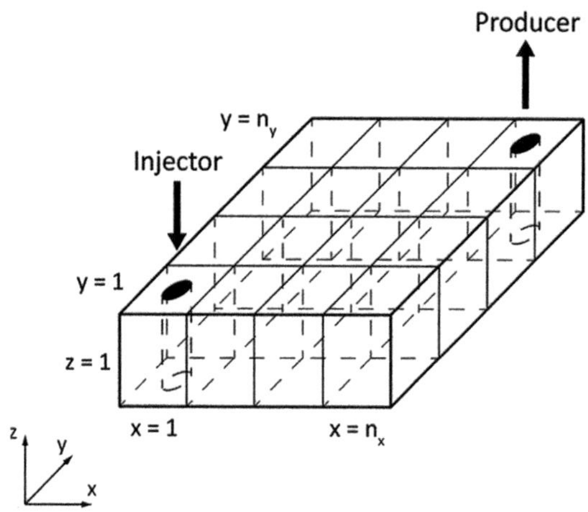

Fig. 2.20 A quarter of the five-spot model used for simulations [45].

2.12 Modeling and simulations

For modeling of EOR processes in a 2D oil field, various well patterns might be used, among which five-spot pattern is more popular. In this pattern, an injection well is placed at the center of square with four production wells at the corners. Most of the time, a quarter of this pattern is used in academic studies and reservoir simulations, as shown in Fig. 2.20. The selected model comprises a reservoir with a domain of Ω and specified reservoir characteristics including porosity (φ) and permeability (k). These parameters might be assumed constant or of specified distribution. For a simple case, the flow is incompressible and isothermal. The numerical simulations are performed in the domain of $n_{x\times}$ n_y blocks. The gravitational force is considered negligible in comparison to viscous force. In the polymer flooding process, the injected polymer solution is followed by a water bank injection to sweep the oil into production wells. To simulate such a process, a combination of nonlinear parabolic/elliptic partial differential equations with a series of algebraic relations is required. In this system, the fluids and rock physical properties are presented, including relative permeabilities, capillary pressure, rock wettability, phase viscosities, residual phase saturations, IPV, viscoelastic properties, chemical degradation, and adsorption [45–47].

For the simulation of the polymer flooding process, no modification is performed on oil flow equations because polymer solutions have no effect on it. On the other hand, the modification should be performed on the flow equation of water to consider the change in water viscosity and relative permeability during polymer injection. It is assumed that the reduction of permeability is proportional to the retained polymer in porous media. The permeability reduction is permanent. Moreover, the formation volume factor and density of water are not influenced by the concentration of polymer [20].

$$\varnothing\frac{\partial S_w}{\partial t}-\vec{\nabla}\frac{K_{rw}K}{\mu_{w.eff}R_K}\left(\vec{\nabla}P_w-\rho_w\vec{g}\right)=0 \tag{2.45}$$

Moreover, the conservation equation is also coupled to the above equation that represents the amount of polymer dissolved in water.

$$\varnothing_{IPV}\frac{\partial S_w C_p}{\partial t}+\varnothing\frac{\partial C_p^{Ad}}{\partial t}+\left[\vec{\nabla}\frac{K_{rw}K}{\mu_{p.eff}R_K}\left(\vec{\nabla}P_w-\rho_w\vec{g}\right)\right]C_p=0 \tag{2.46}$$

where R_K is the permeability reduction factor, C_p^{Ad} is polymer adsorption, K_{rw} shows the water relative permeability, K states permeability, C_p is the polymer solution concentration, $\varnothing$ denotes the porosity, S_w represents saturation of water, $\varnothing_{IPV}$ represent the inaccessible pore volume, $\mu_{w\cdot eff}$ states the effective water viscosity, P_w is the pressure of water, ρ_w is the density of water, and $\mu_{p\cdot eff}$ is the effective polymer viscosity. Other functionalities for water and polymer viscosity are

$$\mu_{p\cdot eff}=\left(\mu m\left(C_p\right)\right)^{\omega}\cdot\mu_p^{1-\omega} \tag{2.47}$$

$$\mu_{w\cdot e}=\left(\mu m\left(C_p\right)\right)^{\omega}\cdot\mu_w^{1-\omega} \tag{2.48}$$

$$\frac{1}{\mu_{w\cdot eff}}=\frac{1-\overline{C}}{\mu_{w\cdot e}}+\frac{1-\overline{C}}{\mu_{p\cdot eff}} \tag{2.49}$$

where $\mu m(C_p)$ denotes the viscosity of the fully mixed polymer solution, $\mu_{w.\ e}$ represents the partially mixed water viscosity, $\mu_{w\cdot eff}$ represents the effective water viscosity, $\mu_{w\cdot eff}$ states the effective polymer solution, μ_p shows the polymer solution viscosity at the highest polymer concentration, μ_w gives the viscosity of water, and $\overline{C}$ denotes the effective concentration of the injected polymer presented as follows:

$$\overline{C} = \frac{C_p}{C_{p,max}} \tag{2.50}$$

where C_p represents the polymer concentration and $C_{p,\ max}$ denotes the maximum concentration of the injected polymer [20]. Studying polymer adsorption is performed using the Langmuir isotherm by the following equation, or by a set of polymer adsorption data vs. concentration of polymer.

$$C_p^{Ad} = \frac{aC^m}{1 + bC} \tag{2.51}$$

where C_p^{Ad} states the concentration of the adsorbed polymer, C shows the polymer concentration in the water adjacent to the reservoir rock, m represents the concentration dependence, a denotes the adsorption coefficient, and b is the adsorption coefficient.

$$R_k = 1 + (RRF - 1)\frac{C_p^{Ad}}{C_p^{Ad\ max}} \tag{2.52}$$

RRF shows the residual resistance factor and $C_p^{Ad\ max}$ denotes the maximum concentration of the adsorbed polymer [20].

2.12.1 Analytical methods

Buckley-Leverett equation is introduced as a case for the general mass conservation equation. This model includes the following assumptions: (a) the model is 1D, homogeneous, and isotropic, (b) the fluids are immiscible and incompressible, (c) there are at least two phases present in the porous media, (d) all the model parts are in local equilibrium, (e) there is no capillary pressure, (f) there is no dispersion in the model, and (g) adsorption isotherm has negative curvature which just depends on one component. Following these assumptions, the mass conservation equation would lead to:

$$\varnothing\frac{\mathrm{dS_j}}{\mathrm{dt}} + \frac{\mathrm{d}\,\vec{\mathrm{u_j}}}{\mathrm{dx}} = 0 \tag{2.53}$$

In this equation $\varnothing$ denotes the porosity, S_j and $\vec{u_j}$ represent the saturation and superficial velocity of phase j [35].

Commercial reservoir simulators are applied for simulation of polymer flooding process including IMEX by CMG, ECLIPSE by Schlumberger, and UTChem. The latter is an academic reservoir simulator applied for chemical flooding processes, which is developed by the University of Texas at Austin. The distinct difference of the stated simulators is their method for modeling the viscosity. In ECLIPSE, there is a black oil package called E100, that is, three-phase, 3D, fully implicit reservoir simulator. This package is capable of modeling the important parameters in the polymer flooding process. In this simulator, the effects of hardness and salinity are not considered in the modeling of viscosity. IMEX that is produced by CMG is also capable of modeling polymer injection process. In this approach, nonlinear mixing rules are applied for the calculation of the polymer viscosity in solution. Moreover, the functionality of polymer viscosity with salinity and hardness are not considered in simulations. In UTChem, complex phase behavior of the polymer, physical, and chemical transformations, are applied in simulations. This is a multiphase, 3D, multicomponent simulator. In the simulation of polymer viscosity, Flory-Huggins equation is applied and the salinity and hardness are applied in the modeling, in addition to polymer concentration. Non-Newtonian behavior of the polymer solution is also applied in this simulator. The shear rate is calculated using the capillary bundle model in modified Blake-Kozeny equation. This is applied as the input of the meter equation in which the shear rate is introduced in the equation of polymer solution. In ECLIPSE, two approaches are applied to consider the functionality of shear rate and viscosity. In the first approach, the viscosity of polymer is related to that of water using a tabular input of data. The other approach uses the Herschel-Bulkley model for modeling the viscosity of polymer. Moreover, tabular input of velocity and viscosity is also applied in IMEX [20].

A reservoir simulation was conducted in an anticline reservoir, with 5 km wide and 7 km long (Fig. 2.21). The oil layer was 123.5 m. A gas cap was generated while producing. This composite reservoir consisted of four zones and three faults. This reservoir was simulated and a history matching was conducted for 7 years for bottom hole pressure [48]. In simulation studies, a model is generated and a history matching is performed before conducting different scenarios. History matching for well P1 is presented in Fig. 2.22.

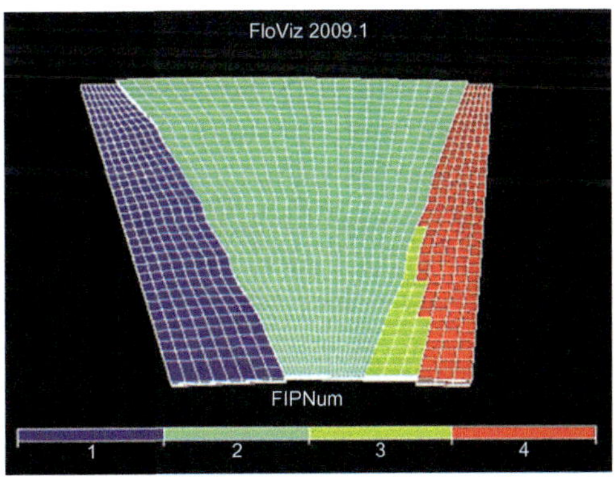

Fig. 2.21 Schematic representation of reservoir model [48].

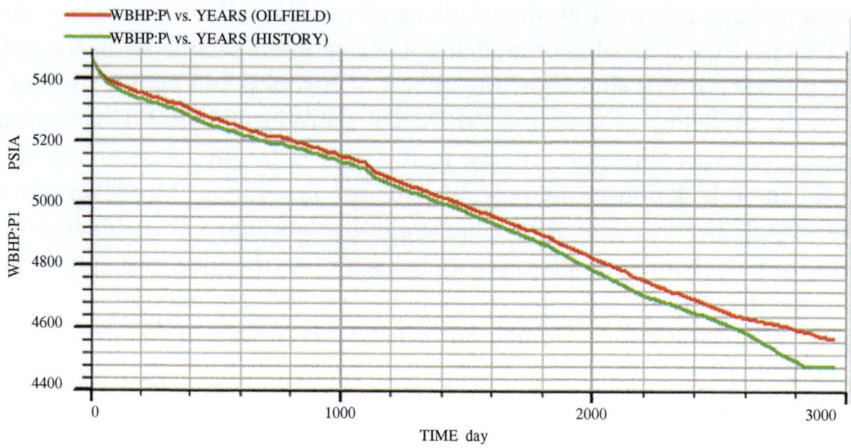

Fig. 2.22 History matching for well P1 [48].

The gas cap expansion and solution gas drive were the primary production mechanisms for 15 years. The maximum value of GOR (10 Mscf/STB) was achieved due to increased gas production [48]. In another zone of the reservoir, aquifer led to production, but some of the wells were closed due to the high value of water cut [48].

2.12.1.1 Water flooding scenario

In this scenario, some injection wells were drilled for water injection. The pressure drop was reduced due to the water injection. After the breakthrough time, increased water production led to the closure of some wells.

2.12.1.2 Polymer flooding scenario

This scenario was conducted to reduce the amount of produced water. Polymer adsorption on porous media led to a reduction in polymer concentration. Various scenarios were considered including (1) injection of polymer with an optimum concentration, (2) polymer injection with concentrations of more than the optimum value, and (3) injection of polymer for three times at optimum concentration. The oil recovery factors obtained in simulation studies are reported in Table 2.3.

Table 2.3 Oil efficiency in three different slug scenarios [48].

Scenario	Oil efficiency
A slug period inject by 7 lb./stb	40.2
A slug period inject by 17 lb./stb	40.8
Three slug period inject by 7 lb./stb	42.9

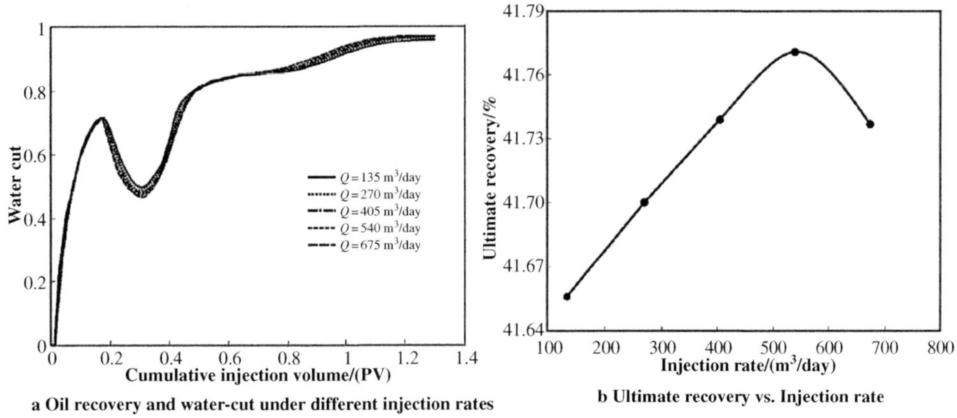

Fig. 2.23 (A) Oil recovery and water cut under different injection rates and (B) ultimate recovery vs injection rate [15].

The effect of the polymer flooding process on increasing the oil recovery factor would be obvious before the polymer breakthrough time. For some cases, this would be 2.6 years in comparison to 4.8 years for polymer breakthrough [7]. In Fig. 2.23, the water cut and recovery factors are presented for various injection rates. At the beginning of the polymer injection process with high rate, the water cut reduces further. For the next water flooding process, at high injection rates, the water cut would increase very fast [15].

The effect of injection rate on ultimate recovery is not significant. This is due to high values of the extensional viscosity to shear viscosity at high flow rates. This would lead to increasing the effective viscosity for the same values of shear viscosity for high flow rates.

The effect of shearing on the molecular structure of the polymer is more significant at high flow rates. This would reduce the shear viscosity. At early stages of polymer flooding, the first effect is more significant. For the next injection process, since the polymer concentration is reduced, the two effects are comparable and the water flows quickly at high flow rates. At high injection rates the injection pressure increases. Since the injection rate has little influence on the development plan of the field, the injection rate should be in accordance with the production demand [15].

Various techniques are utilized to investigate the polymer flooding performance and its effect on cumulative oil production and oil recovery factor. Moreover, its effect on oil viscosity of polymer, interfacial tension, capillary pressure, and most importantly fingering phenomena are studied. One of the most robust techniques includes numerical simulation, which uses the fractional flow theory and other mathematical approaches. These techniques require various information including the geometry and geology of the reservoir, and reservoir rock and the fluid properties. Analysis of these data is a complicated process, which leads to cumulative errors. Therefore, using fast, simple, and accurate techniques for evaluating the performance of polymer flooding is very important.

One of the most applied methods for predicting various parameters in petroleum engineering studies including the polymer flooding process is artificial neural networks (ANNs). This method has been applied for predicting the cumulative oil production and oil recovery factor. Using higher orders of ANN, Chithra Ebaga-Ololo et al. [49] predicted the oil production rate without using a large data set for training the data.

2.13 Upscaling

It is crucial to investigate the flooding mechanisms before the polymer injection. It is impossible to scale up a polymer flooding process before investigating the properties of the polymer [10]. Since the polymer flooding projects are of high cost and risk, all the uncertainties and limitations should be defined before the beginning of the project. Upscaling is a confident technique in which most of the parameters that are upscaled in single phase are applied in full-field simulations. However, the models based on single-phase flow is applied in non-Newtonian two-phase flow models [50].

Most of the simulators can perform simulations for up to 10^6 cells. Other techniques including parallel simulation are also applied to increase the efficiency of calculations. However, with increase in the number of simulations, this technique also becomes inefficient since various geological models are applied for history matching or analyzing the uncertainties. In this technique, a fine-scale model with high resolution is replaced with a coarse one with higher uncertainty. Through upscaling, the properties of each coarse grid are defined in such a way to represent the behavior of the fine model. This concept is presented in Fig. 2.24 [51].

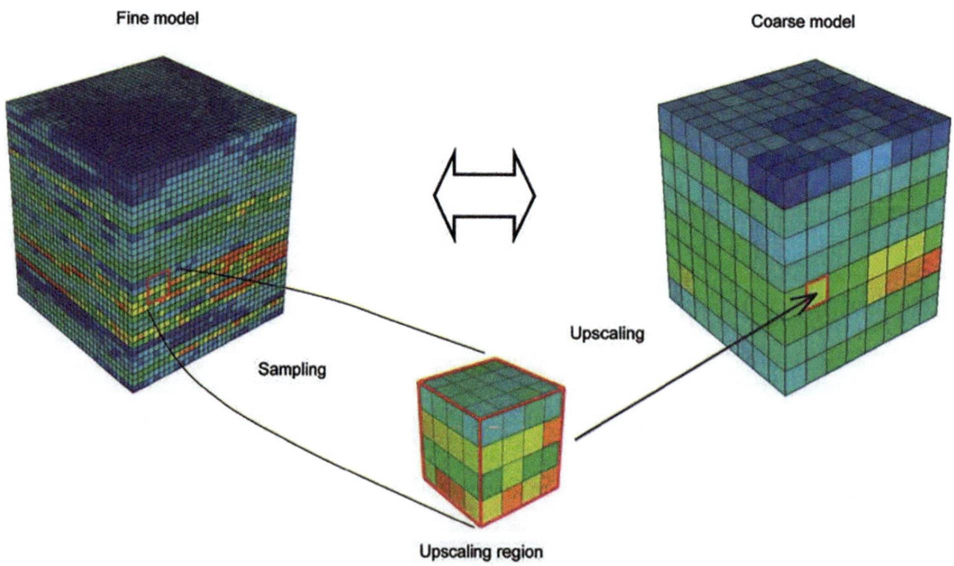

Fig. 2.24 Representation of the upscaling concept [51].

The effective viscosity of polymer solution can be eliminated from equations as for polymers with steady-state concentration it would not change with space. As a result, the mixing parameter ω and effective viscosities of $\mu_{w,eff}$, $\mu_{p,eff}$, are used without upscaling. Important parameters of the polymer flooding process which are upscaled include polymer adsorption c_p^a, reduction factor R_k, and residual resistance factor. With a relation obtained for water flux, the permeability reduction factor was upscaled in all directions. In addition to polymer concentration, the upscaled reduction factor is a function of water saturation, but their functionality is not defined. A two-phase upscaling on relative permeability curves is performed for upscaling the reduction factor. The result would be.

$$k_{rw}^{fine} = \frac{k_{rw}}{R_k(c_p)} \tag{2.54}$$

The upscaled reduction factor was obtained by upscaling the relative permeability curves. This parameter was defined as the water relative permeability with no R_k to upscaled relative permeability, considering the reduction of permeability. The upscaled reduction factor was reduced as water saturation was increased. Polymer adsorption was upscaled using the volume averaging technique:

$$c_a^*\left(c_p^*\right) = \frac{\int \rho_{rock}(1-\varphi)c_a(c_a)dV}{\int \rho_{rock}(1-\varphi)dV} \tag{2.55}$$

The upscaled RRF was reported as a function of R_k and c_p^a.

$$R_k = 1 + (\mathrm{RRF} - 1)\frac{c_p^a}{c_p^{a,\,max}} \tag{2.56}$$

Variations in the reduction factor vs. adsorbed polymer lead to a straight line with slope designated with RRF and $c_p^{a,\,max}$. The RRF is found using the slope of the straight line.

$$RRF^* = \frac{(R_k^* - 1)}{c_p^{a*}} c_p^{a,\,max*} + 1 \tag{2.57}$$

2.14 Laboratory tests and interpretation of the results

The laboratory tests are conducted to study the solubility of the polymer in formation water. Filtration and flow tests are also performed using a viscometer. Other tests are performed to check the mechanical, thermal, and chemical stability of the synthesized polymer. These tests are performed by analyzing the viscosity variations at different conditions. Other tests measure the polymer adsorption on porous media [52].

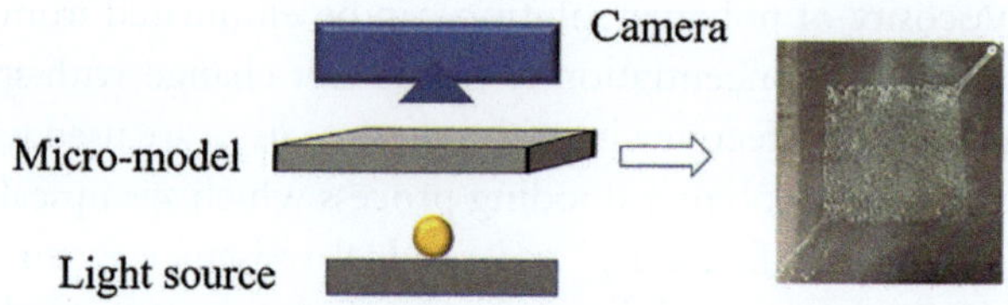

Fig. 2.25 The schematic diagram of etched glass micromodel [13].

2.14.1 Microscopic oil displacement tests

Displacement tests are applied to define the EOR mechanism, which most of them are performed using micromodels. A schematic representation of a micromodel is presented in Fig. 2.25. Micromodels are first saturated with water and then with oil sample. The prepared polymer solution is injected into the micromodel at a constant rate. Then water is injected into the micromodel. Using a camera, dynamic images are taken at different times to record the position and distribution of the residual oil [13]. The main issue in this regard is the complexity of measuring the concentration while hydrocarbon is present [30].

To determine RF and RRF, using a 1D sand pack, the following steps are designed:

(a) The sand pack is saturated with synthetic brine and the absolute permeability and porosity are calculated.
(b) After injection of the synthesized brine, the stabilized pressure drop is recorded (ΔP_b).
(c) Using the stated equations, the RF and RRF are calculated.

To study the mobility control of various polymer solutions, the core flood tests are conducted in 1D sand packs in the following order:

(a) The porosity and absolute permeability are determined.
(b) Oil is injected into the sand pack at a constant rate. The system is then aged for 3 days.
(c) Water is injected to achieve a water cut of 95%.
(d) Polymer is injected with PV of 0.3.
(e) Water is then injected to reach the water cut of 95%.
(f) The pressure variations are recorded during injections and the effluents are collected [27].

2.14.2 Polymer dynamic retention test

Analyzing the effluents of the polymer injection process leads to the calculation of polymer retention. Mass balance technique and polymer frontal breakthrough are the approaches for the calculation of polymer dynamic retention. The latter requires decoupling the effect of IPV. Therefore, this parameter should be specified. In the mass balance technique, the polymer concentration at the inlet and outlet is investigated.

The retained polymer is obtained by subtracting the produced polymer from the injected amount of polymer. The polymer retention is obtained using the following equation:

$$Q = \frac{C_0 V_0 - \sum_{i=1}^{n} C_0 V_0}{w} \tag{2.58}$$

where w stands for dry weight of the core sample, Q represents the polymer retention, V_0 and C_0 denote the volume and initial concentration, respectively. C_i and V_i show concentration and volume of effluent i, respectively. There are static and dynamic approaches to estimate adsorption. In static technique, the adsorbent is diffused. The polymer concentration is determined in the liquid phase. Adsorption is calculated using the following formula in which a decrease in polymer concentration is observed.

$$G = \frac{(C_0 - C) \times V}{m} \tag{2.59}$$

where m shows the mass of the solid particle, G states the adsorption, C_0 and C denote initial and equilibrium polymer concentrations, respectively, and V designates the solution volume [30]. In the dynamic approach, the polymer is injected into the core under pressure. For calculating the amount of polymer adsorption, the sample should be analyzed continuously and the initial pressure should be specified. The polymer concentration is checked until it is equal to its initial value. The total amount of adsorption is defined by combining all the samples that have been tested with polymer solution [30].

Tracer is introduced as the best method for defining the IPV and polymer adsorption in terms of accuracy. Salts are defined as a tracer to be used in this approach. However, defining the polymer concentration is a challenging method in the tracer approach either. The spectrophotometric method in defining the polymer concentration does not show an acceptable result since the wavelength in which the tracer is notched is not in accordance with that of polymer capturing. While the ranges for absorption spectra coincide, this method is not efficient. The most accurate techniques for determining the solution concentration include titrimetric analysis, spectrophotometry, conductometry, thermometric approach, gel permeation chromatography, and fractionation method [30].

2.14.3 Measurement of viscoelastic properties

A stress-controlled rheometer is used for defining loss modulus and storage modulus as rheological parameters of the polymer. There are a number of studies that report the rheological parameters of loss and storage modulus of various polymers calculated at constant temperature and different concentrations and frequencies (0.1–100) [13].

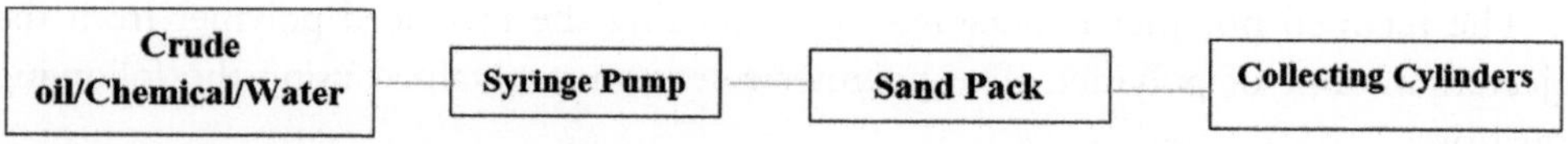

Fig. 2.26 Experimental setup used for single-phase displacement. *(Reproduced from Kakati A, Kumar G, Sangwai JS. Low salinity polymer flooding: effect on polymer rheology, injectivity, retention, and oil recovery efficiency. Energy Fuel 2020;34(5):5715–5732.)*

2.14.3.1 Emulsification tests

Different oil-in-water emulsions are prepared by mixing various ratios of oil and polymers of different concentrations. The prepared emulsions are kept at a constant temperature for 2 min. Then separation efficiency of oil and water is evaluated at a constant temperature [13].

2.14.4 Displacement test for single-phase flow

A study was performed to analyze the effect of salinity on polymer retention and its injectivity in a single-phase flow. The experimental apparatus is illustrated in Fig. 2.26. A syringe pump was utilized to inject the fluid into the sand pack. The accumulators are attached to the inlet of the core holder using a manifold. Pressure variations were studied using a pressure sensor [36].

2.14.4.1 UV analysis of the Effluent

UV-visible spectroscopy was utilized to analyze the effluent water from the single-phase flow tests. The quantitative amount of polymer was obtained using this approach [36].

2.14.4.2 SEM-EDS of flooded sand

SEM-EDS analysis was performed on the sand packs at the end of single-phase flow tests. The samples were dried before the test. SEM micrographs were applied to determine the surface morphology of the sand grains. In this experiment, the elements on the sand pack were analyzed and their mass percent was reported [36].

2.14.4.3 Polymer flooding with low salinity

In this set of experiments, seawater was applied for the preparation of the sand packs to determine the permeability. In most of the reservoirs, water of high salinity is accumulated in the reservoir as connate or initial water. Permeability was calculated, then the sample was saturated with oil and water to calculate the initial saturation. The capillary end effects were avoided by increasing the injection flow rate in saturation tests. This might lead to higher water saturation at the outlet of the sand pack in comparison to other parts of the pack. In this stage, the water saturation is named initial or irreducible water saturation. The remaining amount of oil is calculated using material balance and is called original oil in place (OOIP). The following relations are used to determine the initial water saturation (S_{wi}) and initial oil saturation (S_{oi}) [36].

$$S_{oi} = \frac{OOIP}{PV} \times 100\% \tag{2.60}$$

$$S_{wi} = \frac{PV - OOIP}{PV} \times 100\% \tag{2.61}$$

The residual oil saturation at the end of the test is determined using the following relation. The sand packs were aged to attain the equilibrium between sand grains and fluids.

$$S_{oi} = \frac{OOIP - \text{volume of oil produced}}{PV} \times 100\% \tag{2.62}$$

Polymer degradation might lead to overestimation in determining retention. Moreover, the proper relation between viscosity and polymer concentration must be applied to determine the correct value of polymer retention. If a sufficient amount of brine is not flushed to the sand pack, IPV tend to be overestimated [53].

2.14.4.4 Three-layer oil displacement experiment

The main points of this experiment are as follows:

(1) Initial oil saturation was determined. This was performed by injection of the brine with a flow rate of 0.1 cc/min for 24 h and then with a flow rate of 1 cc/min for 12 h.
(2) Brine was injected to reach the water cut of 90%.
(3) Polymer solution was injected at the rate of 3 cc/min.

Polymer flooding enhances the oil recovery factor with emulsification and mobility control. The relative contribution of each of the parameters is not well defined. In a survey conducted for comparison of emulsification of the two polymers (RH-4 and KLP-6) with HPAM in EOR process, Bottle tests and rheometer were applied to define emulsification properties and viscoelasticity. The mechanism of fluid flow in micromodel was studied using glass micromodel and core tests. Results stated that RH-4 and KLP-6 represented better results in comparison to HPAM with no emulsification property. This study led to an important issue in terms of the application of these polymers in EOR processes [13].

2.15 Field cases

There are various polymer flooding around the world, which affect incremental oil recovery. A number of patents are released by various companies for keeping their income. There are 69 flooding projects conducted in different countries including India, China, Canada, Russia, United States, Argentina, France, Indonesia, and Germany

Thermal/Mechanical, 308				
Immiscible Gas Injection, 40	Miscible Gas Injection, 215	Polymer Flooding , 53	Surfactant Flooding, 16	Microbial, 4

Fig. 2.27 Summary of EOR projects performed worldwide. *(Reproduced from Rellegadla S, Prajapat G, Agrawal A. Polymers for enhanced oil recovery: fundamentals and selection criteria. Appl Microbiol Biotechnol 2017;101(11):4387–4402.)*

from 1996 to 2010. A summary of these projects is presented in Fig. 2.27. The knowledge gained in a project could be applied in another field to increase oil production.

Most of these projects led to enhanced oil production. Some difficulties encountered in the polymer flooding project include polymer degradation, polymer adsorption, and reduced injectivity [14].

One of the polymer flooding projects was performed in the Mangala field to increase the incremental oil recovery. In this strategic project, 74 million barrels of oil are expected to be added to the field production by 2040, which is a more than 6% increase in the oil recovery factor. This successful experience was applied in other fields [54]. Another study was conducted in the Dos Cuadras field in California with low to poor injectivity of polymer. In this project, hydrogen peroxide was used for wellbore treatment. The total injection rate attained was 16,000 bpd in comparison to the anticipated value of 20,000 bpd [55]. This problem was also reported in Eliasville Caddo Unit in Texas that led to various patterns for unequal polymer slugs. The injectivity loss was about 70% [55].

2.15.1 Polymer flooding in the Tambaredjo field, Suriname

The first case was polymer flooding in a heavy oil reservoir located in the Tambaredjo field in Suriname. This pilot plant consisted of three injection and nine production wells. The oil viscosity varies in the range of 1260–3057 mPa s. The average heterogeneous reservoir permeability was 4D. The prepared polymer solution with a concentration of 100 ppm and viscosity range of 45–50 mPa s was injected into the reservoir. The fingering phenomena were observed since the polymer viscosity was lower than that of oil. By increasing the polymer concentration, the performance of this process was enhanced. In all, 10%–60% of the injected polymer was produced from nine production wells. The water cuts were reduced and oil production rates were enhanced. Horizontal fractures were observed while injection of the polymer solution. The oxygen content during injection and mixing processes was in the range 3–8 ppm [56].

2.15.2 Polymer flooding in the Marmul field, Oman

In the Marmul field, the polymer flooding process was performed in 1986 with undertaking results. This field is located in the Sultanate of Oman. It is a sandstone reservoir with thickness 65 ft, permeability 8–25 D, and porosity 26%–34%. The water viscosity was 0.64 mPa s and oil viscosity ranged from 40 to 120 mPa. Since the oil price was low at the time, this field was not expanded till 2010. This denotes the sensitivity of the EOR processes to the price of oil [56].

The polymer flooding pilot plant was conducted with a five-spot pattern with well spacing of 140 m. After water flooding of 0.23 PV, the polymer was injected at 0.63 PV. Then post water flushing was performed at 0.34 PV. The injected polymer solution was HPAM with a viscosity of 15 mPa s at reservoir conditions and water salinity of 600 ppm. The mobility ratio was 2. At the end of this process, the oil saturation was increased and OOIP reached 59%. Large-scale polymer flooding was performed in 2010 with 27 injection wells in 4 inverted five-spot pattern, 20 inverted nine spot, and three patterns of horizontal wells. The total injection rate in the field was 82,000 STB/D. The injection rates were under the fracture pressure of the reservoir. Incremental oil recovery and decrease in water cut were reported [56].

2.15.3 Polymer flooding using high MW and high concentration polymer—Daqing field, China

Three pilot tests were conducted in the Daqing field with high concentration and high MW polymer. The injected concentration and molecular weight of the polymer were 2000 mg/L and 25 million Da, respectively. The reservoir permeability and thickness were in the range of 625–912 mD and 10.1–17.9 m, respectively. The well spacing was in the range 237–250 m. After polymer flooding tests, the water cut decreased by 25%–30%. After a successful experience in this scale, a large-scale test was performed. The surface area was 7.75 km^2 with OOIP of 296,838 million m^3. The number of injection and production wells were 204 and 238, respectively, with well spacing of 125 m and permeability of 652 mD. The concentration of the injected polymer was 2030 mg/L with an injection of 0.61 PV. The incremental oil recovery obtained was more than 10% [56].

2.15.4 Gudao field cases

This is an unconsolidated reservoir with a depth of 1230 m. The production started in 1975. Reservoir characteristics are reported in Table 2.4. The initial oil recovery factor was 16.5% with a water cut of 97%. The viscosity of the crude oil was high and the injected water was not able to mobilize it. The reservoir temperature and salinity were low [57].

A pilot plant study was performed in this field in 1992 with OOIP of 1,650,000 tons. Accordingly, the oil production rate enhanced from 120 to 330 ton/day. The water cut

Table 2.4 Typical reservoir parameters for Gudao field [57].

Reservoir depth (m)	1230
Reservoir thickness (m)	7.9–30.5
Original reservoir pressure (MPa)	12.5
Reservoir temperature (°C)	71
Reservoir porosity	0.335
Reservoir permeability (mD)	1800
Oil density at surface (g/mL)	0.92
Oil viscosity at surface (cP)	1475–3875
Oil viscosity in situ (cP)	50–150
Produced water salinity (mg/L)	3900

was reduced from 98% to 68%. The oil production rate was reduced since the water cut was augmented. In this process, 181,000 tons of oil were recovered. After the success of this process, the polymer injection process was performed in more units of this field [57].

In another unit (ZENZ), a polymer flooding process was implemented in 1998. The OOIP was 11,850 thousand tons, which increased from 420 to 1000 ton/day in 1999. The water cut reduced from 95% to 82%. More than 490,000 tons of incremental oil was recovered [57].

In ZEB unit, polymer flooding was initiated in 2005. This process consisted of 56 injection wells and 79 production wells. 0.59 PV of polymer was injected to the reservoir. The production rate was increased with the maximum value of 820 ton/day. The water cut reduced from 95% to 88%. The incremental produced oil was 853,000 tons [57].

At Dong unit, polymer-surfactant flooding was initiated in 2008. The original oil in place was 14,670 thousand tons. This plant consisted of 55 injection and 84 production wells, respectively. Till the end of 2012, 0.4 PV of the polymer was injected. The polymer solution viscosity ranged from 25 to 35 cP. The water cut was reduced from 97% to 84% and the oil production reached 872 ton/day from 187 ton/day. In total, 3.7% incremental oil recovery was obtained. Post water flooding was performed in 2015 [57].

2.15.5 East Bodo reservoir, Alberta Canada

This is a sandstone reservoir in Canada with a permeability of 1000 mD and oil viscosity in the range of 600–2000 cP. The polymer flooding in horizontal wells was performed in 2006. The quality of the water to prepare polymer solution was a challenging issue in this reservoir. The maximum viscosity of polymer solution was 1 cP in the concentration of 1500 ppm. The water was then changed to freshwater and the viscosity of the polymer solution was reached to 60 cP. The injection rate of the polymer solution was 200 m^3/day with a pressure of 6000 kPa [58].

2.15.6 Turkey case study

The Bati Raman field in the southeast of Turkey has oil with API gravity of 10–15 and an initial RF of 1.5%. CO_2 injection was performed to increase the recovery factor to 5%. The polymer solution was injected to improve the oil recovery factor. A total of 10,000 barrels of the polymer solution was injected into three injection wells. After a week, the CO_2 injection was continued. After 3 months, the produced oil from 16 production well was increased. The polymer solution total cost was reported to be USD 445,000 [58].

2.15.7 Oman case study

The Marmul field is located in Southern Oman with OOIP of 390 million m^3 viscous oil. The reservoir temperature was 46°C with a depth of 2001 ft. The oil viscosity ranged from 80 to 110 cP. A small pilot plant was developed in 1986–88 with a five-spot pattern. The OOIP was 190,000 m^3. Water preflush was performed with 0.23 PV. PAM solution with a flow rate of 500 m^3/day and concentration of 1000 ppm was injected into the reservoir at 0.63 PV. Finally, water postwash was performed with 0.34 PV. The oil recovery factors after these stages were 12%, 46%, and 59%. This project was not extended, since cheaper oil extraction was conducted from other discoveries [58].

2.16 Injection scheme

An important parameter that influences the economics of the polymer flooding process is the injectivity of the solution which is also uncertain. Various techniques are available for the prediction of the polymer flooding process. The injectivity of the polymer is a deterministic parameter for the injection time and production duration, all of which have a great impact on the economics of the process [59]. Prediction of the polymer injectivity is not an easy task and various injectivity tests are performed to obtain trustworthy results. These tests are expensive and would delay the start of the real project and are not able to predict the variations in polymer injectivity with time. It is possible to history match the results of this test with simulation outputs based on the reservoir parameters and polymer characteristics [59].

Increasing the polymer viscosity leads to a reduction in its injectivity and the production rate. Some important parameters that affect polymer injectivity include: (1) the existence of debris, (2) the viscoelastic behavior of polymer, and (3) mechanical degradation of the polymer. The presence of debris and hydration of the polymer would lead to formation damage in the near-wellbore region. Polymer injectivity problems led to the development of the filter tests to evaluate the formation damage [60]. The volume of the injected particles has no effect on injectivity. If the injection was performed above the formation parting pressure, increasing the size of the fracture would not be harmful and might lead to enhancement in fluid injectivity and production efficiency [60]. For a constant injection pressure, the injectivity is reduced, while the polymer viscosity is increased. Increasing the injection pressure is confined since it might harm the reservoir rock integrity. So, achieving an optimum condition in terms of mobility reduction and production rate is a challenge [55].

2.16.1 Strategies for injection rates

Voidage replacement ratio (VRR) is defined as the ratio of reservoir barrels for the injected fluid to those of produced fluids. This parameter is analyzed vs. time to evaluate the pressure maintenance by water or gas injection. This parameter is used by various regulatory agencies to define the effect of field development on the management of energy resources. The injection rate for polymer flooding tests depends on the regulations of the country. For instance, the Alberta Energy Regulator (AER) defines VRR of 1 to keep the reservoir pressure. The injection pressure is limited to 90% of formation partition pressure. In medium or light oils, VRR values of more than 1 are a well-defined concept for reservoir management. However, in heavy oil reservoirs, this concept is not efficient for water flooding where VRR values of less than 1, could be beneficial. The optimum value for VRR in water flooding projects for heavy oil reservoirs could be effective in polymer flooding performance and should be evaluated in economic and technical feasibility studies in heavy oil reservoirs [1]. Reducing the injectivity as a common problem in polymer flooding projects leads to reducing the VRR value. Due to the higher viscosity of the polymer in comparison to water, injectivity is reduced. Moreover, most of the injection wells are fractured during water flooding, before the polymer injection begins. Some strategies to reduce the injectivity problems are presented as follows:

- Using flow back injectors
- Using stimulation techniques
- Using horizontal wells
- Reduction of polymer concentration
- Increasing the number of shots during perforation [1].

2.17 Operation problems

Adsorption of the polymer on formation is irreversible which would lead to formation damage. Recovery of the remaining oil in the formation is very hard since the formation permeability is reduced. The reduction of polymer injectivity is a function of two parameters: reduction of permeability and increasing the viscosity. Hydraulic fracturing might be used to improve the injectivity. Polymer flooding projects would lead to the reduction of suspended particles in water, but oil concentration would increase, considerably. The presence of polymer in the produced water leads to stabilization of water-in-oil emulsions that would cause difficulty in the separation of oil in the produced water since it should meet the specifications of high-quality water for reinjection into the reservoir [52].

The concentration of the injected polymer solution is another important parameter in the polymer flooding process. For high concentrations of polymer solution, post water injection is hard to break through the injected polymer solution. In solutions with low concentrations, the recovery factor would not enhance significantly. Successful projects of polymer flooding have been implemented with slug sizes of 7%–33% PV. If the EOR package was designed properly, injectivity problems due to pore plugging would be removed [14].

Microbial polymer degradation: This would happen due to the production of hydrogen sulfide and the action of microbes. Bactericides are added to the solution to prevent the degradation of polymers. The acidity is controlled by the addition of nitrates that lead to the accumulation of sulfide and heterotrophic-oxidizing nitrate reducing (so NRB and hNRB) which reduces the growth of SRB population. On the other hand, nitrates might lead to microbial corrosion. So, the amount of the injected nitrate should be optimized to reduce the risk of corrosion in equipment. One way to reduce the chance of sulfide production is using sulfate-free water to prepare the polymer solution [14].

In situ polymer degradation: the presence of the concentrated Fe^{+2} might lead to polymer degradation. To avoid this, brine is injected into the reservoir for the precipitation of Fe^{+2}, prior to polymer injection. This would also be beneficial for corrosion inhibition in flow lines and equipment [14].

Viscosity reduction in reservoir: At high temperature and salinity conditions and in the presence of divalent ions, the viscosity of polymer solutions reduces. The polymer viscosity could be maintained by reducing the hardness of the medium. Injection of freshwater would be beneficial in this situation. Thermo-viscosifying polymers are so beneficial in high temperature and high salinity reservoirs and have efficient results in EOR projects. Recently developed biopolymers are another group that has good functionality in low concentration and high viscosity [14].

Inaccessible pore volume: Almost 30% of the connected pore volumes are not available for the polymer solution and they do not have access to all the pore volumes This is due to the higher velocity of the polymer in comparison to the solution. This leads to more contact between polymer and pore spaces of the small area in comparison to the solvent. One approach is to reduce the inaccessible pores in bio-plugging, in which polymers are moved to occupy the inaccessible pores and steer the oil to the production wells.

Another important parameter that is effective in a successful polymer flooding project is the selection of the appropriate polymer, which otherwise would lead to plugging. Polymers selected for the EOR process have shear-thinning behavior in which the polymer viscosity is reduced as the shear rate is increased. In the near-wellbore region, the flow rates are high, which leads to a reduction in apparent viscosity. At low flow rates, the shear viscosity is significant [61].

Mechanical degradation is another induced condition that has occurred in high flow rates that creates significant mechanical stress. The polymer viscosity is reduced by breaking the polymer chain. Other important parameters affecting the mechanical degradation include pore tortuosity, polymer-rock, and polymer-fluid interactions, pore geometry, and polymer physicochemical properties. Various studies have revealed that mechanical degradation is independent of the length and mostly occurs at the injection point [61]. The mechanical degradation was studied on HPAM, and the results revealed that since this polymer is submitted to some degree of mechanical degradation before its injection, the final mechanical degradation is minimized. Moreover, polymers with high molecular weight show more resistance to flow. This means that high elongational stresses, which lead to breaking the molecules, result in degradation [61].

2.18 Well pattern

There are a number of guidelines for designing a polymer flooding project. These guidelines are based on the geological properties in various parts of the field. Important parameters in optimizing the well pattern include permeability ratio, connectivity factor, and well spacing. The connectivity factor is referred to the ratio of the accessed pore volumes to the total PV in the field. For the values less than 70%, the incremental oil recovery by polymer flooding is reduced. To control this parameter, the polymer molecular weight and well pattern must be optimized. There are a number of numerical simulations that describe the functionality of well pattern with incremental oil recovery. Based on the results, the incremental oil recovery is almost the same for nine-spot, five-spot, and four-spot patterns. However, the five-spot pattern is preferred due to its connectivity and suitable injection [5].

In infill drilling, the locations are designated using probabilistic quality maps. Optimization of the production and injection wells is more challenging. Using probabilistic incremental NPV maps, the locations of production and injection wells are determined. The uncertainties of operating parameters and geometries are optimized [62].

The considerations for polymer flooding in field scale include the type of the well, polymer concentration, size of polymer slug, and injection and production rates. Based on the field conditions, the operating parameters are optimized and the well type is selected. The horizontal and vertical well schemes exist for polymer flooding projects. Vertical wells have low cost and injection is performed in various layers in the reservoir, but there might be problems in the injectivity of these wells. On the other hand, the productivity of horizontal wells is higher, injectivity problems are less, and more sweeping efficiencies are obtained. The reported recovery factors are also higher in horizontal well

[62]. To select the horizontal vs vertical wells, two parameters should be considered including (1) the near-wellbore effects such as viscoelasticity of polymers and the shear-thinning behavior, (2) displacement efficiency including the increased oil production and the costs in polymer flooding process for the vertical and horizontal wells [62].

2.19 Surface facilities

Based on the polymer state which might be powder or emulsion, various approaches are considered for its transport. For powder form polymer, it is dissolved in water and pumped for injection to the reservoir. The polymer is stored in the hopper to be distributed to nitrogen-blanketed polymer slicing unit (PSU) with a dosing screw. This solution is then fully hydrated and dissolved using a maturation tank. Then it is diluted for the target reservoir [2].

The skid mounted is more applicable for onshore. Based on the field conditions, pumps, hydration, and polymer dissolution systems are placed inside the containers with 20–40 ft length. The main equipment is blanket with nitrogen to avoid polymer degradation due to oxygen entrance. In these systems, the polymer is quickly hydrated using slicing units and gels and fish eyes are not formed. For the case of high-pressure water, the configuration is presented in Fig. 2.28 [2].

For the low-pressure water, the configuration presented in Fig. 2.29 is recommended for proper hydration of polymer.

For the emulsion form polymer, the W/O emulsions are converted to O/W, using static mixers. This is less expensive in comparison to the powder form of polymer. However, the tank volume for the keeping of emulsion is twice that of the powder form.

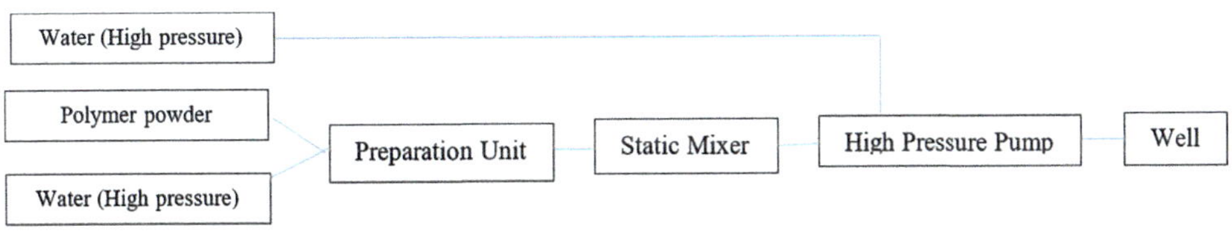

Fig. 2.28 Configuration for polymer preparation when high-pressure water is available. *(Reproduced from Thomas A. Polymer flooding. In: Chemical enhanced oil recovery (cEOR); 2016.)*

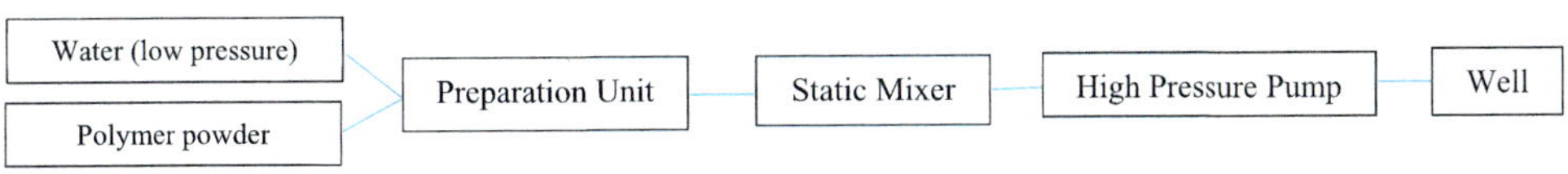

Fig. 2.29 Configuration for polymer preparation when low-pressure water is available. *Reproduced from Thomas A. Polymer flooding. In: Chemical enhanced oil recovery (cEOR); 2016.)*

As stated before, using nitrogen or various scavengers is essential for minimizing chemical degradation. There are guidelines available to minimize the contaminants and mechanical degradation in surface facilities and injection lines while the chokes are present. The contaminants might also be removed.

In low-pressure operation, the rotary pumps and progressive cavity pumps are applied. On the other hand, diaphragm and plunger pumps are recommended. The maximum fluid velocity should be lower than a critical value for the fluid flow in agitator or pipelines. An important parameter for the amount of degradation is well completion. A number of perforations and their density must be considered. The fluid flow is effective on the potential of sharing. Usually, 12 shots/foot is recommended [2]. The minimum equipment that is required for the polymer flooding process are injection pumps, valves, pipes, water treatment, metering facilities, and mixing tools. Most of the processors are conducted at low pressures.

The basic equipment are as follows:

(1) Remote well pads: These are applied in small units in the field. They are composed of producers and injectors with different required facilities.

(2) Pipeline corridor: A passage is provided for pipelines to connect CPF and well pads and processing terminal and vale pads. In production fluid lines, the production is gathered from various well pads and flows to the processing terminal. On the other hand, injection water lines conduct the water in well pads to be used in the subsurface injection. The concentrated polymer solution is transferred to the well pads using mother solution lines where the water is added for dilution, before the injection.

(3) Processing terminal: This is the backbone of the whole process. In this unit, the products are processed to meet specifications. The separated and raw water are processed to satisfy the specific occasions of water. Other facilities are also used in this terminal to completely control this operation.

(4) Central polymers facility (CPF): This is the main unit for the polymer injection process and is equipped with utility and facility stations. In this unit, the prepared solution with desired specifications is transferred to different pads [63].

2.19.1 Polymer friendly choke valves

These valves including coiled configuration or vortex pressure reducers (VPR) are the nonshearing chokes that create accurate and stable control on the polymer injection process with the minimum mechanical degradation. Injection parameters including pressure, water composition, polymer type, and temperature are effective parameters to design these valves. VPR unit is composed of:

- A number of orifices (5–7) that provide the total pressure drop.
- A number of globe valves for tuning the flow rate or pressure drop in this device [4].

In this technology, a specified pressure drop is created without shearing effects. In this technique, the modules are connected in series to the main tube, each of which having the same diameter but a varying length. The pressure drop is modified by changing the overall tubing length for the solution. The modules might be closed or opened with a valve [4].

2.20 Economics and feasibility study of polymer flooding processes

The average value for the polymer consumption for an economic polymer flooding project is reported 0.76 lbm/barrel of the recovered oil. Based on the studies, 400 ppm PV of the injected polymer leads to 6.7% increase in oil recovery. In this study, the saturation of the connate water was 0.25 and the OOIP was 0.75 PV. Since the cost of the polymer was $1.5/lbm, the average cost of the polymer was 4.35 $/bbl for the incremental oil. In this study, the operating and facility costs were not included [52]. The most important factors that designate the overall cost of a polymer flooding process include operating and chemical costs, the price of crude oil, environmental issues, process effectiveness, capital investment, and taxation. Factors that depend on the process are based on the chemicals selected for the process. For example, the price of xanthan gum solutions is higher than that of polyacrylamide, but it is more unstable because polyacrylamide is derived from crude oil and its price is not constant. Moreover, there are other costs including biocides, water, and oxygen scavengers which depend on the field age, well depth and location, and its relative position [14].

Various analyzes on the polymer flooding projects reveal that the total cost is a combination of independent and dependent costs on polymer injectors or producers. These parameters would lead to a net present value (NPV) of zero. This is applied in determining the minimum economic field size (MEFS). These parameters are applied to decide on the projects that are economically viable. For an EOR project, there are various costs including drilling new wells and surface facilities which should be covered by the incomes from the EOR project. Moreover, the excess operating expenditures (OPEX) should also be included. If the field is not large enough, the additional capital expenditures (Capex) are not easy to recover. Enough incremental oil should be produced to make the project economically viable [64].

2.20.1 Robust number of patterns

For economic studies in the polymer flooding process, the minimum number of wells should be specified through which the NPV = 0. If there are not enough patterns, large changes should be applied in incremental oil to achieve the NPV = 0. Therefore, a term called "minimum number of robust patterns" is introduced, in which the required incremental oil is not changed while oil production is less than the anticipated value for oil production in a specified pattern. This parameter states that lower oil production in one pattern, could be compensated by other patterns, and this incremental value is produced by other specified patterns. The minimum number of patterns is not a function of oil price. Only the pattern-dependent and pattern-independent costs are deterministic. The minimum number of patterns is specified using the following analytical approach for NPV.

$$NPV = I \cdot f_o \cdot c \cdot \frac{P}{E} \cdot n_p - \left(\alpha + n_p \cdot \beta\right) \tag{2.63}$$

$$\alpha = f_A \cdot A \tag{2.64}$$

$$\beta = f_B \cdot B \tag{2.65}$$

where α denotes the total costs which are independent of a number of patterns. β represents the total costs per pattern. NPV defines the difference between discounted incomes and discounted costs. Incomes are calculated based on the incremental oil per pattern (I), price of oil (P), number of polymer patterns (n_p), and exchange ratio (E)$\cdot c$ is 6.2898. Total costs are divided into pattern-dependent (B) and pattern-independent (A) costs. Parameters f_A and f_B define discounting effect on the costs. These are the total discounted and total undiscounted costs ratios. a_n and b_n show the yearly costs. d_n states the discount factor of year n in this process. In this time-dependent equation, the polymer rheology, its adsorption, heterogeneity of the reservoir rock, and other properties of the reservoir are considered. The response times and decline curves are obtained from pilot data. The costs and income are separately discounted by f_A and f_B for the costs, and f_o for incomes. The production and incomes are related by the oil price at constant P.

$$f_A = \frac{\sum_{n-1}^{n=25} a_n \cdot d_n}{\sum_{n=1}^{n=25} a_n} \tag{2.66}$$

$$f_B = \frac{\sum_{n-1}^{n=25} b_n \cdot d_n}{\sum_{n=1}^{n=25} b_n} \tag{2.67}$$

$$f_o = \frac{\sum_{n-1}^{n=25} i_n \cdot d_n}{\sum_{n=1}^{n=25} i_n} \tag{2.68}$$

For NPV of zero, a relation between oil per pattern (I) and n_p is introduced. γ is represented as follows [64]:

$$I = \frac{\alpha}{\gamma \cdot n_p} + \frac{\beta}{\gamma} \tag{2.69}$$

$$\gamma = c \cdot \frac{P}{E} \cdot f_o \tag{2.70}$$

The following example analyzes the economic potential of a polymer injection process for increasing the oil recovery factor. The capital costs consisted of M\$1000 for polymer mixing equipment and M\$1000 for drilling each of the horizontal wells. The operating cost for each of the horizontal wells was \$1500, the processing cost of the outlet streams per m^3 of water and oil was \$3 and \$25, respectively. The injection costs per m^3 for water and polymer were \$3 and \$10, respectively. The oil price was considered \$240/m^3. The gas solution drive and water flooding recovered 6% and 23% OOIP, respectively. On the other hand, the polymer injection with the concentration of 1500 ppm led to a recovery of 51% OOIP. All of the scenarios led to positive cash flow as reported in Table 2.5. The final oil recovery for the injection on 200 m well spacing was more than that of 100 m pattern. The incremental value of NPV for water injection was M\$5200; however, those for the polymer injection were M\$ 7100 and M\$13700 for the 200 m and 100 m well spacing, respectively [65]. The economic indicators are reported in Table 2.6.

Table 2.5 Oil recovered at economic limit [65].

Case	Oil prod. E3m3	%OOIP	Years
Primary	50	6	3
Waterflood (200 m)	213	23	22
Poly. flood (200 m)	479	51	44
Poly. flood (100 m)	457	49	17

Table 2.6 Economic indicators [65].

Case	Capital cost M\$	Op. costs M\$	Net revenue M\$	NPV M\$ at 15%
Primary	2000	1400	6600	5500
Waterflood (200 m)	2000	21,000	21,200	10,700
Poly. flood (200 m)	3000	45,000	61,600	12,600
Poly. flood (100 m)	4000	33,000	50,200	19,200

In this simple example, the functionality of polymer flooding in comparison to water injection is clearly demonstrated, which is more than double the net income. Moreover, the well spacing is an important parameter in economic studies of polymer flooding processes. Closer well spacing led to increased NPV and the test duration became shorter [65].

References

[1] Manrique E, Ahmadi M, Samani S. Historical and recent observations in polymer floods: an update review. CT&F—Ciencia, Tecnología y Futuro 2017;6:17–48.

[2] Thomas A. Polymer flooding. In: Chemical enhanced oil recovery (cEOR). IntechOpen; 2016.

[3] Hayatolgheibi SH, Ameli F. A mechanistic study on the behavior of injected hydrogels in porous media for enhanced oil recovery. J Oil, Gas Petrochem Technol 2020;7(1):75–87.

[4] Flavien G, Christophe R, Lionel L, Antoine T. Offshore polymer EOR injection philosophies, constrains and solutions. In: SPE improved oil recovery conference. Society of Petroleum Engineers; 2020.

[5] Wei B. Advances in polymer flooding. In: Viscoelastic and viscoplastic materials, 1. IntechOpen; 2016.

[6] Farhadi A, Primera A, Aponte J, Centeno M. Feasibility study on offshore polymer flooding, forecasting production through integrated asset modelling, a technical and economic approach. SPE Review London; 2015.

[7] Lu X, Puckett D, Xu J, Li Y. Polymer flood in offshore viscous oil reservoirs: implementation, performance and reservoir management. In: IOR 2019–20th European symposium on improved oil recovery. 2019. European association of geoscientists & engineers; 2019. p. 1–14.

[8] Rock A, Hincapie RE, Tahir M, Langanke N, Ganzer L. On the role of polymer viscoelasticity in enhanced oil recovery: extensive laboratory data and review. Polymers 2020;12(10):2276.

[9] Liu J-F, Feng J-Y, Yang S-Z, Gang H-Z, Mu B-Z. The recovery of viscosity of HPAM solution in presence of high concentration sulfide ions. J Petrol Sci Eng 2020;195:107605.

[10] Scott AJ, Romero-Zerón L, Penlidis A. Evaluation of polymeric materials for chemical enhanced oil recovery. Processes 2020;8(3):361.

[11] Alfazazi U, Thomas NC, Alameri W, Al-Shalabi EW. Experimental investigation of polymer injectivity and retention under harsh carbonate reservoir conditions. J Petrol Sci Eng 2020;107262.

[12] Corredor-Rojas LM, Hemmati-Sarapardeh A, Husein MM, Dong M, Maini BB. Rheological behavior of surface modified silica nanoparticles dispersed in partially hydrolyzed polyacrylamide and xanthan gum solutions: experimental measurements, mechanistic understanding, and model development. Energy Fuel 2018;32(10):10628–38.

[13] Kang W, Sarsenbekuly B, Turtabayev S, Yang H, Zhu T, Aidarova S, et al. Study on the influence of emulsification property of functional polymers on enhanced oil recovery and its mechanism. J Petrol Sci Eng 2020;185:106627.

[14] Rellegadla S, Prajapat G, Agrawal A. Polymers for enhanced oil recovery: fundamentals and selection criteria. Appl Microbiol Biotechnol 2017;101(11):4387–402.

[15] Wang J, Liu H. A novel model and sensitivity analysis for viscoelastic polymer flooding in offshore oilfield. J Ind Eng Chem 2014;20(2):656–67.

[16] Kamal MS, Sultan AS, Al-Mubaiyedh UA, Hussein IA. Review on polymer flooding: rheology, adsorption, stability, and field applications of various polymer systems. Polym Rev 2015;55 (3):491–530.

[17] Scott AJ, Penlidis A. Designing optimal terpolymers for enhanced oil recovery (polymer flooding). Ind Eng Chem Res 2020;59(16):7426–37.

[18] Cuthiell D, Kissel G, Jackson C, Frauenfeld T, Fisher D, Rispler K. Viscous fingering effects in solvent displacement of heavy oil. J Can Pet Technol 2006;45(07).

[19] Rez-Morejó JJ, Bertin H, Omari A, Hamon G, Cottin C, Morel D, et al. A new approach to polymer flooding: effects of early polymer injection and wettability on final oil recovery. SPE J-Richardson 2019;1–12.

[20] Alshawaf M. Experimental, numerical and economical analysis of polymer floods in stratified reservoirs. PhD thesis. Imperial College London; 2018.
[21] Luo H, Mohanty KK, Delshad M. Modeling and upscaling unstable water and polymer floods: dynamic characterization of the effective viscous fingering. SPE Reserv Eval Eng 2017;20(04):779–94.
[22] Taber J, Martin F, Seright R. EOR screening criteria revisited—part 2: applications and impact of oil prices. SPE Reserv Eng 1997;12(03):199–206.
[23] Taber JJ, Martin F, Seright R. EOR screening criteria revisited-part 1: introduction to screening criteria and enhanced recovery field projects. SPE Reserv Eng 1997;12(03):189–98.
[24] Bengar A, Moradi S, Ganjeh-Ghazvini M, Shokrollahi A. Optimized polymer flooding projects via combination of experimental design and reservoir simulation. Petroleum 2017;3(4):461–9.
[25] Ranjbar M, Rupp J, Pusch G, Meyn R. Quantification and optimization of viscoelastic effects of polymer solutions for enhanced oil recovery. In: SPE/DOE enhanced oil recovery symposium. Society of Petroleum Engineers; 1992.
[26] Azad MS, Trivedi JJ. Quantification of the viscoelastic effects during polymer flooding: a critical review. SPE J 2019;24(06):2731–57.
[27] Zhu S, Shi L, Wang X, Liu C, Xue X, Ye Z. Investigation into mobility control mechanisms by polymer flooding in offshore high-permeable heavy oil reservoir. Energy Sources A: Recovery Util Environ Eff 2020;1–14.
[28] Gao J, Zhang G, Wang L, Xue X, Ding L, Li X, et al. Properties of a novel imbibition polymer with ultra-high wetting ability as a fracturing fluid system. New J Chem 2020;44(2):492–504.
[29] Barati R, Liang JT. A review of fracturing fluid systems used for hydraulic fracturing of oil and gas wells. J Appl Polym Sci 2014;131(16).
[30] Mirsayanova E, Ilyasov I, Cheremisin A, Evseeva M, Cheremisin A. Numerical simulation of polymer flooding. In: SPE Russian petroleum technology conference. Society of Petroleum Engineers; 2020.
[31] Wei B, Romero-Zerón L, Rodrigue D. Oil displacement mechanisms of viscoelastic polymers in enhanced oil recovery (EOR): a review. J Pet Explor Prod Technol 2014;4(2):113–21.
[32] Olajire AA. Review of ASP EOR (alkaline surfactant polymer enhanced oil recovery) technology in the petroleum industry: prospects and challenges. Energy 2014;77:963–82.
[33] Sarkar S. Evaluation of alkaline, surfactant and polymer flooding for enhanced oil recovery in the norne E-segment based on applied reservoir simulation. Institutt for petroleumsteknologi og anvendt geofysikk; 2012.
[34] Farajzadeh R, Wassing B, Lake L. Insights into design of mobility control for chemical enhanced oil recovery. Energy Rep 2019;5:570–8.
[35] Ding L, Wu Q, Zhang L, Guérillot D. Application of fractional flow theory for analytical modeling of surfactant flooding, polymer flooding, and surfactant/polymer flooding for chemical enhanced oil recovery. Water 2020;12(8):2195.
[36] Kakati A, Kumar G, Sangwai JS. Low salinity polymer flooding: effect on polymer rheology, injectivity, retention, and oil recovery efficiency. Energy Fuel 2020;34(5):5715–32.
[37] Al-Shakry B, Shaker Shiran B, Skauge T, Skauge A. Polymer injectivity: influence of permeability in the flow of EOR polymers in porous media. In: SPE Europec featured at 81st EAGE conference and exhibition. Society of Petroleum Engineers; 2019.
[38] Skauge A, Zamani N, Gausdal Jacobsen J, Shaker Shiran B, Al-Shakry B, Skauge T. Polymer flow in porous media: relevance to enhanced oil recovery. Colloids Interfaces 2018;2(3):27.
[39] Gaillard N, Giovannetti B, Leblanc T, Thomas A, Braun O, Favero C. Selection of customized polymers to enhance oil recovery from high temperature reservoirs. In: SPE Latin American and Caribbean petroleum engineering conference. Society of Petroleum Engineers; 2015.
[40] Delamaide E, Zaitoun A, Renard G, Tabary R. Pelican lake field: First successful application of polymer flooding in a heavy oil reservoir. In: SPE enhanced oil recovery conference. Society of Petroleum Engineers; 2013.
[41] Wang J, Dong M. Optimum effective viscosity of polymer solution for improving heavy oil recovery. J Petrol Sci Eng 2009;67(3–4):155–8.
[42] Guo Z, Dong M, Chen Z, Yao J. A fast and effective method to evaluate the polymer flooding potential for heavy oil reservoirs in Western Canada. J Petrol Sci Eng 2013;112:335–40.

[43] Xu L, Xu G, Yu L, Gong H, Dong M, Li Y. The displacement efficiency and rheology of welan gum for enhanced heavy oil recovery. Polym Adv Technol 2014;25(10):1122–9.
[44] Zhu S, Zhu S, Shi L, Zhang H. Simulation and analysis of polymer dispersed dissolution process on offshore platforms. IOP Conf Ser Earth Environ Sci 2020;565:012045 [IOP Publishing].
[45] Druetta P, Picchioni F. Influence of the polymer properties and numerical schemes on tertiary oil recovery processes. Comput Math Appl 2020;79(4):1094–110.
[46] Ameli F, Moghbeli MR, Alashkar A. On the effect of salinity and nano-particles on polymer flooding in a heterogeneous porous media: experimental and modeling approaches. J Petrol Sci Eng 2019;174:1152–68.
[47] Jamali A, Moghbeli M, Ameli F, Roayaie E, Karambeigi M. Synthesis and characterization of pH-sensitive poly (acrylamide-co-methylenebisacrylamide-co-acrylic acid) hydrogel microspheres containing silica nanoparticles: application in enhanced oil recovery processes. J Appl Polym Sci 2020;137(12):48491.
[48] Salehi MM, Hekmatzadeh A, Sajjadian VA, Masoumi M. Simulation of polymer flooding in one of the Iranian oil fields. Egypt J Pet 2017;26(2):325–30.
[49] Ebaga-Ololo J, Chon BH. Prediction of polymer flooding performance with an artificial neural network: a two-polymer-slug case. Energies 2017;10(7):844.
[50] Pongthunya P. Upscaling in polymer flooded reservoirs. PhD thesis. Imperial College London; 2013.
[51] Danilova A. Upscaling polymer flooding using MRST. NTNU; 2016.
[52] Sheng JJ, Leonhardt B, Azri N. Status of polymer-flooding technology. J Can Pet Technol 2015;54 (02):116–26.
[53] Wang D, Li C, Seright RS. Laboratory evaluation of polymer retention in a heavy oil sand for a polymer flooding application on Alaska's North slope. SPE J 2020;25(04):1842–56.
[54] Jain S, Parekh B. Enhanced oil recovery-polymer flooding: surface transfer facility design, graphical approach for hydraulics and tackle induced vibrations. In: SPE annual technical conference and exhibition. Society of Petroleum Engineers; 2017.
[55] Torrealba V, Hoteit H. Improved polymer flooding injectivity and displacement by considering compositionally-tuned slugs. J Petrol Sci Eng 2019;178:14–26.
[56] Sheng JJ. Polymer flooding—fundamentals and field cases. In: Enhanced oil recovery field case studies. Elsevier; 2013. p. 63–82.
[57] Gao C, Shi J, Zhao F. Successful polymer flooding and surfactant-polymer flooding projects at Shengli oilfield from 1992 to 2012. J Pet Explor Prod Technol 2014;4(1):1–8.
[58] Gao CH. Scientific research and field applications of polymer flooding in heavy oil recovery. J Pet Explor Prod Technol 2011;1(2–4):65–70.
[59] Delamaide E. How to use analytical tools to forecast injectivity in polymer floods. In: SPE Europec featured at 81st EAGE conference and exhibition. Society of Petroleum Engineers; 2019.
[60] Seright RS, Seheult JM, Talashek T. Injectivity characteristics of EOR polymers. In: SPE annual technical conference and exhibition. Society of Petroleum Engineers; 2008.
[61] Al-Shakry B, Skauge T, Shaker Shiran B, Skauge A. Impact of mechanical degradation on polymer injectivity in porous media. Polymers 2018;10(7):742.
[62] Sieberer M, Clemens T, Peisker J, Ofori S. Polymer flood field implementation-pattern configuration and horizontal versus vertical wells. In: IOR 2019–20th European symposium on improved oil recovery. 2019. European Association of Geoscientists & Engineers; 2019. p. 1–21.
[63] Jain S, Vasavada K. Polymer flooding: surface facility design for mitigating induced vibrations and on-field performance validation. In: Abu Dhabi International Petroleum Exhibition & Conference. Society of Petroleum Engineers; 2018.
[64] Sieberer M, Jamek K, Clemens T. Polymer-flooding economics, from pilot to field implementation. SPE Econ Manag 2017;9(03):51–60.
[65] Wassmuth F, Green K, Hodgins L, Turta A. Polymer flood technology for heavy oil recovery. In: Canadian international petroleum conference. Petroleum Society of Canada; 2007.

CHAPTER 3

Enhanced oil recovery using surfactants

Sabber Khandoozi[a,b], Amin Sharifi[a,b], and Masoud Riazi[a,b]
[a]Enhanced Oil Recovery (EOR) Research Centre, IOR/EOR Research Institute, Shiraz University, Shiraz, Iran
[b]Petroleum Engineering Department, School of Chemical and Petroleum Engineering, Shiraz University, Shiraz, Iran

3.1 Overview

The petroleum industry has been relying on enhanced oil recovery (EOR) to produce the demanding amount of oil. To fulfill this aim, different EOR approaches have been developed and implemented such as CO_2 injection, low salinity water flood (LSWF), and chemical flooding methods [1–3]. Theoretically, chemical enhanced oil recovery (CEOR) is divided into surfactant, polymer, and alkaline flooding. Uren and Fahmy [4] initially unveiled surface-active agents, i.e., surfactants, as an EOR agent in 1927.

In years gone by, surfactants have been used as an active agent to produce extra oil in some of the reservoirs requiring EOR. The diversity of mechanisms has enticed many experts to turn their efforts to this promising CEOR method. In concrete terms, due to the complexity of structure, surfactant provides a wide range of applicability for different conditions, i.e., oil type, brine composition, reservoir condition, rock mineralogy, etc. However, akin to other EOR methods, surfactant flooding needs to be evaluated practically and economically beforehand. For this reason, many studies in academia and industry have been performed to determine the applicability of diverse surfactants under different circumstances [1].

Surfactants can enhance oil recovery of reservoirs through different mechanisms such as wettability alteration and interfacial tension (IFT) reduction. Depending on the type of reservoir rock, oil, water, and water-oil interaction, the desirable mechanism(s) and surfactant(s) should be selected. This issue is so critical that any mistake in the surfactant screening process not only fails the project performance but also can result in significant formation and even surface facility damage. In turn, recognition of these aspects is a prerequisite for each study.

In this chapter, it is attempted to review in part all the key aspects of surfactant EOR. In this regard, the results of some worthwhile research items as well as industrial practices are presented. Besides, authors strive to share their intimate knowledge regarding surfactants roles. To begin with, a roadmap for surfactant flooding is shown in Fig. 3.1. As can be seen, for the implementation of surfactant flooding at field scale, a number of key steps need to be followed. Hence, when a surfactant fails to succeed in each step, a review of previous steps and even changing of surfactant seems to be inevitable. In this chapter, the

Chemical Methods
https://doi.org/10.1016/B978-0-12-821931-7.00007-9

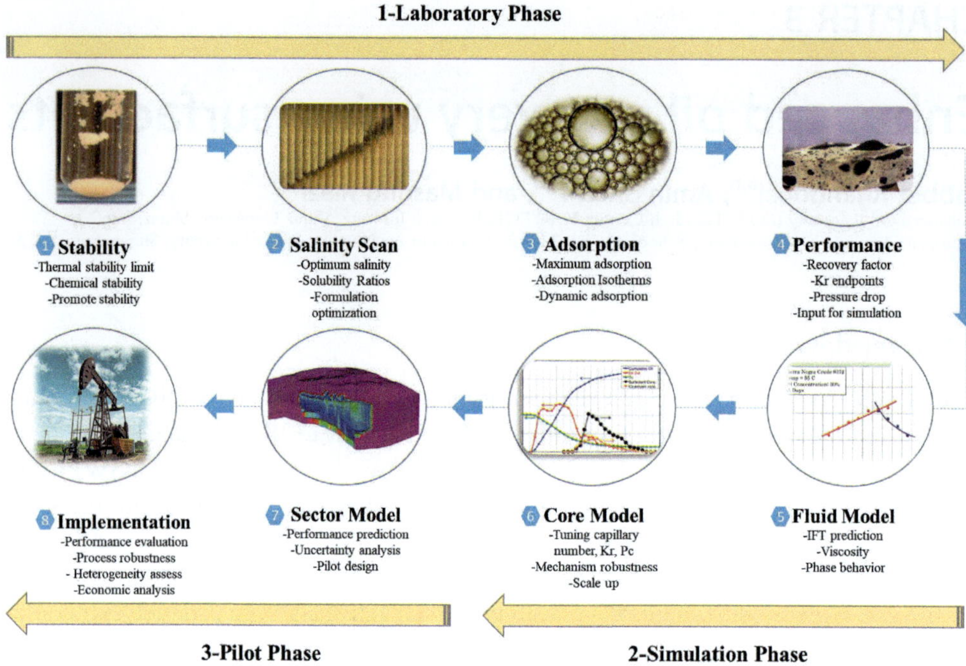

Fig. 3.1 Roadmap to design and implement a surfactant-flooding project.

main emphasis is on addressing all parts of the mentioned roadmap; however, to embark on it, initially, some basics such as surfactant types are reviewed. In this way, authors give precedence to cutting-edge issues regarding EOR via surfactants.

3.2 Types of surfactants

The hydrophilic nature of a surfactant monomer is the key property to categorize surfactants. According to the charge of the polar head group, four categories are recognized: anionic, cationic, zwitterionic, and nonionic surfactants [5, 6], being as follows:

3.2.1 Anionic

The negative charge of the hydrophilic portion is the main property of the anionic surfactant class. Due to their availability and low cost, anionic surfactants are generally known as the most widely used surfactants in the petroleum industry [7, 8]. Anionic hydrophilic groups commonly consist of sulfate, sulfonate, carboxylate, and phosphate groups. Their typical formulation and structure are represented in Table 3.1 and Fig. 3.2, respectively [5, 7, 9].

Table 3.1 Typical structures of anionic head groups.

No.	Anionic surfactants	Formulation	Examples
1	Sulfate	$ROSO_3^-X^+$	SDS
2	Sulfonate	$RSO_3^-X^+$	SDBS
3	Carboxylate	$RCOO^-X^+$	AEC
4	Phosphate	$ROPO(OH)O^-X^+$	AEP

R: Alkyl chain usually carbon number: 8–16. X: Alkaline metal counter ion.

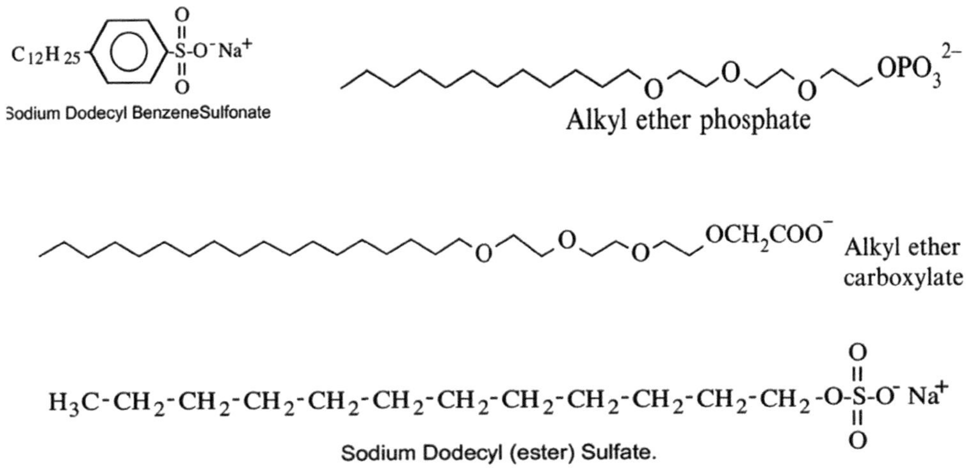

Fig. 3.2 Schematic structure of anionic surfactants.

Anionic surfactants, especially petroleum types, are not tolerant at the harsh environment (salinity and temperature) in which their tolerance based on salt type can be ordered as follow [10]:

Sulfonate-Sulfate < Phosphate < Carboxylate

Today, anionic surfactants can be tailored for a different range of reservoir conditions, based on their hydrophobic (chain length and branching) and hydrophilic properties.

3.2.2 Cationic

The positive charge ion of the polar group (mostly nitrogen) is the distinguishing characteristic of cationic surfactants in which negative counter ions are usually chloride. Quaternary ammonium-based surfactants and amine types are the most common cationic types [10, 11]. On the other hand, alkyl trimethyl ammonium chloride and dialkyl dimethyl ammonium chloride are two common classes of cationic surfactants, used by the oil industry. Schematic structures of two main cationic classes are depicted in Fig. 3.3.

Cationic surfactants are the third major surfactant types used in the petroleum industry. Besides that, they are usually compatible with zwitterionic and nonionic ones [12].

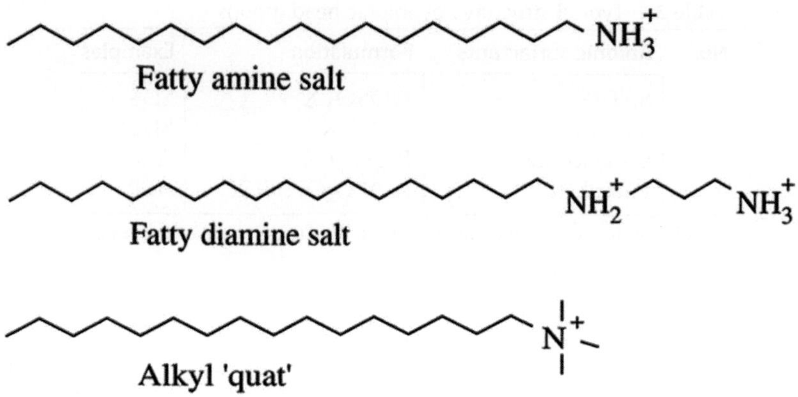

Fig. 3.3 Schematic structure of some typical cationic surfactants.

Generally, cationic surfactants are more expensive than anionic ones. It is noteworthy that they are preferred for carbonate rocks since they are adsorbed infinitesimally on the carbonates due to the positive charge of rocks at normal pH conditions. More importantly, they are generally stable with chemical, salinity, and pH variation [13].

3.2.3 Zwitterionic

Zwitterionic/amphoteric surfactants have both positive and negative charged groups concurrently in their molecular structure. Ammonium and carboxylates are the most common moieties in amphoteric surfactant monomers [10]. Their active charge types are specifically a function of solution pH in such a way that it sustains the anionic state at the alkaline condition and the cationic state at low pH. It also behaves akin to a nonionic surfactant at isoelectric point (point of zero charge) [10, 13]. Betaine and Imidazoline are two common classes of amphoteric surfactants. In addition, some typical examples of these classes are shown in Fig. 3.4.

Zwitterionic surfactants are not commonly used in the oil industry, mainly due to their cost. On the other hand, owing to their tolerance at high electrolyte and hardness conditions as well as ensuring compatibility with other surfactant types, they can be used as a cosurfactant in the chemical EOR process.

3.2.4 Nonionic

This class of surfactants does not have any charge type and the head group size is usually larger than the tail group [8]. Major nonionic head groups are polyether and polyhydroxy units, consisting mainly of ethylene oxide, being referred to as ethoxylated surfactants. Hydrophobic tails are alcohols, alkyl phenol, fatty acids, sorbitans, and fatty amines. Fig. 3.5 shows the structure of typical nonionic surfactants.

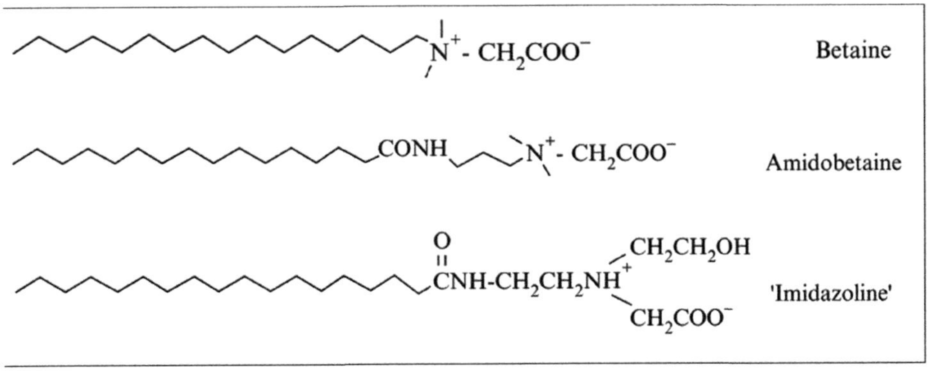

Fig. 3.4 Schematic representation of typical zwitterionic surfactants.

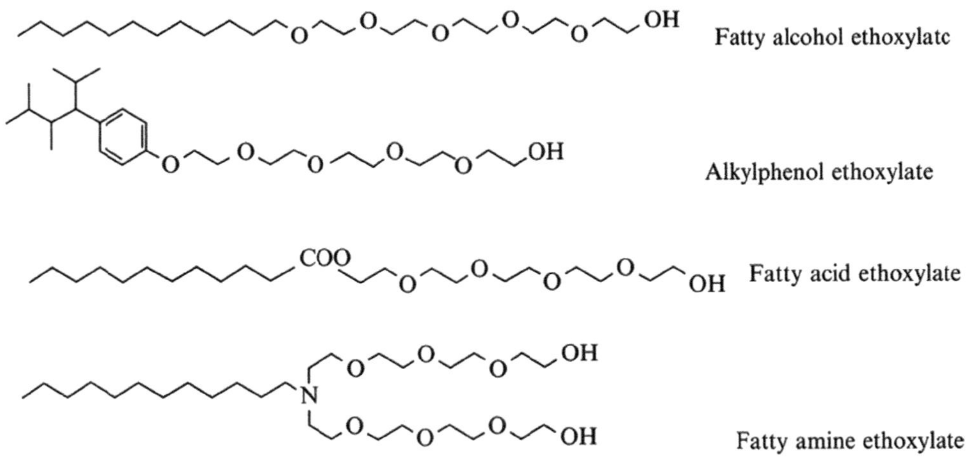

Fig. 3.5 Schematic representation of typical nonionic surfactants.

Following anionic surfactants, nonionic ones are the second-largest class in the oil industry and usually compatible with other surfactant types. Moreover, they are tolerant at high salinity conditions and a wide range of hardness but generally sensitive to temperature [10]. Besides, nonionic surfactants are usually used as a cosurfactant to enhance the phase behavior of solution and interfacial properties of other surfactant types.

3.3 Chemicals used in surfactant flooding

Since the 1900s, surfactants have been used as promising CEOR agents to unleash remaining oil in the reservoirs. Extensive works have been carried out in the lab and field media to design and tailor surfactants for specific conditions of brine salinity, temperature,

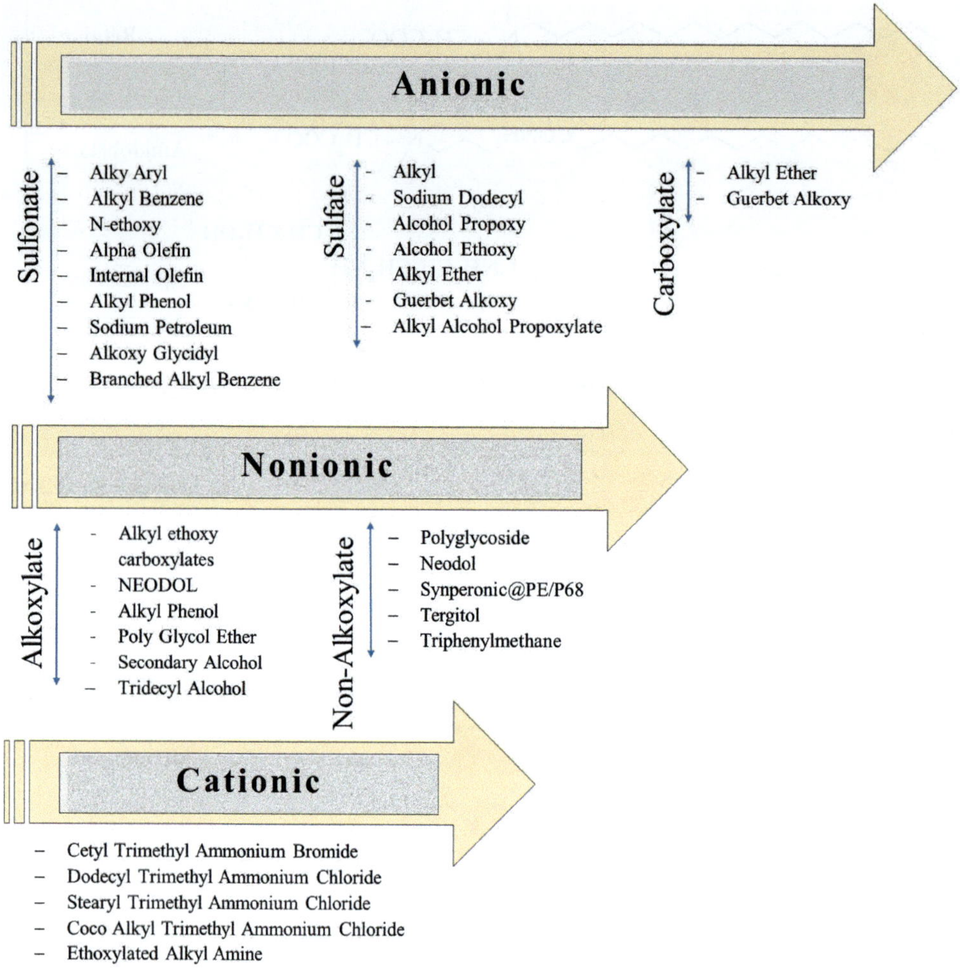

Fig. 3.6 Typical surfactants used in CEOR.

oil composition, and rock mineralogy. The most common surfactant types used in CEOR are shown in Fig. 3.6.

3.4 Thermal and aqueous stability

Reviewing the stability of the aqueous solution is a key step of any surfactant flooding process. A stable surfactant slug without any cloudiness and phase separation is a primary requirement for the surfactant-screening phase at laboratory study (Fig. 3.7). In this sense, surfactant sustains a retention time in the reservoir prior to commencement of its activity in the interfacial area. Hence, structural properties should be maintained during this time

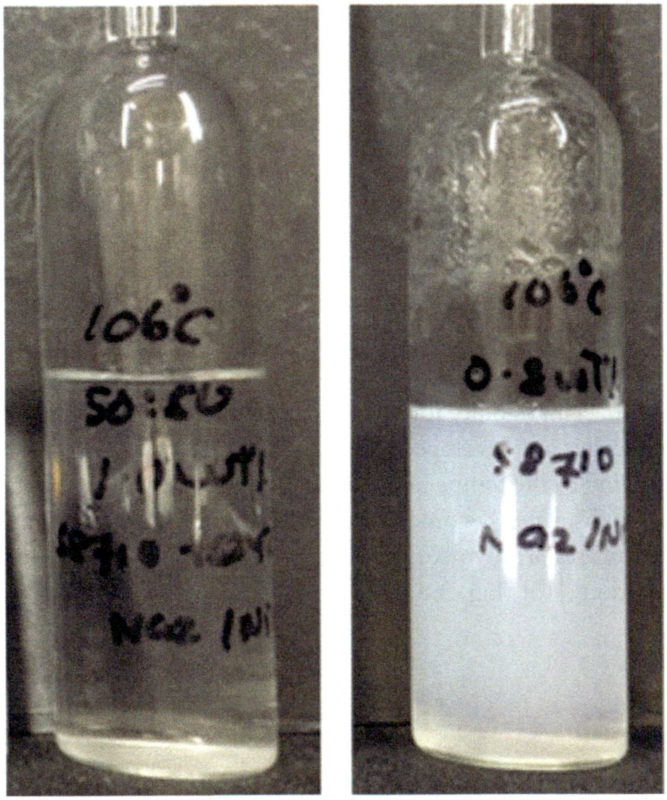

Fig. 3.7 Surfactant stability test as a screening criterion via examining the transparency of solutions, Left: stable solution and Right: cloudy (milky) unstable solution [14].

interval and also the later application period. Two main factors, challenging surfactants stability, are salinity and temperature. To assess surfactant stability, some tests such as stability tests are usually recommended. A stability test is usually conducted in two stages: aqueous stability and thermal stability limits. The more a solution is transparent, the more stable it is.

3.4.1 Aqueous and chemical stability

Any precipitates or phase separation causes trapping and heterogeneous propagation of injected slug, which finally results in the reduction of surfactant flooding performance. Therefore, solution transparency in a specific range of salinity is a preliminary requirement for any surfactant flooding process. The upper limit for aqueous stability is usually up to injected salinity and no issue will be raised by oil addition [1]. Furthermore, the oxygen content is a crucial parameter increasing surfactant decomposition rate and therefore in the lab phase, anerobic condition should be provided like in the reservoir. In addition, high pH can also decrease hydrolysis following surfactant decomposition [6].

3.4.2 Thermal stability

Temperature usually affects the performance of a surfactant solution especially in reservoir condition since it can adversely reduce the chemical stability of the solution by increasing decomposition. On the other hand, high temperature enhances surfactants' hydrolysis, making adsorption faster especially in some anionic surfactants such as sulfate types [6]. Cloud point is a typical approximation for thermal stability in the screening phase.

To improve surfactant stability for a wide range of salinities and temperatures, different classes of surfactants are developed. Sulfate-based surfactants can be selected for stability issues according to the temperature of 60°C since above this temperature, C-S-O bonds hydrolyze and sulfates become unstable [7]. Internal olefin sulfonates (IOS) and alcohol—alkoxy sulfates are two surfactant classes being able to tolerate a wide range of reservoir conditions [15] (Fig. 3.8).

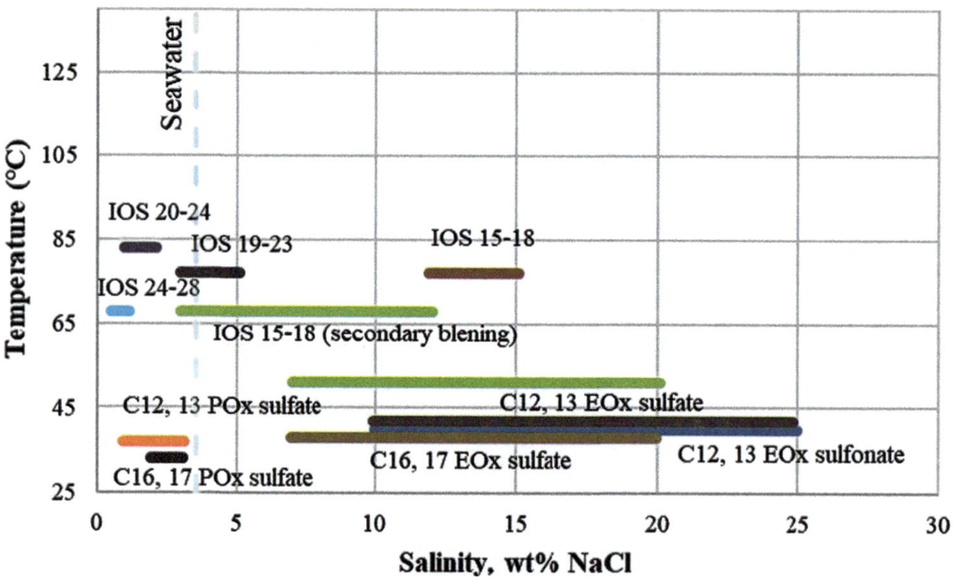

Fig. 3.8 Sulfonate surfactants used for a wide range of salinities and temperatures. *(Redrawn after Barnes JR, et al. Application of internal olefin sulfonates and other surfactants to EOR. Part 1: structure-performance relationships for selection at different reservoir conditions. In: SPE improved oil recovery symposium, 2010, Society of Petroleum Engineers.)*

3.5 Optimum salinity

An oil/brine/surfactant system can form a middle phase microemulsion within a special range of salinity. Winsor [16] was the first who categorized microemulsions as type I (oil in water), type II (water in oil), and type III (bi-continuous oil and water in a third phase). The salinity at which the middle phase microemulsion formed, contains an equal volume of oil and brine, being known as the optimal salinity [17] (Fig. 3.9). IFT between microemulsion/oil and water/microemulsion is minimum at this salinity. To determine optimum salinity, it is observed that if the volume of oil (V_o) and water (V_w) per unit volume of surfactant (V_s) is calculated and plotted, the intersection of V_o/V_s and V_w/V_s is the optimum solubilization ratio (σ^*) [18]. A typical example of optimum salinity, solubilization ratios, and Winsor types are represented in Fig. 3.10. At optimum salinity, not only IFT, but also surfactant loss, microemulsion viscosity and equilibration time are minimum defining success criteria needed for maximum oil recovery in a typical surfactant flooding (Fig. 3.11).

Different parameters can change optimum salinity, mainly defined by phase behavior tests. Solairaj [20] predicted the effect of these parameters on the optimal salinity via a correlation which can be used to predict surfactant structure for the optimized formulation:

$$LOG(S^*) = a_1 E_O + a_2 N_{PO} + a_3 N_{EO} + a_4\left(T - T_{ref}\right) + a_5 N_c + a_6 N_c{}^* PO + C \quad (3.1)$$

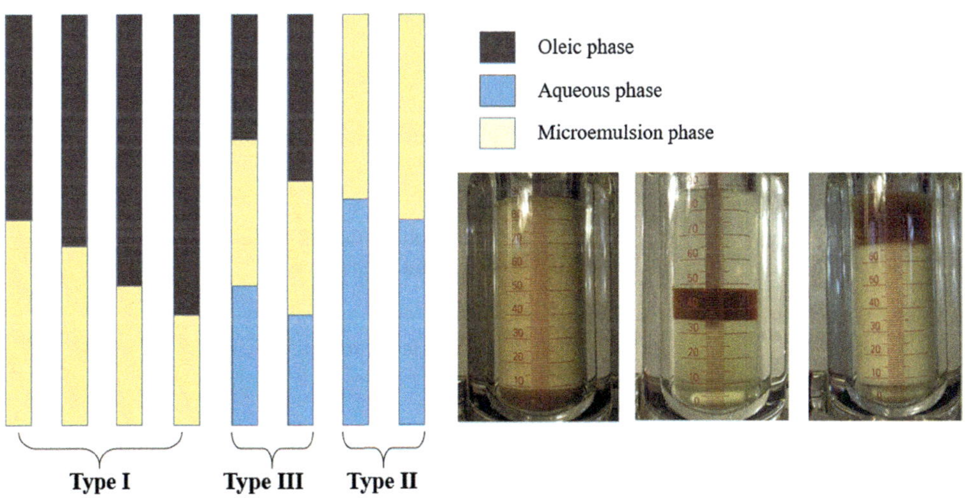

Fig. 3.9 Schematic diagram of different Winsor types for recognizing optimum salinity. *(Redrawn after Barnes JR, et al. Development of surfactants for chemical flooding at difficult reservoir conditions. In: SPE symposium on improved oil recovery, 2008. Society of Petroleum Engineers.)*

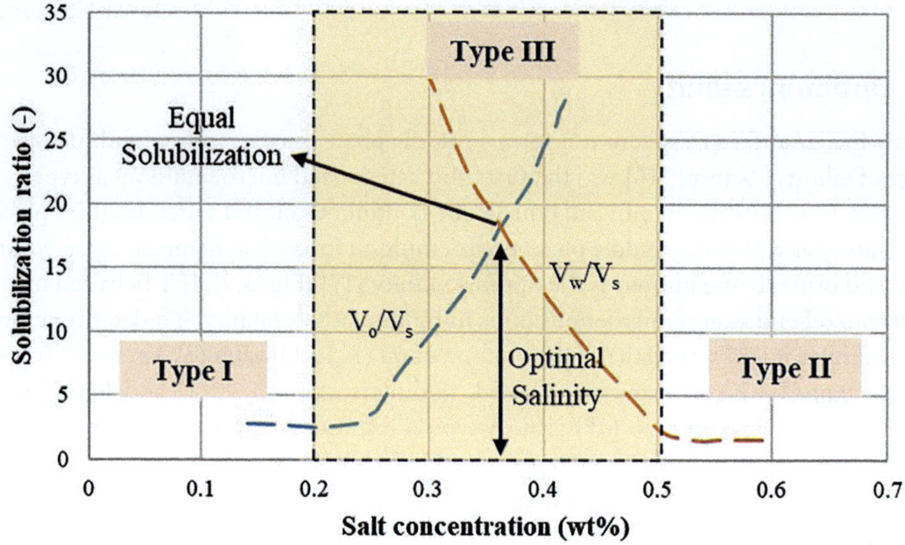

Fig. 3.10 Typical optimum salinity window in line with Winsor types, redrawn after [19].

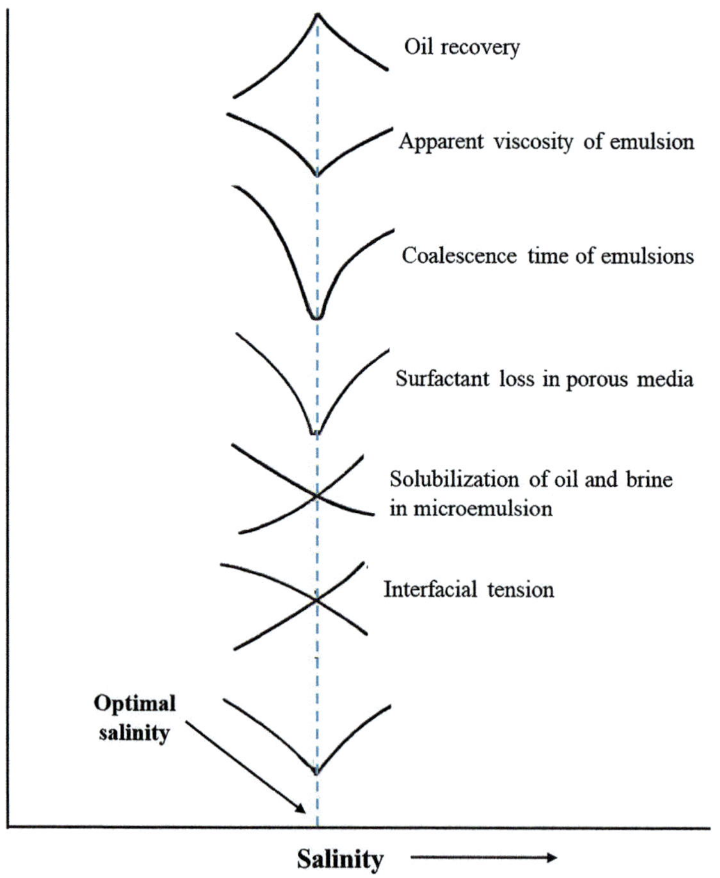

Fig. 3.11 Critical phenomena usually happen at optimum salinity. *(Redrawn after Shah DO. Surface phenomena in enhanced oil recovery, 1981.)*

where a_1, a_2, a_3, a_4, and a_5 are regression coefficients and C is the intercept. N_c is the mole average weighted carbon number of the surfactants mixture. E_o is the equivalent alkane carbon number (EACN) of oil. N_{PO} is mole average weighted propylene oxide (PO) number in the surfactant mixture, N_{EO} is mole average weighted ethylene oxide (EO) number in the surfactant, T is the temperature of interest (°C), T_{ref} is reference temperature and LOG(S^*) is the common logarithm of the optimum salinity, S^* (ppm). Main parameters affecting optimum salinity are temperature, surfactant structure, oil characteristics, and cosolvents.

3.5.1 Temperature

Phase behavior, especially solubility and optimal salinity, significantly depends on the solution temperature. Optimal salinity increased with temperature which can be interpreted as reducing gel, precipitates, and any viscous phases [21]. Optimal salinity trend with respect to temperature can change in the following order [22]:

(a) If formulation contains only Krafft point, optimal salinity will increase with temperature (Fig. 3.12).

(b) If EO and PO groups are introduced in the structure (cloud point effect is appeared), optimal salinity will decrease by temperature elevation (Fig. 3.12).

3.5.2 Surfactant structure

Every parameter changing HLB (hydrophilic-lipophilic balance) can cause a variation in optimal salinity. Hydrophobic properties (length, branching, and ethoxylation) and hydrophilic head types especially at anionic surfactants can significantly change optimal

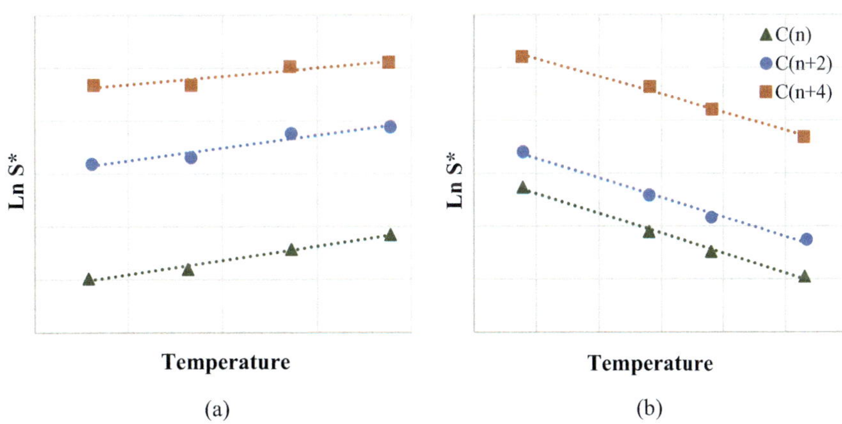

Fig. 3.12 Effect of temperature on optimal salinity: (A) formulation with Krafft point and (B) formulation with cloud point for three different alkanes (*n* designates the number of carbon in the structure of surfactant). *(Redrawn after Pinnawala Arachchilage GW, et al. Evaluating the effect of temperature on surfactant phase behavior and aqueous stability to forecast optimum salinity at high temperature. In: SPE improved oil recovery conference, 2018, Society of Petroleum Engineers.)*

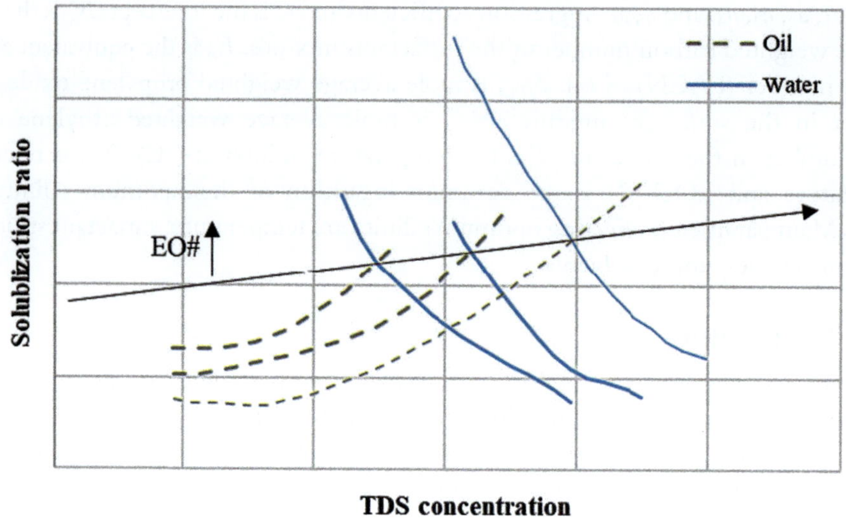

Fig. 3.13 Effect of ethylene oxide (EO) group on optimal salinity. *(Redrawn after Yang HT, et al. Low-cost, high-performance chemicals for enhanced oil recovery, In: SPE improved oil recovery symposium, 2010, Society of Petroleum Engineers.)*

salinity [23]. For instance, increasing hydrophobic length decreases optimal salinity. To tailor a surfactant for special salinity conditions, functional groups such as hydrophilic ethylene oxide (EO) and propylene oxide (PO) numbers are a priority [24]. Indeed, PO and EO have the opposite effect on optimal salinity and solubilization ratios. EO contribution into surfactant structure, in up to 10 spots, can increase optimal salinity without significant change in the solubilization ratio [25]. An example of the EO effect on optimum salinity and solubilization ratio is shown in Fig. 3.13.

3.5.3 Oil characteristic

Crude oil properties can be usually modeled using the equivalent alkane carbon number concept (EACN) [26]. It has been found that optimum salinity increases with EACN and a typical example is shown in Fig. 3.14.

3.5.4 Cosolvent

Generally, alcohol as the most common co-solvent in surfactant flooding can increase optimum salinity via enhancing the solubility of surfactant in the aqueous phase. However, experiment results reveal that high molecular weight alcohols such as n-pentanol can decrease optimum salinity significantly whereas the solubilization ratio increases considerably [27]. On the other hand, low molecular weight alcohols cause higher optimal salinity when added to the surfactant solution. It is also believed that secondary alcohols (such as SBA) have the same effect on optimal salinity as the primary alcohol with one less

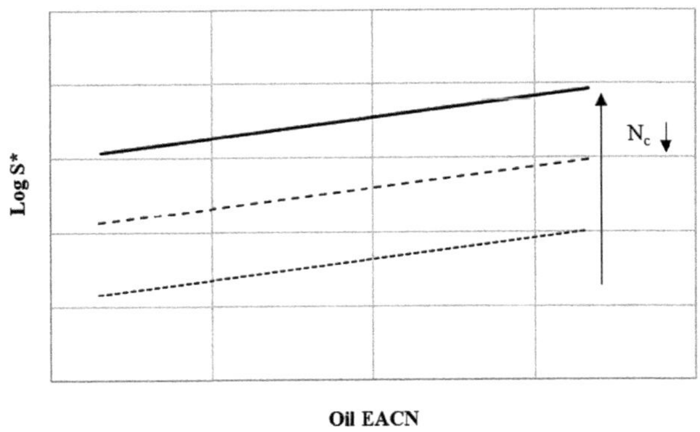

Fig. 3.14 Oil EACN effect on the optimum salinity at different mole average weighted carbon numbers. *(Redrawn after Pinnawala Arachchilage GW, et al. Evaluating the effect of temperature on surfactant phase behavior and aqueous stability to forecast optimum salinity at high temperature. In: SPE improved oil recovery conference, 2018, Society of Petroleum Engineers.)*

carbon (such as n-pentanol). Similarly, tertiary alcohols (such as tert-amyl alcohol) does the same with respect to the secondary alcohol with one less carbon (such as SBA) in their alkyl chain [28].

3.6 Mechanisms

3.6.1 IFT reduction

Surfactant as the main chemical agent to recover trapped oil, initially was introduced to reduce interfacial tension [4]. At very low concentration of surfactants, surfactant molecules remain as monomers where there is an equilibrium with those molecules adsorbed at the interface film [18]. Increasing surfactant concentration results into the aggregation of monomers, which finally forms micelles. Micelle is formed at the concentration termed critical micelle concentration (CMC) above which monomer concentration remains constant and only the number of micelles will increase. IFT is generally reduced as far as CMC is attained then it remains constant whereas more micelles are formed. Micelle with an inner nonpolar environment can swallow oil droplets and increase oil solubility [18]. CMC is normally at the range of 0.00001–0.0001 mol/lit [4]. Adsorption of surfactant monomers on the brine-hydrocarbon surface can decrease interfacial tension up to a specific concentration of monomers. Beyond the CMC, micelles form and no significant change happens to the IFT [8]. It is well established that ultralow IFT (around 0.001 mN/m) is required to attain a high capillary number and mobilize trapped oil [16]. When the surfactant is added to the injected fluid, it reduces the interfacial tension and resistance capillary force, leading to elongation of trapped oil droplets (Fig. 3.15).

Ultralow value of IFT is usually observed only when dispersed-crystalline middle phase appeared between brine and hydrocarbon [30, 31].

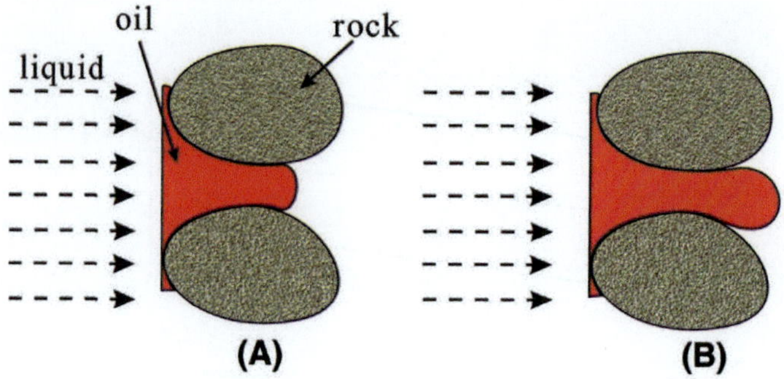

Fig. 3.15 Schematic representation of trapped oil (A) and elongated oil owing to surfactant usage (B) [29].

3.6.2 Wettability alteration

Nearly 60% of proven oil reserves are estimated to be in the carbonate reservoirs where oil wetness, heterogenity and low permeability contribute to a low production rate [32]. Changing wettability from oil-wet to intermediate/water-wet proved to be the most promising production mechanism in carbonate rocks especially low permeability ones. This process promotes the spontaneous imbibition where the recovery mechanism is mainly controlled by capillary forces. Three hypotheses are proposed for the wettability alteration process via surfactants: coating, cleaning, and micellar solubilization.

Coating: Adsorption of anionic surfactant molecules on the oil-wet rock surface can be accomplished by the interaction of the hydrophobic portion of surfactant with organic oil components on the surface [33]. This can make a surfactant double layer regarding polar head group is dissolved in water and create a thin water layer between oil and rock surface. Although it activates the imbibition process, this surfactant double layer is not stable and permanent since the hydrophobic bond is not strong [33]. Thus, this mechanism is a function of anionic surfactant hydrophilicity. Schematic representation of this mechanism is shown in Fig. 3.16.

Cleaning: It is well recognized that negatively charges components of oil, including carboxylic groups, adsorbing to the rock surface, are responsible for oil wetness [34]. Electrostatic and hydrophobic interaction between surfactant monomers and anionic organic materials through the formation of ion-pair are expected to cause wettability change [33]. Standness and Austad [33] stated that after desorption, ion-pair is dissolved in oil phase or in micelle and rock surface becomes water wet (Fig. 3.17). In turn, this mechanism is a strong function of diffusion rate, surfactant structure, and solution temperature. As electrostatic interaction leads to the desorption of organic materials, the cationic surfactant is the most common wettability modifier in the cleaning process of carbonates.

Micellar solubilization: Adsorbed anionic molecules of oil can be solubilized into the surfactants micelle during surfactant invasion into the pore spaces [35]. Kumar [35]

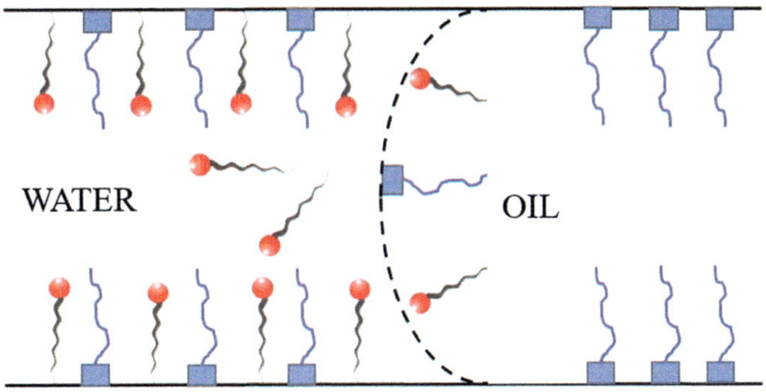

Fig. 3.16 Schematic representation of coating process during wettability alteration via surfactants; surfactant monomers *(red color symbols)* and carboxylates *(blue color symbols)*. *(Redrawn after Zhou Y, et al. A comprehensive review of emulsion and its field application for enhanced oil recovery. Energy Sci Eng 2019;7(4):1046–58.)*

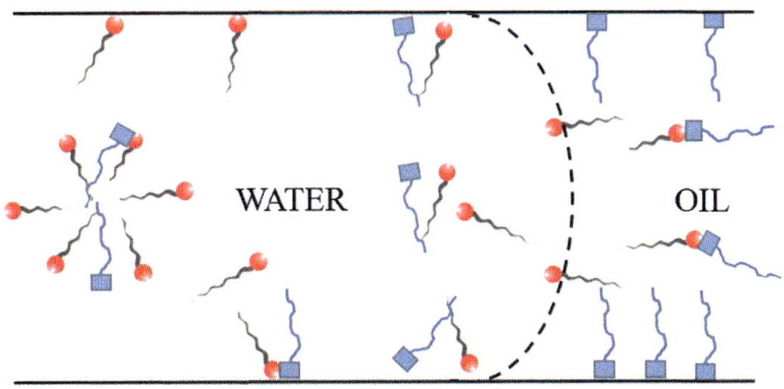

Fig. 3.17 Schematic representation of cleaning process during wettability alteration via surfactants; surfactant monomers *(red color symbols)* and carboxylates *(blue color symbols)*. *(Redrawn after Zhou Y et al. A comprehensive review of emulsion and its field application for enhanced oil recovery. Energy Sci Eng 2019;7(4):1046–58.)*

reckoned that anionic surfactants (propoxylated sulfate) are usually more prone to solubilization mechanism than cationic (DTAB) in carbonate rocks. According to the experimental results this mechanism has a faster rate than the ion-pair mechanism usually playing role in opposite charge surfactants [35].

3.6.3 Emulsification

Some surfactant solutions have a significant emulsification effect on crude oil. Under the oil-water flow shear, the surfactant system can quickly emulsify, disperse, and strip the crude oil on the rock surface, forming oil-in-water (II-) emulsion. This emulsion can increase solution viscosity leading to profile control potential by improving the oil-water mobility ratio. Meanwhile, being charged by the surfactant adsorbing on them, the oil droplets cannot easily stick back to the formation rock surface and thus may flow to the producing well with the surfactant solution.

3.7 Emulsion formation and treatment

Effective surfactant flooding is not a matter of principal but rather a question of the economy. Surfactant loss by retention is a determinant process threatening the economy and performance of a surfactant-flooding project. This retention may be controlled by emulsion properties such as solubility and viscosity. Any precipitate, crystalline phase, or even colloids in the emulsion phase may cause surfactant chromatographic separation. Cosolvents are generally used in the chemical formulation to improve phase behavior, reduce emulsion viscosity, and promote surfactant-polymer compatibility [24]. Alcohols, as the main cosolvents, can increase surfactant solubility and modify emulsion properties via eliminating any crystalline and/or viscous gel phase. The main characteristics of renowned alcohols in CEOR have been depicted in Table 3.2.

3.7.1 Microemulsion rheology

A mixture of oil, brine, and surfactant can form a thermodynamically stable phase under special conditions, called microemulsion [36]. Microemulsion rheology affects phase separation time, pressure gradient, and even sweep efficiency in heterogeneous porous media depending on physicochemical properties. Viscosity, as the representative parameter of microemulsion rheology, does not influence the trapping/mobilization of oil droplets but contributes to the displacement of residual oil [37].

Two main issues dealing with microemulsion rheology are high viscosity and non-Newtonian behavior which may result in significant surfactant retention, high-pressure gradient, high separation time, and low sweep efficiency. Microemulsion rheology can be determined by assessment of the following parameters [38]:

- **Hydrophobe structure:** Hydrophobe branching can be used to decrease microemulsion viscosity and equilibrium time [24].

Table 3.2 Alcohols used in the CEOR and their main properties.

Solvent type / properties	Key points	Remarks
Common types	IPA, IBA,SBA, EGBE, TEGBE	Effective for sandstone and carbonate
Structure	Higher chain length, more viscosity	Cost issue
Concentration	More solvent, lower equilibrium time, and emulsion viscosity	Lower solubilization ratio

IPA, isopropanol alcohol; IBA, isobutyl alcohol; SBA, sec-butanol alcohol; EGBE, ethylene glycol monobutyle ether; TGBE, triethylene glycol monobutyle ether.

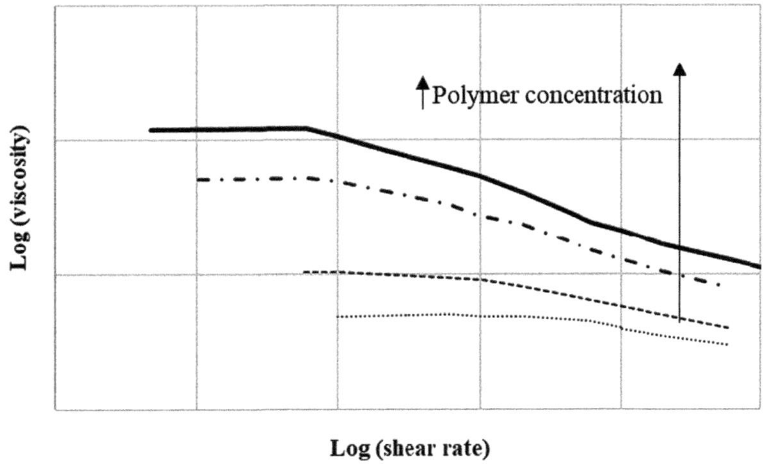

Fig. 3.18 Effect of a typical polymer concentration on microemulsion viscosity. *(Redrawn after Al Kalbani M, et al. Impact of polymer EOR and salinity on barium sulphate scale risk. in SPE international oilfield scale conference and exhibition. 2018. Society of Petroleum Engineers.)*

- **Co-Surfactant:** Orderly arrangement of surfactant molecules (the main reason for crystalline phase or high viscosity) can be also disrupted using a dissimilar or branched co-surfactant including IOS [38].
- **Polymer:** Viscosity change with polymer addition is a function of the microemulsion type. The polymer can increase viscosity when microemulsion of types I and III are present (Fig. 3.18) and it seems that no significant change is predicted for microemulsion type II. The polymer can also decrease coalescence or equilibrium time when added to a specified concentration [39]. Fig. 3.19 shows the effect of polymer on the microemulsion samples. It is clear that samples with polymer are so close to the equilibrium while those without polymer are far from equilibrium state [40].

- **Cosolvent:** The most common method to reduce microemulsion viscosity is solvent (such as IBA and TEGBE) contribution into surfactant formulation [38, 41]. An example of solvent effect on the microemulsion viscosity is depicted in Fig. 3.20. It is worth to note that care must be taken for alcohol selection since using some alcohols can change the ultralow IFT condition leading to less prize of CEOR. In this sense, a general procedure for alcohol screening as illustrated in Fig. 3.21 can be devised to lower this risk.

3.8 Surfactant retention

While in practical terms, surfactant remained in solution for IFT reduction or wettability alteration is critical for the success of a surfactant EOR, high initial concentrations of

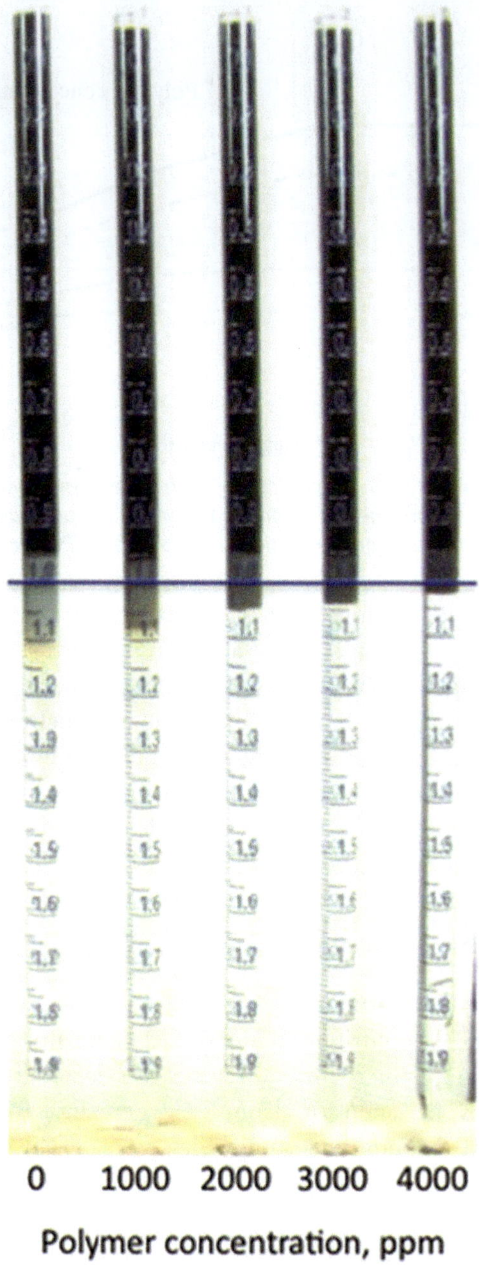

Fig. 3.19 Effect of polymer presence on microemulsion formation [40].

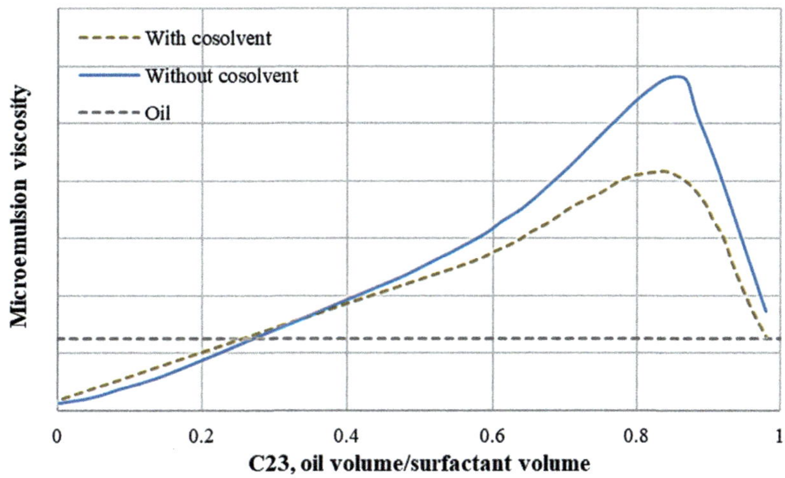

Fig. 3.20 Solvent effect on the microemulsion viscosity. *(Redrawn after Mohan K. Alkaline surfactant flooding for tight carbonate reservoirs. In: SPE annual technical conference and exhibition, 2009, Society of Petroleum Engineers.)*

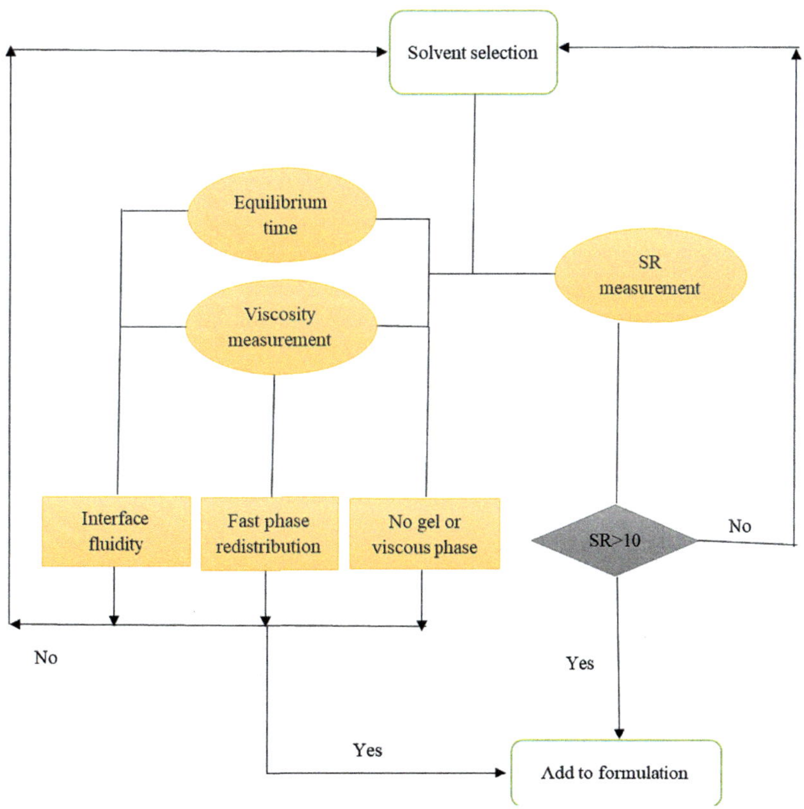

Fig. 3.21 Procedure to screen solvents for higher performance in CEOR.

surfactants are not commercially viable. In other words, there has been a trade-off between economic and practical aspects of each surfactant EOR, termed as surfactant retention. Hence, prior to any surfactant application, it is essential that the surfactant loss in the porous media should be defined. Generally, surfactant loss or retention in reservoirs can be ascribed to adsorption, precipitation, and phase trapping [1, 42]. Indeed, the effects of these three phenomena in the porous media can end up in the reduction of technical and economic feasibility of EOR for a certain surfactant. The higher the surfactant retention, the more the sum of money squandered and the less the efficiency attained [43–45].

Currently, although differentiation between these effects at reservoir conditions seems to be implausible, obtaining knowledge regarding each one at lab scale is within our grasp. In turn, related lab tests allow reservoir engineers to preevaluate different surfactants and consider how to deal with the issue of surfactant retention. In the remaining parts of this section, three forms of surfactant retention are investigated.

3.8.1 Surfactant adsorption

The amount of adsorbed surfactant is usually measured through static and dynamic adsorption tests [1, 3, 42] It is noteworthy that merely static tests could demonstrate the maximum amount of surfactant adsorption directly in which the adsorption amount is usually reported in mass of adsorbed surfactant per unit mass of rock (mg surfactant/g of rock) [3].

Surfactant adsorption on rock surfaces relies on various parameters such as surfactant type, rock mineralogy, type, and concentration of electrolytes [46]. Besides that, surfactant concentration is another key factor. In this regard, finding surfactant adsorption with respect to equilibrium concentration is an inevitable step to define any surfactant flooding performance.

When surfactant adsorption is plotted versus equilibrium surfactant concentration in log-log scale, the diagram can be divided into four regions (see Fig. 3.22). In region I, the adsorption increases linearly with concentration obeying Henry's law [47]. In region II, due to the surface aggregation of surfactants, a sharp increase in adsorption is observed. The surface aggregates are known as "Hemi-micelles" forming between critical hemimicellar concentration (HMC) and critical micelle concentration (CMC), followed by a gradual increase in region III. Finally, in region IV, the diagram is plateaued. However, in some cases, a downward trend in region IV is envisaged. Although surfactant adsorption causes loss of surfactant in the flooding process, it may change wettability using the coating process.

To correlate surfactant adsorption with equilibrium surfactant concentration, some proposed models are usually fitted to lab results. In the next section, this issue is elaborated.

3.8.1.1 Adsorption models

In simple words, an adsorption model is an equation correlating the equilibrium surfactant adsorption to the equilibrium surfactant concentration at a specific temperature and

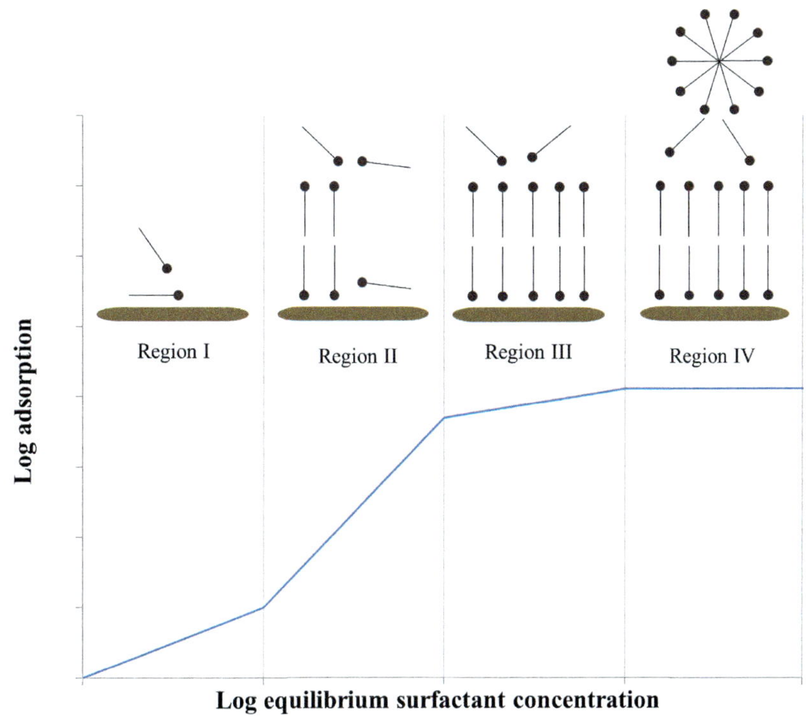

Fig. 3.22 Adsorption vs equilibrium surfactant concentration (log-log scale): changes of surfactant adsorption with surfactant concentration in four regions. *(Redrawn after Paria S, Khilar KC. A review on experimental studies of surfactant adsorption at the hydrophilic solid-water interface. Adv Colloid Interface Sci 2004;110(3):75–95.)*

pressure. Recognition of apt adsorption models allows reservoir engineers to characterize adsorption [48]. Furthermore, in some simulators, these adsorption models have been equipped in such a way that adsorption can be defined through model parameters rather than the table insertion. In this section, two of the renowned adsorption models are described: Langmuir and Freundlich models.

Langmuir model

The most common adsorption model is the Langmuir model expressed as follows [49]:

$$q_e = \frac{q_0 K_{ad} C_e}{1 + K_{ad} C_e} \tag{3.2}$$

where C_e and q_e are in order concentration of surfactant and adsorbed surfactant at equilibrium. K_{ad} and C_e are Langmuir adsorption constant (adsorption capacity) and concentration of surfactant, respectively. Additionally, q_0 and q_e define initially adsorbed surfactant and adsorbed surfactant at equilibrium, respectively.

Table 3.3 Type of adsorption through the value of R_L.

No.	R_L	Adsorption
1	$R_L > 1$	Unfavorable
2	$0 < R_L < 1$	Favorable
3	$R_L = 1$	Linear
4	$R_L = 0$	Irreversible

(Data from Nimibofa A, Augustus N, Donbebe W. Modelling and interpretation of adsorption isotherms. Chemistry 2017; 3039817. p. 2020.)

In this model, it is presumed that the adsorption film is monomolecular [49]. The Langmuir adsorption model can be also rearranged as in the following equation:

$$\frac{C_e}{q_e} = \frac{1}{K_{ad} q_{max}} + \frac{C_e}{q_{max}} \tag{3.3}$$

where q_{max} is the maximum adsorption capacity. Other parameters are as defined for Eq. (3.2). As $\frac{C_e}{q_e}$ is plotted versus $\frac{1}{C_e}$, a straight line with slope equal to $1/q_0 K_{ad}$ and intercept of $1/q_e$ is formed.

In Langmuir adsorption, R_L, being a dimensionless number, is termed as the separation factor expressed as follows [48]:

$$R_L = \frac{1}{1 + K_{ad} C_0} \tag{3.4}$$

where K_{ad} is Langmuir constant and C_0 is the initial concentration of surfactant. This number can be used to categorize surfactants based on their adsorption behavior. Classification of adsorption via the value of R_L has shown in Table 3.3.

Freundlich model

In this model, it is assumed that surfactant adsorption on the solid surface is multilayer. This model is appropriate for adsorption on heterogeneous surfaces [48]. The following equation is used for this model:

$$q_e = K_f C_e^{1/n} \tag{3.5}$$

where C_e and q_e are in order concentration of surfactant and adsorbed surfactant at equilibrium. K_f and $1/n$ designate adsorption capacity and adsorption intensity, respectively.

This equation can be rearranged as follows:

$$\log q_m = \log K_f + \frac{1}{n} \log C_e \tag{3.6}$$

Therefore, as $\log q_m$ is plotted versus $\log C_e$, the slope of the trend line equals to $\log K_f$ and intercept of this line equals $\frac{1}{n} \log C_e$.

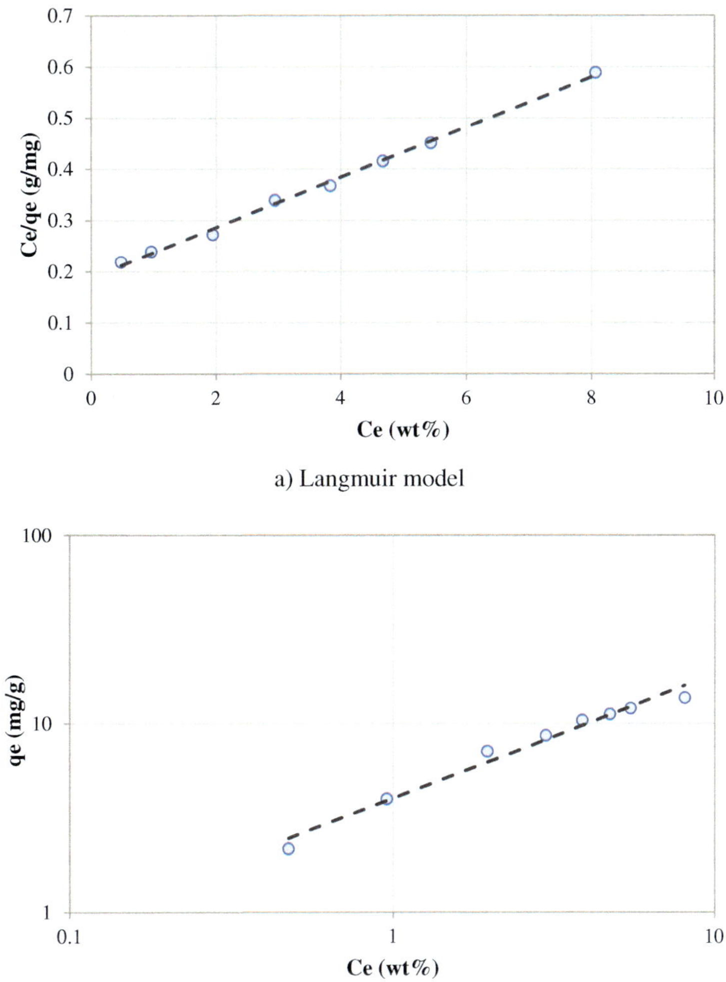

Fig. 3.23 Comparison of (A) Langmuir and (B) Freundlich adsorption isotherm models for a typical lab result. *(Data from Barati A, et al. Adsorption of a new nonionic surfactant on carbonate minerals in enhanced oil recovery: experimental and modeling study. Chem Eng Res Des 2016;105:55–63.)*

In Fig. 3.23, these two models are matched on the results of static adsorption of a certain surfactant [50]. As can be seen, the Langmuir model matches the lab results better than the Freundlich model.

3.8.2 Phase trapping

Although salinity increases surfactant loss on the rock, it is believed that surfactant-rich-phase trapping can be the dominant mechanism [37]. The microemulsion of type II (+)

phase usually has less mobility (higher viscosity) than the excess brine phase and chase water. In addition, interfacial tensions in the Winsor type II (+) are not as low as in the Type III phase environment. Therefore, surfactant-rich, microemulsion phases can be trapped in the porous media during a chemical flood much like oil is trapped in rock during waterflooding, especially in the rear mixing zone [37, 51]. Some investigators believe that surfactant phase trapping is a complex mechanism including hydrodynamic trapping, phase partitioning, and mechanical trapping [1]. All these conditions usually depend on the multiphase flow conditions.

3.8.2.1 Surfactant partitioning

The process of the transmission of surfactant from an aqueous to the oleic phase and vice versa is termed surfactant partitioning. In Fig. 3.24, a schematic representation of surfactant partitioning has been shown. As surfactants excessively partition from an aqueous phase to an oleic one, the surfactant has no potential to affect surface activities whether the fluid-fluid interface or fluid-rock interactions. Consequently, it can result into high-pressure gradient issue. Hence, to avert this problem, the critical role of surfactant partitioning in surfactant EOR on different processes needs to be considered: Surfactant phase behavior, oil recovery, etc. [52].

In technical terms, partitioning coefficient (k_p), a measure of surfactant partitioning, can be defined as follows:

$$k_p = \frac{C_{oil}}{C_{water}} \tag{3.7}$$

In this equation, C_{oil} and C_{water} represent the surfactant concentration in oil and water phases, respectively.

When it comes to surfactant partitioning, factors such as temperature and salinity are foremost among pertinent parameters [53]. At low temperatures, surfactant mobility is negligible because of the low rate of surfactant solubility and high viscosity of surfactant, resulting in lower surfactant partitioning. While, at higher temperatures, the transfer rate

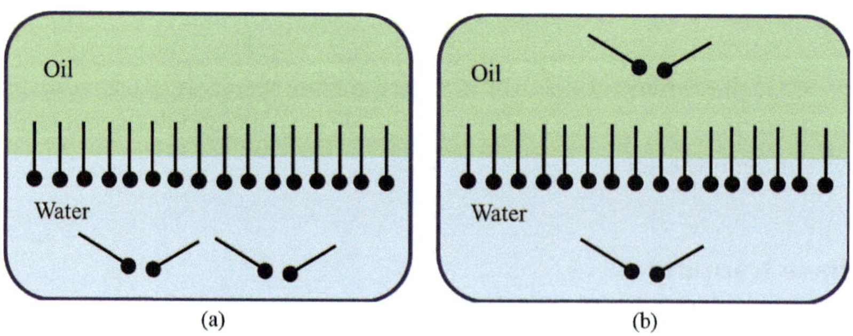

Fig. 3.24 Schematic representation of surfactant partitioning: (A) without partitioning and (B) surfactant partitioning into oleic phase.

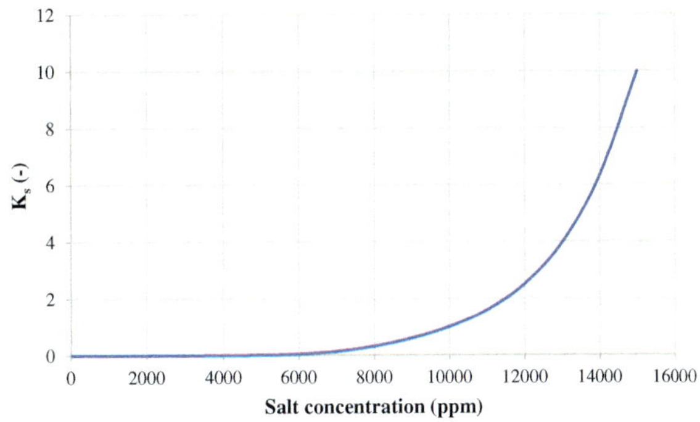

Fig. 3.25 Partitioning coefficient as a function of salinity according to Eq. (3.8), $C_{seop}=10{,}000$ ppm and $0<C_{se}<15{,}000$ ppm.

of surfactant molecules in the aqueous phase increases leading to transportation of them to the oil-water interface [54]. On the other hand, the presence of ions can increase the amount of surfactant in the oil phase; in this regard, it should be noted that divalent ions compared to monovalent ions have more capability [55].

Besides the experimental evaluation of surfactant partitioning, a correlation is also recommended to obtain surfactant partitioning as follows:

$$K_s = \begin{cases} 10^{2\left(C_{se}/C_{seop}-1\right)}, C_{se} > C_{seop} \\ 10^{2\left(1-C_{seop}/C_{se}\right)}, C_{se} < C_{seop} \end{cases} \tag{3.8}$$

where C_{se} and C_{seop} represent effective salinity and effective optimum salinity, respectively.

By considering 10,000 ppm for C_{seop}, partitioning coefficient can be plotted versus salt concentration as shown in Fig. 3.25 (C_{se} ranged from 0 to 15,000 ppm). This plot can be input in different numerical simulators.

There are some experimental methodologies which can be used to determine partitioning coefficients such as the high-performance liquid chromatography (HPLC) and solubility ratio method [56]. For instance, in the solubility approach, the steps mentioned below need to be followed to attain partitioning coefficient [55]:

- Initial aqueous surfactant solution is provided.
- The solution is placed in a separating funnel and the crude oil is placed on top of it (It is attempted to preserve the volume ratio for all measurements).
- Two phases are kept for a certain period of time (48 hours) to allow the surfactant solution to diffuse from the aqueous to the oleic phase.
- Finally, the partitioning coefficient can be defined as follows [55, 57]:

$$k_p = \frac{C_{oil}}{C_{water}} = \frac{V_{water}}{V_{oil}}\left(\frac{C_{water,0} - C_{water}}{C_{water}}\right) \tag{3.9}$$

In this equation, C_{oil} and C_{water} represent the surfactant concentration in oil and water phases, respectively. In addition, V_{water} and V_{oil} are volume of aqueous and oil phases, respectively. Partitioned surfactant in the oil phase is usually obtained using the material balance in which initial surfactant concentration in the aqueous phase ($C_{water,\ 0}$) is set, then correspondingly, equilibrium surfactant concentration in the aqueous phase after partitioning (C_{water}) is recorded.

3.8.3 Surfactant precipitation

Electrostatic interaction between surfactant monomers and counter ions (monovalent and multivalent) in the aqueous phase is a basic phenomenon that contributes to the precipitation of ionic surfactants [58]. In this phenomenon, the interaction of monovalent ions (especially Na^+) with anionic surfactant (sulfonate for example) produces a colloidal solution which can finally cause the onset of surfactant precipitation by increasing salinity. In the first mechanism, precipitation can occur below or above CMC. As Al-Hashim et al. (1988) described, micelle phase transformation (cylindrical, lamellar) of a petroleum sulfonate just prior to precipitation takes place [59] (Fig. 3.26). This description is summarized in the following reactions:

$$mNa^+ + nR^- \Leftrightarrow M^{-(n-m)} \tag{3.10}$$

$$(n-m)Na^+ + M^{-(n-m)} \Leftrightarrow (NaM)_n \tag{3.11}$$

where Na^+ and R^+ are counter ion and surfactant molecules, respectively. n is the number of monomers in the micelle and M represents the number of Na^+ ions attached to the micelle. M is used to show micelle with an ionic net charge of (n-m). By increasing salinity (Na^+), the equilibrium of Eq. (3.11) changes and produces micelle with Na^+ attached which finally goes into liquid crystalline and leads to precipitation [59].

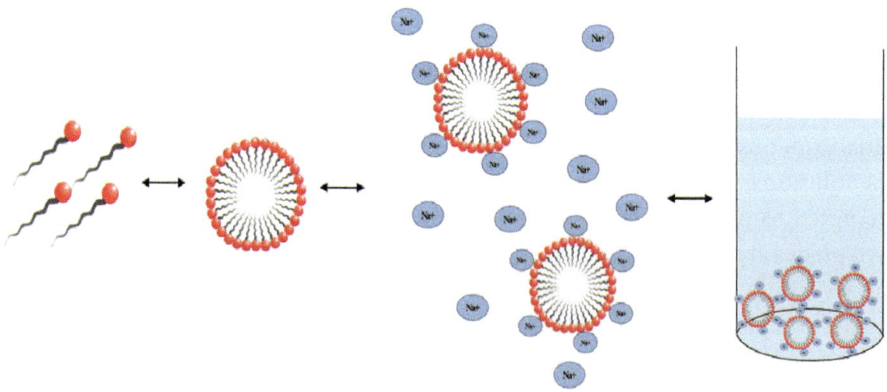

Fig. 3.26 Surfactant precipitation via micelle complex formation

3.8.4 Measurement of surfactants loss

Totally, two sets of experiments are proposed to estimate retention in CEOR [3]: Static and dynamic retention. These experiments are explained in the following section.

3.8.4.1 *Static retention*

This section of review involves the estimation of retention isotherms via crushed samples. Initially, the crushed samples need to be sifted to have a particle size less than certain values [60]. On the other hand, a solution containing water and surfactant is provided. Then, the sifted solid materials are added to the water. Then, the amount of adsorption can be calculated as follows [60]:

$$\Gamma = \frac{\left(C_i - C_{eq}\right).m_{surf}}{m_{abs}.S_{BET}} \tag{3.12}$$

where C_i and C_{eq} are in order the initial and the equilibrium concentration of the surfactant in wt%, m_{surf} is the mass of the solution in g, m_{abs} is the mass of the rock in g, and S_{BET} denotes the specific area of the rock in m^2/g. The amount of surfactant before and after rock exposure is usually obtained through liquid chromatography.

For a single-component surfactant, the shape of isotherm as well as curve of retention versus surfactant equilibrium concentration will give information regarding the type and mechanism of adsorption. After all, this type of experiment provides no information about the other states of surfactant loss (i.e., precipitation and phase partitioning).

3.8.4.2 *Dynamic retention*

Aside from static retention, dynamic retention is the other key evaluation process that needs to be deemed as well. Obviously, dynamic retention is a more pragmatic approach compared to static retention as oil and pore geometry are also considered. Generally, a standard dynamic retention is conducted experimentally through the following approaches: [61]:

(a) Determination of effluent surfactant ratio versus cumulative injection volumes (Fig. 3.27).
(b) Circulation of about 20 pore volume of solution through the core for 24 hours.
(c) Extraction of the surfactant from the core with an efficient solvent.

Applying the first method, the injection is continued as far as the effluent surfactant ratio is less than one (Fig. 3.27). This process can be tedious in cases where the injection rate is low or surfactant adsorption is high. Similarly, for the other two methods, it is argued that when the adsorption equilibrium is established.

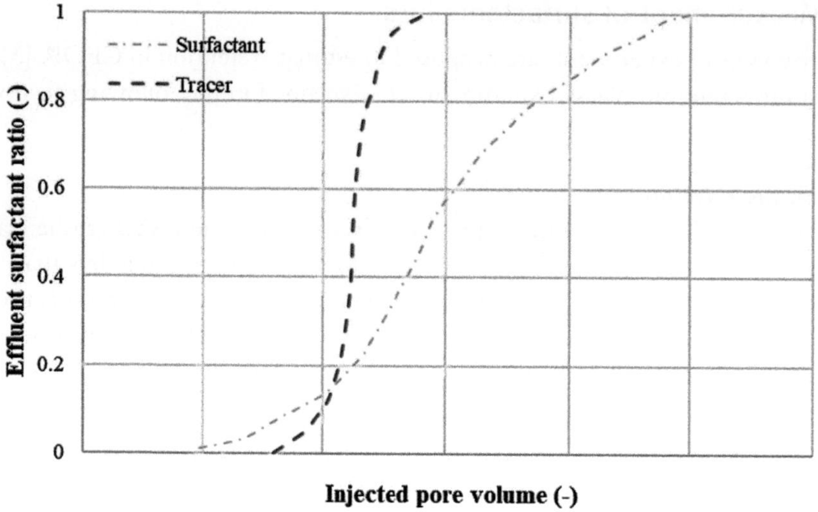

Fig. 3.27 Schematic comparison of the effluent surfactant ratio and the tracer ratio at different injected pore volumes. *(Redrawn after Skjæveland SM, Kleppe J. SPOR monograph, Recent advances in improved oil recovery methods for north sea sandstone reservoirs. Norwegian Petroleum Directorate, Norway, 1992.)*

3.8.5 Modeling and simulations

Modeling and simulation of surfactant flooding are the other issues which need to be investigated properly. In this sense, it is anticipated that the modeling of surfactant flooding would be complex thanks to the versatile mechanisms in this process. It should be noted that active mechanisms could vary from one type of surfactant to another one. For instance, in a surfactant with only wettability alteration capability and the other one with IFT reduction the modeling approach is different. Therefore, the modeling requires distinct consideration of each mechanism to address surfactant EOR holistically.

During these years, experts put forward different methodologies to model surfactant as well as water and rock behavior during surfactant injection. In time, the applied approach to simulate surfactant flooding has evolved.

At present, diverse numerical simulators in petroleum engineering support the simulation of surfactant flooding: ECLIPSE, CMG, UTCHEM, MRST, etc. In this regard, the properties of different simulators are compared in Table 3.4 [62,63].

Among different software packages, UTCHEM is supreme because of capturing all the proposed mechanisms. Yet, the tedious simulation process is a hurdle in applying it to field scales. In addition, in terms of geologic issues, this software lack some prerequisites. Therefore, its application can be currently restricted to the core or well models.

Aside from the simulation software selection, there are some inputs required for simulation. These data are shown in Fig. 3.28. As mentioned, due to the difference in

Table 3.4 Comparison of different numerical simulators in simulation of surfactant EOR [62,63].

No.	Surfactant module	UTCHEM	CMG-STARS	ECLIPSE
1	Microemulsion viscosity	✓	✗	✗
2	Interfacial tension	✓	✓, Tabular format	✓, Tabular format
3	Phase behavior	✓	✗	✗
4	Surfactant adsorption	✓	✓	✓
5	Ion exchange effect	✓	✓	✓
6	Effective salinity window	✓	✗	✗

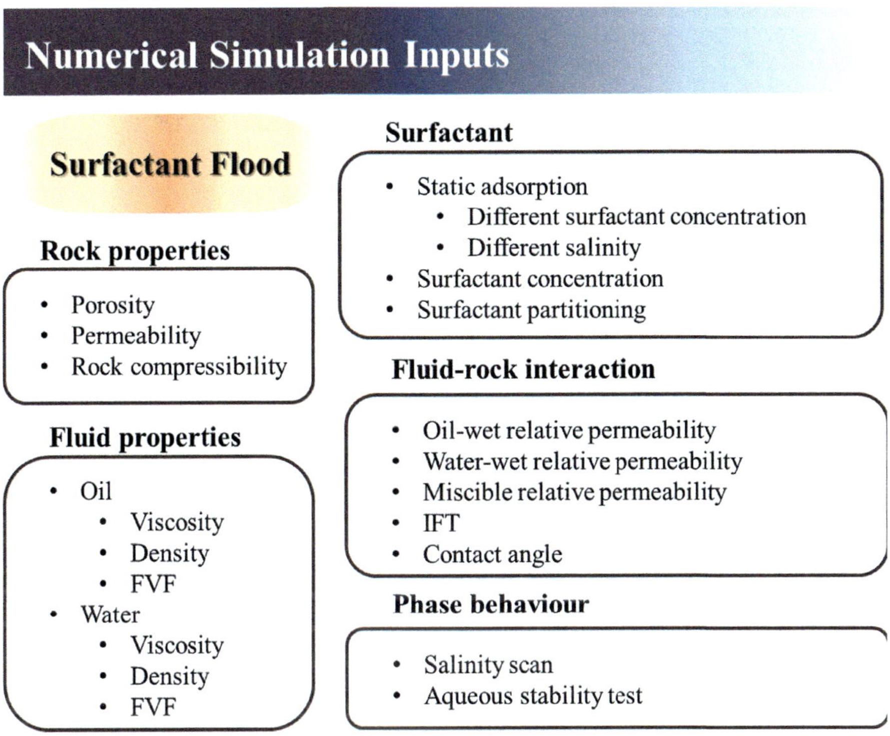

Fig. 3.28 Required inputs for simulation of surfactant flood.

modeling approaches of surfactant flooding by different simulators, the process of usage of these data would differ from one to the others.

In some simulators such as ECLIPSE and CMG, the surfactant is assumed as a single component in the water phase being able to be transferred to the oil phase via partitioning. Therefore, the presence of microemulsion cannot be directly considered and there is no third phase for microemulsion. However, in STARS-ME and UTCHEM, microemulsion can be defined as a separate phase.

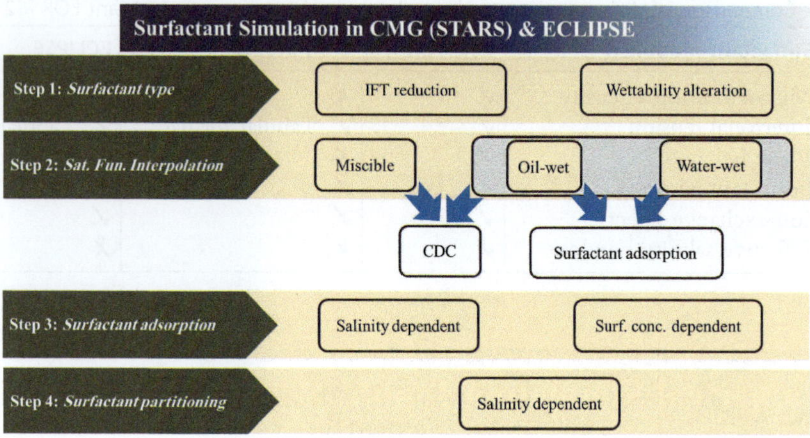

Fig. 3.29 Surfactant EOR simulation in software such as CMG (STARS) and ECLIPSE.

Aside from the details of surfactant considerations in STARS and ECLIPSE software packages, its simulation can be embraced into four definite sections (Fig. 3.29):

Step 1: Three relative permeabilities can be considered for a reservoir rock type. This number can be reduced to two with respect to the surfactant type. For instance, for a surfactant with the capability of wettability alteration and having no effect on IFT reduction, elimination of miscible relative permeability is out of the question. In oil-wet fractured reservoirs where wettability alteration is desirable, these kind of surfactants are usually opted.

Step 2: When IFT reduction is the dominant mechanism, miscible relative permeability curves are considered. In this regard, relative permeability curves for each phase are obtained through interpolation of miscible and initial relative permeabilities. To achieve this goal, a capillary desaturation curve is required in which at each capillary number the value of residual saturation for each phase is estimated.

On the other hand, as the surfactant is able to alter wettability, the interpolation between oil-wet and water-wet immiscible relative permeabilities is performed. This process can be done based on different parameters. However, according to the preceding studies, surfactant adsorption on the rock surface is an appropriate parameter to interpolate immiscible relative permeabilities.

Step 3: As explained in previous sections, surfactant adsorption is a function of salinity, surfactant concentration, rock mineralogy, etc. In this issue, presuming other factors are identical in all grids of a particular rock type, it is considered that surfactant adsorption is merely a function of surfactant concentration and salinity.

Step 4: For surfactant partitioning, a table relating partitioning coefficient to salinity is required. For this purpose, Eq. (3.8) can also be applied.

3.8.6 Relative permeability

As explained earlier, surfactant type can determine its role in the reservoir. For a typical surfactant with the ability to reduce IFT, reduction of capillary forces leads to increasing capillary number and reducing residual saturation. This usually happens when the concentration of surfactant increases (see Fig. 3.30). On the other hand, with the IFT reduction, the relative permeability curves tend to become straight lines (Fig. 3.31).

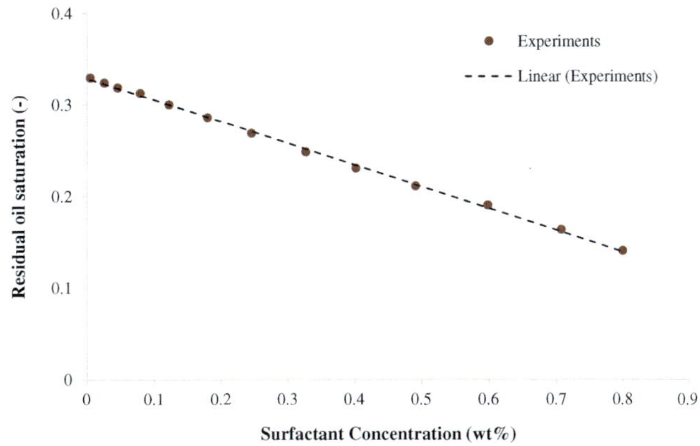

Fig. 3.30 Effect of a typical surfactant concentration on residual oil saturation. *(Data from Ahmadi MA, Shadizadeh SR. Spotlight on the new natural surfactant flooding in carbonate rock samples in low salinity condition. Sci Rep 2018;8(1):1–15.)*

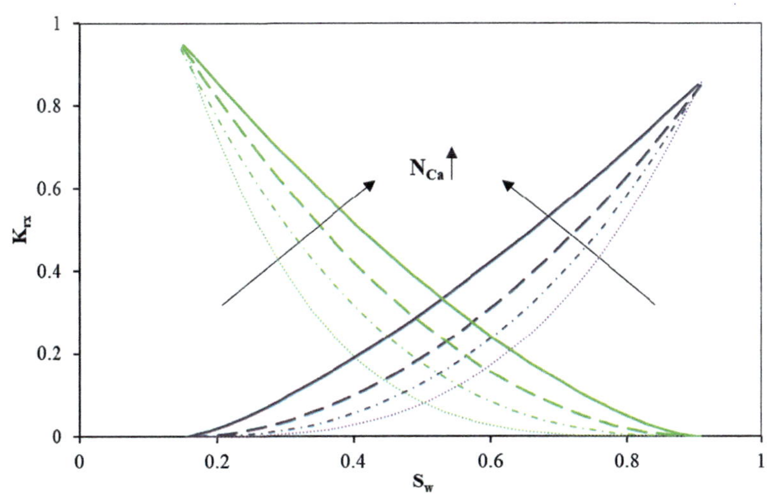

Fig. 3.31 Effect of the capillary number on relative permeability curves. *(Redrawn after Guo H, et al. Proper use of capillary number in chemical flooding. J Chem 2017;2017:1–11.)*

In the simulation, main surfactant features (mechanisms) determining the method of interpolation between relative permeabilities are: wettability alteration and IFT reduction.

With regard to the wettability alteration, two relative permeability curves are considered: oil-wet and water-wet. Indeed, the interpolation of these two sets of relative permeability is performed via an interpolation parameters, being a function of surfactant concentration (or adsorption).

In addition, with respect to the IFT reduction, initially, the surfactant is considered as a component in the water phase in which by changing of surfactant concentration as well as salinity in each grid block, IFT is changed and estimated. Then, capillary number is calculated. Afterwards, using capillary desaturation curves, residual saturation or interpolation parameters are attained. Next, via these parameters, interpolation between relative permeabilities can be performed. It should be noted that, in this case, relative permeability sets are related to immiscible and miscible conditions.

3.8.7 Capillary desaturation curve (CDC)

To reduce residual oil saturation, one certain method is increasing capillary number, which can be accomplished through reducing IFT and increasing velocity or viscosity. The capillary number can be obtained through the following equation demonstrating the ratio of viscous to capillary forces:

$$N_c = \frac{u\mu}{\gamma} \tag{3.13}$$

where N_c is the capillary number, u is the velocity of displacing fluid, μ is the viscosity of displacing fluid, and γ is the interfacial tension between the displacing and displaced fluids.

To define the effect of capillary number on residual saturation, a capillary desaturation curve is used [64]. In simple words, a capillary desaturation curve is a plot of residual saturation of the wetting or nonwetting phase versus the capillary number. Of course, in some cases, this form of capillary number has been replaced with others [65–67]. However, this form of the equation is more popular than the others due to its simplicity [42].

CDC is an important curve in designing surfactant formulation since myriad surfactants can decrease IFT followed by the increase of the capillary number and the decrease of residual saturation.

There are some contributing factors depicting the shape of CDC [68]: wettability (see Fig. 3.32) and pore size distribution (see Fig. 3.33). As can be seen in Fig. 3.32, wetting fluid has the lowest residual saturation in addition to having the highest critical capillary number. In Fig. 3.33, typical carbonates with wider pore size distribution show the lowest critical capillary number and highest residual saturation at reservoir condition [68]. Thus, it is anticipated that each rock type in a specified reservoir has different decline slopes and critical points of CDC.

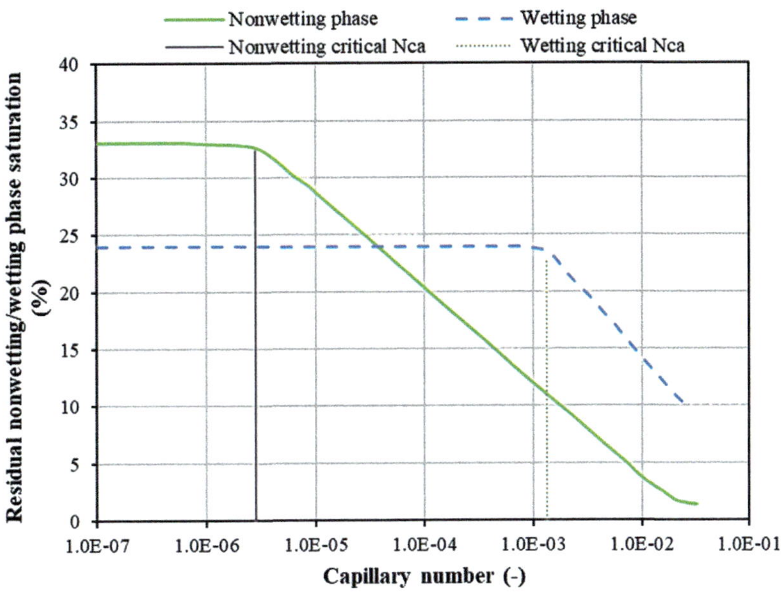

Fig. 3.32 Schematic representation of the effect of wettability on residual saturation of wetting and nonwetting phase. *(Redrawn after Lake LW, et al. Fundamentals of enhanced oil recovery, 2014.)*

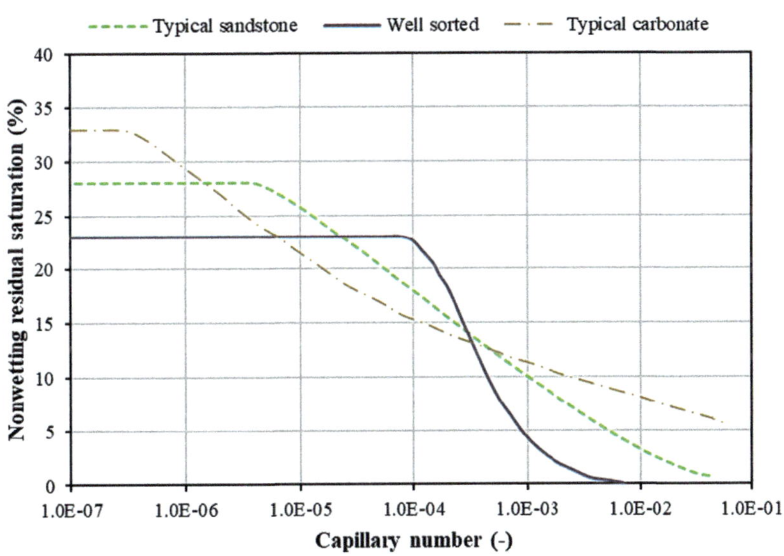

Fig. 3.33 Schematic representation of the effect of pore size distribution on capillary desaturation curve. *(Redrawn after Lake LW, et al. Fundamentals of enhanced oil recovery, 2014.)*

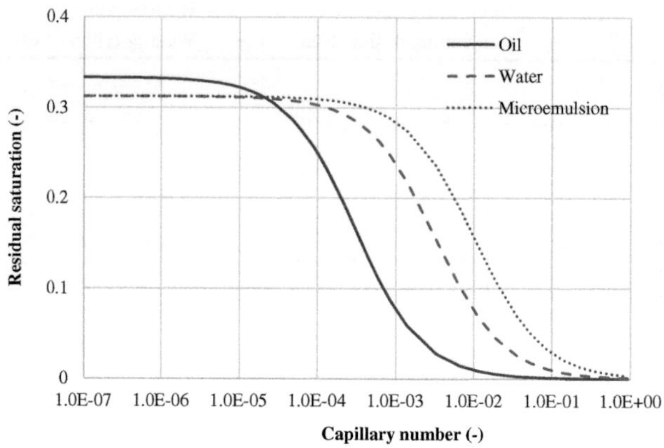

Fig. 3.34 Schematic representation of CDC in the presence of three phases of oil, water, and microemulsion. *(Redrawn after Delshad M. Chemical enhanced oil recovery, 2014. [PowerPoint presentation]. Retrieved from NISOC.)*

Aside from the effects of different parameters on each CDC, it is important to attain CDCs for simulation purposes. This is because, in numerical simulation software, the interpolation between oil-wet and miscible relative permeabilities is performed via CDC. In addition, for each rock type, i.e., CDC; different relative permeabilities (at least three sets) need to be provided. However, as attaining relative permeability via experiments are time- and money-consuming , usually in two states, these curves are plotted.

As microemulsion is the most wetting phase, its critical capillary number is the highest, and the highest residual saturation is expected (see Fig. 3.34). In this sense, water and oil follow this trend with respect to the wetness of the rock, i.e., oil-wet and water-wet. In Fig. 3.34, as the reservoir is water-wet, CDC of water phase is observed between the microemulsion and oil CDCs.

After all, the CDCs generated at lab conditions can be applied in numerical simulators. Indeed, interpolation of relative permeabilities with the immiscible relative permeability is based on the CDC. In this regard, for simulation purposes of surfactant flooding at different scales, it is a prerequisite.

3.9 Upscaling

In addition to the importance of lab and simulation results at the core scale, the applicability of this data for pilot, reservoir, and field scales is the other major issue. It is essential to have identical behavior in both scales. However, these scales differ significantly from the core scale in terms of the type of dominant forces, the magnitude of forces, etc. In turn, these contributing factors should be deemed in experiment design to be representative of large-scale conditions.

Different forms of upscaling equations for the surfactant process can be found in the literature. In Table 3.5, some of them have been tabulated. Initially, Mattax and Kyte [69]

Table 3.5 Scaling groups used for surfactant imbibition.

No.	Scaling group	Reference
1	$t_D = \frac{\sigma\sqrt{k/\varnothing}}{\mu L_c^2} t$	[69]
2	$t_{DCij} = \left[\frac{\sigma_i \cos(\theta_i) F_{si}}{\mu_{wi}} \sqrt{\frac{K_i}{\varnothing_i}}\right] \times t_{ij}$	[70]
3	$t_D = c^2 \left[\frac{M_e^* P_c^* (S_{wf} - S_{wi})}{\varnothing L_a^2}\right] t$	[71]
4	$t_D = c^2 \frac{kk_{re}^*}{\varnothing \mu_e} \frac{P_c^* (S_{wf} - S_{wi})}{L_a^2} t$	[72]

shed the light on this concept in petroleum engineering. In this sense, a correlation was put forward. This correlation was derived for the imbibition of water in a single reservoir matrix block. Following researches have pursued this target via developing this equation or proposing new ones in different circumstances.

Scaling groups are useful in the projection of reservoir behavior in model scale via prototype scale results. In simple words, plotting different target parameters such as recovery factor versus these scaling groups gives a uniform shape. In turn, where variables change (i.e., those changing from core to field scale), the same curve is expected to attain.

To further clarify this issue, the methodology proposed by Li and Horne [72] for surfactant imbibition into a matrix block is elaborated. In the dimensionless time equation, t_D is dimensionless time, c is the ratio of the gravity force to the capillary force, $\frac{kk_{re}^*}{\mu_e}$ is the effective mobility of the two phases, $\varnothing$ is porosity, P_c^* is the capillary pressure at S_{wf}, L_a is the characteristic length, S_{wf} is the water (or wetting phase) saturation behind the imbibition front, S_{wi} is the initial water saturation in the media, and t is time. As can be seen, different parameters affecting imbibition process has been considered in this dimensionless number. Results show that plotting the normalized recovery versus this dimensionless time ends up in the identical recovery curves [72].

Aside from the imbibition process, some scaling groups have been proposed for surfactant flooding [73, 74]. In flooding, the process is more sophisticated than imbibition. Consequently, instead of one scaling group, some scaling groups have to be identical in the model and prototype. Totally, two categories of scaling groups can be obtained via inspectional and dimensional analysis [75]. While inspectional analysis is based on governing equations, dimensional analysis is founded on dimensions of pertinent parameters. In Fig. 3.35, recovery curves of two cases with identical scaling groups at different produced pore volumes are compared. The similar results in these two cases show the desirability of the selected scaling groups.

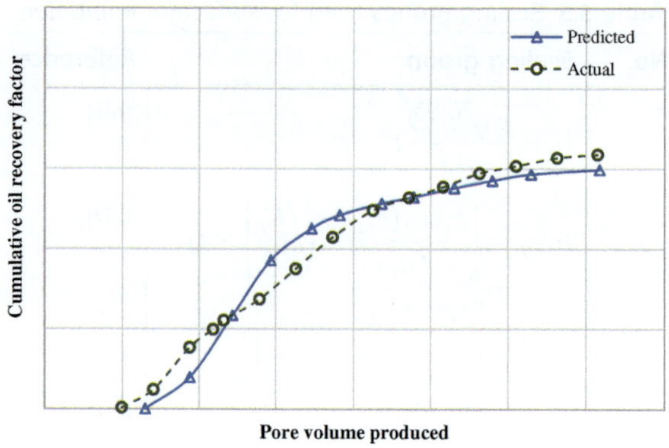

Fig. 3.35 Comparison of predicted and actual recoveries. *(Redrawn after Thomas S, Ali S, Thomas N. Scale-up methods for micellar flooding and their verification. In: Annual technical meeting, 2000. Petroleum Society of Canada.)*

3.10 Screening criteria

Like other EOR approaches, some screening criteria for surfactant flooding can be found in the literature. These criteria act as yardsticks for the initial evaluation of oil reservoirs characteristics for surfactant EOR without any preceding experience of surfactant EOR in the target reservoir. In many cases, this initial evaluation can give insights to the companies in issues such as possible obstacles and the chance of success. For instance, by comparing the reservoir temperature with previous successful surfactant EORs or screening tables, it seems to be so high that with the current knowledge, the surfactant cannot stand at reservoir condition. In Table 3.6, a handful of criteria reported in different studies have shown [76–87]. With the advent of new methodologies in the synthesis of surfactants and technology improvement in surfactant flooding, these criteria have been developed during these years. In the following, some of these criteria are investigated:

- **Oil composition, viscosity, and API gravity**

  These parameters are important in the phase behavior of the applied surfactant since they affect the optimum salinity range in addition to partitioning.
- **Formation type**

  Most of the surfactant flooding projects were applied in sandstone reservoirs, whereas in carbonate rocks the number is lower. This is because anionic surfactants adsorption in carbonates reservoirs is higher.
- **Formation water salinity**

  Formation water salinity, in particular, the concentration of divalent ions is essential for screening surfactants. However, it is worthy to note that many reservoirs witness waterflooding prior to surfactant flooding. Thus, the reservoir salinity would be around the injection water salinity resulting in mitigation of the problem of high formation water salinity. Injection water salinity needs to be near the optimum salinity to

Table 3.6 Summary of screening criteria for parametric screening method.

No.	Ref.	API	Oil viscosity (cP)	Res. thickness (ft)	Perm. (mD)	Depth (ft)	Temp (°F)	Lith.	Form. TDS (ppm)	Divalent (ppm)	Clay	Oil Sat. (fraction)	Aquifer	Gas cap
1	[76, 77]	>20	<35	NC	>10	<9000	<200	Sandstone preferred	–	–	–	–	–	–
2	[78]	22–39	2.63–15.6	NC	50–60	625–5300	122–155	Sandstone	–	–	–	–	–	–
3	[79]	≥25	≤30	–	≥20	–	≤250	–	≤200,000	–	–	0.20	–	–
4	[80]	>25	<20	–	>20	NC	<200	Sandstone	50,000	1000	Low	>0.25	None to Weak	None to Weak
5	[81]	–	–	–	>50	<6561	<176	Sandstone	Low	Low	–	>0.30	Weak	Weak
6	[82]	–	<40	–	>40	–	<200	Sandstone and carbonate	<100,000	–	–	–	–	–
7	[83]	>25	100	–	>20	<9000	<200	Sandstone preferred	100,000	–	–	0.3	–	–
8	[84]	–	<150	–	>50	–	<158	Sandstone	50,000	1000	Low	0.35	None	None
9	[85]	>20	<35	–	>100	500–9000	<200	–	<200,000 if Tr < 140 <50,000 if Tr > 140	–	–	>0.45	–	–
10	[86]	NC	<35	–	>10	NC	<200	Sandstone	<50,000	<100	Low	>0.3	Weak	Weak
11	[87]	>12	<200	>10	>10	>500	<250	–	<250,000			>0.2	–	–

demonstrate effective production. Optimum salinity is a function of the type of crude oil, surfactant structure, etc.

- **Reservoir temperature and depth**

  As at high reservoir temperature, the feasibility of surfactant degradation as well as incompatibility are increased, thus surfactant EOR is proposed to be performed at lower temperatures. However, as explained in Section 3.4.2 (**Thermal stability**), with progress in the field of surfactant design, at present, there are a handful of surfactants that are resistant to higher temperatures.
- **Formation permeability**

  This criterion is based on the previous projects. In fact, surfactants' limitation with regard to permeability has decreased. In fact, this limitation can be traced back to the lack of data or primitive structure of surfactants rather than a practical issue. Consequently, at present, with the advent of new technologies regarding surfactant synthesis and wider field application of surfactants, surfactant EOR can be a promising proposition for lower permeabilities.
- **Oil saturation**

  As reviewed in previous sections, surfactant flooding can reduce oil saturation. Hence, more oil saturation in the reservoir is favored for surfactant flooding. In other words, the surfactant is exposed to more amount of oil leading to more oil production.
- **Aquifer and gas cap**

  Although surfactant flooding is primarily performed to produce oil through active mechanisms, another goal of this method is associated with pressure maintenance [86]. As a result, aquifer and gas caps because of pressure maintenance can diminish their capability.

As explained previously, under high temperature and/or high salinity conditions, this method is usually not recommended. This is because most of the surfactants do not have enough stability in harsh conditions leading to degradation. However, in the recent years, some studies were performed to pave the way for using surfactants in these conditions.

From the authors' viewpoint, these criteria cannot be taken into account at special conditions, e.g., for carbonate reservoirs since they are usually associated with sandstones.

On the other hand, there are many software such as SelectEOR, EORgui, SWORD, and EORt (EOR S&P) using the diverse database to rank different EOR methodologies. The major problem regarding available screening software is the limitations regarding new confirmed technologies. In fact, the database of each software is restricted to those field scales results. As a result, the new shifting in benchmarks is not considered.

Example 3.1. Surfactant parametric EOR screening.

A team of reservoir engineers was charged to perform surfactant EOR screening for a reservoir. The properties of this reservoir have been shown in Table 3.7. Please as a reservoir engineer perform surfactant EOR screening using screening criteria proposed by Sheng [86] and Izadi et al. [87].

Table 3.7 Reservoir properties of pilot under study.

No.	Parameter (unit)	Field A
1	Rock type (–)	Sandstone
2	Depth (ft)	1800
3	Temperature (°C)	28
4	Reservoir pressure (psi)	1000
5	Porosity range (%)	10–20
6	Permeability range (mD)	10–500
7	Oil API gravity (–)	37
8	Oil viscosity (cP)	9
9	Formation salinity (ppm)	55,000–60,000

Data from Izadi M, et al. Assessing productivity impairment of surfactant-polymer EOR using laboratory and field data. In: SPE improved oil recovery conference, 2018. Society of Petroleum Engineers.

Table 3.8 Comparison of the reservoir properties of pilot under study with EOR screening criteria.

			EOR screening approach	
No.	Parameter (Unit)	Field A	[86]	[87]
1	Rock type (–)	Sandstone	✓	✓
2	Depth (ft)	1800	✓	✓
3	Temperature (°C)	28	✓	✓
4	Reservoir pressure (psi)	1000	✓	✓
5	Porosity range (%)	10–20	✓	✓
6	Permeability range (mD)	10–500	✓	✓
7	Oil API gravity (–)	37	✓	✓
8	Oil viscosity (cP)	9	✓	✓
9	Formation salinity (ppm)	55,000–60,000	✗	✓

Solution:

As can be seen in Table 3.8, the results of comparing Field A properties with two EOR screening approaches have been shown. From the results, it can be inferred that this pilot is appropriate for surfactant flooding; however, formation salinity has failed to meet [86] EOR screening limitations. Of course, it is worth mentioning that the previous advent of technology in synthesizing surfactants pushed the salinity criterion. In turn, [87] EOR screening criterion for salinity seems to be more reliable.

3.11 Field cases

Utilizing surfactant as an EOR method is not restricted to the lab scale. Indeed, this method has been broadly applied in many cases at large scales whether for pilot or field-scale purposes. Some further cases can be found in Ref. [1]. Only a few cases exist that merely surfactant is used in field scales. In other words, surfactants tend to be used with other chemical agents such as polymer and alkaline. In this sense, in this section, the emphasis is on some cases with these properties.

3.11.1 Cambridge Minnelusa field

McAdams, Roux, and associates discovered Cambridge Field in 1989. A handful of data related to this field is shown in Table 3.9. Apart from this information, the oil production mechanism in this field is rock and fluid expansion with a zero initial GOR [88]. A combination of surfactant, polymer and alkaline was used to flood into this field.

To reduce the risk of field application, a wide range of laboratory tests were performed [88]. In this regard, diverse formulations were examined. Laboratory results depicted that a surfactant concentration of 0.1 wt% is desirable to achieve optimal phase behavior. Finally, results showed that using the selected formulation 30% more oil can be produced compared to the numerical simulation waterflood predictions.

3.11.2 Gudong field

Gudong oil field is located in the north-east of Shandong province in China [89]. This field was discovered in 1984, developed in 1985 [89]. The properties of the southern Ng5–6 layer of the West 7 Block of Gudong Oilfield has shown in Table 3.10. Shengli Oilfield Company performed a surfactant-polymer EOR operation at pilot scale of this field [90].

Table 3.9 Summary of Cambridge Minnelusa field data.

No.	Parameter (Unit)	Value
1	Pore volume (Mbbl)	7117
2	OOIP (MSTB)	4875
3	Peak primary oil production rate (BOPD)	489
4	Reservoir temperature (°F)	132
5	Average reservoir thickness (ft)	28.7
6	Oil formation volume factor (rbbl/STB)	1.03
7	Average porosity (%)	18
8	Average permeability (mD)	845
9	Initial water saturation (%)	31.6
10	Initial reservoir pressure (psia)	1792
11	Oil API (–)	20
12	Oil viscosity (cP)	31
13	Initial GOR (scf/STB)	0

Data from Vargo J, et al. Alkaline-surfactant-polymer flooding of the Cambridge Minnelusa field. In: SPE Rocky Mountain regional meeting, 2000. Society of Petroleum Engineers.

Table 3.10 Summary of Gudong field data.

No.	Parameter (unit)	Value
1	Reservoir area (km^2)	0.94
2	Initial oil in place (ton)	277×10^4
3	No. of injection wells (−)	9
4	No. of producing wells (−)	16
5	Average permeability (mD)	1320
6	Formation water salinity (mg/L)	8207
7	Ca^{2+}+Mg^{2+} concentration (mg/L)	321
8	Oil viscosity (cP)	45
9	Reservoir temperature (°C)	68
10	Initial reservoir pressure (Mpa)	12.7
11	Crude oil acid number (mgKOH/g)	3.11

Data from Zhijian Q, et al. A successful ASP flooding pilot in Gudong oil field. In: SPE/DOE improved oil recovery symposium, Society of Petroleum Engineers, 1998; Li Z, et al. A successful pilot of dilute surfactant-polymer flooding in Shengli Oilfield. In: SPE improved oil recovery symposium, OnePetro, 2012.

Initial experimental evaluations showed that to get the IFT of around 10^{-2} mN/m at the reservoir condition 0.3 wt% of the selected surfactant (Mahogany sulfonate) is required [90].

Shengli Oilfield Company performed the dilute surfactant-polymer flooding in Gudong Field in 2003 [90]. The effect of performing this EOR method in the pilot on oil rate and water cut is in a way that, while oil production rate increased sharply, water cut decreased remarkably.

3.11.3 Semoga field

This field was discovered in 1996, located in the Rimau block in South Sumatra [91]. Generally, this field is comprised of three different formations: Telisa Formation (tight sandstone), Baturaja Formation (carbonate), and Talang Akar Formation (sandstone). Surfactant flooding was carried out in the Baturaja Formation. Surfactant stimulation was performed in 2009 in two wells to reduce water cut in these wells. To do this, nonionic surfactants that showed an acceptable decrease in contact angle were chosen. Through a huff-n-puff operation, surfactant solution was an injection in two wells, following 7 days of soaking time to allow the surfactant to attain its high performance. Some of the core characteristics of this field is shown in Table 3.11. In some wells, water cut witnessed a downward trend depicting the success of surfactant EOR in this field.

Table 3.11 Summary of Gudong field data (Baturaja formation).

No.	Parameter (unit)	Value
1	Proven volume (acre-ft)	317,856
2	Net pay thickness (ft)	77
3	No. of injection wells (−)	28
4	No. of producing wells (−)	82
5	No. of shut-in wells (−)	17
6	Oil production rate peak (BOPD)	36,200

Data from Rilian NA, Sumestry M, Wahyuningsih W. Surfactant stimulation to increase reserves in carbonate reservoir "a case study in semoga field". In: SPE EUROPEC/EAGE annual conference and exhibition. 2010. Society of Petroleum Engineers.

References

[1] Sheng JJ. Enhanced oil recovery field case studies. Gulf Professional Publishing; 2013.
[2] Al Shalabi EW, Sepehrnoori K. Low salinity and engineered water injection for sandstone and carbonate reservoirs. Gulf Professional Publishing; 2017.
[3] Green DW, Willhite GP. Enhanced oil recovery. 2nd ed. vol. 6. Henry L. Doherty Memorial Fund of AIME, Society of Petroleum Engineers; 2018.
[4] Uren LC, Fahmy E. Factors influencing the recovery of petroleum from unconsolidated sands by waterflooding. Trans AIME 1927;77(01):318–35.
[5] Salager J-L. Surfactants types and uses. In: FIRP booklet; 2002. p. 300.
[6] Shosa JD, Schramm LL. Surfactants: fundamentals and applications in the petroleum industry. Palaios 2001;16:614.
[7] Becher P. Anionic surfactants: physical chemistry of surfactant action (surfactant science series, volume 11). E. H. Lucassen-Reynders, editor. Marcel Dekker, Inc., New York and Basel, 1981. xi+412 pp. $55.00. J Dispers Sci Technol 1982;3:211–2.
[8] Green DW, Willhite GP. Enhanced oil recovery. vol. 6; 1998.
[9] Blake TD, Everett DH, Haynes JM. Wetting. In: Surfactants. London: Academic Press; 1984. p. 221–75.
[10] Holmberg K, Jönsson B, Kronberg B, Lindman B. Surfactants and polymers in aqueous solution. West Sussex, England: John Wiley & Sons; 2003.
[11] Tadros TF. Applied surfactants: principles and applications. John Wiley & Sons; 2006.
[12] Holmberg K. Natural surfactants. Curr Opin Colloid Interface Sci 2001;6:148–59.
[13] Tadros TF. An introduction to surfactants. De Gruyter; 2014.
[14] Zulkifli NN, et al. Evaluation of new surfactants for enhanced oil recovery applications in high-temperature reservoirs. J Pet Explor Prod Technol 2020;10(2):283–96.
[15] Barnes JR, et al. Application of internal olefin sulfonates and other surfactants to EOR. Part 1: structure-performance relationships for selection at different reservoir conditions. In: SPE improved oil recovery symposium. Society of Petroleum Engineers; 2010.
[16] Stegemeier G. Mechanisms of entrapment and mobilization of oil in porous media. In: Improved oil recovery by surfactant and polymer flooding. Elsevier; 1997. p. 55–91.
[17] Shah DO. Improved oil recovery by surfactant and polymer flooding. Elsevier; 2012.
[18] Healy RN, Reed RL, Stenmark D. Multiphase microemulsion systems. Soc Pet Eng J 1976;16(03):147–60.
[19] Dang C, et al. Modeling and optimization of alkaline-surfactant-polymer flooding and hybrid enhanced oil recovery processes. J Petrol Sci Eng 2018;169:578–601.
[20] Solairaj S, et al. New correlation to predict the optimum surfactant structure for EOR. In: SPE improved oil recovery symposium. Society of Petroleum Engineers; 2012.

[21] Puerto MC, Reed RL. A three-parameter representation of surfactant/oil/brine interaction. Soc Pet Eng J 1983;23:669–82.
[22] Pinnawala Arachchilage GW, et al. Evaluating the effect of temperature on surfactant phase behavior and aqueous stability to forecast optimum salinity at high temperature. In: SPE improved oil recovery conference. Society of Petroleum Engineers; 2018.
[23] Skauge A, Palmgren O. Phase behavior and solution properties of ethoxylated anionic surfactants. In: SPE international symposium on oilfield chemistry. Society of Petroleum Engineers; 1989.
[24] Levitt D, et al. Identification and evaluation of high-performance EOR surfactants. In: SPE/DOE symposium on improved oil recovery. Society of Petroleum Engineers; 2006.
[25] Yang HT, et al. Low-cost, high-performance chemicals for enhanced oil recovery. In: SPE improved oil recovery symposium. Society of Petroleum Engineers; 2010.
[26] Cayias J, Schechter R, Wade W. Modeling crude oils for low interfacial tension. Soc Pet Eng J 1976;16(06):351–7.
[27] Puerto M, Gale W. Estimation of optimal salinity and solubilization parameters for alkylorthoxylene sulfonate mixtures. Soc Pet Eng J 1977;17(03):193–200.
[28] Jones S, Dreher K. Cosurfactants in micellar systems used for tertiary oil recovery. Soc Pet Eng J 1976;16(03):161–7.
[29] Zhou Y, et al. A comprehensive review of emulsion and its field application for enhanced oil recovery. Energy Sci Eng 2019;7(4):1046–58.
[30] Hall AC. Interfacial tension and phase behavior in systems of petroleum sulfonate/brine/n-alkane. Colloids Surf 1980;1:209–28.
[31] Puig JE, et al. On interfacial tensions measured with alkyl aryl sulfonate surfactants. Soc Pet Eng J 1979;19:71–82.
[32] Roehl PO, Choquette PW. Carbonate petroleum reservoirs. Springer Science & Business Media; 2012.
[33] Standnes DC, Austad T. Wettability alteration in chalk: 2. Mechanism for wettability alteration from oil-wet to water-wet using surfactants. J Petrol Sci Eng 2000;28:123–43.
[34] Thomas MM, Clouse JA, Longo JM. Adsorption of organic compounds on carbonate minerals: 1. Model compounds and their influence on mineral wettability. Chem Geol 1993;109:201–13.
[35] Kumar K, Dao EK, Mohanty KK. Atomic force microscopy study of wettability alteration by surfactants. SPE J 2008;13:137–45.
[36] Bourrel M, Schechter RS. Microemulsions and related systems: formulation, solvency, and physical properties. Editions Technip; 2010.
[37] Shah DO. Surface phenomena in enhanced oil recovery. USA: Springer; 1981.
[38] Walker D, et al. The impact of microemulsion viscosity on oil recovery. In: SPE improved oil recovery symposium. Society of Petroleum Engineers; 2012.
[39] Suniga PT, Fortenberry R, Delshad M. Observations of microemulsion viscosity for surfactant EOR processes. In: SPE improved oil recovery conference. Society of Petroleum Engineers; 2016.
[40] Zhang G, Yu J. Effect of commonly used EOR polymers on low concentration surfactant phase behaviors. Fuel 2021;286:119465.
[41] Sahni V, et al. The role of co-solvents and co-surfactants in making chemical floods robust. In: SPE improved oil recovery symposium. Society of Petroleum Engineers; 2010.
[42] Sheng JJ. Modern chemical enhanced oil recovery: theory and practice. Gulf Professional Publishing; 2010.
[43] Glover C, et al. Surfactant phase behavior and retention in porous media. Soc Pet Eng J 1979;19(03):183–93.
[44] Kamari A, et al. Reliable method for the determination of surfactant retention in porous media during chemical flooding oil recovery. Fuel 2015;158:122–8.
[45] Yassin MR, et al. Prediction of surfactant retention in porous media: a robust modeling approach. J Dispers Sci Technol 2014;35(10):1407–18.
[46] Austad T. A review of retention mechanisms of ethoxylated sulfonates in reservoir cores. In: SPE international symposium on oilfield chemistry. Society of Petroleum Engineers; 1993.
[47] Paria S, Khilar KC. A review on experimental studies of surfactant adsorption at the hydrophilic solid–water interface. Adv Colloid Interface Sci 2004;110(3):75–95.

[48] Nimibofa A, Augustus N, Donbebe W. Modelling and interpretation of adsorption isotherms. Chemistry 2017;3039817:2020.
[49] Langmuir I. The adsorption of gases on plane surfaces of glass, mica and platinum. J Am Chem Soc 1918;40(9):1361–403.
[50] Barati A, et al. Adsorption of a new nonionic surfactant on carbonate minerals in enhanced oil recovery: experimental and modeling study. Chem Eng Res Des 2016;105:55–63.
[51] Gupta SP, Trushenski SP. Micellar flooding-compositional effects on oil displacement. Soc Pet Eng J 1979;19:116–28.
[52] Graciaa A, et al. The partitioning of complex surfactant mixtures between oil/water/microemulsion phases at high surfactant concentrations. J Colloid Interface Sci 1983;93(2):474–86.
[53] Miller CA, Neogi P. Interfacial phenomena: equilibrium and dynamic effects. vol. 139. CRC Press; 2007.
[54] Belhaj AF, et al. Experimental investigation of surfactant partitioning in pre-CMC and post-CMC regimes for enhanced oil recovery application. Energies 2019;12(12):2319.
[55] Kedar V, Bhagwat SS. Effect of salinity on the IFT between aqueous surfactant solution and crude oil. Pet Sci Technol 2018;36(12):835–42.
[56] Hodges G, et al. A comparison of log K ow (n-octanol–water partition coefficient) values for non-ionic, anionic, cationic and amphoteric surfactants determined using predictions and experimental methods. Environ Sci Eur 2019;31(1):1–18.
[57] Belhaj AF, et al. Partitioning behaviour of novel surfactant mixture for high reservoir temperature and high salinity conditions. Energy 2020;198:117319.
[58] Stellner KL, Scamehorn JF. Surfactant precipitation in aqueous solutions containing mixtures of anionic and nonionic surfactants. J Am Oil Chem Soc 1986;63:566–74.
[59] Al-Hashim HS, et al. Adsorption and precipitation behaviour of petroleum sulfonates from Saudi Arabian limestone. J Petrol Sci Eng 1988;1:335–44.
[60] Jian G, et al. Characterizing adsorption of associating surfactants on carbonates surfaces. J Colloid Interface Sci 2018;513:684–92.
[61] Skjæveland SM, Kleppe J. SPOR monograph, Recent advances in improved oil recovery methods for north sea sandstone reservoirs. Norway: Norwegian Petroleum Directorate; 1992.
[62] Goudarzi A, Delshad M, Sepehrnoori K. A chemical EOR benchmark study of different reservoir simulators. Comput Geosci 2016;94:96–109.
[63] Goudarzi A, Delshad M, Sepehrnoori K. A critical assessment of several reservoir simulators for modeling chemical enhanced oil recovery processes. In: SPE reservoir simulation symposium. Society of Petroleum Engineers; 2013.
[64] Dombrowski H, Brownell L. Residual equilibrium saturation of porous media. Ind Eng Chem 1954;46 (6):1207–19.
[65] Boom W, et al. On the use of model experiments for assessing improved gas-condensate mobility under near-wellbore flow conditions. In: SPE annual technical conference and exhibition. Society of Petroleum Engineers; 1996.
[66] Boom W, et al. Experimental evidence for improved condensate mobility at near-wellbore flow conditions. In: SPE annual technical conference and exhibition. Society of Petroleum Engineers; 1995.
[67] Garnes J, et al. Capillary number relations for some north, sea reservoir sandstones. In: SPE/DOE enhanced oil recovery symposium. Society of Petroleum Engineers; 1990.
[68] Lake LW, et al. Fundamentals of enhanced oil recovery. Society of Petroleum Engineers; 2014.
[69] Mattax CC, Kyte J. Imbibition oil recovery from fractured, water-drive reservoir. Soc Pet Eng J 1962;2 (02):177–84.
[70] Gupta A, Civan F. An improved model for laboratory measurement of matrix to fracture transfer function parameters in immiscible displacement. In: SPE annual technical conference and exhibition. Society of Petroleum Engineers; 1994.
[71] Li K, Horne RN. A general scaling method for spontaneous imbibition. In: SPE annual technical conference and exhibition. Society of Petroleum Engineers; 2002.
[72] Li K, Horne RN. Generalized scaling approach for spontaneous imbibition: an analytical model. SPE Reserv Eval Eng 2006;9(03):251–8.

[73] Thomas S, Ali S, Thomas N. Scale-up methods for micellar flooding and their verification. In: Annual technical meeting. Petroleum Society of Canada; 2000.

[74] Srivastava R, Huang S, Dyer S. Scaling criteria for micellar flooding experiments. In: Annual technical meeting. Petroleum Society of Canada; 1994.

[75] Khandoozi S, et al. Inspectional and dimensional analyses for scaling of low salinity waterflooding (LSWF): from core to field scale. J Petrol Sci Eng 2020;189:106956.

[76] Taber JJ, Martin F, Seright R. EOR screening criteria revisited—part 1: introduction to screening criteria and enhanced recovery field projects. SPE Reserv Eng 1997;12(03):189–98.

[77] Taber J, Martin F, Seright R. EOR screening criteria revisited—part 2: applications and impact of oil prices. SPE Reserv Eng 1997;12(03):199–206.

[78] Al Adasani A, Bai B. Analysis of EOR projects and updated screening criteria. J Petrol Sci Eng 2011;79 (1–2):10–24.

[79] Crespi I. Enhanced oil recovery: report on a focus group discussion. Princeton, NJ (USA): Mathematica Policy Research; 1978.

[80] Brashear J, Kuuskraa V. The potential and economics of enhanced oil recovery. J Petrol Tech 1978;30 (09):1,231–9.

[81] Carcoana AN. Enhanced oil recovery in Rumania. In: SPE enhanced oil recovery symposium. Society of Petroleum Engineers; 1982.

[82] Mortis G. Enhanced oil recovery. Oil Gas J 2002;100:43–7.

[83] Goodlett G, et al. The role of screening and laboratory flow studies in EOR process evaluation. In: SPE Rocky Mountain regional meeting. Society of Petroleum Engineers; 1986.

[84] Al-Bahar MA, et al. Evaluation of IOR potential within Kuwait. In: Abu Dhabi international conference and exhibition. Society of Petroleum Engineers; 2004.

[85] Dickson JL, Leahy-Dios A, Wylie PL. Development of improved hydrocarbon recovery screening methodologies. In: SPE improved oil recovery symposium. Society of Petroleum Engineers; 2010.

[86] Sheng J. Status of surfactant EOR technology. Petroleum 2015;97–105.

[87] Delshad M. Chemical enhanced oil recovery. NISOC; 2014. [PowerPoint presentation]. Retrieved from.

[88] Vargo J, et al. Alkaline-surfactant-polymer flooding of the Cambridge Minnelusa field. In: SPE Rocky Mountain regional meeting. Society of Petroleum Engineers; 2000.

[89] Zhijian Q, et al. A successful ASP flooding pilot in Gudong oil field. In: SPE/DOE improved oil recovery symposium. Society of Petroleum Engineers; 1998.

[90] Li Z, et al. A successful pilot of dilute surfactant-polymer flooding in Shengli Oilfield. In: SPE improved oil recovery symposium. OnePetro; 2012.

[91] Rilian NA, Sumestry M, Wahyuningsih W. Surfactant stimulation to increase reserves in carbonate reservoir "a case study in semoga field". In: SPE EUROPEC/EAGE annual conference and exhibition. Society of Petroleum Engineers; 2010.

CHAPTER 4

Alkaline flooding

Ali Rahimi[a], Morteza Jami[b], Hassan Divandari[c], and Mehdi Safari[d]
[a]Department of Petroleum Engineering, Marvdasht Branch, Islamic Azad University, Marvdasht, Iran
[b]School of Petroleum and Chemical Engineering, Universiti Teknologi Brunei, Bandar Seri Begawan, Brunei Darussalam
[c]Department of Petroleum Engineering, Shahid Bahonar University of Kerman, Kerman, Iran
[d]Department of Petroleum Engineering, Curtin University, Miri, Malaysia

4.1 Introduction

The alkaline flooding process is a chemical flooding method (CEOR) that is used to increase oil recovery. The process includes the injection of alkaline chemicals into the porous medium to react with the crude oil organic acid, produce an in situ surfactant, and assist oil recovery. The surfactant produced can reduce interfacial tension (IFT) and increase the recovery of trapped oil in the reservoir.

This chapter will discuss the fundamental and theoretical aspects of alkaline flooding and describe the microscopic and macroscopic functions of alkaline agents in the flooding method. For this purpose, first, the common chemicals used in alkaline flooding are presented and their properties and reactions with oil will be discussed. Then the alkaline phase behavior during flooding and the three-phase diagram will be presented. In this context, the reaction between alkaline agents, water, rock, and oil will be described. In order to understand the role of alkaline better, the active mechanism in the porous media and alkaline function in a microscopic and macroscopic scale will be explained. The required conditions for using alkaline in an EOR process will also be explicated. The simulation and modeling process of alkaline flooding is an important process which will be discussed in this chapter. Finally, the results of a few alkaline flooding projects, economic and environmental feasibility, and field operation problems are presented.

4.2 Commonly used alkaline agents

There are three common alkali agents used in alkaline water flooding, sodium hydroxide (NaOH), sodium carbonate (Na_2CO_3), and sodium silicate ($NaSiO_4$), although other alkaline agents such as $NaHCO_3$, Na_4SiO_4, NH_3, $NaBO_2$, NH_4OH, and KOH are also used as in both experimental and field applications. All of these chemical agents are able to increase the water pH [1–3]. Changing the pH is a vital factor in order to form an in situ surfactant within the reservoir [4]. After an alkali agent such as sodium hydroxide is added to water, it breaks into sodium and hydroxide ions, in which hydroxide is known as a basic factor for increasing pH. Similarly, sodium silicate is another alkaline chemical agent

Chemical Methods
https://doi.org/10.1016/B978-0-12-821931-7.00010-9

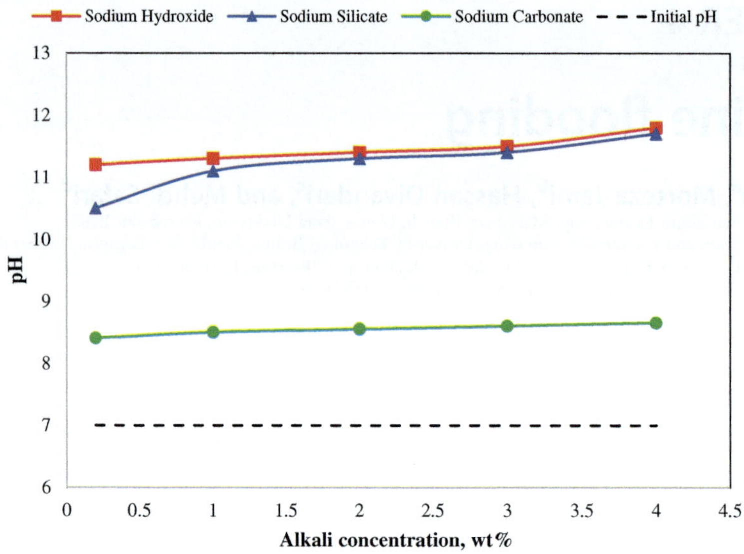

Fig. 4.1 Effect of different alkali agents' concentrations on pH. *(Modified from Fakher S, Abdelaal H, Elgahawy Y, Imqam A. A characterization of different alkali chemical agents for alkaline flooding enhanced oil recovery operations: an experimental investigation. SN Appl Sci 2019;1:1–11.)*

that will dissociate into sodium and silicate. The most important feature of using sodium silicate as an alkali agent is that it is much easier to inject into the reservoir due to its having a liquid form in ambient conditions. The third common alkaline chemical agent is sodium carbonate, which has the ability to generate both acid and alkaline when reacted with water. Fig. 4.1 illustrates the impact of these three alkali agents on water pH with different concentrations. All of these measurements were performed at the initial condition of pH=7 for all cases [5].

These results showed that after adding the alkali chemicals, the pH of the solution has increased to 64% for sodium hydroxide, 62.5% for sodium silicate, and 21% for sodium carbonate. Researchers [6, 7] proved that the pH decreases when crude oil and an alkaline solution are in contact for a longer time period (Fig. 4.2). They also claimed that the pH reduction has a direct relationship with OH^- consumption by acidic components of the in situ crude oil. The higher the acidic components of crude oil, the higher the consumption of OH^-, and therefore lower the pH values.

4.3 Alkaline reaction

4.3.1 Reaction with oil

The main influence of alkaline agents on crude oil is in situ surfactant production and emulsification [8]. Sheng described and linked the production of in situ surfactants to

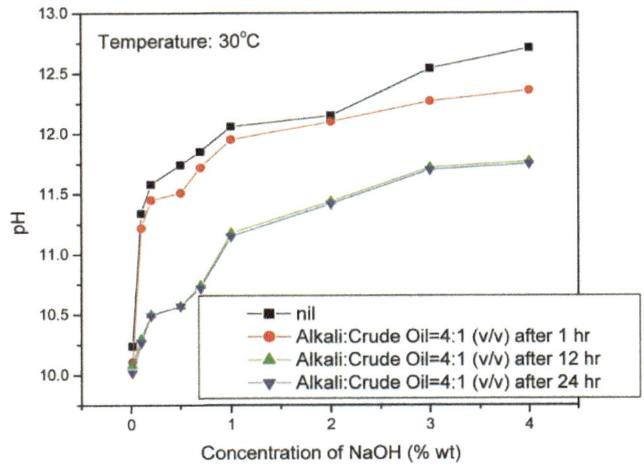

Fig. 4.2 pH of the alkali/aqueous phase of reservoir fluid after alkali addition at various NaOH concentrations [6].

the reaction between injected alkaline and saponifiable components or naphthenic acids in crude oil. This reaction is called the "saponification reaction." This reaction happens when alkali chemicals produce high concentrations of anionic hydroxide (OH^-) in the aqueous phase resulting in high pH [9, 10].

In aqueous solutions with neutral pH, the crude oil acid is insoluble enough to be produced during the flooding process. Naphthenic acid is the nominal mixture for some nonspecific mixtures of carboxylic cyclohexyl and cyclopentyl acids with a molecular weight of 120–700 [11]. When this acid is exposed to an alkaline solution, it reacts to generate a water-soluble anionic surfactant (A^-). Fig. 4.3 shows the reaction of an acid with an alkaline in a porous medium. The components of this hydrocarbon acid (HA) will dissociate from into components the components shown in the following reactions. These components which are produced in the presence of an alkaline at the water-oil interface are soluble due to the production of an anionic surfactant (A^-) [12].

$$HA_o \leftrightarrows HA_w \quad \text{(molecular acid between the aqueous phase and oleic phase)} \tag{4.1}$$

$$HA_w = A^- + H^+ \quad \text{(overall reaction)} \tag{4.2}$$

Therefore, the NaOH solution will be as follows [12]:

$$NaOH \rightarrow Na^+ + OH^- \tag{4.3}$$

The exact nature of hydrocarbon acid (HA_w) is not clearly known because it depends on the type of crude oil [11]. For example, some crude oils have a lower fatty acid and produce a less in situ surfactant when reacting with an alkaline. Eq. (4.4) shows the

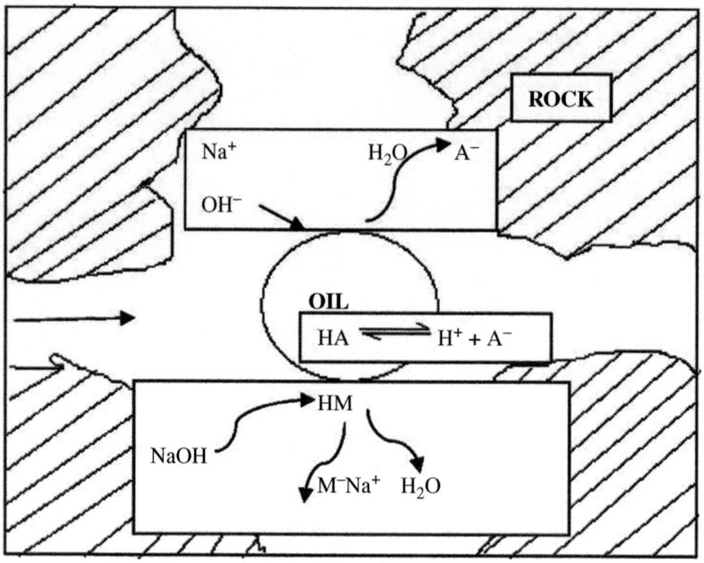

Fig. 4.3 Schematic representation of interaction between an alkaline solution and saponifiable components [6].

dissociation constants for Eqs. (4.1) and (4.2), and Eq. (4.5) presents the distribution coefficient of the molecular acid:

$$K_A = \frac{c_{H^+} \cdot c_{A^-}}{c_{HA_w}} \tag{4.4}$$

$$K_D = \frac{c_{HA_o}}{c_{HA_w}} \tag{4.5}$$

where c is a molar concentration.

The acid dissociation constant controls the pH and, as a result, the surfactant hydrolysis, which controls the distribution coefficient of acid solubility in the water. It should be noted that salt does not affect the chemical balance. Based on the fact that acids are mostly in the oil phase, it is anticipated that the production of anionic surfactants will not be high. The water dissociation constant is expressed as follows:

$$K_w = c_{H^+} \cdot c_{OH^-} \tag{4.6}$$

Assuming a constant concentration of water, an increase in c_{OH^-} concentration is equivalent to a decrease in c_{H^+} and vice versa where the pH of the water is derived as follows [12]:

$$pH = -\log H^+ \tag{4.7}$$

By Eq. (4.4), the amount of anionic surfactant is obtained as follows:

$$c_{A^-} = \frac{K_A \cdot c_{HA_w}}{c_{H^+}} \tag{4.8}$$

And by combining Eqs. (4.5) and (4.8):

$$c_{A^-} = \frac{K_A \cdot c_{HA_o}}{K_D \cdot c_{H^+}} \tag{4.9}$$

And by considering Eq. (4.6), which is related to water, with Eq. (4.9):

$$c_{A^-} = \frac{K_A \cdot c_{HA_o} \cdot c_{OH^-}}{K_D \cdot K_w} \tag{4.10}$$

Now if we define the expression K as follows:

$$K = \frac{K_D \cdot K_w}{K_A} \tag{4.11}$$

Therefore, Eq. (4.10) is rewritten as follows:

$$c_{A^-} = \frac{c_{HA_o} \cdot c_{OH^-}}{K} \tag{4.12}$$

For an alkaline solution containing NaOH in water, the following equation holds:

$$c_{Na^+} + c_{H^+} = c_{A^-} + c_{OH^-} \tag{4.13}$$

where the acid is hydrolyzed by NaOH as follows:

$$HA_o + NaOH \leftrightarrows NaA + H_2O \tag{4.14}$$

where Na^+ is the amount of sodium isolated from NaOH. At high pH values, c_{OH^-} has a small value. By combining the above equations, Eq. (4.15) is generated for calculating the anionic surfactant [12]:

$$c_{A^-} = \frac{1}{2}\left[\frac{c_{Na^+}}{1 + K/c_{HA_o}} + \sqrt{\left(\frac{c_{Na^+}}{1 + K/c_{HA_o}}\right)^2 + \frac{4K_w c_{HA_o}}{K}}\right] \tag{4.15}$$

Eq. (4.15) could be simplified at high pH as follows:

$$c_{A^-} = \frac{c_{Na^+}}{1 + K/c_{HA_o}} \tag{4.16}$$

Therefore, for a constant acid concentration in the oil phase and for a given pH or equivalent at a given c_{Na^+}, Eqs. (4.15) and (4.16) show that K is regulating the amount of active surface components in the aqueous phase. The balance constant in Eq. (4.14) is $1/K$.

Application of the acid hydrolysis chemistry requires a measure of the equilibrium constants of K_A, K_D, and K_W. The dissociation constant for water is dependent on the reservoir temperature. Table 4.1 presents the distribution or partition coefficient (K_D), dissociation constants for water (K_W), and acid (K_A) by DeZabala et al. [12].

Table 4.1 Distribution or partition coefficient (K_D), dissociation constant for water (K_W) and acid (K_A) from DeZabala et al. [12].

Parameter	K_D	K_W	K_A
Constant value (mol/dm^3)	10^4	5×10^{-14}	10^{-10}

For a better understanding of the reaction between oil and alkaline, the reaction between acidic oil and an NaOH alkaline solution will be analyzed here. For example, by adding NaOH to water, it dissociates into OH^- and Na^+ ions as shown in Eq. (4.3). The carboxyl in crude oil is hydrolyzed to parental carboxylic acid and an alcohol when exposed to OH^- ions on the contact surface. This reaction is known as saponification as it is the source for soap creation from fats or crude oil glycerol triesters [13–16]. Fig. 4.4A shows the three-step reaction of carboxylic acid and its hydrolysis by OH^- and finally the formation of the parent carboxylic acid and an alcohol. Subsequently, the hydrolyzed acid reacts with Na^+ ions and inhibits surfactant production where this step is shown in Fig. 4.4B [13–18].

4.3.2 Reaction with water

Alkalis come into contact with formation or connate water after injection into a porous medium. Subsurface formations and associated connate water generally contain monovalent and divalent ions, so it is very likely that the injected alkaline fluid reacts with rocks and reservoir fluids [19]. When divalent ions, for example, Ca^{2+} and Mg^{2+}, of the connate water encounter alkaline, they react with the produced OH^-. The following example shows the reaction between the calcium ion and OH^-:

$$Ca^{2+} + 2OH^- \rightleftharpoons Ca(OH)_2 \tag{4.17}$$

The solubility of $Ca(OH)_2$ in aqueous solution is 0.077 mg/L; therefore, it precipitates after Ca^{2+} reacts with the in situ surfactant [20]. The reaction of these ions with alkaline depends on the temperature, pH, and concentration of ions [9].

4.3.3 Reaction with rock

There are various types of reaction between the rock and an alkaline solution and it is difficult to analyze the details due to the high lithology variety of rocks. For example, when alkaline encounter clays, they try to reach equilibrium with the new environment. On the other hand, clays exchange ions with the aqueous phase due to their very large surface area. The common reaction of clay minerals with an alkaline solution is when the hydrogen ions of the clay surface react with OH^- ions of the alkaline solution and lower

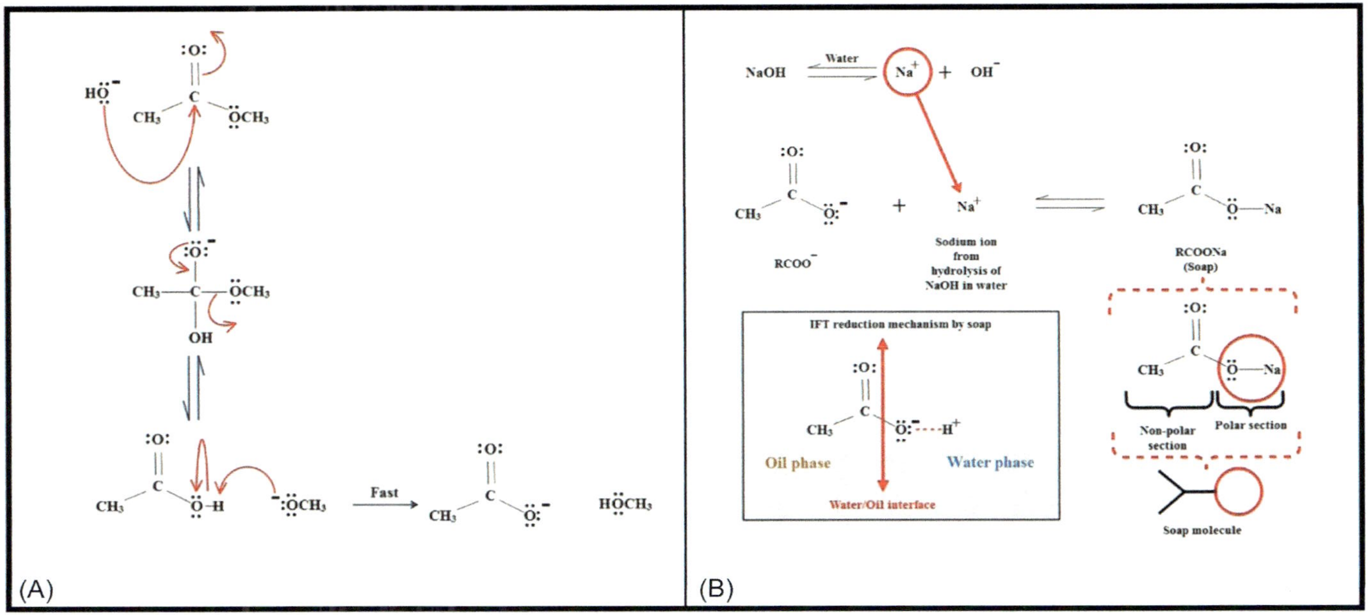

Fig. 4.4 Reaction between alkaline and acidic oil: (A) saponification mechanism and (B) IFT reduction mechanism by NaOH [13].

the pH. So the added alkaline will be consumed through the movement of alkaline water in the reservoir which is shown in the following Eq. [4.9]:

$$H - X + Na^+ + OH^- \leftrightarrow Na - X + H_2O \tag{4.18}$$

where X represents the mineral-base exchange sites.

For example, ion exchange of Na^+ and Ca^{2+} is as follows:

$$2Na - X + Ca^{2+} \leftrightarrow Ca - X + 2Na^+ \tag{4.19}$$

Besides hydrogen, clays have divalent ions, for instance, calcium and magnesium. When alkaline water comes in contact with clays, calcium or magnesium ions of the clay surface exchange with the sodium ions of the alkaline solution, which can cause the calcium or magnesium to deposit [9]. The reaction of this type of cation exchange can be shown by the following equations:

$$Ca - X_2 + 2Na^+ + 2OH^- \leftrightarrow 2(Na - X) + Ca(OH)_2 \tag{4.20}$$

$$Ca - X_2 + 2Na^+ + CO_3^{2-} \leftrightarrow 2(Na - X) + CaCO_3 \tag{4.21}$$

Ion exchange is a rapidly reversible process, but alkaline rock dissolution is a long-standing irreversible kinetic process that can occur. The number of possible reactions between rock and alkaline is very high because of the high variety of rocks mineralogy [9].

Fig. 4.5 shows the time τ_B in pore volumes for NaOH to elute from a Wilmington sand core alkaline injection. Fig. 4.5 predicts that the lower the injected caustic concentration, the longer it will take for breakthrough. For example, at a pH of 11 and 13, alkaline appear in a production well after 10 and 2 PV [12].

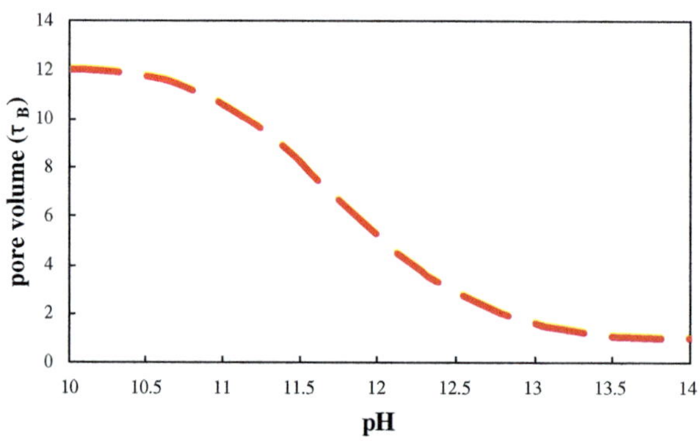

Fig. 4.5 Pore volumes for caustic versus concentration [12].

Table 4.2 Alkaline consumption in different mineral [11].

Mineral	Interaction with alkaline
Gypsum and kaolinite	Highest
Illite, montmorillonite, zeolite, and dolomite	Temperate
Chlorite and feldspar	Fairly low
Quartz	Fine
Quartz and insignificant for calcite	Lowest

Ehrlich and Wygal [21] studied various minerals and found that clay alkalis consumption was very high when rock and alkaline come into contact, while it was low for dolomite, calcite, and quartz. Sheng [11] also summarized the amount of alkaline consumption of minerals as shown in Table 4.2.

Shen and Chen [22] prepared a list of alkaline consumption from high to low for various minerals as follows:

Gypsum ≫ **Montmorillonite** ≫ **Kaolinite** ≫ **Illite** ≫ **Plagioclase** ≫ **Microcline** ≫ **Quartz** ≫ **Mica** ≫ **Dolomite** ≫ **Calcite**

Other reservoir variables can also affect the consumption of alkaline. For example, a higher temperature will increase the alkaline consumption. Another factor that has a great effect on the alkaline consumption on the surface of the rock is the contact time. In summary, the three most important and influential factors affecting alkaline consumption are rock mineralogy, pH, and temperature and contact time [23].

4.4 Mechanisms

The alkaline solution interacts with the acidic oil, resulting in the formation of an in situ surfactant. The in situ surfactant is then absorbed at the crude oil-brine-rock interface, assisting the oil in moving. The alkaline flooding process improves the sweep efficiency based on four major mechanisms: (1) IFT reduction, (2) wettability alteration, (3) emulsification, and (4) saponification. Saha et al. [24] summarized the major mechanisms of the alkaline flooding process as shown in Fig. 4.6. Cooke et al. [23] found that alkaline flooding formed the oil bank in porous media by activating the IFT reduction and emulsification mechanisms. Similar studies also confirmed that emulsion formation improves the oil recovery during alkaline flooding [25, 26]. In the following sections, the alkaline process governing mechanisms are discussed for a better understanding of alkaline flooding.

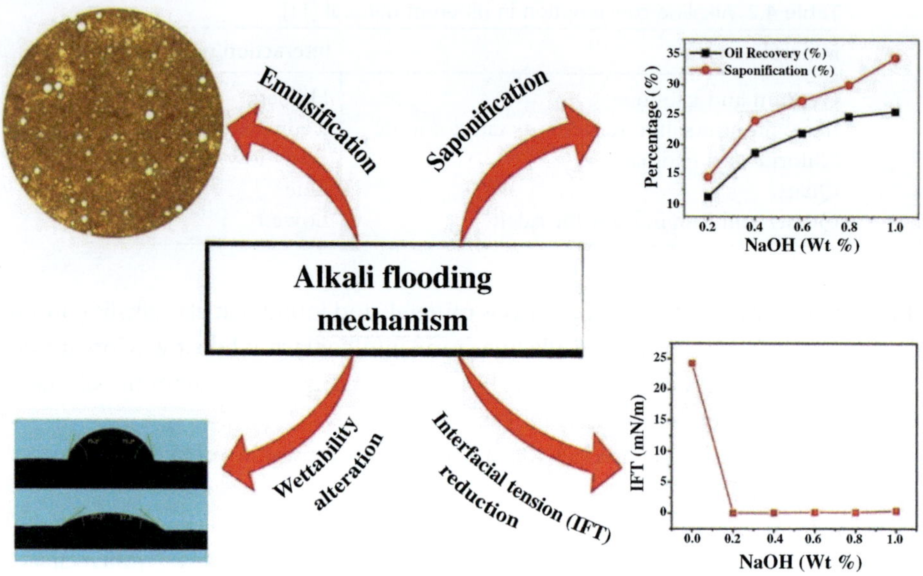

Fig. 4.6 Mechanism governing on alkaline flooding [24].

4.4.1 Saponification

Saponification is one of the processes during alkaline injection that occurs at the interface of the alkaline solution and crude oil. At this interface, the acid oil is hydrolyzed in the presence of alkaline and, as a result soap is produced which ultimately decreases the IFT [20]. Saha et al. [24] investigated the degree of neutralization and saponification of acid oil for the alkaline solution. They used NaOH in different concentrations where the total acid amount of oil was measured pre- and postinjection of the alkaline solution. Fig.4.7 depicts the neutralization and saponification of crude oil at various NaOH concentrations. These results showed that at an NaOH concentration of 0.2wt%, only 14.6% of the oil acid group participated in saponification and neutralization reactions. The neutralization and saponification increased by increasing the alkaline concentration and reached up to 34.5% at 1.0wt% NaOH. The reason for this increase is the higher Na^+ ions concentration. This shows that nearly 75% of the acid group remained unconverted.

4.4.2 IFT reduction

After the alkaline solution is injected into the oil reservoir, the solution flows through the porous medium and comes in contact with oil. Under suitable conditions (the alkaline efficiency is described in other sections), the acidic oil reacts with the alkaline and produces an in situ surfactant, which is produced at the water-oil interface. Like the artificial surfactant, the produced in situ surfactant accumulates on the interface and diffuses into

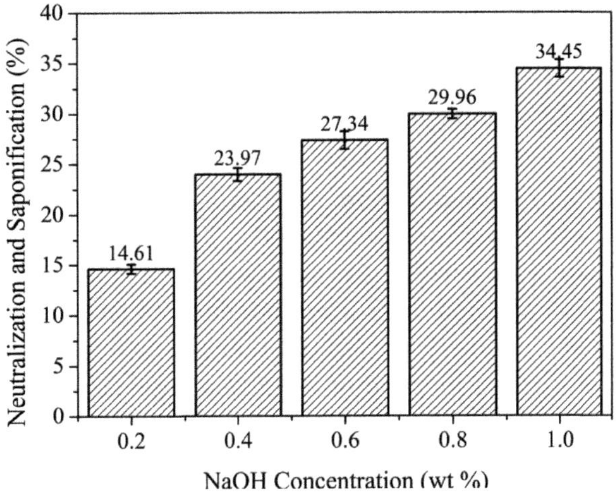

Fig. 4.7 Neutralization and saponification of acid oil at different NaOH concentrations [24].

the aqueous and oil phases, the distribution of which depends on their preference for the aqueous and oil phases. As a result, the interfacial tension decreases and eventually reaches equilibrium [7].

By using the spinning drop tensiometer, Rubin and Radke [27] concluded that the lowest interfacial tension observed is indicative of the minimum achievable reservoir equilibrium value. Several studies [28, 29] have shown that the laboratory achieved equilibrium interfacial tension is used for alkaline injection in the field conditions. McCaffery [29] stated that the laboratory obtained interfacial tension could not be a good representative of the real reservoir condition because the interfacial tension is obtained shortly after alkaline and oil contact. Jennings et al. [30] also found that in a very low IFT condition in a potentially water-wet core, residual oil can be emulsified in situ and move within the moving alkaline solution, and trapped again by pore throats which are too narrow to allow the oil emulsion droplets to pass. This can not only reduce the water mobility but also improves the crude oil sweep efficiency [31].

Although there are different opinions about the interfacial tension in the alkaline injection process, many studies state that the IFT will decrease during this process. Li et al. [32] explained that increasing the alkaline concentration decreases the IFT due to increased diffusion and reaction, while Nasr-El-Din et al. [33] suggest that interfacial tension decreases to a minimum point and then increases again. The increase in IFT beyond a critical concentration is related to the reduction in the ionic species at high alkali concentration because of micelles formation and the compression of the electric double layer at high ionic strength [34].

Many researchers have shown that time is a significant factor in all EOR processes [35–38]. In particular, time is an important factor in the interface reaction between crude oil and the alkaline solution when alkaline reacts with oil to produce an in situ surfactant. The produced surfactant moves between the surface and the bulk of oil and water, and the amount of reaction between the surface and the aqueous phase of the oil determines the changes in the interfacial tension between the alkaline solution and the oil. After a certain time (equilibrium time), the interfacial tension becomes constant, which is said to be in equilibrium [20]. The equilibrium interfacial tension is known as the best choice for the actual interfacial tension in the reservoir [39]. However, as alkaline properties change during the interaction with reservoir lithology, it is important to evaluate the minerals. Pei et al. [7] calculated the interfacial tension of two different types of oil as a function of interaction time with the alkaline solution and oil type. They have shown that the interfacial tension is a function of the alkalinity (pH value). For example, after NaOH solution injection, the IFT will decrease as a function of time for NaOH concentrations less than 0.6 wt% and reaches a minimum value. With the passage of time, the IFT will increase. At higher concentrations of NaOH, the interfacial tension stays almost constant after initial reduction. This means that, at high alkalinity, the interfacial tension is almost constant after the initial reduction, while for sodium carbonate the interfacial tensile behavior was a function of time, similar to the low concentration of sodium hydroxide. It is quite obvious that the reason for the reduction in the interfacial tension is a function of time in alkaline injection due to the production of surfactants on the water-oil interface [40]. In general, the more compact the surfactant pack is at the surface between the acidic oil and the alkaline solution, the greater the initial reduction in interfacial tension as a function of time [7, 41]. Generally, after the collision of the alkaline solution and acidic oil, a surfactant is rapidly produced on the alkaline solution and oil interface, where the diffusion speed of the in situ surfactant inside the aqueous and oil phases (bulk) is lower than the chemical reaction between the alkaline and acidic oil, which produces an in situ surfactant accumulation at the oil-alkaline interface and creates a sharp decrease in the IFT in the early stages. Accordingly, the chemical reactions will continuously weaken, and the diffusion of the surfactant into the bulk begins to dominate [42].

Fig. 4.8 shows the variations of dynamic IFT (DIFT) and equilibrium IFT (EIFT) for the alkaline solution and heavy oil based on the results of a research by Pei et al. [43]. According to Fig. 4.8A, DIFT was the lowest value for 4% concentration, and at this concentration, after 10 min, the interfacial tension decreased to 10^{-3}, but as time passed, it increased again. Fig. 4.8B displays the results of the interfacial tension of the alkaline in Fig. 4.8A after reaching equilibrium. According to Fig. 4.8B, the interfacial tension first decreases to the minimum point and then increases where the EIFT reaches the minimum point at a concentration of 0.8% [43]. Chan et al. also obtained similar EIFT results and showed that the interfacial tension of the NaOH solution and oil initially decreases to the minimum point and then increases with increasing alkaline concentration [44]. These results are also consistent with the work of Yen et al. [45], Jennings [30], and McCaffery

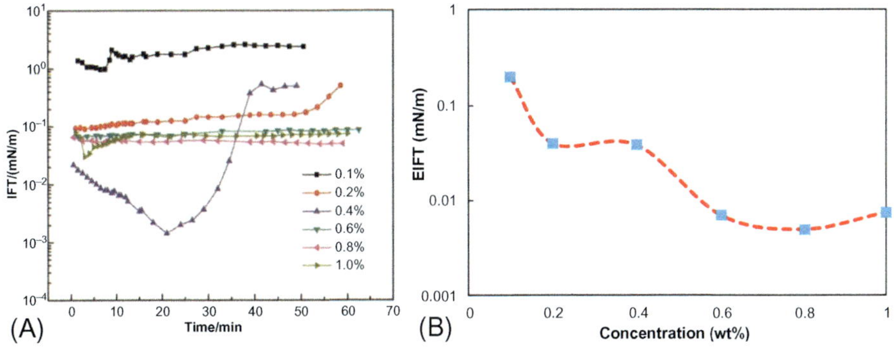

Fig. 4.8 DIFT and EIFT versus concentration and time during alkaline injection [43].

Table 4.3 The variation of IFT for different crude oils at various pH and water chemistry [46].

Crude type	Asphaltenic and resinous agents	Water chemistry	IFT vs pH
Light crude	Basic functional group in resin fractions	Distilled water Brine	Increase Decrease
Medium crude	Acidic functional groups	Distilled water Brine	Decrease More decrease
Heavy crude	High indigenous surface active agents	Distilled water Brine	Decrease More decrease

[29] who relate the increased interfacial tension to the formation of undissociated (inactive) soap at high concentrations of alkaline which could decrease the charge density [44].

Lashkarbolooki and Ayatollahi [46] measured the IFT of three light, medium, and heavy crude oils at different pH and water chemistry conditions. The comparison of the IFT of these oils and their asphaltenic and resinous equivalent oils showed that several factors, including the amount of asphaltene and resin content, aqueous salinity, and pH, could influence the IFT. Table 4.3 summarizes these findings.

These IFT tests show that asphaltene and resin content, salinity, and having a high acid number, decrease the IFT in the freshwater, and the IFT will furthermore decrease when brine is used. IFT in the combination of light crude and freshwater is an exception.

IFT reduction between the alkaline solution and crude oil will result in several interfacial events, for example, emulsification that increases the oil recovery. Generally, after the water flooding process, the remaining oil will become completely immobile due to a strong interfacial tension between oil and water. The injection of alkaline, and consequently the reduction of interfacial tension, makes the remaining oil mobilize and begin

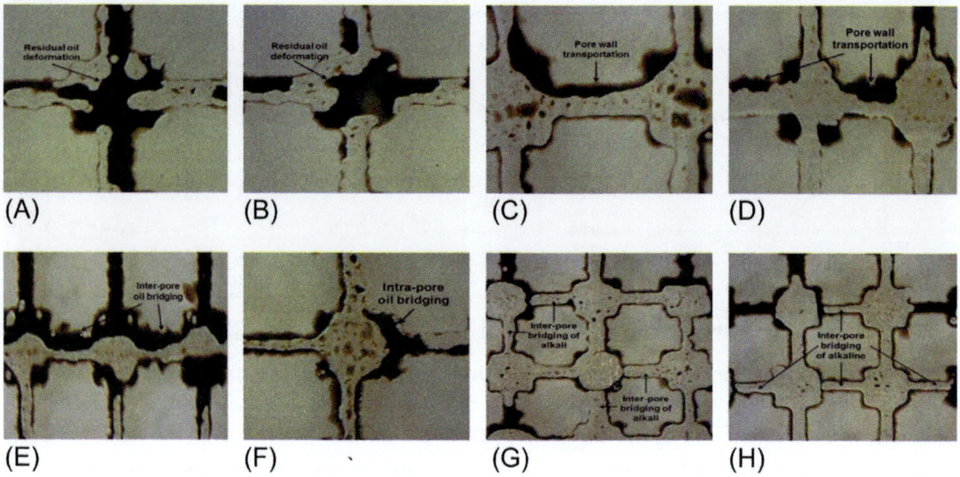

Fig. 4.9 Behind mechanism of IFT reduction in alkaline flooding: (A and B) deformation of residual oil, (C and D) pore-wall transportation, (E) interpore oil bridging, (F) intrapore oil bridging, and (G and H) interpore bridging of the alkaline solution [47].

to move through the pores. Therefore, one of the hidden mechanisms in reducing interfacial tension is the residual oil deformation. The remaining oil generally transforms into long and narrow oil films. In an oil-wet reservoir, oil adsorbs to the pore wall and, as a result, is transferred along the pore wall, resulting in the production of oil. This mechanism is called "pore wall transportation." Maintaining this process at low interfacial tension causes the remaining oil to move and connect between several adjacent pores. This mechanism is called "interpore bridging." This type of bridge may also occur inside a hole called intrapore oil bridging [47]. Furthermore, during alkaline injection, the solution passes through the pore body due to its larger diameter and less capillary force. As a result of the reduction of capillary force, the alkaline solution is mixed through the alkaline stream of the pores. Therefore, by increasing the flooding area, more volume of residual oil will be recovered, which is called the interpore bridging of the alkaline solution. Fig. 4.9 shows these mechanisms during an alkaline injection process in a laboratory micromodel [47].

4.4.3 Wettability alteration

Wettability is one of the most important factors influencing liquid distribution and flow in reservoir pores [48–50]. The change in wettability from hydrophobic to hydrophilic can increase oil recovery. This mechanism can be applied to oil-wet reservoirs because the residual oil in water-wet systems is generally immobile and discontinuous as compared to oil-wet systems [51].

Mungan [52, 53] showed that alkaline flooding is able to change the rock wettability from oil-wet to water-wet where the relative water permeability increased after alkaline flooding. Kumar et al. [54] discovered that wettability can be controlled by asphaltene adsorption and that anionic surfactants can remove these adsorbed components from the mineral surface, causing the rock surface to become water-wet.

Wettability change from oil-wet to water-wet can improve the oil recovery. This is due to the fact that the nonwetting phase occupies the large pores and the wetting phase occupies the small pores. By changing the wettability, the small pores will be occupied by nonwetting fluid while the large pores will be filled with wetting fluid. The resulting fluid redistribution as the phases try to move to their natural state would make both phases vulnerable to recovery via viscous forces [47, 55]. The enhancing oil recovery by wettability alteration from oil-wet to water-wet happens because of two mechanisms: (1) the effect of relative permeability, which decreases the mobility ratio of a displacement and (2) a change in the capillary distribution curve [47].

Fig. 4.10A and B shows water flooding with the Na_2CO_3 solution in an oil-wet glass micro-model. According to this figure, the wettability of some pore-walls has changed to water-wet, whereas some other pore-walls remained oil-wet, although the remaining oil-wet surfaces have also changed from strong oil-wet to weak oil-wet. Fig. 4.10C shows the alteration in wettability of the glass surface for NaOH flooding. Detailed microscopic analysis shows that NaOH flooding is more effective in altering wettability from oil-wet to water-wet as shown in Fig. 4.10C. By using NaOH, a complete wettability alteration happened in many areas and nearly all the pore-walls became water-wet, while in sodium carbonate flooding there was partial wettability alteration. Therefore, it can be concluded that more severe wettability alteration will cause more efficient oil recovery [47].

There are other cases where the effect of alkaline is the opposite, which is the change of wettability from water-wet to oil-wet. Cooke et al. [23] pointed to this mechanism where they observed the increase in oil production due to the wettability change from water-wet to oil-wet. In this mechanism, the residual crude oil as a nonwetting phase is converted into a continuous wetting phase in the porous media. In this method, the goal

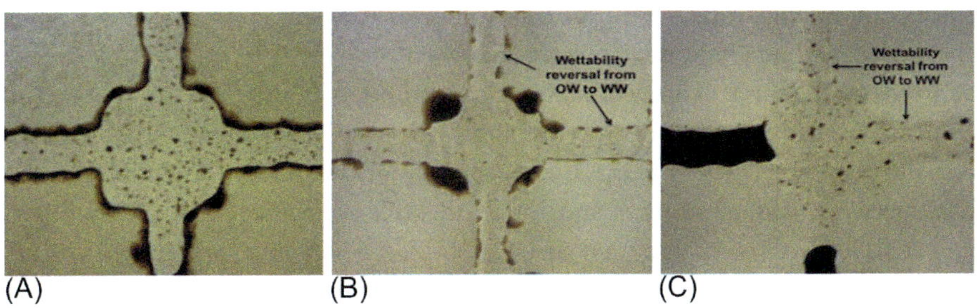

Fig. 4.10 Wettability alteration: (A) after water flooding, (B) sodium carbonate injection, and (C) sodium hydroxide injection in an oil-wet micromodel [47].

is to increase the capillary number, by reducing the IFT and emulsifying water in oil. The production of soap in the alkaline flooding process reduces the interfacial tension and creates water in oil emulsion. These emulsion droplets tend to block the flow and create a high-pressure gradient in the area where they formed. This high-pressure gradient ultimately overcomes the capillary forces, which have already reduced with low surface tension and which decrease the residual oil saturation furthermore.

Wettability alteration to partially oil-wet is the other governing mechanism in alkaline flooding. The acidic oil reaction with an alkali produces the in situ surfactant and, as a result, alters wettability to preferentially oil-wet [47].

In a laboratory study, Gong et al. [56] used quartz to evaluate changes in wettability during alkaline flooding. They observed that the wettability of the rock altered from water-wet to oil-wet in the presence of the alkaline solution, whereas the contact angle of the rock changed from 24 to 167 degrees by adding 0.5 wt% sodium carbonate solution to the distilled water. The addition of NaOH to the solution created the same effect and changed the wettability to strong oil-wet (167 degrees).

Buckley et al. [57] listed four mechanisms responsible for changing the rock wettability: (1) polar interaction, (2) surface precipitation, (3) acid-base interaction, and (4) ion binding.

Polar interaction is a reaction between the rock surface charges and the opposite charges of the crude oil chemicals. This occurs when there is no water film between the oil and the reservoir rock surface. At pH values around 7, negatively charged sandstone surfaces interact with positively charged components, while positively charged carbonate surfaces interact with negatively charged chemicals of the oil phase. Nevertheless, the polar components of crude oil usually have a nonpolar component that is typically a long chain of hydrocarbon. This long chain moves to the rock surface, creating a layer of nonpolar components and changing it to oil-wet. The precipitation of these components on the rock surface is primarily dependent on the solubility of heavy components of oil in asphaltene. The mechanism of wettability alteration happens when oil is a weak solvent for its asphaltene. As a result, the asphaltene cannot dissolve properly and deposit on the reservoir rock pore surface. The rock pore surface becomes oil-wet due to the presence of nonpolar components of heavy oil on the surface of the rock.

The acid/base interaction of acidic or alkaline components can totally or partially ionize the atoms on the surfaces of carbonate and sandstone rocks through the aqueous phase. In sandstones, negative surface charges interact with positive charges of the ionized base. In contrast, positive charges on the surface of carbonate rock react with ionized acid. Bonding between surface charges and ionized acid/base occurs via a water film. The water-wettability will be maintained through a fixed water film, which can collapse in the unsteady state of the water film, letting the acid or base to interact directly with the rock surface charges, and consequently change the wettability to oil-wet. Generally, ionic bonding plays the main role in changing the sandstone's wettability.

Ion binding has an important role in sandstone wettability alteration. Ions (primarily divalent ions in the aqueous phase such as calcium and magnesium ions) act as a bridge between the negative charges of the sandstone surface and the negative charges of the ionized acid from the oil phase in this mechanism. This bridge happens in the aqueous phase as a base-acid interaction. In the carbonate rocks, the presence of divalent ions due to repulsion between rock surface ions and divalent ions has no effect on the change in wettability at the rock surface.

By analyzing electrostatic forces, Liu et al. [58] inspected the mechanism of wettability alteration in a heavy oil-brine-sand system. They inferred that if the injected solution has both Na_2CO_3 and Mg^{2+} ions, the wettability of the sandstone could shift from water-wet to preferentially oil-wet by magnesium ion binding. The decrease in zeta-potential value caused by Mg^{2+} addition to the heavy oil-alkaline solution-sand system confirmed the combination of Mg^{2+} and ionized organic acids at the oil-water interface. Furthermore, the decrease in the zeta-potential value of sand in the water phase suggested that Mg^{2+} was also adsorbed on sand surfaces, weakening the electrostatic forces. The decrease in repulsive electrostatic forces between oil droplets and sand surfaces contributed to the sand's wettability changing from water-wet to more oil-wet.

A water-oil emulsion is produced at the displacement front by injecting alkaline into the porous medium, during which the emulsion is located in the middle of the pore spaces and the oil phase remains on the grain surfaces. The alkaline solution replaces the oil phase on the surfaces as the injection continues. This is due to the pore walls' affinity to become wet with the aqueous phase. As chemicals continue to be injected, the wettability of some pores changes from water-wet to weakly oil-wet due to the interactions between the alkaline surface and the rock surface. Fig. 4.11A and B shows the flooding of Na_2CO_3 and NaOH in a water-wet glass micro-model. These figures illustrate the formation of an oil film on the pore walls.

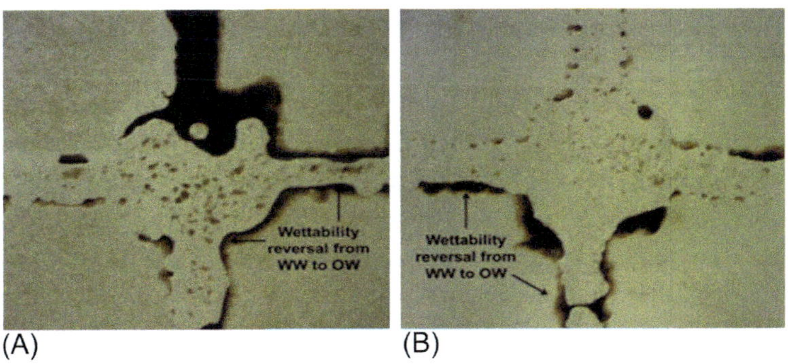

Fig. 4.11 Wettability alteration after: (A) sodium carbonate injection and (B) sodium hydroxide injection in a water-wet micromodel [47].

The in situ surfactant adsorbs on the quartz through the electrostatic or hydrogen bond interactions between the hydrophilic head group of the surfactant and the quartz surface. When an in situ surfactant adsorbs on the quartz, the quartz surface wettability will change because the hydrophobicity groups of the surfactants are resolvable in the oil phase [56]. Liu et al. [59] examined the effects of alkaline-surfactant flooding on the wettability alteration of bitumen-treated glass surfaces and concluded that the synergistic impact of anionic surfactants and alkaline environments is extremely successful in mobilizing the hydrocarbon and improving the bitumen-treated glass surface's water-wettability.

4.4.4 Emulsification

Researches have shown that alkaline flooding is able to increase oil recovery in porous media by the emulsification mechanism [23, 30, 60]. The production of emulsion by alkaline is also dependent on the interfacial tension of oil-water, which means that the reduction of the interfacial tension will increase the production of emulsion. Emulsion stability is also related to the film between water and oil.

Two types of emulsions have been proposed for the injection of alkaline into a porous medium. Some researchers have suggested that the development of oil-in-water (O/W) emulsions is a likely mechanism for oil extraction [61, 62]. Oil emulsion in water is produced in two ways:

- Emulsification and entrainment
- Emulsification and entrapment

In the emulsification and entrainment mechanism, the reduction of the interfacial tension between the alkaline solution and crude oil causes a complicated interfacial phenomenon and, as a result, an oil emulsion will form in water. Assuming a model, when oil is emulsified in water, the oil is involved with water and removed from the system. Low acid number, low salinity, and high pH are the required parameters for this process. In this mechanism, the oil water emulsion size is generally less than the diameter of the pore throat [7, 60, 63–65].

Meanwhile, in emulsification and entrapment, the oil is emulsified into the water phase and oil droplets generally detached from the reservoir pore walls, which results in oil recovery improvement. This mechanism requires high alkaline pH, moderate oil acid number, low water salinity, and low concentration of alkali. The O/W emulsion size is usually larger than the pore-throat size [7, 63, 65]. Generally, when the concentration of alkaline in the system is low, adding a surfactant dominates the O/W emulsion mechanism in the oil displacement process.

The second type of emulsion as an alkaline process mechanism has been introduced as the water/oil (W/O) emulsion [62, 66]. Ma [67] and Liu [68] showed that the viscosity of W/O emulsion changes from 2000 to 20,000 cP where the emulsion viscosity depends on the ratio of water/oil and alkaline concentration in the system. In addition, the W/O emulsion improves the sweep efficiency by decreasing the mobility of the water phase, blocking the high permeability water conduits, and diverting the injected water into the nonswept region [43, 47, 69].

In general, the W/O emulsion mechanism is predominant when high alkaline concentrations are used. W/O emulsion has a much higher viscosity compared to the O/W mechanism, which can effectively block the water conduits to improve back-and-forth efficiency. This method can be highly effective in heavy oil reservoirs [70, 71].

The formation of a water-in-oil emulsion also results in a large nonuniform pressure differential in a limited region at the emulsion front. The pressure gradient must overcome the capillary force, which has decreased as the IFT has decreased, causing the crude oil to be displaced in the pore space [47].

The water-in-oil emulsion was created by Pei et al. [71] in a sand pack during the alkaline flooding process. They measured the droplet size with "Scope Image Plus" software during alkaline injection. Fig. 4.12 shows the changes of the droplet size of water-in-oil emulsions in different alkaline concentrations. According to this figure, increasing alkaline concentration will increase the droplet size and the average droplet sizes produced by emulsion in 0.1%, 0.2%, 0.4%, 0.6%, 0.8%, and 1.0% NaOH concentrations are measured at 1.012, 1.354, 1.637, 1.864, 2.016, and 2.785 μm, respectively.

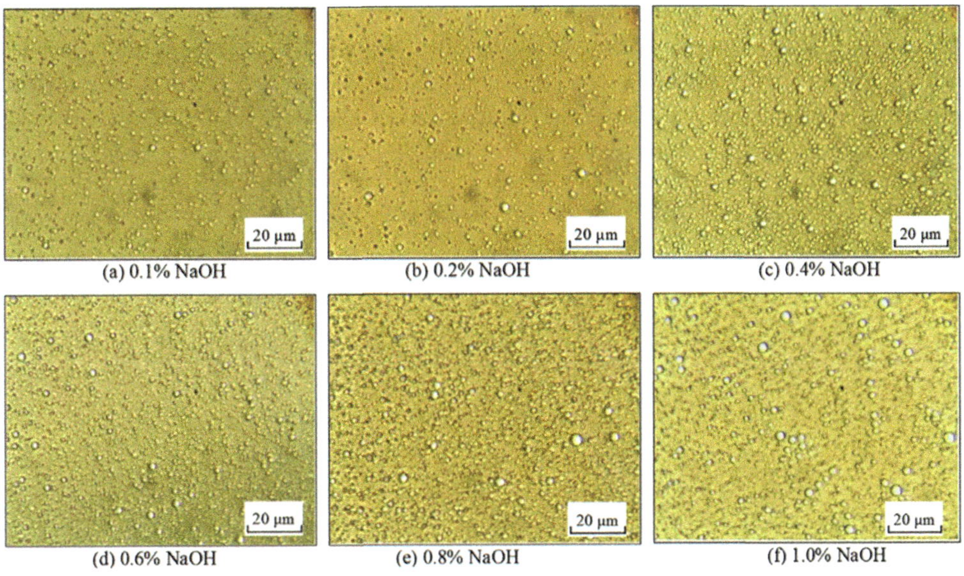

Fig. 4.12 Photomicrograph of water droplet size in oil emulsion: with NaOH concentrations of: (A) 0.1%, (B) 0.2%, (C) 0.4%, (D) 0.6%, (E) 0.8%, and (F) 1.0% [71].

Emulsion quality is defined as the volume percent of the dispersed aqueous phase. They discovered that when the alkaline concentration rises, the quality of the emulsion and the size of the droplets will increase [71].

4.5 Effect of reservoir condition on alkaline process

Temperature and pressure are two important reservoir factors affecting the alkali injection process during operation and later in the reservoir [72]. This section will summarize these effects on the alkaline injection process. Xie et al. investigated the impact of temperature and pressure on the wettability of the reservoir. They showed that temperature affected the wettability of the rock, while pressure does not have much effect on wettability [73]. On the other hand, as mentioned in Section 4.3.3, temperature increases the alkaline consumption due to the absorption of alkaline on the rock surface, which consequently increases the change in wettability, while the interfacial tension changes with temperature for alkaline solutions were in contrast to wettability. According to Saha et al. [24] and Ge et al. [64], increasing temperature increases the interfacial tension of the alkaline solution and oil. Although Ge et al.'s [64] experiments at 50°C for Binnan heavy oil showed that the interfacial tension decreased slightly, in general the interfacial tension increases as a function of temperature in the solution. Fig. 4.13 illustrates the effect of temperature on the IFT between the alkaline solution and oil. The reaction between the alkaline solution and the crude oil is likely to be accelerated at higher temperatures. These changes have a negative effect on the formation of water emulsion within the oil phase, which is the main mechanism of alkaline flooding [43, 69]. According to Fig. 4.14, pressure increase reduced oil recovery at all temperatures, which is due to the reduction of intermolecular distance [74]. By increasing the pressure of the oil and alkaline system, the distance between the molecules decreases and the intermolecular forces increase. This

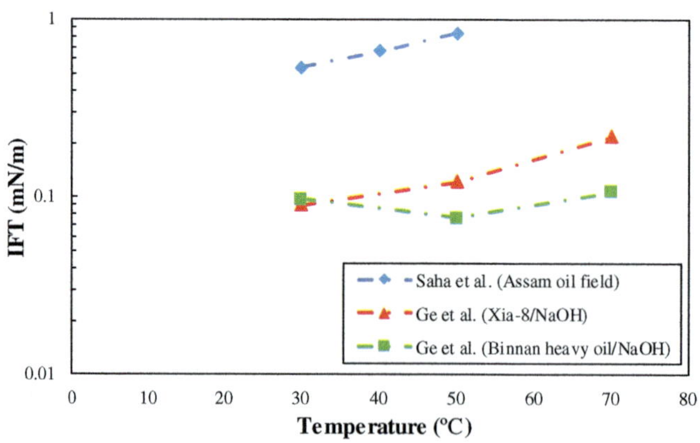

Fig. 4.13 IFT between alkaline/oil versus temperature (Saha et al. [24] and Ge et al. [64]).

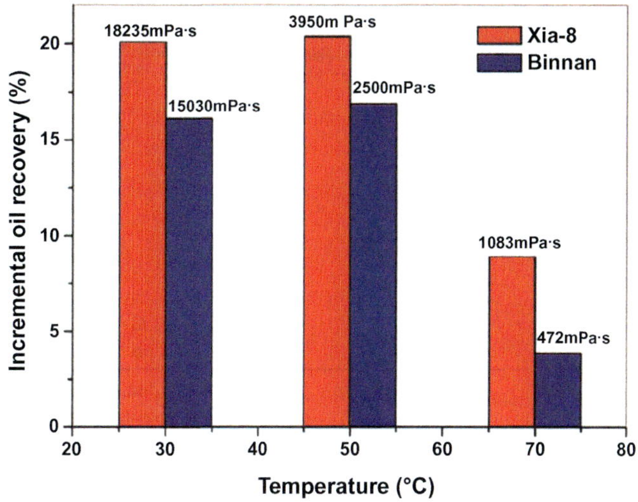

Fig. 4.14 The effect of temperature and pressure on incremental oil recovery during alkaline flooding [64].

creates a strong film on the surface that reduces the solubility of the oil and water, making the emulsion harder to form.

4.6 Geology and lithologic variation of reservoir

The geological, lithological, and structural variations have a significant impact on the wettability and permeability heterogeneity of the reservoir. These heterogeneities are recognized as the most common cause of the low success rate of flooding and achieved recovery efficiency [75].

The highest impact is caused by both horizontal and vertical reservoir heterogeneity, particularly variations in permeability. The secondary structures, such as fault barriers or joint sets as high permeability channel flow system, can restrict the flooding fluid to sweep the oil or divert the fluid in particular channels and bypass the main reservoir. In some cases, the primary depositional settings such as particular permeable channels or layers, lateral pinch outs, and facies change have a significant impact on enhanced oil recovery.

Several researches demonstrated that raising the pH through using alkaline chemicals can change the rock wettability [21, 51, 76, 77]. The nonwetting phase occupies the larger pores in most porous media in their initial wetting state, while the wetting phase occupies the smaller pores. By reversing the wettability of the reservoir rock, the wettability of larger pores changes from oil-wet to water-wet which will enhance the oil recovery. Alkaline flooding can react with the reservoir in several ways depending on the rock type and its mineralogy. These interactions are recognized by Somerton and Radke [78] as surface exchange and hydrolysis, congruent and incongruent dissolution reactions, and development of insoluble salts by reacting with ions in the fluid or ion exchange with minerals [79].

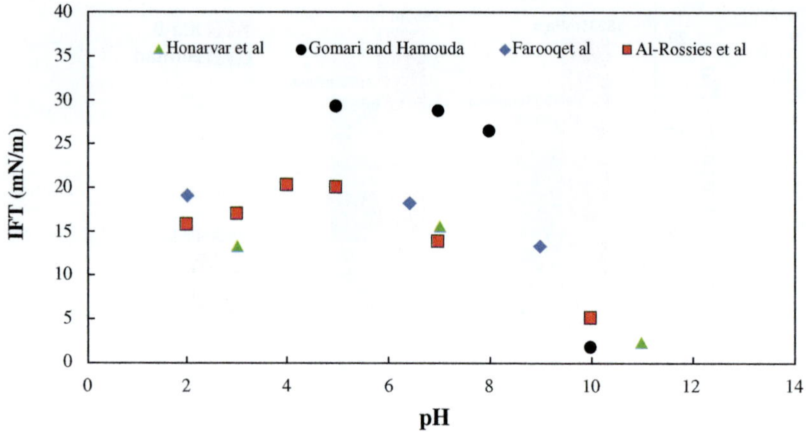

Fig. 4.15 pH effect on the IFT in a brine/crude oil system. *(Source by Honarvar B, Rahimi A, Safari M, Rezaee S, Karimi M. Favorable attributes of low salinity water aided alkaline on crude oil-brine-carbonate rock system. Colloids Surf A Physicochem Eng Asp 2020;585. https://doi.org/10.1016/j.colsurfa.2019.124144; Gomari KAR, Denoyel R, Hamouda AA. Wettability of calcite and mica modified by different long-chain fatty acids (C18 acids). J Colloid Interface Sci 2006;297:470–9; Farooq U, Tweheyo MT, Sjøblom J, Øye G. Surface characterization of model, outcrop, and reservoir samples in low salinity aqueous solutions. J Dispers Sci Technol 2011;32:519–31; and Al-Rossies AAS, Al-Anazi BD, Paiaman AM. Effect of pH-values on the contact angle and interfacial tension. Nafta 2010;61:181–6.)*

4.7 Effect of pH

Many researches show that changes in pH are almost the most important parameter in alkaline flooding. As shown in Section 4.2, pH changes are dependent on the alkaline concentration and the type of alkaline. In this section, the effect of pH on the alkaline flooding mechanism is investigated. The interfacial tension changes are shown in Fig. 4.15 as the governing mechanism for alkaline injection versus pH in the basic environment. Although the acidic environment is not discussed in advance in this chapter, the comparison with the alkaline environment is presented in Fig. 4.15. As Fig. 4.15 shows, the environment changes from acidic to neutral and then to alkaline, which reduces the interfacial tension. At pH values of 10–11, the interfacial tensions measured by Al-Rossies et al. [80], Gomari and Hamouda [81], and Honarvar et al. [20] are reported as 5.2, 1.9, and 2.39 mN/m, respectively. These measurements are performed using the pendant drop and ring method, where the interfacial tension measurement in alkaline flooding by the Spinning Drop Technique gives more accurate results. However, a significant change in interfacial tension in the alkaline environment was observed compared to a neutral and acidic environment. Rudin and Wasan [82] also investigated the effect of pH changes by NaOH on interfacial tension. The results of their experiments were in good agreement with Fig. 4.15, although at a pH above 11 the interfacial tension increases, which could be due to a heavy surface hydrolysis at high pH where the presence of Na^+ ions at a high pH suddenly decreases. Therefore, the process of interfacial tension can be divided into three parts; first, it decreases from neutral to pH around 10; then at pH values of 10–11, the interfacial tension will drop to a minimum, and finally at a pH above 11, it starts to increase.

The solution pH can also affect the wettability of the host rock. As the pH increases, electrostatic changes occur at the rock surface. For example, calcite has a positive charge at pH values less than 9 and, as a result, the adsorption of oil components with a negative charge is likely on the surface of the rock and the system is oil-wet. This means that lowering the pH reduces the repulsive forces between the oil components and the calcite surface, which results in increased oil-wetness of the rock surface [50, 83]. According to various studies, increases in pH on the shale surface cause the stability of the water film on the rock surface. The stability of the water film on the shale surface is lower at a neutral pH and, as a result of the alkaline water flooding, the system changes to a strong water-wet environment [84, 85]. Najafabadi et al. [86] also observed the change of wettability on the surface of dolomitic rock under an alkaline (high pH) condition, although Zhang et al. [83] observed different results for the wettability in the $NaBO_2$ and $NaCO_3$ solution at the same pH (11.5). In this regard, an increase in oil recovery was observed with increasing pH and wettability alteration on the rock surface [87]. Table 4.4 shows the effect of high pH on wettability in various studies.

4.8 Salinity effect on alkaline flooding

The injected fluid and reservoir brine salinity affect the alkaline flooding performance and are considered as important factors in enhanced oil recovery. The formation water and seawater as injectable base fluids contain several monovalent and divalent ions. The most important ions in the formation water and seawater are Na^+, K^+, Ca^{2+}, Mg^{2+}, Cl^-, and HCO_3^-. Water salinity can also affect the oil-water interface properties and change the surface charges and reservoir fluid equilibrium. The ion concentration in the aqueous phase changes its bonds and, consequently, interface changes, according to the change in water and oil interface properties [72, 88, 89].

Suleimanov et al. [90] investigated the impact of water salinity on oil recovery at a sandstone reservoir during alkaline flooding. They observed that divalent ions react with alkaline and produce precipitates, as a result a combination of low salinity water and alkaline solution recover more oil compared to high salinity water and the alkaline solution. On the other hand, removing divalent ions can also raise the oil recovery in high salinity water used in alkaline flooding. Honarvar et al. [20] showed that precipitation increases with increasing divalent ions due to the reaction between Ca^{2+} and OH^-.

$$Ca^{2+} + 2OH^- \rightleftharpoons Ca(OH)_2 \tag{4.22}$$

Table 4.4 The pH effect on wettability [50].

Alkaline	Rock type	Brine	Experimental test	Result	Ref.
NaOH-HCl (0.1–0.4 M)	Bera sandstone	– Freshwater – 5500 ppm KCl – Formation brine	– Contact angle – Field test	– By increasing NaOH, the contact angle reduced from (135–170 degrees) to (40–70 degrees) – Injection of NaOH modifies the wettability of oil-wet – Wet formation to be more water-wet	Leach et al. [87]
NaOH 1.25 N	Berea sandstone	Water	Core flood	– After the caustic flooding, the wettability changed to more water-wet – The oil recovery improvement by NaOH reduces by higher clay content or acoustic consumption	Ehrlich and Wygal [21]
NaOH-Na_2CO_3-$NaBO_2$	Texas cream limestone	Water surfactant	– Dynamic contact angle – CT-aided visualization of in situ imbibition	– $NaBO_2$ demonstrate a higher potential to change the wettability of carbonate rocks compare to other alkalis – $NaBO_2$ has high tolerance to elevated divalent ion concentration of the brines	Zhang et al. [83]

The $Ca(OH)_2$ solubility in the aqueous solution is 0.077 mg/L; therefore, it precipitates after Ca^{2+} reacts with OH^-. In addition, according to the results of Honarvar et al. [20], in carbonate reservoirs, the pH of the solution decreases with increasing divalent ions and also the presence of in situ surfactants reduces the attraction of the oil component on a carbonate rock. As a result, the wettability of carbonate rocks at high salinity and high concentration of divalent ion (Ca^{2+}) will change toward the oil-wet. Li et al. [91] calculated the IFT between oil and the alkaline solution in the presence of monovalent and divalent ions. The results show that the divalent ions increase the IFT while the monovalent ions act as inactive ions [20]. However, the reaction of the alkali and divalent ion and formation of the precipitates may be the target and, as a result, water flow is diverted to less permeability zones, improving sweep efficiency [92, 93]. If this mechanism is the key goal of the alkaline injection, then sodium silicate should be used because the reaction between sodium silicate and calcium or magnesium produces the most insoluble precipitants and reduces permeability the most [19, 93].

Fig. 4.16 illustrates the microimages of emulsions for Zhuangxi-106 heavy oil at different salinities and NaOH solution according to Ge et al.'s [64] results. The results show that the water salinity has a significant effect on the resulting emulsion types. The oil/water emulsion forms at low salinity water while the water/oil emulsion mostly occurs in high salinity. These two types of emulsions can appear at the same time when the salinity is 0.6% [64].

Saha et al. [24] showed that Na^+ ions reduce the IFT between oil and the alkaline solution. The measured IFT changed from 2.4×10^{-1} to 5.84×10^{-1} mN/m with

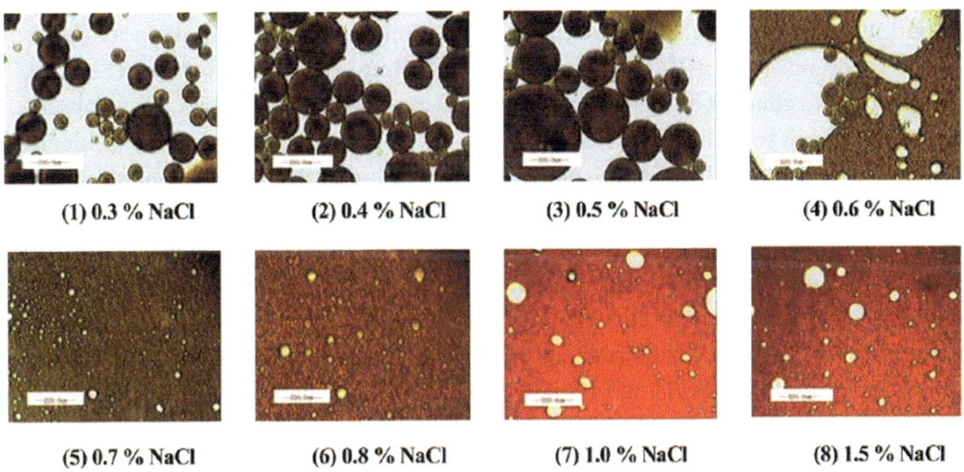

Fig. 4.16 Photomicrographs of heavy oil emulsions and the alkaline solution with different salinity concentrations [64].

decreasing salinity. When the Na^+ ions increase at the interface, the hydrolysis of the acid component also increases and more in situ surfactant will be produced and, as a result, the IFT will reduce [94, 95].

As a general result, the presence of a monovalent cation in the alkaline solution can help the in situ production, while in the active ion cases it prevents the in situ surfactant from accessing the carbonate rock. In addition, divalent ions are recognized as a disturbing factor in alkaline injection due to the precipitation produced.

4.9 Effects of oil composition on alkaline flooding

As mentioned earlier, the alkaline solution is also called an artificial surfactant because it can produce an in situ surfactant in the reservoir by reacting with acidic components of crude oil [69]. Hence, the total acid number (TAN) of crude oil is defined as a crucial index for alkaline flooding in petroleum reservoirs [96].

According to the literature [97–99], the crude oil acid number is a vital factor for alkaline flooding but is not the only factor. In fact, the crude oils with high acid number respond very well to alkaline flooding, whereas there are few successful cases reported for crude oil containing a low acid number. The reaction of alkali compounds with acidic components of crude oil leads to the reduction of interfacial tension between water and crude oil, change in rock wettability and creating oil-water emulsion, which all can improve oil recovery. Cooke et al. [23] found that alkaline water is capable of recovering the residual oil left behind by conventional water flooding when having acidic oils. Jennings et al. [30] showed that caustic solutions (~0.05–0.50 wt% NaOH) could move residual oil from an oil-wet core by reducing the IFT low enough. The success of this mechanism largely depends on the existence of enough surface active organic acid concentration in crude oil, to lower the IFT, improve oil recovery, and to decrease the produced water-oil ratio (WOR) during the flooding process. Hornof et al. [100] investigated the effect of linoleic acid in crude oil as shown in Fig. 4.17A. They observed that when there is some linoleic acid in paraffin oil, the oil recovery reduced considerably. As mentioned earlier, the IFT can decrease only at high alkaline pH values when the acidic components are present in crude oil [101]. It has been found that the alkaline flooding process can act effectively on heavy crude oils due to their higher acidic components content [68, 102]. A crude oil would be considered as an acid when the total acid number (TAN) of the sample exceeds 0.5 mg KOH/g [103]. Dong et al. [104] found that with a TAN of 0.19 mg KOH/g, the crude oil IFT and brine can be dropped to below 0.1 mN/m by adding NaOH and Na_2CO_3 to the brine. This will change the interfacial tension and causes some interfacial instability.

The crude oil IFTs at different pH were linked to the asphaltene and resin content of the crude since these polar species exhibit more amphiphilic behavior when they are charged. Lashkarbolooki and Ayatollahi [46] measured the IFT of three light, medium,

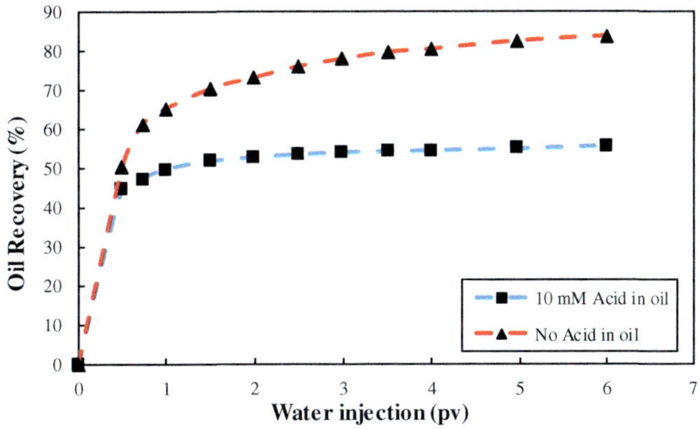

Fig. 4.17 Effect of addition of linoleic acid on the recovery of paraffin oil by water flooding [100].

Table 4.5 Extracted results of alkali flooding in oil recovery with different scenarios [105].

Flooding scenarios	**Concentration of alkali**	**Oil recovery**		
		NaOH	**$NaHCO_3$**	**NaOH + 5 oleic acid number in oil phase**
Continuous flood	0.05	0.676	0.666	0.700
	0.20	0.698	0.686	0.733
	0.50	0.716	0.709	–
Intermittent flood	0.05	0.792	0.782	0.760
	0.20	0.820	0.812	0.720
	0.50	0.690	0.696	–

and heavy crude oils at different pH and water chemistry conditions. They found that the IFT trend of asphaltene and resin fractions at different pH is similar to that found for crude oil except for the light crude oil with an ultralow asphaltene content.

Bortolotti et al. [105] studied the effect of irregular alkali flooding in oil-wet fractured carbonate reservoirs to assess the oil recovery effectiveness. Based on the laboratory results, they found that the 1-week alternating alkali flooding process has shown about 10% greater oil sweep efficiency compared to the conventional continuous EOR. Table 4.5 was tabulated according to the outcomes of the NaOH, $NaHCO_3$, and NaOH +5.0 acid number of oleic acid flooding in oil recovery. Table 4.5 shows that NaOH flooding displayed considerably high oil recovery compared to $NaHCO_3$ (weaker alkali) flooding, especially in an alternating flooding scenario. Furthermore, increasing acid concentration in the oil phase can produce soap in place precipitation and formation even in low alkali concentrations. Thus, more limitations should be considered when there is a high acid concentration in crude oil reservoirs.

Sheng [106] was curious to understand which fraction of acid components could react with the alkaline solution to generate soap, so he utilized a numerical simulation study. He concluded that only a fraction of acid components could be affected in soap generation. In addition, he found that the acid number does not have a specific effect on incremental oil recovery for three reasons: (1) different approach to determine the acid number; (2) more acid content might not be able to produce soap; and (3) the real circumstances might not have the proper salinity to produce soap. However, Ge et al. [64] believed that heavy oil with high acid numbers could react remarkably with the alkaline solution in order to generate an in situ surfactant, which leads to improved oil recovery. They carried out a number of experiments to understand the influence of crude acid number, brine salinity, type and concentration of alkaline, and other parameters of the sand pack. These parameters and their impacts are listed in Table 4.6. In other words, oil acid number is a prominent factor in alkali flooding, as a high organic acid number in heavy oil can greatly and quickly reduce the IFT as a result of formation of an in situ surfactant on contact with alkali agents. Afterward, the interfacial event causes water droplets creation in the oil phase, which leads to increasing the displacement front viscosity, consequently decreasing viscous fingering, diminishing the channeling of water, and enhancing the oil recovery of heavy oil. Fig. 4.18 illustrates the effect of acid number on oil recovery. Additionally, alkali flooding can intensify emulsion stability

Table 4.6 Summary of sand pack flooding tests [64].

Run no.	Heavy oil	Initial oil saturation (%)	Chemical formula	Oil recovery (%)		
				Water flooding	Alkaline flooding	Total
1	Zhuangxi 106	83.50	NaOH (0.5%) +NaCl (0.5%)	38.60	12.40	51.00
2	Chenzhuang	85.20	NaOH (0.5%) +NaCl (0.5%)	33.70	12.90	44.60
3	Binnan	83.70	NaOH (0.5%) +NaCl (0.5%)	39.00	16.90	55.90
4	Xia-8	82.70	NaOH (0.5%) +NaCl (0.5%)	37.60	20.40	58.00
5	Zhuangxi 106	85.30	NaOH (0.5%) +NaCl (1.0%)	37.50	11.90	49.40
6	Zhuangxi 106	84.60	NaOH (0.5%) +NaCl (1.25%)	38.20	13.30	51.50
7	Zhuangxi 106	86.50	NaOH (0.5%) +NaCl (1.5%)	38.4	11.8	50.20
8	Zhuangxi 106	86.40	NaOH (0.5%) +NaCl (2.0%)	36.90	10.20	47.11

Table 4.6 Summary of sand pack flooding tests—cont'd

Run no.	Heavy oil	Initial oil saturation (%)	Chemical formula	Oil recovery (%)		
				Water flooding	Alkaline flooding	Total
9	Chenzhuang	82.50	NaOH (0.5%) + NaCl (1.0%)	37.60	14.40	52.00
10	Chenzhuang	84.70	NaOH (0.5%) + NaCl (1.25%)	37.80	13.80	51.60
11	Chenzhuang	83.30	NaOH (0.5%) + NaCl (1.5%)	38.10	16.50	54.60
12	Chenzhuang	87.10	NaOH (0.5%) + NaCl (2.0%)	38.30	15.40	53.80
13	Xia-8	83.40	NaOH (0.5%) + NaCl (0.5%)	34.70	20.10	54.80
14	Xia-8	85.90	NaOH (0.5%) + NaCl (0.5%)	43.30	8.90	52.20
15	Xia-8	82.90	NaOH (0.5%) + NaCl (0.5%)	37.80	16.10	53.90
16	Xia-8	87.10	NaOH (0.5%) + NaCl (0.5%)	42.90	3.90	46.80
17	Zhuangxi 106	85.00	NaOH (0.25%) + NaCl (0.5%)	39.80	7.20	47.00
18	Zhuangxi 106	87.50	NaOH (1.0%) + NaCl (0.5%)	38.70	15.60	54.30
19	Zhuangxi 106	84.50	NaOH (1.25%) + NaCl (0.5%)	37.60	16.20	53.80
20	Zhuangxi 106	86.10	NaOH (1.5%) + NaCl (0.5%)	38.50	15.40	53.90
21	Zhuangxi 106	85.00	Na_2CO_3 (0.5%) + NaCl (0.5%)	39.80	5.40	45.10
22	Zhuangxi 106	83.90	Na_2CO_3 (0.75%) + NaCl (0.5%)	37.20	7.60	46.80
23	Zhuangxi 106	84.10	Na_2CO_3 (1.0%) + NaCl (0.5%)	38.40	13.20	51.60
24	Zhuangxi 106	86.10	Na_2CO_3 (1.5%) + NaCl (0.5%)	39.20	12.60	51.80
25	Zhuangxi 106	85.40	Na_2CO_3 (2.0%) + NaCl (0.5%)	37.90	12.90	50.80

because of the reaction with acidic components in crude oil to generate an in situ surfactant. The polarity of S^-, N^-, and O^- complex molecules in crude oil (interface activity) increases by adding alkali to the solution. In other words, a higher solution pH causes more reaction of crude oil acidic components with alkali agents that can improve interface activity [107].

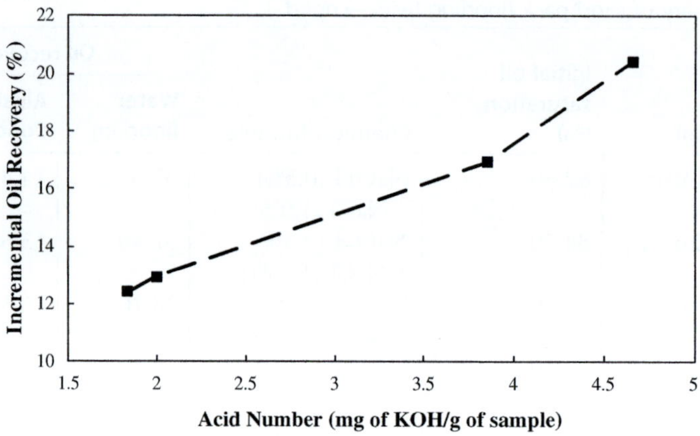

Fig. 4.18 Effect of crude oil acid number on oil recovery [64].

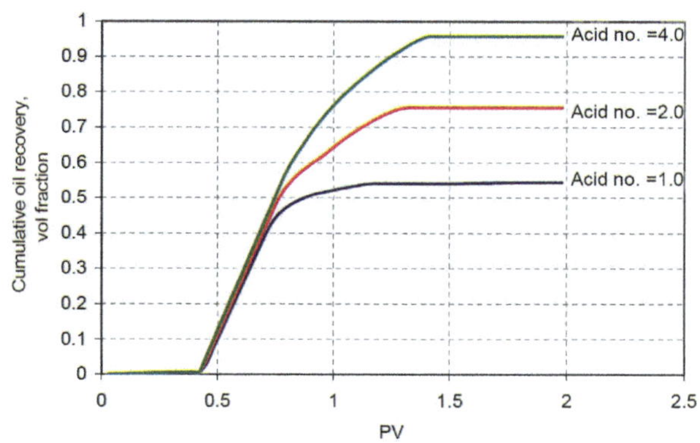

Fig. 4.19 Impact of acid number on crude oil recovery [108].

Mohammadi [108] investigated the effects of different ranges of acid numbers in crude oil systems. The results are summarized in Fig. 4.19, which shows that increasing the acid number to 2 and 4 can increase oil recovery to 75% and 95% OOIP, respectively. This is because more soap is formed as a result of the higher acid number. In addition, certain types of crude oil with acidic components are favorable to generate soap when interacting with alkali agents, so a high acid number of oil is a vital factor for alkaline flooding. Hence, light oils are not preferable cases for alkaline flooding due to the low acid numbers, which leads to less in situ soap formation. Recently, Arekhov et al. [109] studied the impact of high and low TAN oils on the IFT and recovery factor. They provided a schematic chart (Fig. 4.20) which explains the IFT variation for high TAN and low TAN oils. Fig. 4.20 illustrates the addition of alkali to high TAN oil, which causes considerable reduction in the IFT and reaches a low equilibrium value. Furthermore, they utilized these low and high TAN oils to inject into nonaged core plugs.

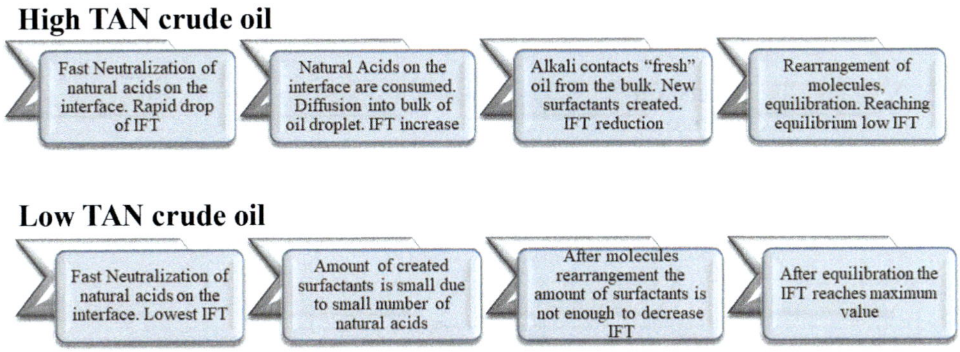

Fig. 4.20 Flowchart illustrating the IFT alteration process for high and low TAN crude oils [109].

4.10 Ternary diagram in alkaline flooding

A ternary phase diagram displays the possible phases and their equilibrium depending on the composition of a mixture of three components at constant temperature and pressure. Fig. 4.21 illustrates a ternary phase diagram including a single phase, two phases, and three phases that are separated by tie lines [110].

Chen et al. [111] displayed the effect of Alkali/Surfactant/Polymer (ASP) solution viscosity in a ternary phase diagram as illustrated in Fig. 4.22. According to this plot, the viscosity of the solution increases by gradually adding in the polymer. Furthermore, the alkali plays as a restrictive agent and limits the contribution of polymer solution to viscosity enhancing. This is because the addition of alkali agents reduces the cross-linking probability of polymer chains with decrease in molecular hydraulic radius.

Winsor [112] was one of the first researchers who discovered phase behavior, especially for microemulsion phases. He introduced three types of emulsion formation: solubilized oil in water (type I), solubilized water in oil (type II), and a three-phase system containing the stabilized microemulsion phase (type III). Type II has been also classified into two subtypes, II(+) and II(−). The plus and minus indicate the slope of tie-lines in the ternary diagram. The phase composition calculation and their effects of salinity on phase behavior for all three microemulsion cases are shown in Fig. 4.23. de Castro Dantas et al. [113] investigated the impact of acid microemulsion flooding of carbonate reservoirs and proposed a pseudo-ternary phase diagram using *n*-bitanol/alkconat L90, acid solution, and kerosene of the microemulsion system. Their proposed phase equilibrium was matched with the Winsor phase behavior classification. According to Fig. 4.24, acid concentration on the aqueous phase did not have a great effect on the constructed system. Thus, they announced a favorable composition with a low surfactant, low oil concentration by varying only the acid concentration for the recovery process. Consequently, they used four systems only with different acid concentrations of 1.5%, 5%, 10%, and 15% HCl to evaluate the recovery factor. Accordingly, they understood that a higher acid

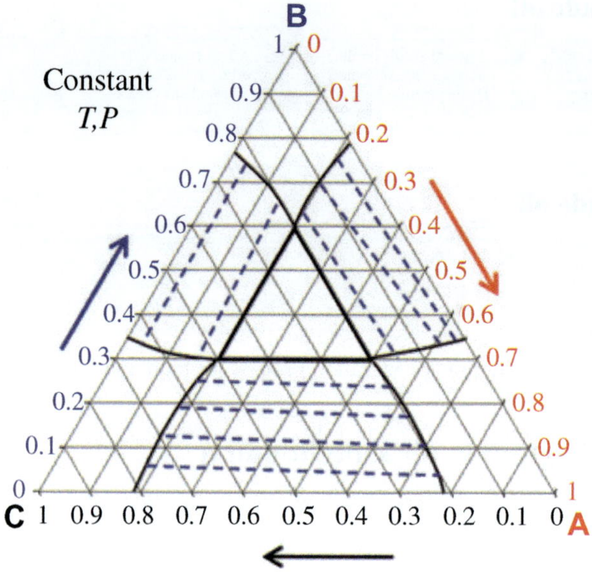

Fig. 4.21 A schematic ternary phase diagram for a mixture of A, B, and C components. Phase boundaries are shown with *black lines*. Regions within the *black lines* display similar types of phase behavior. The composition of phases in equilibrium is shown by *dashed blue lines*. Axis A of the ternary diagram denotes a composition that is quantified in mole or weight fractions. The point (0.3, 0.4, 0.3) is at the middle of the small triangle at the center of the diagram and is located by following the *red* diagonal 60-degree line at *red* 0.3 and the horizontal line at *blue* 0.4 or any combination of two of the coordinates (A, B, C) [110].

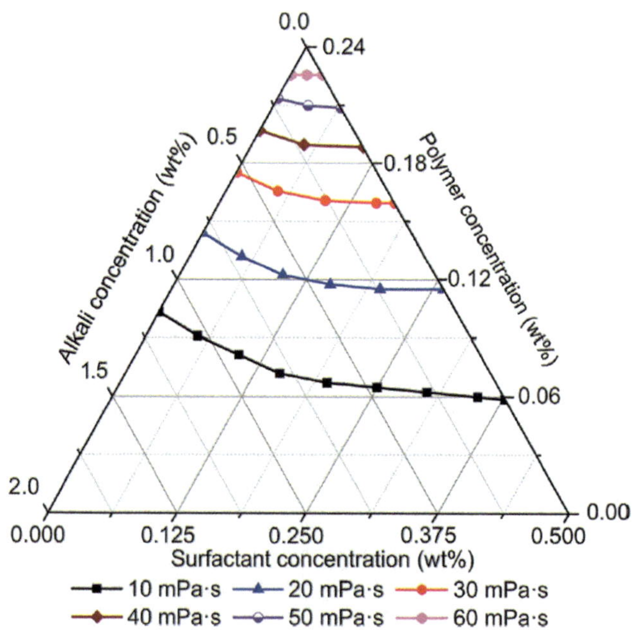

Fig. 4.22 Ternary phase diagram ASP solution viscosity [111].

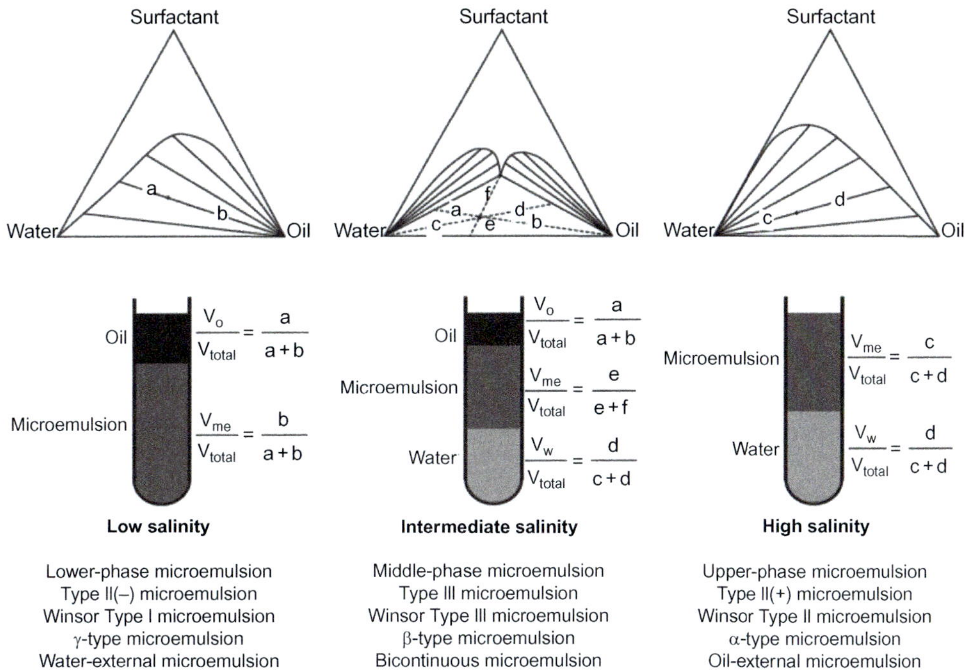

Fig. 4.23 Three major types of microemulsion and the impact of salinity on phase behavior [11].

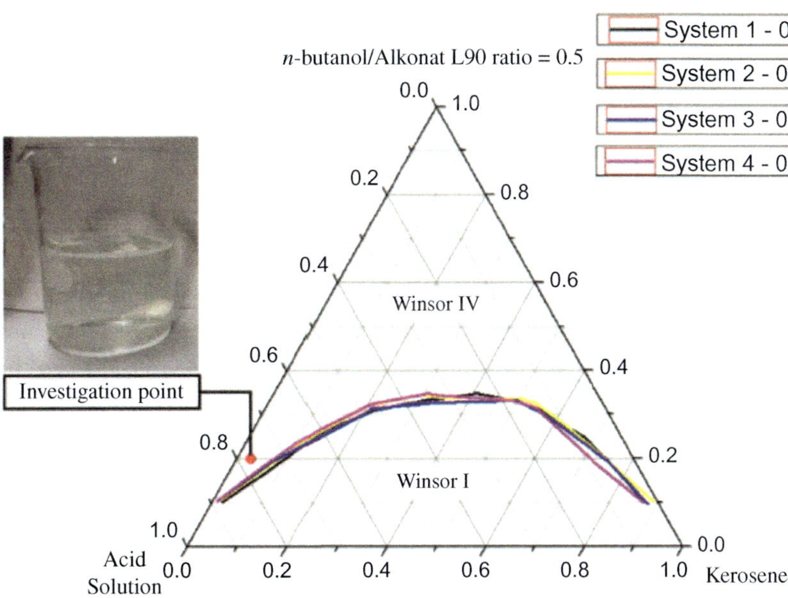

Fig. 4.24 Pseudo-ternary diagram of a microemulsion system with various HCl concentrations [113].

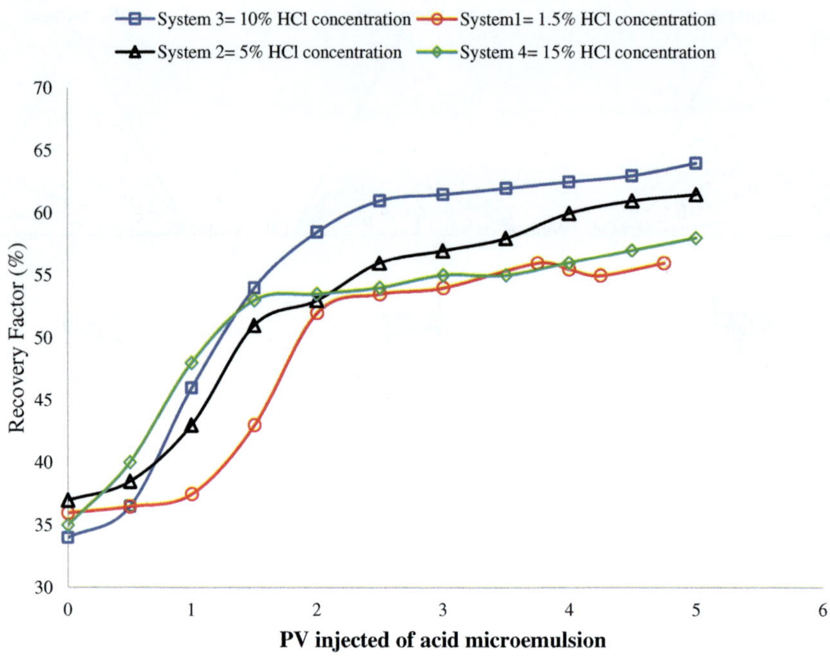

Fig. 4.25 Recovery factor of different systems with different PV injected of acid microemulsion [113].

concentration could dissolve carbonate formation, which leads to simultaneously increasing the permeability of the low permeable zones and connecting isolated zones, ultimately improving the oil recovery (Fig. 4.25).

4.11 Success rate and screening criteria

Alkaline flooding is a complex process and its field success depends on several variables including oil type, reservoir heterogeneity, brine chemistry, and some operational procedures. Historically, a large number of enhanced oil recovery projects have failed to achieve their targets. The common reason was the poor oil sweep efficiency due to the lithological heterogeneity of the reservoir and misunderstanding of the reservoir properties such as oil saturation. In many cases, the injected flooding fluids, for example, alkaline fluids, follow a certain path of more permeable zones and fail to sweep the oil from the main reservoir, which results in poor oil recovery and loss of the flooding agent [11].

In order to design a successful enhanced recovery and considering that we cannot take all of the reservoir variables and field operation aspects into account, there is a great need for laboratory measurements to understand the impact of alkaline flooding on a particular reservoir. Following these lab tests, the process will be tested on a pilot scale. It is extremely important to plan and execute a detailed monitoring program and testing during the pilot test. These data will help to design the flooding process more accurately and avoid failure [114]. Table 4.7 shows the oil and reservoir condition for the alkaline flooding process.

The acid number of crude oil is one of the most essential factors in alkali flooding. As mentioned before, this variable characterizes the quantity of natural soap that can be produced by the aqueous alkali agent. The amount of potassium hydroxide (KOH in milligrams) necessary to neutralize 1 g of crude oil is known as the acid number [12]. By utilizing an in situ surfactant produced by the interaction of alkalis with natural organic acids in oil, alkaline flooding enhances crude oil recovery. Many researches show that the crude oil carboxylic acids and sometimes phenolics and porphyrins are responsible for saponification and act as surfactants [115, 116].

4.12 Displacement efficiency in alkaline process

The alkaline solution is injected into the porous media and leads to improved oil recovery by a two-stage process:

First stage: the residual oil is mobilized by configurational changes like emulsification and wettability alteration. The organic acids in the residual oil reacted with the alkaline solution and produced the in situ surfactant. The in situ surfactant adsorbs on the oil-water interface to lower the interfacial tension and in some cases causes spontaneous emulsification. In addition, the in situ surfactant absorbed on the rock surface and change the wettability and as a result, change the configuration of the remaining oil [28].

Second stage: the macroscopic production characteristics of the mobilized oil phase is modified. In other words, displacement efficiency through mobility control is improved by a decrease in the water mobility and the overall recovery efficiency may be increased [28].

Fig. 4.26 presents the residual oil versus in situ surfactant concentration in an alkaline solution. The change in residual oil versus in situ surfactant concentration is expressed using the deZabala equation (4.12) as follows:

$$S_{\text{or}} = a_1 + a_2 \exp\left(\frac{-C_{\text{A}}}{a_3}\right) \tag{4.23}$$

where C_{A} is the in situ surfactant versus mol/dm^3, a_1, a_2, and a_3 are constants and equal to 0.05, 0.35, and 1.18×10^{-3} mol/dm^3, respectively. It is assumed that the system for chemical displacement modeling is a linear, homogeneous, porous medium of uniform initial saturation and composition and constant temperature; immiscible, mobile, and incompressible fluid phases of oil and water [12]. The system is also assumed to have a

Table 4.7 Oil and reservoir condition for alkaline flooding process.

Crude oil properties				Reservoir rock characteristics				
Gravity (API)	Viscosity (cP)	Oil composition	Oil saturation	Formation lithology type	Net thickness (ft)	Average reservoir permeability (mD)	Reservoir depth (ft)	Reservoir temp (°F)
13–35	<200	Some Organic acids	Above water flood residual	Sandstone preferred	N.C.	>20	<9000	<200

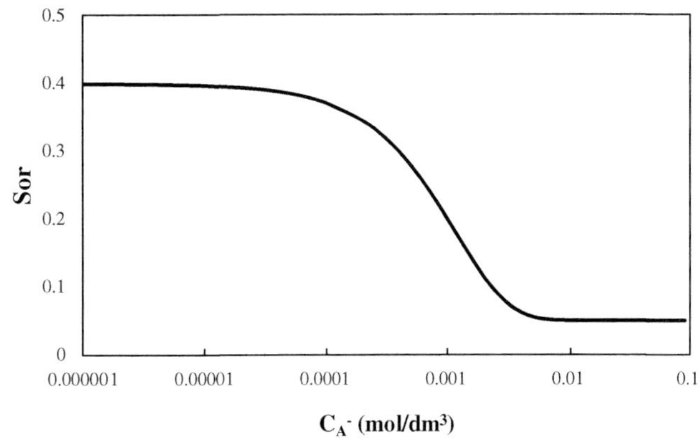

Fig. 4.26 The effect of in situ surfactant concentration on residual oil [12].

stable displacement front with no viscous fingering and a local and instantaneous thermodynamic equilibrium; negligible dispersion and capillary-pressure forces [12].

Residual oil displacement with an alkaline solution was investigated by Donaldson and Crocker [117] where they used 1000-ppm solutions of sodium hydroxide, a nonionic surfactant, and a mixture of the alkaline and surfactant solutions. The results of the sand packed test showed that only individual alkaline or surfactant injection improved the oil recovery by 1%–3% while the alkaline and surfactant mixture removed virtually all of the residual oil. Based on several published literature, Mungan [114] has concluded that the minimum IFT can be obtained with an NaOH concentration range of 0.05%–0.2% with an equivalent pH value of 12.5. the capillary number is about 10^{-6} for conventional water floods, which must be increased to at least 10^{-4}. Table 4.8 shows the capillary number of alkaline floods as provided by Castor et al. [28]. Therefore, alkaline flooding can provide the critical capillary number required for enhancing oil recovery.

4.13 Combined flooding processes

One of the most important factors in alkaline-surfactant flooding is achieving a low IFT [118, 119], which can be accomplished using various alkaline chemicals and dilute

Table 4.8 The capillary number for different alkaline flood [28].

Alkaline	Capillary number
Distilled water	7.5×10^{-6}
NaOH	1.3×10^{-3}
NaOH/NaCl	1.2×10^{-4}
$Na_2B_4O_7$/NaCl	7.2×10^{-5}
$Na_2B_4O_7$	4.2×10^{-4}

surfactant techniques [120]. Adding a small volume of surfactant to the alkaline water can reduce the interfacial tension more than can either the surfactant or alkali agents alone [121].

Nelson et al. [122] first showed that the in situ soap, as an anionic surfactant, has its own optimum salinity, usually different from reservoir salinity, and they introduced it as "Cosurfactant Enhanced Alkaline Flooding." A proper blend of the alkali agent and synthetic surfactant can create a more effective process. This combined process later developed into an alkaline-surfactant-polymer (ASP) process, which in recent years by merging the advantages of the alkali, surfactant, and polymer is considered as the most effective chemical flooding technique in enhanced oil recovery. Several successful field ASP enhanced recovery tests in the United States, China, and India have proved its higher efficiency [123–130].

Most researchers conclude that there are several important factors impacting the alkaline-surfactant flooding method including IFT reduction at low surfactant concentration, wettability alteration, low adsorption of surfactant by alkali, and mobility control [68]. However, the mechanisms involved with alkaline-surfactant flooding are still not entirely understood.

One of the major controlling factors of IFT decrease of the alkaline-surfactant-oil system is the amount of organic acid in the oil [131]. Rudin and Wasan [131] found that adding an alkali to the surfactant flooding solution will increase the interfacial tension, while adding an alkali to reservoirs with intermediate to high acid concentrations can result in an ultralow interfacial tension.

It has also been found that the alkaline-surfactant systems can achieve a highly low IFT, which is the main reason for its successful enhanced oil recovery, and the presence of an alkali can reduce the consumption of the surfactant through reducing the surfactant adsorption [77, 118, 132–139].

4.14 Simulation and modeling

4.14.1 Simulation

Enhanced oil recovery numerical simulation has an important role in recovery evaluation. Simulation can save cost, time, correct the process to maximize recovery, and minimize the problems and complexities before the actual process executed in the field. Due to these facts, most operators consider numerical modeling as the first necessary stage of EOR processes. The established model can provide the capacity to evaluate different variables in alkaline flooding.

One of the first numerical models was developed by Pope and Nelson [140]. They created a 1D numerical simulator for surfactant oil recovery. This simple simulator has evolved into a 3D simulator that is capable of modeling various chemical flooding [8, 141–143].

Arhuoma et al. [144] investigated the numerical simulation of alkaline flooding to improve heavy oil recovery. NaOH was used as an alkali agent with two different concentrations for comparison purposes. In these simulations, the development of in situ W/O emulsion and reduction in the IFT were concluded because of the reaction between the alkaline flooding solution and reservoir acidic crude oil. In addition, they considered alkali, water, oil, and W/O emulsion as four principal components during the simulation process. The results of this simulation, cumulative oil production, and pressure drops for alkaline flooding at 0.2% and 0.8% NaOH concentration are depicted in Fig. 4.27. It would be obvious that absolute matching occurred between measured and simulated data. Consequently, resistance to water flow and formation of in situ W/O emulsion leads to an increasing pressure drop, while a higher NaOH concentration causes an increase in cumulative oil production.

Wang et al. [8] specifically studied increasing heavy oil recovery using alkali agents in experimental and simulation conditions. Fig. 4.28 represents experimental and simulation

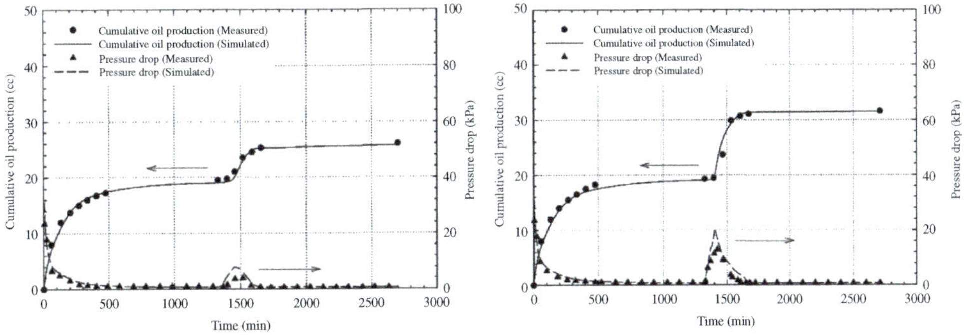

Fig. 4.27 Simulated and measured cumulative oil production and pressure drop at 0.1% (left) and 0.8% NaOH (right) concentration [144].

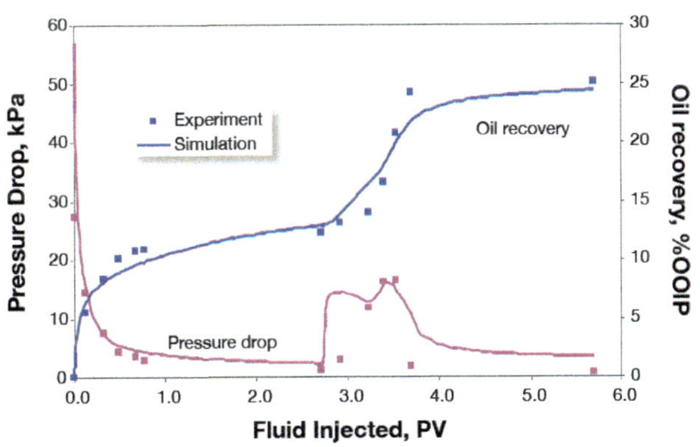

Fig. 4.28 Experimental and simulation of cumulative oil recovery and pressure drop as a function of injected solution (0.3 wt% NaOH + 0.3 wt% Na_2CO_3) [8].

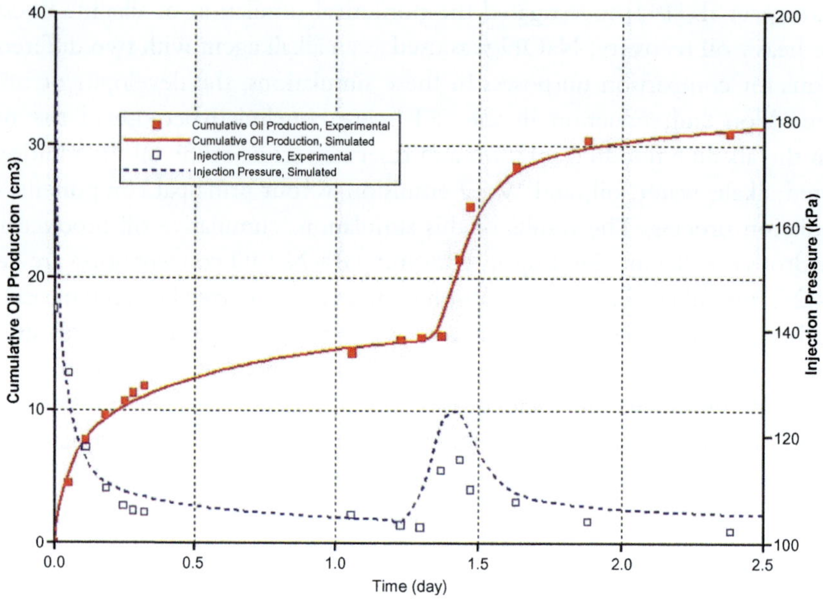

Fig. 4.29 Experimental and simulation of cumulative oil recovery and pressure drop as a function of alkaline/surfactant injected fluid [145].

of pressure drop and cumulative oil recovery as a function of injected fluid (0.3 wt% NaOH + 0.3 wt% Na_2CO_3). Thus, improving cumulative oil recovery could be observed in order to swiftly rise in pressure drop. This is due to the generation of W/O emulsion, which leads to the increasing pressure drop during alkaline injection and, as a result, reducing water mobility. This will conclude a higher cumulative oil recovery. In addition, Wang and Dong [145] changed the alkaline solution to an alkaline/surfactant solution as an injected fluid. While Fig. 4.29 depicts the exact trend of Fig. 4.28, both the pressure drop and cumulative oil recovery illustrated in Fig. 4.28 are higher. The reason behind this is when there is a surfactant in solution, some oil can enter the continuous water phase and move out in the form of O/W emulsion, which can be trapped by smaller pore throats, so this emulsion can dilute the mobility of the water phase to increase oil recovery due to diversion of the water phase toward unswept zones.

4.14.2 Modeling

There are many commercial reservoir simulators commonly used in the industry, such as CMG STARS (Computer Modeling Group Ltd), ECLIPSE (Schlumberger), VIP and Nexus (Halliburton), and UTCHEM (University of Texas). There are also several simulators which have been developed by major oil companies as an in-house software such as CHEARS by Chevron and Empower by ExxonMobil.

Most of the EOR projects are a combination of alkaline, surfactant, and polymer injection. Alkaline flooding simulation is generally difficult as it involves complex chemical reactions with the formation and other fluids. Sheng [11] lists the major chemical reactions which are generally subject to modeling:

- Reaction between alkali agent and crude oil acidic components and in situ soap generation
- Reaction between injected alkali and formation water (precipitation) and formation minerals (ion exchange and dissolution)
- Impacts of products of these chemical reactions on transport phenomena
- Interactions of alkali agent with surfactant and polymer if they are involved

Generally, the modeling process starts with a program (e.g., EQBATCH) to estimate and evaluate the initial state (equilibrium state) of the reservoir based on the formation and brine composition, crude oil acid number, and water and oil saturations [11]. These equilibrium data will be used by the alkaline model (e.g., UTCHEM) for further EOR processing.

The EQBATCH can also determine the injected solution and formation brine compatibility, equilibrium composition, and the subsequently injected fluid pH once the different electrolytes are added [11].

Bhuyan developed and validated the geochemical model for several cases [146, 147], while some researches [147, 148] utilized it for modeling the ASP process, which is called UTCHEM. Later, this model was promoted by adding fluid species, reactions, and any number of elements [149, 150]. However, they neglected the effect of soap on the phase behavior in their simulation. Furthermore, Mohammadi et al. [151] considered the influence of soap in phase behavior in their experiments and then modeled it using Hand's equation [152] in UTCHEM, as we can see cumulative oil production, oil cut, pressure drop, and pH as a function of pore volumes in Figs. 4.30–4.32. They concluded and observed a significant coordination between measured and modeled (UTCHEM-EQBATC) data.

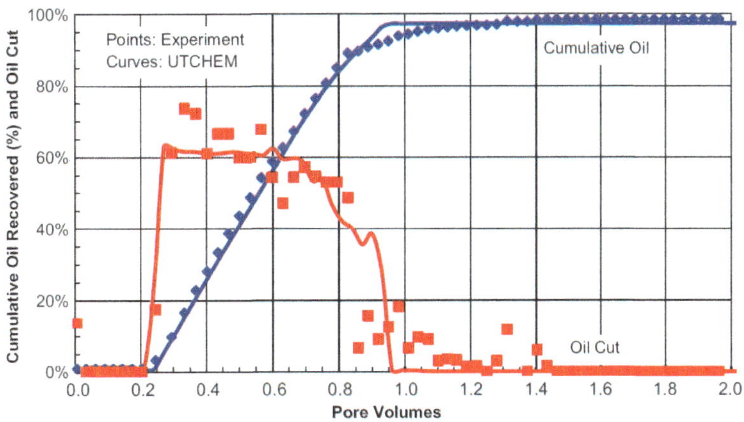

Fig. 4.30 The simulated and measured oil recovery and oil cut comparison for ASP flooding [151].

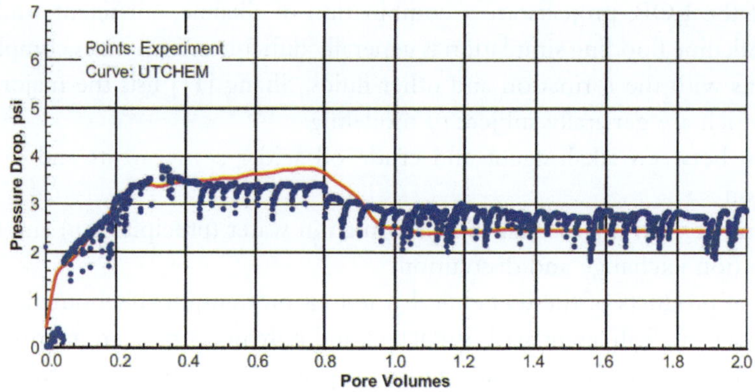

Fig. 4.31 The simulated and experimental pressure drops comparison [151].

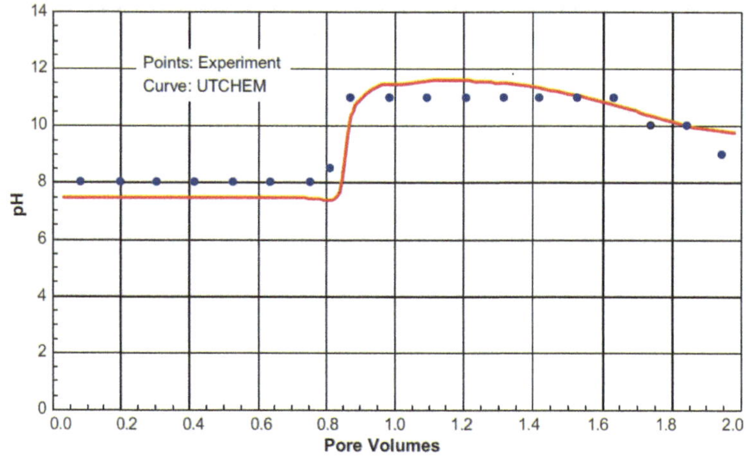

Fig. 4.32 Experimental and modeled values of effluent pH as a function of pore volumes [151].

Korrani et al. [153] used the UTCHEM-IPhreeqc model in their works. IPhreeqc, which uses geometry calculation of Charlton and Parkhurst [154], coupled with the UTCHEM reaction related process. They expressed the UTCHEM algorithm after coupling with IPhreeqc in Fig. 4.33. They used Mohammadi et al.'s [151] experimental data to simulate it using the UTCHEM-IPhreeqc model and compared it with the Mohammadi (UTCHEM-EQBATC) model. Figs. 4.34 and 4.35 display the UTCHEM-IPhreeqc simulation and its comparison to the UTCHEM-EQBATC model. According to these figures, it is obvious that the UTCHEM-IPhreeqc model has a more comparable match than the UTCHEM-EQBATC model for both pressure drops and cumulative oil recovery. This was because the UTCHEM-IPhreeqc model totally overcomes the specific assumptions and restrictions intrinsic in the UTCHEM-EQBATC model.

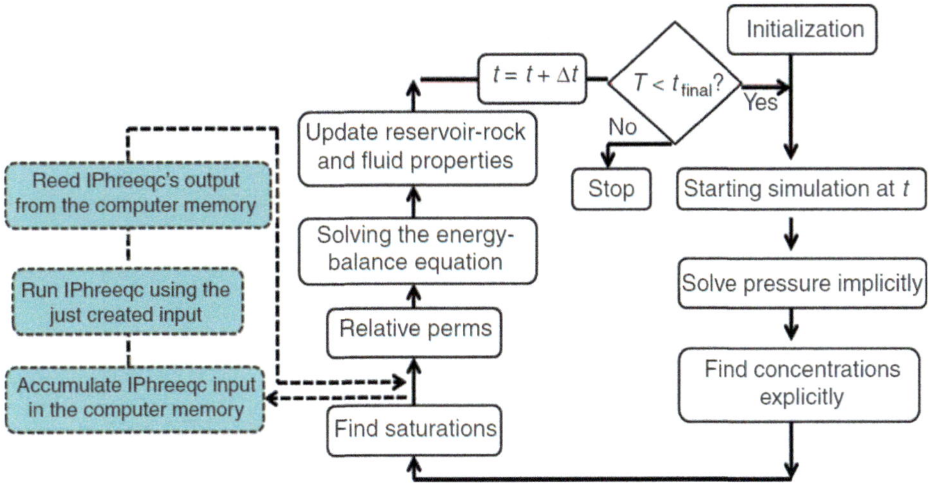

Fig. 4.33 Simplified UTCHEM flowchart after coupling with IPhreeqc [153].

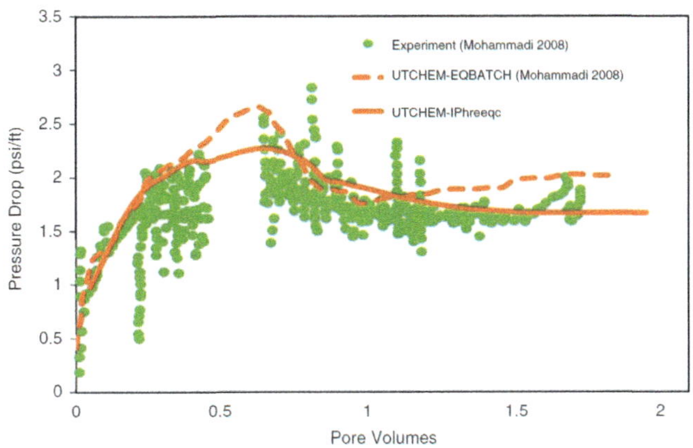

Fig. 4.34 The results of simulation using UTCHEM-IPhreeqc and UTCHEM-EQBATC against pressure drops measured data [153].

Dang et al. [155] proposed the ASP model using Nghiem et al.'s [156] reservoir geometry model. They considered polymer properties, alkaline consumption, microemulsion phase behavior, in situ soap, and aqueous and mineral reactions in the ASP model. The Korrani and Mohammadi models [154, 157] used the explicit modeling of Type III, while the pseudo two-phase approach (without the explicit modeling of Type III) was utilized by Dang et al. [157]. Accordingly, the ASP model has shown considerable matching among the new models with experimental data, UTCHEM-EQBATC, and UTCHEM-IPhreeqc for cumulative oil recovery and pressure drop purposes even when there is no third liquid phase according to Figs. 4.36 and 4.37.

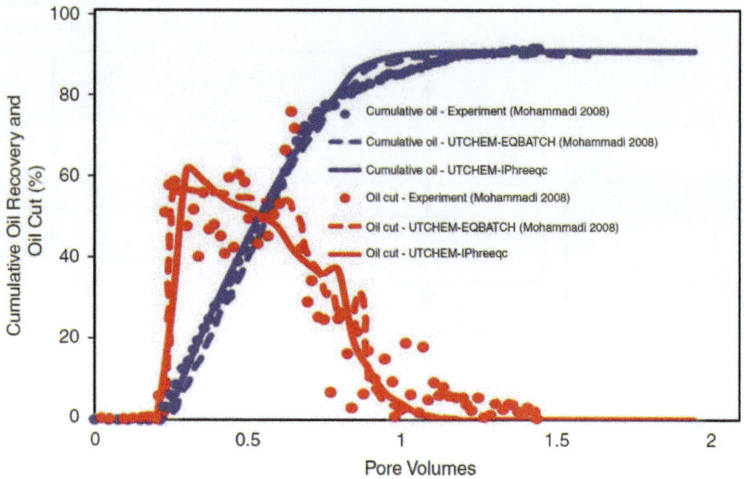

Fig. 4.35 The plot of simulation results with UTCHEM-IPhreeqc and UTCHEM-EQBATC against cumulative oil recovery and oil cut recorded data [153].

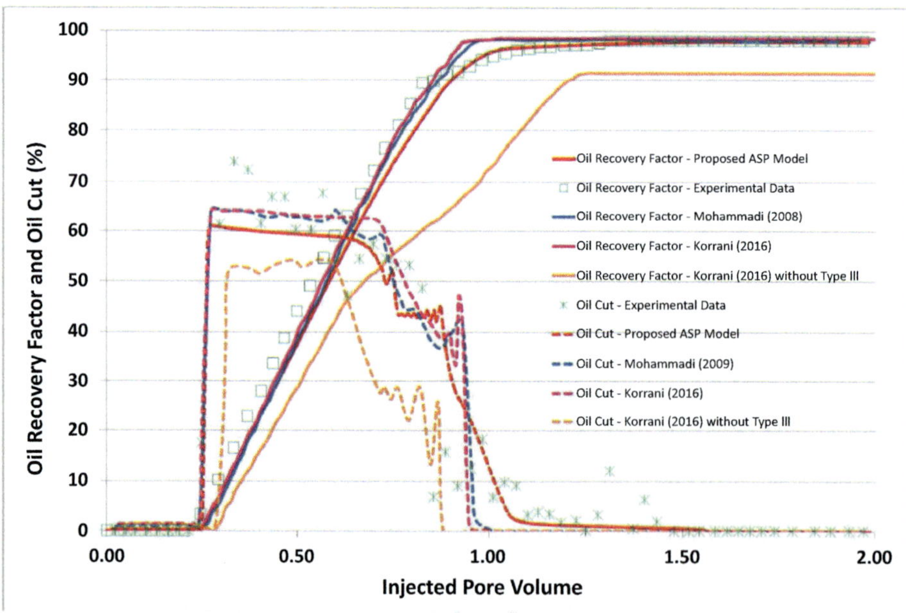

Fig. 4.36 Measured experimental and simulated data of oil recovery factor and oil cut plot with acidic crude oil [157].

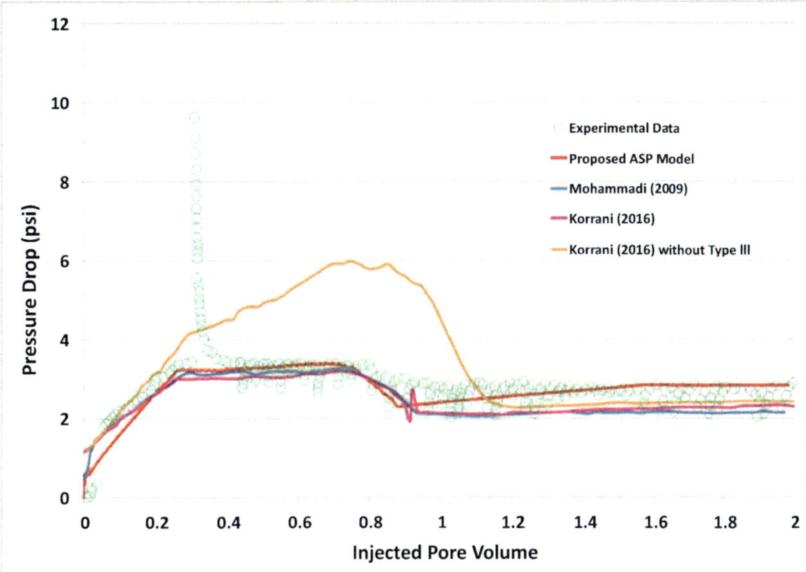

Fig. 4.37 Experimental and simulated pressure drops cut with acidic crude oil [157].

4.14.3 Mathematical formulation of chemical reactions and equilibrium state

Bhuyan [146] described variables identification in the process including "N" elements from formation and water composition following the definition of "K" solid species, "J" fluid species, "I" matrix adsorbed cations, and "M" micelle-associated cations all made up of N elements. To determine the unknown $(K+J+I+M)$ equilibrium concentrations, four equations $(K+J+I+M)$ are required. These independent equations are as follows:

Mass balances of elements, providing N equations of the following form:

$$C_n^t = \sum_{j=1}^{J} h_{nj} C_j + \sum_{k=1}^{K} g_{nk} C_{k(s)} + \sum_{i=1}^{I} f_{ni} \overline{C}_i + \sum_{m=1}^{m} e_{nm} \overline{\overline{C}}_m \quad \text{for} \quad n = 1, \ldots, N \tag{4.24}$$

where C_n^t is the total concentration of component n; C_j is the concentration of fluid species j; $C_{k(s)}$ is the concentration of solid species k; C_i is the concentration of matrix adsorbed cation i, and C_m is the concentration of micelle-associated cation m, all in mol/L water. The coefficient before each of these concentrations is the stoichiometric coefficient.

From the chemical species of J fluid, select N-independent chemicals randomly such that the concentrations of the remaining $(J-N)$ fluid species can be expressed as the concentrations of these N-independent chemical species using equilibrium relationships of the following equation:

$$C_r = K_r^{eq} \prod_{j=1}^{N} C_j^{W_{rj}} \quad \text{for} \quad r = (N+1), \ldots, J \tag{4.25}$$

For example, $C_{H_2CO_3} = K^{eq}_{H_2CO_3} C^2_H C_{CO_3}$. In Eq. (4.25) and the following equations, the W variable represents the exponent of concentration. There is a solubility product constraint for each solid which is defined as follows:

$$K_k^{sp} \geq \prod_{j=1}^{N} C_j^{W_{rj}} \quad \text{for} \quad k = 1, \ldots, K \tag{4.26}$$

The solubility product constants, K_k^{sp}, are explained based on the concentrations C_j of the independent chemical species. In the case of a solid absence, the corresponding solubility product constraint is the inequality and if the solid is present, the constraint is the equality. For example, $K_{CaCO_3}{}^{sp} \geq [Ca_3^{2-}][CO_3^{2-}]$ for the presence of solid $CaCO_3$.

For adsorbed cations on the matrix (I), there are ($I-1$) independent exchange equilibrium relationships as stated in Eq. (4.27):

$$K_r^{ex} = \prod_{j=1}^{N} C_j^{W_{pj}} \prod_{i=1}^{I} \overline{C}_i^{W_{pi}} \quad \text{for} \quad p = 1, \ldots, (I-1) \tag{4.27}$$

where C_j is the concentration in fluid and C_i is the concentration of adsorbed cations on the matrix in mol/L. W_{pj} and w_{pi} are the exponents for the concentrations C_j and C_i, respectively. As shown in the following example, these exponents will be negative if the species is on the left side, and positive if the species is on the right side of the equilibrium equation.

$$Ca^{2+} + 2\overline{Na^+} \xleftrightarrow{K^{ex}_{Ca-Na}} \overline{Ca^{2+}} + 2Na^+ \tag{4.28}$$

Then

$$K^{ex}_{Ca-Na} = \frac{\left[\overline{Ca^{2+}}\right][Na^+]^2}{\left[\overline{Na^+}\right]^2[Ca^{2+}]} = [Ca^{2+}]^{-1}[Na^+]^2\left[\left[\overline{Na^+}\right]^{-2}\left[\overline{Ca^{2+}}\right]\right] \tag{4.29}$$

There is one electroneutrality condition which would be given by

$$Q_v = \sum_{i=1}^{I} z_i \overline{C}_i \tag{4.30}$$

where Q_v is the cation-exchange capacity [CEC in mol/L pore volume (PV)]; C_i is the concentration of cation i adsorbed on the matrix [in mol/L pore volume (PV)], and z_i is the charge of the adsorbed cation i.

In the same way, for M cations associated with surfactant micelles there are ($M-1$) cation exchange (on micelle) equilibrium relations of the following equation:

$$K_q^{ex} = \prod_{j=1}^{N} C_j^{w_{qj}} \prod_{m=1}^{M} \overline{C}_m^{w_{qm}} \quad \text{for} \quad q = 1, \ldots, (M-1) \tag{4.31}$$

And the electroneutrality conditions for the total micelles will be defined by the following equation:

$$C_{A^-} + C_{s^-} = \sum_{m=1}^{M} \overline{\overline{z}}_m \overline{\overline{C}}_m \tag{4.32}$$

where C_{A^-} is the in situ generated surfactant and C_{s^-} is the injected surfactant concentrations, C_m is the concentration of cation m adsorbed on micelles in mol/L, and z_m is the charge of the adsorbed cation m.

As a result, we managed to have N mass balances, $(J-N)$ aqueous reaction equilibrium relations, K solubility-product-constant equations, $(I-1)$ cation-exchange on matrix equilibrium relations, as well as one electroneutralinity condition, $(M-1)$ cation-exchange on micelle equilibrium relations and one electroneutralinity condition for the micelles, which provides all of the $(J+K+I+M)$-independent equations to solve the same number of concentration unknowns. For the detailed calculation algorithm, see Bhuyan [146].

4.14.4 Mathematical formulation of alkaline flooding

To express an isothermal flow problem, we start with the following material balance equations of N components [158]:

$$\begin{aligned}
&\frac{\partial(\phi C_n^t)}{\partial t} + \frac{\partial}{\partial x}\sum_{j=1}^{n_p}\left[C_{nj}u_{xj} - \phi S_j\left(D_{xxnj}\frac{\partial C_{nj}}{\partial x} + D_{xynj}\frac{\partial C_{nj}}{\partial y} + D_{xznj}\frac{\partial C_{nj}}{\partial z}\right)\right] \\
&\quad + \frac{\partial}{\partial y}\sum_{j=1}^{n_p}\left[C_{nj}u_{yj} - \phi S_j\left(D_{yynj}\frac{\partial C_{nj}}{\partial y} + D_{yznj}\frac{\partial C_{nj}}{\partial z} + D_{yxnj}\frac{\partial C_{nj}}{\partial x}\right)\right] \\
&\quad + \frac{\partial}{\partial z}\sum_{j=1}^{n_p}\left[C_{nj}u_{zj} - \phi S_j\left(D_{zznj}\frac{\partial C_{nj}}{\partial z} + D_{zynj}\frac{\partial C_{nj}}{\partial y} + D_{zxnj}\frac{\partial C_{nj}}{\partial x}\right)\right] \\
&= r_n, \quad n = 1, \ldots, N
\end{aligned} \tag{4.33}$$

where D_{xx}, D_{yy}, and D_{zz} are the elements of dispersion tensor; n_p is the number of phases; S_j is the saturation of phase j; u is the Darcy velocity; ϕ is the porosity; and r_n is the source (+) or sink (−) term.

The overall mass continuity equation has been prepared by summarizing the conservation equations over all components:

$$\phi c_t \frac{\partial p}{\partial t} + \vec{\nabla} \cdot \sum_{j=1}^{n_p} \vec{u}_j \sum_{n=1}^{N_V} (1 + c_n \Delta p) C_{nj} = q, \tag{4.34}$$

$$q = \sum_{n=1}^{N_V} q_n, \tag{4.35}$$

where N_v is the total number of volume-occupying components, q_n is the specific volumetric rate of either injection (+) or production (−) of component n, c_n is the compressibility of volume-occupying component n, c_t is the total compressibility, and p is the pressure.

In summary, the solution process steps are as follows:

1. First, solve Eq. (4.34) for pressure.
2. After determining pressure, calculate fluxes.
3. Solve Eq. (4.33) for the total concentrations of N components. Then by using the reaction equilibrium equations as described before, calculate all the concentrations.
4. Determine the phase concentrations, phase saturations, and other physical and transport properties required to solve for pressure for the next time level.

4.15 Application of machine learning

Reservoir behavior and characterization, planning of field development, and production prediction could be depicted by constructing mathematical models, examining field information, and performing experimental mensuration. Therefore, numerical simulation can be used to adequately provide the needs of reservoir engineers. However, numerical simulation becomes useless and time-consuming when the problem size extends [159]. Recently, artificial intelligence (AI) technologies have attracted enormous interest due to a rapid response speed and for promoting conventional problems containing a large spectrum of reservoir engineering [160, 161]. Hence, many advanced machine learning algorithms such as artificial neural networks (ANN), fuzzy logic (FL), genetic algorithm (GA), response surface model (RSM), particle swarm optimization (PSO), and support vector machine (SVM) have been utilized in many oil and gas reservoirs [162–165].

Fig. 4.38 illustrates how reservoir engineering problems can be solved using forward-looking intelligence systems. Prediction of field response can be done with forward-looking models using project design parameters and reservoir characteristics as input.

There is also a reverse AI-model which always utilizes the field response data as input unlike forward-looking models. An inverse ANN model capable of determining porosity, permeability, and thickness distribution of oil reservoirs was founded by Ramgulam et al. [166]. Fig. 4.39 illustrates the mechanism of the inverse AI-model which was proposed by Sun and Ertekin [167] for the polymer flooding process.

AI-models have also been widely used in chemical EOR projects. Ramakrishnan and Wasan [168] introduced an EOS to model IFT reduction between aqueous and oleic phases as a function of alkali concentration. More researchers [169–171] developed EOS to model the phase behavior of surfactant systems with consideration in the presence of alkali agents. Reduction in IFT as a variable is a challengeable agent to develop the ANN model for the alkaline/polymer process. Hence, a table of IFT as a function of

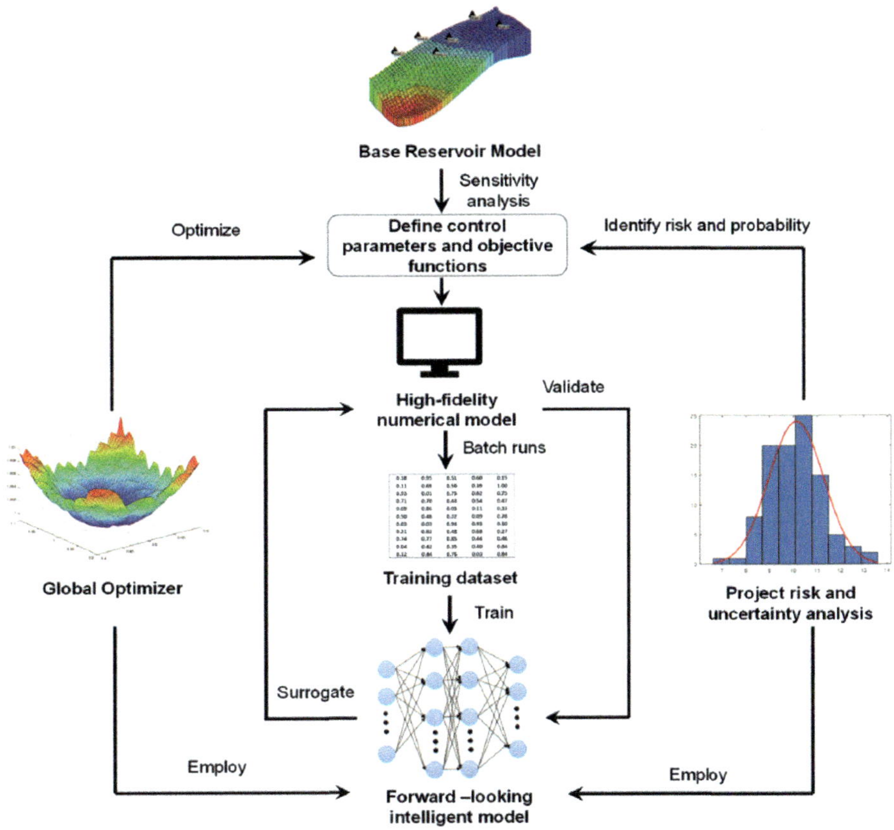

Fig. 4.38 A common workflow that uses forward-looking intelligent models to achieve computationally intensive reservoir engineering analysis [159].

NaOH concentration must be generated because of the generalization of the ANN model with IFT effects, so generation of the relationship between NaOH concentrations against the IFT can occur by implementing an EOS as illustrated in Fig. 4.40.

Sun [172] generalized the phase behavior of the ASP flooding process using the HLD-NAC equation of state established by Ghosh and Johns [171]. The procedure of this model was based on using HLD-NAC EOS, which is shown in Fig. 4.41. He then validated the model with the experimental results of Mohammadi et al. [151] as can be seen in Fig. 4.42. Sun [172] also used forward-looking and inverse ANN models for the chemical EOR process with a difference in the injected fluid type. Table 4.9 shows the summary of forward-looking and inverse ANN models for chemical EOR processes. According to Table 4.9, forward-looking models had more promising results than inverse ANN models in average relative errors. This is due to the forward-looking model depicts the expert system that effectively matches the data

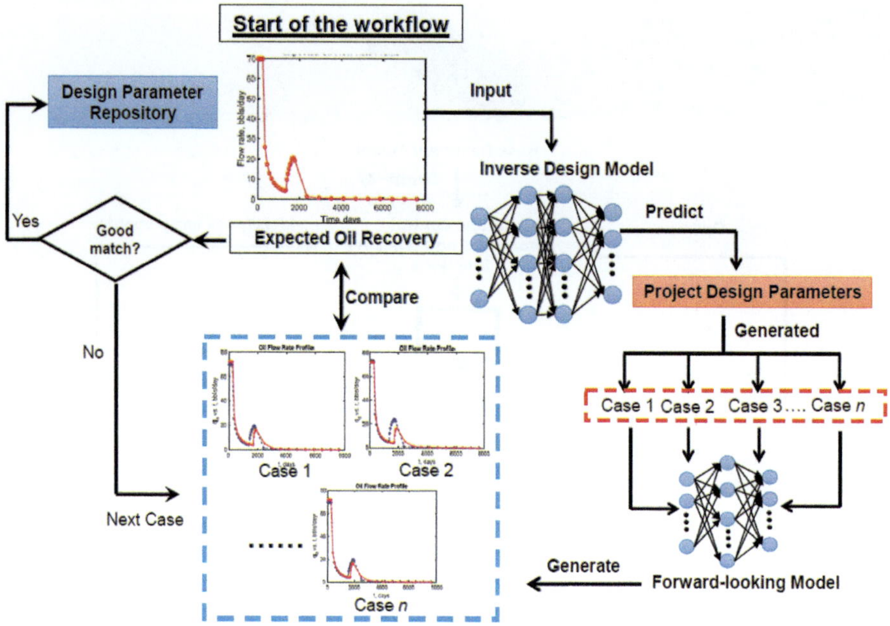

Fig. 4.39 Inverse project design workflow [159].

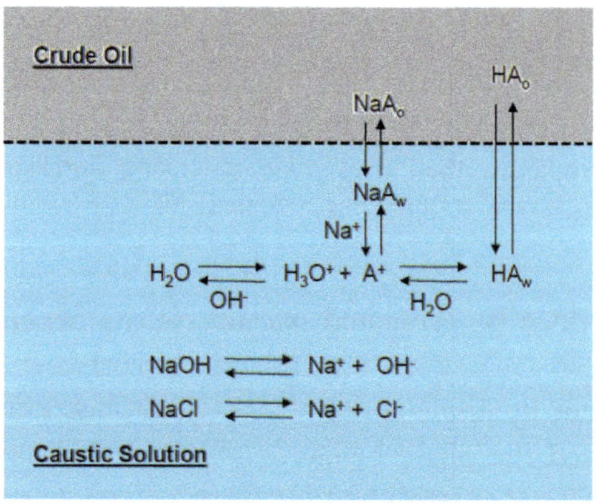

Fig. 4.40 Chemical equilibriums of a caustic solution-oil system. *(Reproduced from Ramakrishnan TS, Wasan DT. A model for interfacial activity of acidic crude oil/caustic systems for alkaline flooding. Soc Pet Eng J 1983;23:602–12.)*

structure presented in the forward-looking problems because of the low error values. In addition, using a different chemical type composition and slug size are multiple ways of accessing a particular predicted oil recovery, whereas the inverse design problem moves toward being nonuniqueness as the chemical formulation of the EOR process becomes more complex.

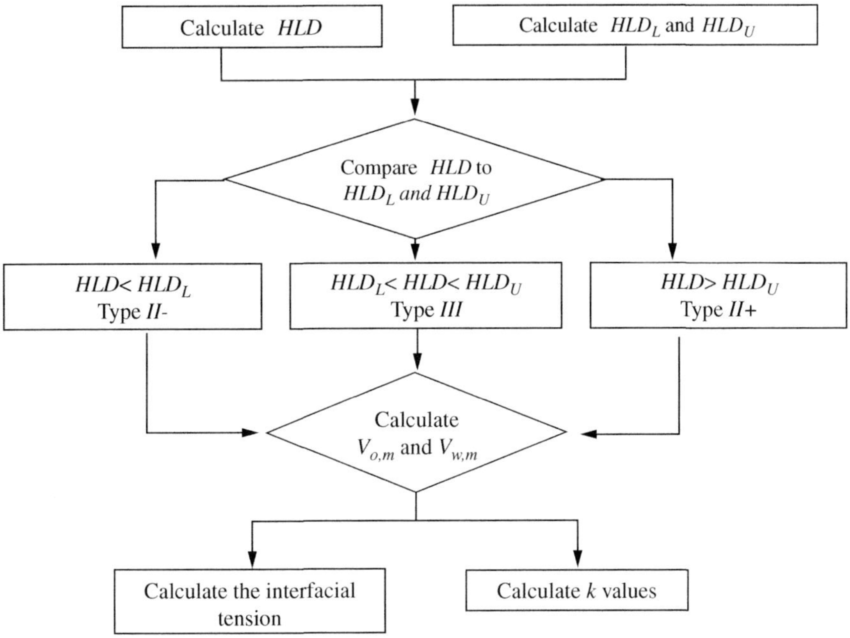

Fig. 4.41 Flash calculation flowchart using HLD-NAC EOS [172].

Table 4.9 General summary of the ANN models for chemical EOR processes [172].

ANN model	Injection type	Hidden layers	Transfer function	Error (%)
Forward-looking	P	60	<*logsig*>	5.83
	AP	67	<*logsig*>	7.52
	SP	37	<*logsig*>	7.81
	ASP	54	<*logsig*>	9.06
Inverse design	P	75	<*logsig*>	13.06
	AP	99	<*logsig*>	20.36
	SP	82	<*logsig*>	17.77
	ASP	36	<*logsig*>	36.66

ANN models are qualified to forecast the slug response time during the project timeline, which presents as an incremental oil production and produced water profiles. Fig. 4.43 indicates a comparison of a numerical model and a forward-looking ANN prediction for both the oil and water flow rate profile with a difference in average relative errors for the ASP flooding process. It is obvious that the best and average quality cases showed an excellent match rather than the high-fidelity (worth case) model, because the majority of the nonlinearities of the problem are as a result of the chemical slug injection presents, while the presence likelihoods of cases with high relative errors are very low [172].

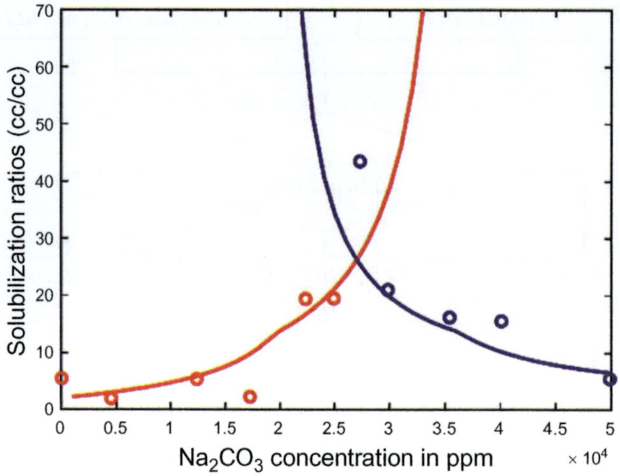

Fig. 4.42 Validation of the flash calculation result of an oil-brine-surfactant-Na_2CO_3 using HLD-NACEOS (line) with experimental data (pointed) [172].

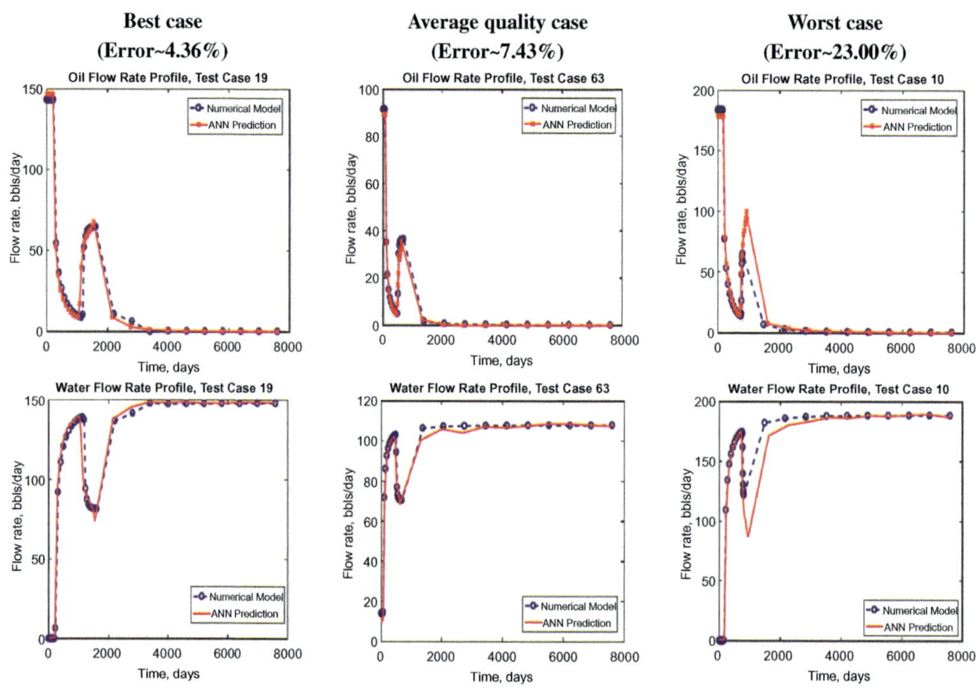

Fig. 4.43 Comparison of a numerical model and a forward-looking ANN prediction in: (A) oil flow rate and (B) water flow rate with variations of average relative errors [172].

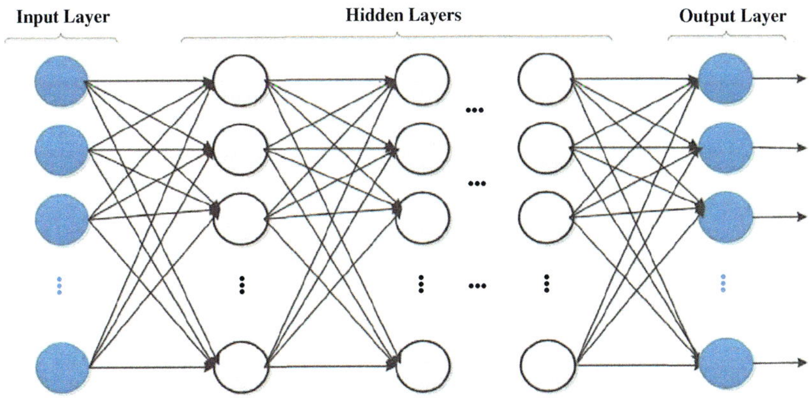

Fig. 4.44 Schematic representation of basic multilayer neural network [178].

There are also other algorithms which are used in the EOR process including FA [173–175], killer Whale algorithm (KWA) [176], and PSO [2, 177]. Hence, Dang et al. [178] deployed a multilayer neural network (ML-NN) technique in terms of the Levenberg-Marquardt algorithm to carry out the *N*-dimensional interpolation of relative permeability. Fig. 4.44 depicts a schematic representation of a basic multilayer neural network which consists of: (1) an input layer (*D*-dimensional input parameters), (2) any number of hidden layer (HL), and (3) an output layer (OL) with the number of nodes. Thus, according to laboratory data, interpolation of the low salinity water flooding process using ML-NN was performed to demonstrate the relative permeability curves for high salinity water flooding (HSW) and low salinity water flooding (LSW). K_{ro} and K_{rw} curves were generated using ML-NN with the NN configuration of 1 HL containing three neurons as shown in Fig. 4.45.

4.16 Surveillance and monitoring of alkaline flooding

The alkaline enhanced oil recovery consists of a series of complex mechanisms and because of the lack of effective field implementation, it is difficult to take the field operation aspects into account. For this reason, the laboratory tests and data are the best tool to conduct detailed pilot designs.

During the flooding pilot testing, a robust observation and monitoring program is exceptionally important. These data can determine the go/no go business decision based on the pilot testing performance. This is one of the most important outcomes of pilot testing for the entire field and if the go decision is taken, it will support the design of the upcoming operation and field implementations.

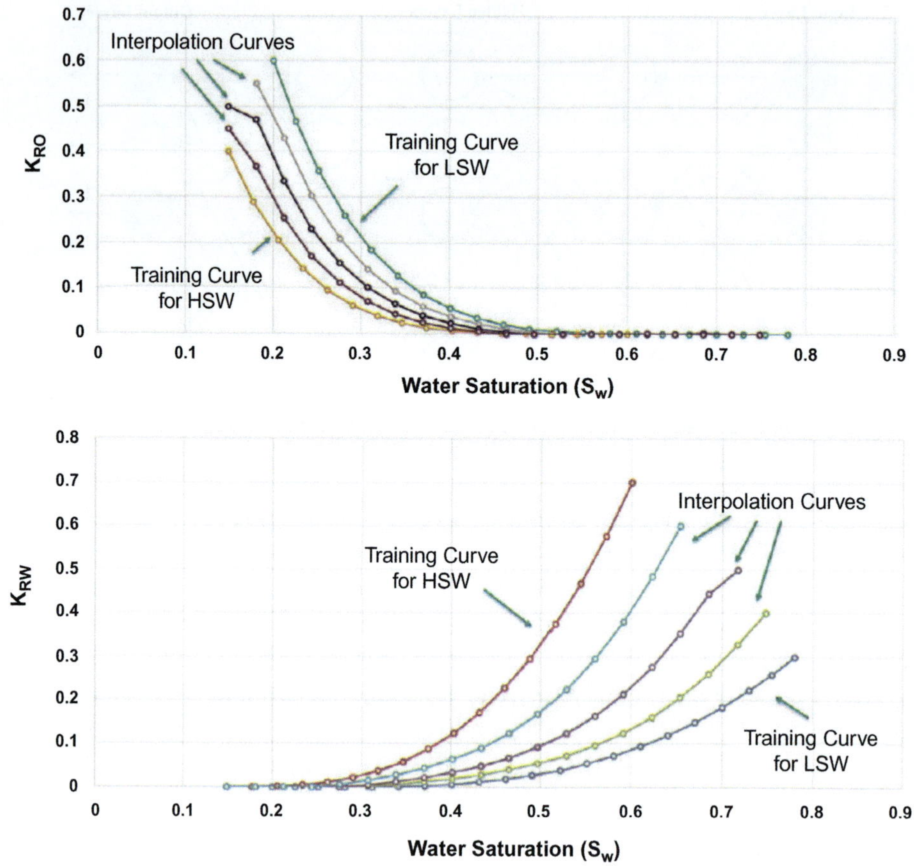

Fig. 4.45 Implementation of ML-NN for relative permeability interpolation for the LSW process [178].

Sheng [11] recommended a surveillance and monitoring program for an alkaline flooding project which is presented in Table 4.10.

4.17 Application conditions of the alkaline flooding project

The success of an alkaline flooding project depends on the wells and reservoir variables and criteria [11]. Some of these criteria are as follows:

- Injection wells should be placed in the oil zone and not in the surrounding brine saturated aquifer. This is to reduce alkali consumption to react with divalent ions. Sometimes the injection of alkaline fluid is beneficial to form precipitates if the injector is in the water saturated zone and brine has high divalent content. This may mitigate the bottom water coning.
- There is no reservoir gas cap.

Table 4.10 Suggested observation and monitoring activities in an alkaline water flooding.

Production wells tests and data	Frequency	Injection wells tests and data	Frequency	Observation wells	Frequency
Production rate	Every day	Fluid injection rate	Every day	Well head pressure	Every day
Water cut	Every day	Wellhead pressure	Every month	Bottom hole pressure (BHP)	Every day
Reservoir pressure	Every quarter	Bottom-hole pressure (BHP)	Every month	Sample analysis	Every day
Bottom-hole pressure (BHP)	Every quarter	Injection profiles	Every quarter		
Pressure buildup tests	Every quarter	Composition of injected water	Every quarter		
Production profiles	Every quarter	Composition analysis	Every week		
Composition of Produced water	Every week	Alkali concentration	Every day		
Composition analysis	Every week	Mechanical impurity analysis	Every quarter		
Alkali concentration	Every week				
Tracer content	Every week				
Produced hydrocarbon composition	Every week				

- When the reservoir has high acid number oils, alkaline flooding can be performed at all stages.
- For the reservoir with low acid number oils, it is recommended to have alkaline flooding in the earlier stages and the residual oil saturation should be higher than 0.4.
- Alkaline flooding has no temperature restriction.
- The clay content of the reservoir should not exceed 15%–25% as the chemical reaction and ion exchange with clay will consume the alkaline content
- Reservoir permeability must be more than 100 mD.
- Oil viscosity has to be below 50–100 cP. Nevertheless, the injection of the alkaline-surfactant into a very high viscous oil reservoir has been not only the center of attention but also quite successful.
- Formation water salinity has to be below 20%, and the divalent species in the injection water should be below 0.4%.

For designing an alkaline flooding project, the following facts should also be considered:
- Generally, the alkaline material consumption in the field is higher than that in the laboratory tests. This is because the alkalis contact time of alkalis with reservoirs in the field is much longer compared to that in the lab test.
- The oil recovery factor is generally lower in the field compared to the laboratory test results.
- The alkaline chemical flooding will be more effective if combined with surfactant flooding, polymer flooding, hydrocarbon gas injection, or thermal recovery.
- An alkaline injection has the potential to create scale complications in the reservoir, wellbore, and surface facilities.
- The contact of the alkaline solution with oil may create stable emulsions which can increase the cost of produced fluids treatment.

A. Appendix

A.1 Solution

Solutions are homogeneous mixtures of a solvent (e.g., water) and one or more solutes, the substance that dissolves in the solvent. The solute can be a solid, liquid, or gas which is dispersed uniformly in the solvent. An aqueous solution is generally referred to the solution that has solute dissolved in water.

Dissolution of a solute in a solvent can happen through two stages:

- The bounds between the particles in the solute are broken
- Solvent particles are surrounded by the solute particle

This process is called solvation, which means all solute particles are totally surrounded by solvent molecules (Fig. A.1).

For aqueous solutions, solute (water)-solvent interactions are referred to as hydration and the interaction between the solvent and solute is called solvation.

It is important to note that the chemistry of the solute and solvent will not change and there are no atom rearrangements.

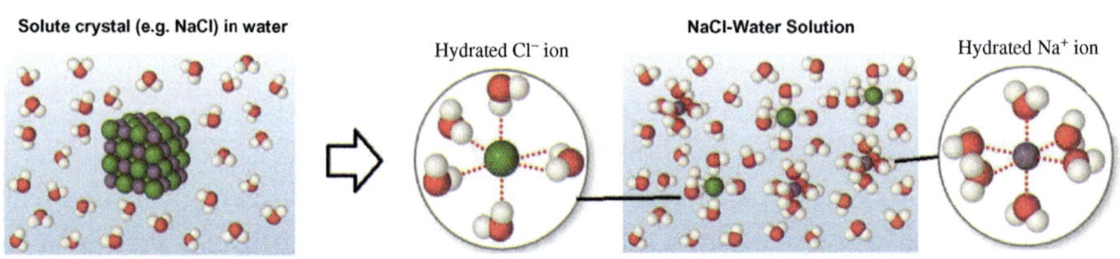

Fig. A.1 Solvation and hydration process.

Concentration of solutions:

There are two main ways to express a solution: the concentration of a solution and the amount of solute per amount of solvent or solution. It is important to note that the amount of solute always comes first.

(A) Amount of solute per amount of solution

Concentration of a solution is generally defined as the amount of solute in the total volume of solution. The solute amount and the volume of solution can be shown as the mass, volume, or number of particles, which is shown in Table A.1.

(B) Amount of solute per amount of solvent

The concentration of a solution can also be shown as the dissolved solute amount per unit of solvent (Table A.2).

Water as the major liquid on Earth can dissolve a wide range of substances. Water is known as the "Universal Solvent" because it can dissolve more substances than any other liquid.

Water's chemical composition and physical attributes are such that they make it one of the best solvents. Water molecules are made up of one oxygen atom and two hydrogen atoms with covalent bonds. The electrons in the covalent bonds are not shared equally; these electrons are more attracted to the oxygen atom than to the hydrogen atoms. This unequal attraction of electrons makes the O—H bonds polar. Therefore, the oxygen atom has a partially negative charge while the hydrogen atoms have a partial positive charge.

Table A.1 Types of solute amount per amount of solution method.

Concentration term	Amount of solute to amount of solution	Concentration units
Mass/mass % % m/m or % w/w	$\frac{\text{Mass of solute in solution}}{\text{Total mass of solution}} \times 100$	g/100 g
Mass/volume % m/v % or w/v %	$\frac{\text{Mass of solute in solution}}{\text{Total volume of solution}} \times 100$	g/100 mL
Volumes/volume % v/v %	$\frac{\text{Volume of solute in solution}}{\text{Total volume of solution}} \times 100$	mL/100 mL
Parts per million (ppm)	$\frac{\text{Mass of solute in solution}}{\text{Total mass of solution}} \times 10^6$	ppm mg/kg, μg/g
	$\frac{\text{Mass of solute in solution}}{\text{Total volume of solution}} \times 10^6$	ppm mg/L, μg/mL
Parts per billion (ppb)	$\frac{\text{Mass of solute in solution}}{\text{Total volume of solution}} \times 10^9$	ppb, μg/L
Mole fraction (X)	$\frac{\text{Mole of solute}}{\text{Total mole of solution}}$	%
Molarity (M)	$\frac{\text{Mole of solute}}{\text{Total liters of solution}}$	mol/L molar, M

Table A.2 Types of amount of solute per amount of solvent method.

Concentration type	Solute amount/solvent amount	Concentration unit
Weight ratio %	$\frac{\text{Mass of solute}}{\text{kg of solvent}}$	g/100 g solvent
Molality	$\frac{\text{Mole of solute}}{\text{kg of solvent}}$	mol/kg molal, m

Note:
- Molarity and molality are equal for very dilute aqueous solutions as 1 L of H_2O weighs 1 kg.
- Choice of units generally depends on the purpose.
- Mole fraction: used for gases and vapor pressures of solutions.
- Molarity is commonly used since volumes of solutions are easier to measure.
- Molality is independent of temperature.
- Percent by mass is also independent of temperature and does not need the value of the molar mass.

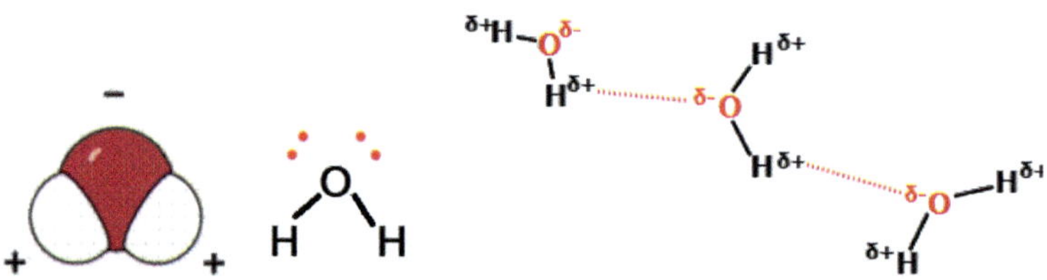

Fig. A.2 Water molecule structure and polarity. The hydrogen bond connecting the water molecules is presented as a *red dotted line* and the covalent bond between the oxygen and hydrogen atoms is shown as a *solid black line*.

The oxygen of the water molecule has a lone pairs, a pair of nonbonding electrons that repel each other more than the repulsion between lone pairs and bonding pairs or between the two bonding pairs of electrons. This pushes the O—H bonds closer to each other and creates a nonsymmetrical bent molecule. The difference in the electronegativity of the oxygen and hydrogen atoms does not cancel each other out and, as a result, a molecule of water is a permanent net dipole molecule. This polarity allows many other types of molecules to attract to the water molecule (Fig. A.2).

The hydrogen bonds in the water molecule are stronger than the attractive forces of most polar molecules (dipole-dipole interactions) and nonpolar molecules (dispersion or London forces), which causes a higher melting and boiling point than expected. That is why water is a liquid at temperature and pressure to support life on Earth. The strength of hydrogen bond makes water a universal solvent that dissolves many substances.

A.2 Dissolving ionic compounds in water

The dissolution of ionic compounds such as salts in water is due to disruption of their ionic bonds caused by the attraction of ions to the opposite charge side of the water

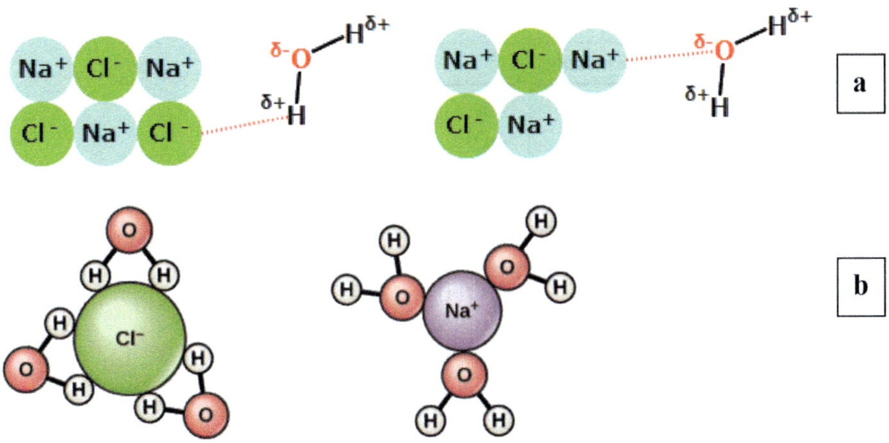

Fig. A.3 (A) Ion-dipole interaction (which is presented as a *red dotted line* between the Cl^- ion and the hydrogen atom of the water molecule with a partially positive charge (shown as $H\delta^+$) and similarly between Na^+ and of the oxygen atom of the water molecule with a negative charge, $O\delta^-$) and (B) dissociation and hydration of NaCl in water.

molecules. Material dissociation happens when atoms or groups of atoms separate from their molecules and produce ions. One of the best examples is the dissolution of sodium chloride (NaCl) in water (Fig. A.3). The dissolution starts with the dissociation of NaCl into Na^+ and Cl^- ions. These separated ions are then surrounded by spheres of water molecules (hydration). The sodium ion is in fact surrounded by the partially negatively charged oxygen of the water molecules and the chlorine ion is surrounded by the partially positively charged hydrogen of the water molecule.

As the attractive force between the sodium ion (Na^+) and the partially negatively charged oxygen of the water molecule is strong enough, the sodium ion will be dragged away from the salt ionic lattice and ultimately will be totally surrounded by water molecules. The result of the dissolution of solid NaCl (solid) in water [H_2O (liquid)] is known as sodium chloride [NaCl (aq)] aqueous solution.

A.3 Base dissociation in water (ionization)

A.3.1 Dissociation or ionization of a strong base

The hydroxides of Group 1 elements (Alkali metals, MOH), and the hydroxides of Group 2 elements (Alkali Earth metals, $M(OH)_2$), are classified as strong bases. When added to water, they completely dissociate and produce a hydroxide ion and a hydrated metal cation (Table A.3).

After the Group 1 or 2 hydroxide dissolution in water, the undissociated hydroxides will not exist in the solution. These hydroxides will exist as hydrated metal cations and hydroxide ions.

Table A.3 Type of hydroxide groups.

Group 1 metals	
General form	Base → hydroxide ions + cation
General Eq	$MOH \rightarrow OH^-_{(aq)} + M^+_{(aq)}$
Example	$NaOH \rightarrow OH^-_{(aq)} + Na^+_{(aq)}$
	$KOH \rightarrow OH^-_{(aq)} + K^+_{(aq)}$
Group 2 metals	
General Eq	$M(OH)_2 \rightarrow 2OH^-_{(aq)} + M^{2+}_{(aq)}$
Example	$Ca(OH)_2 \rightarrow 2OH^-_{(aq)} + Ca^{2+}_{(aq)}$
	$Ba(OH)_2 \rightarrow 2OH^-_{(aq)} + Ba^{2+}_{(aq)}$

The Group 2 hydroxides are less soluble in water, for example, calcium hydroxide ($Ca(OH)_2$). In these cases, some undissociated $Ca(OH)_{2(s)}$ might stay in the water or deposit at the bottom of the solution. But the species in the solution are all in the form of dissociated $Ca^{2+}_{(aq)}$ and $OH^-_{(aq)}$, not $Ca(OH)_{2(aq)}$.

For instance, for 0.020 mol of a strong water-soluble base, (MOH), dissolved in 1 L of water:

	Base **MOH**	⇨ ⇨	**Hydroxide ion** **OH^-**	+ +	**Hydrated cation** **M^+**
Moles of base before adding to water	0.020 mol		0.0 mol		0.0 mol
Moles of base after adding to water	0.0 mol		0.020 mol		0.020 mol

Similarly, for 0.01 mol of a strong base, $M(OH)_2$, dissolved in 1 L of water:

	Base **$M(OH)_2$**	⇨ ⇨	**Hydroxide ion** **$2OH^-$**	+ +	**Hydrated cation** **M^{2+}**
Moles of base before adding to water	0.020 mol		0.0 mol		0.0 mol/L
Moles of base after adding to water	0 mol		2×0.020 mol		0.020 mol

A.3.2 Dissociation or ionization of a weak base or ionization

Weak bases do not have a complete reaction with water. Ammonia aqueous solution (NH_3), amines ($R\text{-}NH_2$), and phosphine (PH_3) are three common cases of weak bases (Table A.4).

The undissociated weak base in aqueous solution is in equilibrium with its conjugate acid and hydroxide ions.

The concentrations of an undissociated base, hydroxide ions, and a conjugate acid of the base are dependent on the extent of the water and base reaction.

Table A.4 Common examples of bases.

	Brønsted-Lowry definition
Form	**Base + water ⇌ hydroxide ions + conjugate acid**
Equation	$B + H_2O_{(l)} \rightleftharpoons OH^-_{(aq)} + BH^+_{(aq)}$
Ammonia	$NH_3 + H_2O_{(l)} \rightleftharpoons OH^-_{(aq)} + NH^+_{4(aq)}$
Ethanamine (ethylamine)	$CH_3CH_2NH_2 + H_2O_{(l)} \rightleftharpoons OH^-_{(aq)} + CH_3CH_2NH^+_{3(aq)}$
Methanamine (methylamine)	$CH_3NH_2 + H_2O_{(l)} \rightleftharpoons OH^-_{(aq)} + CH_3NH^+_{3(aq)}$

For 0.01 mol of a weak base, R-NH2, dissolving in 1 L of water:

	Base + H_2O	⇌	Hydroxide ion	+	Conjugate acid
Moles of base before mixing with water	0.01 mol		0.0 mol		0.0 mol
Moles of base after mixing with water	0.01 − *n* mol		*n* mol		*n* mol

Note: $n < 0.01$.

A.4 Bases strength comparison

A.4.1 Dilute aqueous solutions electrical conductivity

Strong bases are generally able to completely dissociate in water and produce a high concentration of hydroxide ions and metal cations in solutions. In the case of a weak base, a small amount of base reacts with water and creates a lower amount of hydroxide ions.

The quantity of ions in an aqueous solution controls the electrical conductivity (EC) of the solution, so it is possible to measure the mobile ions available by measuring the EC. The higher the EC of the solution, the more mobile ions that are available.

To measure the relative strength of different monovalent base solutions, the EC of each solution should be measured on condition that a similar concentration for each base is used, and also all tests should be conducted at the same temperature as EC is sensitive to temperature.

It is very important to use the same concentration for different bases because a higher concentration in water creates more hydroxide ions which ultimately affects conductivity. That is why all the electrical conductivity (EC) measurements have to be recorded at the same temperature.

A.4.2 Dilute aqueous solution's pH

A common way to measure the relative strengths of different bases that can accept just one proton is to measure the solution pH. Similar to the electrical conductivity measurement, the same concentration for each base should be used, and all measurements must perform at the same temperature.

If the same concentrations are not selected, then the higher concentrations will show a higher pH. As the temperature affects the base solubility and changes the number of free ions in the solution, all measurements must be performed at the same temperature. Common strong Arrhenius bases are as follows:

• $RbOH$:	Rubidium hydroxide
• $LiOH$:	Lithium hydroxide
• $Ca(OH)_2$:	Calcium hydroxide
• $Sr(OH)_2$:	Strontium hydroxide
• $CsOH$:	Cesium hydroxide
• $Ba(OH)_2$:	Barium hydroxide
• $NaOH$:	Sodium hydroxide
• KOH:	Potassium hydroxide

Practical examples:

Example A.1 Which one produces the weaker base in water: ammonia or NaOH?

a. Aqueous solutions of all elements of Group 1 hydroxides are producing strong bases.
Sodium classifies as a Group 1 Alkali metal, so NaOH is a strong base.

b. Ammonia in aqueous solution will produce a weak base.

c. Therefore, ammonia produces a weaker base compared to sodium hydroxide.

Example A.2 Which one of the following aqueous solutions will show a higher pH at 25°C?

0.0151 mol/L KOH or 0.015 mol/L methanamine (methylamine).

a. As both alkalis only accept one proton and both have the same concentration and temperature, the stronger base will produce a higher pH.

b. Aqueous solutions of Group 1 alkali element hydroxides are strong bases.
Potassium is a Group 1 alkali metal, so it is a strong base.

c. Methanamine (methylamine), which classifies aqueous solutions of amines, is a weak base.

d. The stronger the base, the higher the pH.
The higher pH is awarded to aqueous KOH as a result of its stronger base.

Example A.3 At 25°C temperature, the pH of an NaOH aqueous solution is 9.0, and at the same temperature the pH of an aqueous ammonia solution is 10.5. Determine which of these solutions is the stronger base?

a. Aqueous solutions of all Group 1 hydroxides are strong bases.
NaOH is a strong base due to its Group 1 alkali metal.

b. Aqueous solution of ammonia is a weak base.

c. The stronger base is NaOH.

The NaOH solution against the ammonia solution dilutes more, so the variation in pH could be as a result of it.

A.4.3 Equilibrium constant

Consider an equilibrium reaction that synthesizes ammonia gas from nitrogen ($N_{2(g)}$) and hydrogen ($H_{2(g)}$) gases:

$$N_{2(g)} + 3H_{2(g)} \leftrightarrows 2NH_{3(g)}$$

At the same temperature, three experiments were carried out.

The initial concentration of each species was changed for each experiment and then allowed the system to equilibrate. After equilibrium was reached, the concentrations of all species were measured and the values of the mass action expressions were calculated for the final equilibrium concentrations.

Q is a constant ($Q = K$) at equilibrium and, for example, the value of Q for the particular reaction of Table A.5 at a certain temperature is 6. It is important to notice that at all stages with the initial mixture with only reactant molecules and later stages that include both the reactant and product, when the reaction reaches the equilibrium, Q stays the same.

The K value or the equilibrium constant for a specific reaction at a particular temperature will not change when the equilibrium concentrations of the reactant and/or product molecules are changed.

Therefore, if the concentrations of all components in a given chemical reaction at equilibrium at a particular temperature are known, it is possible to calculate the K value or equilibrium constant value for that particular temperature.

The K value or the equilibrium constant can be used to calculate the reactant and product concentrations for this particular reaction in other experiments at the same temperature.

Table A.5 Numerical examples of mass-action expression.

	Initial concentrations (mol/L)			Equilibrium concentrations (mol/L)			
Experiment	$[N_{2(g)}]$	$[H_{2(g)}]$	$[NH_{3(g)}]$	$[N_{2(g)}]$	$[H_{2(g)}]$	$[NH_{3(g)}]$	$Q = \frac{[NH_{3(g)}]^2}{[N_{2(g)}][H_{2(g)}]^3}$
1.	1.00	3.00	0	0.325	0.975	1.350	$\frac{[1.350]^2}{[0.325][0.975]^3} = 6$
2.	1.000	1.000	0	0.781	0.343	0.438	$\frac{[0.438]^2}{[0.781][0.343]^3} = 6$
3.	1.000	1.000	1.000	0.885	0.655	1.230	$\frac{[1.230]^2}{[0.885][0.655]^3} = 6$

Table A.6 Definition of types of equilibrium constant.

Reaction	Equilibrium constant symbol	Equilibrium constant name
Dissociation of bases (*bases ionization*)	K_b	Base dissociation constants
water self-dissociation (*water auto-dissociation*)	K_w	Ion product for water
Dissociation of acids (*acids ionization*)	K_a	Acid dissociation constants
Dissolving an ionic compound	K_{sp}	Solubility products
Gaseous reactions	K_P	(Calculated using partial pressures not concentrations)

Because of the importance of specific types of reactions, there are special names and symbols for the equilibrium constant in these reactions. Table A.6 summarizes some examples of these reactions.

The base dissociation constant or base ionization constant (K_b) is an equilibrium constant that states the ionization or dissociation of a base.

a. When the *Arrhenius base* (BOH) ionizes or dissociates to form OH^- and B^+:

$$BOH \rightleftharpoons OH^- + B^+$$

$$K_b = \frac{[BH^+][OH^-]}{[B]}$$

b. For a *Brönsted-Lowry base*:

$$B + H_2O \rightleftharpoons BH^+ + OH^-$$

$$K_b = \frac{[BH^+][OH^-]}{[B]}$$

- K_b delivers an indication of the equilibrium status
  - **(i)** The large K_b: the products of the dissociation reaction are favored.
  - **(ii)** The small K_b: the undissociated base is favored.
- K_b provides an indication of the base strength
  - **(i)** The large K_b: the base is mostly dissociated, so the base is strong.
  - **(ii)** The small K_b: the base is not well dissociated, so the base is weak.
- The degree of the base dissociates is usually represented as a percentage:

  Ionization or dissociation percentage $= [OH^-$ at equilibrium$]/[$base initial$] \times 100$
  - **(i)** The base is a strong base if the dissociation percentage $\approx 100\%$.
  - **(ii)** The base is a weak base if the dissociation percentage is small.

Table A.7 Values of the base dissociation constant.

Base name	Dissociation constant, K_b	K_b value
Methanamine: CH_3NH_2	4.4×10^{-4}	Higher
Ammonia: NH_3	1.8×10^{-5}	↓
Hydroxylamine: NH_2OH	9.1×10^{-9}	↓
Phosphine: PH_3	1.0×10^{-14}	Lower

Table A.7 shows the values of the dissociation constant of the base (K_b) for different weak bases *aqueous solutions* at the temperature of 25°C.

Comparing the base dissociation constant values for methylamine and ammonia:

Methanamine: $K_b = 4.4 \times 10^{-4}$ (larger).

Ammonia: $K_b = 1.8 \times 10^{-5}$ (smaller).

Considering the same concentration solutions, for example, 0.1 mol/L at 25°C:

(i) The solution will have more unionized ammonia molecules than unionized methylamine molecules.

(ii) The methylamine solution will have more hydroxide ions compared to the ammonia solution; thus:

- The pOH of the methylamine solution will be lower than the pOH of the ammonia solution,
- The pH of the methylamine solution will be higher than the pH of the ammonia solution.

A.5 pH calculation

A.5.1 Strong bases (alkalis)

Before adding a strong Arrhenius base, the water molecules are in equilibrium with the hydrogen and hydroxide ions:

$$\text{Water}\,(H_2O) \rightleftharpoons \text{hydrogen ions}\,(H^+) + \text{hydroxide ions}\,(OH^-)$$

$$H_2O_{(aq)} \rightleftharpoons H^+{}_{(aq)} + OH^-{}_{(aq)}$$

Generally, in water a low number of H_2O molecules dissociate into hydrogen ($H^+_{(aq)}$) and hydroxide ions ($OH^-_{(aq)}$).

At 25°C the concentration of these ions in water is very low: $[H^+{}_{(aq)}] = [OH^-{}_{(aq)}] \approx 10^{-7}$ mol/L And $K_w = [H^+{}_{(aq)}][OH^-{}_{(aq)}] = 10^{-7} \times 10^{-7} = 10^{-14}$

The above equation can be modified from multiplication to addition by measuring the log10 of all of these ionic compounds:

$$K_w = [H^+{}_{(aq)}][OH^-{}_{(aq)}]$$

$$\log_{10}(K_w) = \log_{10}[H^+{}_{(aq)}] + \log_{10}[OH^-{}_{(aq)}]$$

If we multiply both sides of the equation by -1, then:

$$-\log_{10}(K_w) = -\log_{10}\left[H^+{}_{(aq)}\right] + -\log_{10}\left[OH^-{}_{(aq)}\right]$$

Which produce:

$$-\log_{10}(K_w) = pH + pOH$$

since $pH = -\log_{10}\left[H^+{}_{(aq)}\right]$ and $pOH = -\log_{10}\left[OH^-{}_{(aq)}\right]$
For 25°C water:

$$K_w = 10^{-14}$$

$$\log_{10} K_w = \log_{10} 10^{-14} = -14$$

$$-\log_{10} K_w = -\log_{10} 10^{-14} = 14$$

Therefore: $14 = pH + pOH$
So it is clear that when we add a strong Arrhenius base to water, the base totally dissociates and produces a hydrated metal cation ($M^+_{(aq)}$ or $M^{2+}_{(aq)}$) and hydroxyl anion ($OH^-_{(aq)}$):

$$\text{Base} \rightarrow \text{hydroxide ions} + \text{metal cations}$$

Group 1 alkali metal hydroxides:

$$MOH \rightarrow OH^-{}_{(aq)} + M^+{}_{(aq)}$$

Group 2 alkali earth metal hydroxides:

$$M(OH)_2 \rightarrow 2OH^-{}_{(aq)} + M^{+2}{}_{(aq)}$$

Adding the base to the water will disturb the water dissociation equilibrium:

$$H_2O \rightleftharpoons H^+{}_{(aq)} + OH^-{}_{(aq)}$$

Adding further $OH^-_{(aq)}$ to the water would move the equilibrium to the left, according to Le Chatelier's Principle. As more $OH^-_{(aq)}$ is added, the water dissociation equilibrium mechanism reestablishes equilibrium by reacting to the addition of more $OH^-_{(aq)}$ with some $H^+_{(aq)}$ to reinstate the equilibrium. Solution procedure summary of pH amounts is shown in Table A.8.

Table A.8 Summary of solution procedure.

At 25°C	$[OH^-_{(aq)}]$ mol/L	$[H^+_{(aq)}]$ mol/L	K_w	pH	pOH	$-\log(K_w)$ (pK_w)
Water (before base added)	10^{-7}	10^{-7}	10^{-14}	7	7	14
Aqueous solution (after base added)	$>10^{-7}$	$<10^{-7}$	10^{-14}	>7	<7	14

As a result, increasing the concentration of $OH^-_{(aq)}$ in the water lowers the concentration of $H^+_{(aq)}$, but the water dissociation constant remains constant at 10^{-14}.

So:

$$K_w = [H^+_{(aq)}][OH^-_{(aq)}] = 10^{-14}$$

and

$$-\log_{10}K_w = 14 = pH + pOH$$

To measure pH at a given temperature, the values of K_w or $-\log_{10}K_w$ (pK_w) and pOH or $[OH^-_{(aq)}]$ can be used.

(i) The pH can be determined if the pOH of an alkaline solution is known:

$$pH + pOH = -\log_{10} K_w$$

Therefore:

$$pH = -\log_{10}K_w - pOH$$

So at 25°C:

$$pH = 14 - pOH$$

(ii) We can determine the pH of a solution if we know the concentration of hydroxide ions in it.

$$K_w = [H^+_{(aq)}][OH^-_{(aq)}]$$

By rearranging this equation (formula), we can determine the concentration of hydrogen ions in the aqueous solution:

$$[H^+_{(aq)}] = K_w \div [OH^-_{(aq)}]$$

Thus:

$$pH = -\log_{10}[H^+_{(aq)}] = -\log_{10}(K_w \div [OH^-_{(aq)}])$$

$[H^+_{(aq)}]$ and $[OH^-_{(aq)}]$ must be in mol/L (or M) unit.

Examples:

Example A.1 1.1×10^{-13} mol/L ion concentration of hydrogen (oxidanium or oxonium or hydronium) was used in an aqueous sodium hydroxide solution. What is the pH of the solution?

The question can be solved based on the data that have been given within the question as follows:

a. *Specify the object of the question.*
Determine the pH
pH=?

b. *Write down the required information.*
$[H^+_{(aq)}] = 1.1 \times 10^{-13}$ mol/L

c. *Find out the exact relationship between what you need and what you know.*
Write the pH equation:
$pH = -\log_{10}[H^+_{(aq)}]$

d. *Use the appropriate equation for solving the question by substituting the given values:*
$pH = -\log_{10}[1 \times 10^{-13}] = 12.95$

Example A.2 The pOH of an aqueous calcium hydroxide solution is 4.5. What is the pH of this solution at 25°C?

a. *Specify the object of the question.*
Determine the pH
pH=?

b. *Write down the required information.*
pOH=4.5
Temperature=25°C
so: $K_w = 1.0 \times 10^{-14}$ (from the Data Sheet)
and $-\log_{10}K_w = -\log_{10}10^{-14} = 14$
then: pH+pOH=14
Compute pH: pH=14−pOH

c. *Use the appropriate equation for solving the question by substituting the given values:*
pH=14–4.5=9.5

Example A.3 Compute the pH of 0.45 g of sodium hydroxide at 25°C which has been added to pure water to make 600.0 mL of aqueous solution.

(A) Hydroxide ion concentration determination method:

a. *Specify the object of the question.*
Determine the pH
pH=?

b. *Write down the required information from the question.*
Mass of $NaOH_{(s)}$ = 0.45 g
Volume of ($NaOH_{(aq)}$) solution = 600.0 mL
Convert the volume from mL to L by dividing the volume by 1000
Then: the volume of solution = 600.0/1000 = 0.6000 L

c. *Find out the exact relationship between what you know and what you need.*

(i) Compute the moles of NaOH dissolved in water:

Moles = mass/molar mass

Mass of NaOH = 0.45 g (from the question)

Molar mass of NaOH = 22.99 + 16.00 + 1.008 = 39.998 g/mol (from the Periodic Table)

Moles of NaOH = 0.45/39.998 = 0.0112 mol

(ii) The NaOH dissociation equation in water is:

$NaOH \rightarrow Na^{+}_{(aq)} + OH^{-}_{(aq)}$

(iii) Find the mole ratio (stoichiometric ratio) of NaOH to OH^- ions:

$NaOH:OH^-$ is 1:1 (from the balanced chemical equation)

(iv) Find the moles of OH^- ions using the stoichiometric ratio (mole ratio):

Moles $OH^{-}_{(aq)}$ = moles NaOH = 0.010 mol

(v) Measure the hydroxide ions concentration in solution in mol/L (or molarity):

$[OH^{-}_{(aq)}]$ = molarity = moles/volume in liters

$[OH^{-}_{(aq)}] = 0.0112/0.600 = 0.018$ mol/L

(vi) Write the H^+ ions concentration in the water equation:

$K_w = [H^{+}_{(aq)}]\ [OH^{-}_{(aq)}]$

Therefore: $[H^{+}_{(aq)}] = K_w/[OH^{-}_{(aq)}]$

(vii) Substitute the values of K_w and $[OH^{-}_{(aq)}]$ in the equation:

$[H^{+}_{(aq)}] = 1.0 \times 10^{-14}/0.018 = 5.55 \times 10^{-13}$ mol/L

(viii) Write and consider the pH equation:

$pH = -\log_{10}[H^{+}_{(aq)}]$

d. *Use the appropriate equation for solving the question by substituting the given values:*

$pH = -\log_{10}[5.55 \times 10^{-13}] = 12.25$

(B) The pOH method:

a. *Specify the object of the question.*

Determine the pH

pH = ?

b. *Write down the required information.*

Mass of $NaOH_{(s)}$ = 0.45 g

Volume of $NaOH_{(aq)}$ solution = 600.0 mL = 600.0/1000 = 0.6000 L

c. *Find out the exact relationship between what you know and what you need.*

(i) Compute the moles of NaOH dissolved in water:

Moles = mass/molar mass

Mass NaOH = 0.45 g (from the question)

Molar mass NaOH = 22.99 + 16.00 + 1.008 = 39.998 g/mol (from the Periodic Table)

Moles NaOH = 0.45/39.998 = 0.0112 mol

Write the NaOH dissociation in water equation:

$NaOH \rightarrow Na^{+}{}_{(aq)} + OH^{-}{}_{(aq)}$

(ii) Measure the mole ratio (stoichiometric ratio) of NaOH: OH^- ions:
NaOH:OH^- is 1:1 (from the balanced chemical equation)

(iii) Measure the moles of OH^- ions using the mole ratio (stoichiometric ratio):
Moles of $OH^-_{(aq)}$ = moles of NaOH = 0.0112 mol

(iv) Measure the concentration of OH^- ions in solution in mol/L (or molarity):
$[OH^-_{(aq)}]$ = molarity = moles/volume in liters
$[OH^-_{(aq)}] = 0.0112 \div 0.600 = 0.018\,\text{mol/L}$

(v) Write the pOH equation:
$pOH = -\log_{10}[OH^-_{(aq)}]$

(vi) Add the value for $[OH^-_{(aq)}]$ and solve:
$pOH = -\log_{10}[0.020] = 1.75$

(vii) Write the pH equation as a function of K_w:
$-\log_{10}K_w = pH + pOH$ or $pK_w = pH + pOH$

d. *Use the appropriate equation for solving the question by substituting the given values:*
$-\log_{10}[1.0 \times 10^{-14}] = pH + 1.75$
$14 = pH + 1.75$
$14 - 1.75 = pH = 12.25$

Example A.4 There is 0.11 mol/L NaOH(aq) dissociation at 25°C. What is the $[OH^-]$, pOH and the percentage of dissociation.

1. Determine the base ionization or dissociation equation:

$$NaOH_{(aq)} \rightleftharpoons OH^-_{(aq)} + Na^+_{(aq)}$$

2. Compute the existing species concentrations of the initial and final equilibrium conditions:
NaOH will totally dissociate into OH^- and Na^+ because it is a strong base.
Set up an RICE[a] table as presented in Table A.9 to find the equilibrium concentration of the species in the solution:
$[OH^-] = 0.11\,\text{mol/L}$
3. Compute pOH:$pOH = -\log_{10}[OH^-]$
$pOH = -\log_{10}[0.10] = 0.95$
4. Compute the percentage of dissociation:
Dissociation % = $[OH^-]$/[base initial] × 100
$[OH^-] = 0.11\,\text{mol/L}$

[a] **RICE** stands for:
Reaction
Initial concentration
Change in concentration
Equilibrium concentration

Table A.9 Values of the equilibrium concentration.

Species concentration	$NaOH_{(aq)}$	$\rightleftharpoons$	$OH^-_{(aq)}$	+	$Na^+_{(aq)}$
Initial concentration (mol/L):	0.11		0		0
Change in concentration (mol/L):	−0.11		+0.11		+0.11
Equilibrium concentration (mol/L):	0.11 − 0.11 = 0		0.11		0.11

[NaOH initial] = 0.11 mol/L

%dissociation = 0.11/0.11 × 100 = 100%

Weak base examples:

Example A.1. There is 0.30 mol/L, NH_3(aq) dissociation at 25°C. What is the $[OH^-]$, pOH and the percentage of dissociation?

1. Define the base dissociation equation:

$$NH_{3(aq)} + H_2O_{(l)} \rightleftharpoons NH_4{}^+{}_{(aq)} + OH^-{}_{(aq)}$$

2. Define the base dissociation equilibrium expression:

$$K_b = \frac{[NH_4^+][OH^-]}{[NH_3]}$$

$$1.8 \times 10^{-5} = \frac{[NH_4^+][OH^-]}{[NH_3]}$$

3. Compute the initial concentrations and equilibrium concentrations of all existing species:Let x = moles of NH_3 that dissociate to form NH_4^+ and OH^-
   Construct an RICE table as presented in Table A.10 to find the equilibrium concentration of various component types in the solution:
   NH_3 is a weak base (with small K_b), so it only slightly dissociates. The x value will be very small compared to 0.30.
   Therefore, at equilibrium: $[NH_3] \approx 0.30$ mol/L.
4. Use the expression for the base dissociation by substituting the concentration values:

$$K_b = \frac{[NH_4^+][OH^-]}{[NH_3]}$$

Table A.10 Values of the equilibrium concentration.

Species concentration	NH_3	+	H_2O	$\rightleftharpoons$	NH_4^+	+	OH^-
Initial concentration (mol/L):	0.30				0		0
Change in concentration (mol/L):	$-x$				$+x$		$+x$
Equilibrium concentration (mol/L):	$0.30 - x$				x		x

$$1.8 \times 10^{-5} = \frac{[x]\,[x]}{[0.3]} = \frac{x^2}{0.3}$$

$$x^2 = 1.8 \times 10^{-5} \times 0.3 = 5.4 \times 10^{-6}$$

$$x = \sqrt{5.4} \times 10^{-6} = 2.33 \times 10^{-3}$$

$$[OH^-] = x = 2.33 \times 10^{-3}$$

5. Compute pOH:
   $pOH = -\log_{10}[OH^-]$
   $pOH = -\log[2.33 \times 10^{-3}] = 2.63$
6. Calculate the percentage of dissociation:
   Dissociation percentage $= [OH^-]/[\text{base initial}] \times 100$
   $[OH^-] = 2.33 \times 10^{-3}$
   $[NH_3 \text{ initial}] = 0.30\,\text{mol/L}$
   Dissociation percentage $= 2.33 \times 10^{-3}/0.30 \times 100 = 0.77\%$

Example A.2 What is the $[H^+]$ and pH in a 0.7 mol/L aqueous ammonia solution? At 25°C: $K_b = 1.8 \times 10^{-5}$.

1. Define the base dissociation equation:

$$NH_3 + H_2O \rightleftharpoons NH_4^+ + OH^-$$

2. Define the equations of the base dissociation equilibrium:

$$K_b = \frac{[NH_4^+]\,[OH^-]}{[NH_3]} \text{ and } K_b = 1.8 \times 10^{-5}$$

$$1.8 \times 10^{-5} = \frac{[NH_4^+]\,[OH^-]}{[NH_3]}$$

3. Compute the initial concentration and equilibrium concentration of all existing species:
   Let $x = NH_3$ moles that dissociate to produce NH_4^+ and hydroxyl ion (OH^-)
   Construct a table of RICE as presented in Table A.11 to find the equilibrium concentration of each type in solution.

Table A.11 Values of the equilibrium concentration.

Species concentration	NH_3	+	H_2O	$\rightleftharpoons$	NH_4^+	+	OH^-
Initial concentration (mol/L):	0.70				0.0		0.0
Change in concentration (mol/L):	$-x$				$+x$		$+x$
Equilibrium concentration (mol/L):	$0.70 - x$				x		x

NH_3 is known as a weak base (with small K_b), it shows a small dissociates then x will be very small compared to 0.7.

Therefore, at equilibrium, $[NH_3] \approx 0.7\,\text{mol/L}$

4. Use the expression for the base dissociation by substituting the concentration values:

$$K_b = \frac{[NH_4^+][OH^-]}{[NH_3]}$$

$$1.8 \times 10^{-5} = \frac{[x]\,[x]}{[0.3]}$$

$$1.8 \times 10^{-5} = \frac{x^2}{0.7}$$

$$x^2 = 1.8 \times 10^{-5} \times 0.7 = 1.26 \times 10^{-5}$$

$$x = \sqrt{1.26} \times 10^{-5} = 3.55 \times 10^{-3}$$

$$[OH^-] = x = 3.55 \times 10^{-3}$$

5. Measure the H^+ concentration

At 25°C, the equilibrium constant (K_w) for water dissociation is equal to 10^{-14}

Then: $[H^+]\,[OH^-] = 10^{-14}$

$[H^+] = 10^{-14}/[OH^-]$

By substituting the $[OH^-]$ value:

$[H^+] = 10^{-14}/3.55 \times 10^{-3} = 2.81 \times 10^{-12}$ mol/L

6. Calculate the solution pH:

$pH = -\log_{10}[H^+]$

$pH = -\log[2.81 \times 10^{-12}] = 11.55$

References

[1] Green DW, Willhite GP. Enhanced oil recovery. SPE textbook series, vol. 6. Richardson, TX: Society of Petroleum Engineers (SPE); 1998.

[2] Flaaten A, Nguyen QP, Zhang J, Mohammadi H, Pope GA. ASP chemical flooding without the need for soft water. In: SPE annual technical conference and exhibition. Society of Petroleum Engineers; 2008.

[3] Mandal A. Chemical flood enhanced oil recovery: a review. Int J Oil Gas Coal Technol 2015;9:241–64.

[4] Fortenberry R, Kim DH, Nizamidin N, Adkins S, Pinnawala Arachchilage GW, Koh H, et al. Use of cosolvents to improve alkaline/polymer flooding. SPE J 2015;20:255–66.

[5] Fakher S, Abdelaal H, Elgahawy Y, Imqam A. A characterization of different alkali chemical agents for alkaline flooding enhanced oil recovery operations: an experimental investigation. SN Appl Sci 2019;1:1–11.

[6] Samanta A, Ojha K, Mandal A. Interactions between acidic crude oil and alkali and their effects on enhanced oil recovery. Energy Fuel 2011;25:1642–9.

[7] Pei H, Zhang G, Ge J, Jin L, Ding L. Study on the variation of dynamic interfacial tension in the process of alkaline flooding for heavy oil. Fuel 2013;104:372–8.
[8] Wang J, Dong M, Arhuoma M. Experimental and numerical study of improving heavy oil recovery by alkaline flooding in sandpacks. J Can Pet Technol 2010;49:51–7.
[9] Sheng J. Enhanced oil recovery field case studies. Gulf Professional Publishing; 2013.
[10] Lake LW. Enhanced oil recovery. Englewood Cliffs, NJ: Prentice–Hall; 1989.
[11] Sheng J. Modern chemical enhanced oil recovery: Theory and practice. Gulf Professional Publishing; 2010.
[12] DeZabala EF, Vislocky JM, Rubin E, Radke CJ. A chemical theory for linear alkaline flooding. Soc Pet Eng J 1982;22:245–58.
[13] Esfandiarian A, Azdarpour A, Santos RM, Mohammadian E, Hamidi H, Sedaghat M, et al. Mechanistic investigation of LSW/surfactant/alkali synergism for enhanced oil recovery: fluid–fluid interactions. ACS Omega 2020;5:30059–72.
[14] Carey FA. Carboxylic acid derivatives: nucleophilic acyl substitution. Org Chem 2003;5.
[15] Theodorou V, Skobridis K, Tzakos AG, Ragoussis V. A simple method for the alkaline hydrolysis of esters. Tetrahedron Lett 2007;48:8230–3.
[16] Jencks WP, Carriuolo J. General base catalysis of ester hydrolysis. J Am Chem Soc 1961;83:1743–50.
[17] Stefanidis D, Jencks WP. General base catalysis of ester hydrolysis. J Am Chem Soc 1993;115:6045–50.
[18] Johnson SL. General base and nucleophilic catalysis of ester hydrolysis and related reactions. In: Advances in physical organic chemistry, vol. 5. Elsevier; 1967. p. 237–330.
[19] Sheng JJ. Status of alkaline flooding technology. J Pet Eng Technol 2015;5:44–50.
[20] Honarvar B, Rahimi A, Safari M, Rezaee S, Karimi M. Favorable attributes of low salinity water aided alkaline on crude oil-brine-carbonate rock system. Colloids Surf A Physicochem Eng Asp 2020;585. https://doi.org/10.1016/j.colsurfa.2019.124144.
[21] Ehrlich R, Wygal Jr RJ. Interrelation of crude oil and rock properties with the recovery of oil by caustic waterflooding. Soc Pet Eng J 1977;17:263–70.
[22] Shen MD, Chen Z. Chemical behavior of minerals in alkaline solutions. In: Liu S-G, Lu Z-Y, editors. Advance in petroleum geology. Chengdu: Sichun Science and Technology Press; 1996. p. 144–8.
[23] Cooke Jr CE, Williams RE, Kolodzie PA. Oil recovery by alkaline waterflooding. J Pet Technol 1974;26:1–365.
[24] Saha R, Uppaluri RVS, Tiwari P. Influence of emulsification, interfacial tension, wettability alteration and saponification on residual oil recovery by alkali flooding. J Ind Eng Chem 2018;59:286–96.
[25] Tang M, Zhang G, Ge J, Jiang P, Liu Q, Pei H, et al. Investigation into the mechanisms of heavy oil recovery by novel alkaline flooding. Colloids Surf A Physicochem Eng Asp 2013;421:91–100.
[26] Almalik MS, Attia AM, Jang LK. Effects of alkaline flooding on the recovery of Safaniya crude oil of Saudi Arabia. J Pet Sci Eng 1997;17:367–72.
[27] Rubin E, Radke CJ. Dynamic interfacial tension minima in finite systems. Chem Eng Sci 1980;35:1129–38.
[28] Castor TP, Somerton WH, Kelly JF. Recovery mechanisms of alkaline flooding. In: Surface phenomena in enhanced oil recovery. Springer; 1981. p. 249–91.
[29] McCaffery FG. Interfacial tensions and aging behaviour of some crude oils against caustic solutions. J Can Pet Technol 1976;15.
[30] Jennings Jr HY, Johnson Jr CE, McAuliffe CD. A caustic waterflooding process for heavy oils. J Pet Technol 1974;26:1–344.
[31] Mahdavi E, Zebarjad FS. Screening criteria of enhanced oil recovery methods. In: Fundamentals of enhanced oil and gas recovery from conventional and unconventional reservoirs. Elsevier Inc.; 2018.
[32] Li G-Z, Mu J-H, Li Y, Yuan S-L. An experimental study on alkaline/surfactant/polymer flooding systems using nature mixed carboxylate. Colloids Surf A Physicochem Eng Asp 2000;173:219–29.
[33] Nasr-El-Din HA, Hawkins BF, Green KA. Recovery of residual oil using the alkali/surfactant/polymer process: effect of alkali concentration. J Pet Sci Eng 1992;6:381–401.
[34] Tong M, Liu P, Zhang L, Zhu Y. An improved model for interfacial activity of acidic oil/caustic system for alkaline flooding. In: International meeting on petroleum engineering. Society of Petroleum Engineers; 1986.

[35] Safari M, Rahimi A, Gholami R, Permana A, Siaw Khur W. Underlying mechanisms of shale wettability alteration by low salinity water injection (LSWI). J Dispers Sci Technol 2020;1–9.
[36] Rahimi A, Safari M, Honarvar B, Chabook H, Gholami R. On time dependency of interfacial tension through low salinity carbonated water injection. Fuel 2020;280:118492. https://doi.org/10.1016/j.fuel.2020.118492.
[37] Rahimi A, Honarvar B, Safari M. The role of salinity and aging time on carbonate reservoir in low salinity seawater and smart seawater flooding. J Pet Sci Eng 2020;187. https://doi.org/10.1016/j.petrol.2019.106739.
[38] Divandari H, Hemmati-Sarapardeh A, Schaffie M, Ranjbar M. Integrating synthesized citric acid-coated magnetite nanoparticles with magnetic fields for enhanced oil recovery: experimental study and mechanistic understanding. J Pet Sci Eng 2019;174:425–36.
[39] Divandari H, Hemmati-Sarapardeh A, Schaffie M, Ranjbar M. Integrating functionalized magnetite nanoparticles with low salinity water and surfactant solution: interfacial tension study. Fuel 2020;281:118641.
[40] Chiwetelu CI, Hornof V, Neale GH. Mechanisms for the interfacial reaction between acidic oils and alkaline reagents. Chem Eng Sci 1990;45:627–38.
[41] Pei H, Zhang G, Ge J, Ding L, Tang M, Zheng Y. A comparative study of alkaline flooding and alkaline/surfactant flooding for Zhuangxi heavy oil. In: SPE heavy oil conference Canada. Society of Petroleum Engineers; 2012.
[42] Li J, Wang W, Gu Y. Dynamic interfacial tension phenomenon and wettability alteration of crude oil-rock-alkaline-surfactant solution systems. In: SPE annual technical conference and exhibition. Society of Petroleum Engineers; 2004.
[43] Pei H, Zhang G, Ge J, Jin L, Liu X. Analysis of microscopic displacement mechanisms of alkaline flooding for enhanced heavy-oil recovery. Energy Fuel 2011;25:4423–9.
[44] Chan M, Sharma MM, Yen TF. Generation of surface active acids in crude oil for caustic flooding enhanced oil recovery. Ind Eng Chem Process Des Dev 1982;21:580–3.
[45] Yen TF, Chan M, Jang LK. Some applications derived from petroleum component properties for improved waterflooding studies. In: Proc heavy Oil/EOR contractor annual presentations; 1980. p. 413–26.
[46] Lashkarbolooki M, Ayatollahi S. The effects of pH, acidity, asphaltene and resin fraction on crude oil/water interfacial tension. J Pet Sci Eng 2018;162:341–7.
[47] Mehranfar A, Ghazanfari MH. Investigation of the microscopic displacement mechanisms and macroscopic behavior of alkaline flooding at different wettability conditions in shaly glass micromodels. J Pet Sci Eng 2014;122:595–615.
[48] Wang Y, Xu H, Yu W, Bai B, Song X, Zhang J. Surfactant induced reservoir wettability alteration: recent theoretical and experimental advances in enhanced oil recovery. Pet Sci 2011;8:463–76.
[49] Graue A, Fernø MA, Aspenes E, Needham R. Wettability effects on water mixing during waterflood oil recovery. J Pet Sci Eng 2012;94:89–99.
[50] Mohammed M, Babadagli T. Wettability alteration: a comprehensive review of materials/methods and testing the selected ones on heavy-oil containing oil-wet systems. Adv Colloid Interface Sci 2015;220:54–77.
[51] Wagner OR, Leach RO. Improving oil displacement efficiency by wettability adjustment. Trans AIME 1959;216:65–72.
[52] Mungan N. Interfacial effects in immiscible liquid-liquid displacement in porous media. Soc Pet Eng J 1966;6:247–53.
[53] Mungan N. Certain wettability effects in laboratory waterfloods. J Pet Technol 1966;18:247–52.
[54] Kumar K, Dao EK, Mohanty KK. Atomic force microscopy study of wettability alteration by surfactants. SPE J 2008;13:137–45.
[55] Bradley HB. Petroleum engineering handbook. Richardson, TX: SPE; 1987.
[56] Gong H, Li Y, Dong M, Ma S, Liu W. Effect of wettability alteration on enhanced heavy oil recovery by alkaline flooding. Colloids Surf A Physicochem Eng Asp 2016;488:28–35.
[57] Buckley JS. Mechanisms and consequences of wettability alteration by crude oils (Dissertation). Heriot-Watt University; 1996.
[58] Liu Q, Dong M, Asghari K, Tu Y. Wettability alteration by magnesium ion binding in heavy oil/brine/chemical/sand systems—analysis of electrostatic forces. J Pet Sci Eng 2007;59:147–56.

[59] Liu W, Jin Y, Tan X, Yeung A. Altering the wettability of bitumen-treated glass surfaces with ionic surfactants. Fuel 2011;90:2858–62.
[60] Johnson Jr CE. Status of caustic and emulsion methods. J Pet Technol 1976;28:85–92.
[61] Ma S, Dong M, Li Z, Shirif E. Evaluation of the effectiveness of chemical flooding using heterogeneous sandpack flood test. J Pet Sci Eng 2007;55:294–300.
[62] Dong M, Liu Q, Li A. Displacement mechanisms of enhanced heavy oil recovery by alkaline flooding in a micromodel. Particuology 2012;10:298–305.
[63] Mayer EH, Berg RL, Carmichael JD, Weinbrandt RM. Alkaline injection for enhanced oil recovery-a status report. J Pet Technol 1983;35:209–21.
[64] Ge J, Feng A, Zhang G, Jiang P, Pei H, Li R, et al. Study of the factors influencing alkaline flooding in heavy-oil reservoirs. Energy Fuel 2012;26:2875–82.
[65] Srisuriyachai F. Evaluation of alkali flooding combined with intermittent flow in carbonate reservoir (Dissertation). Alma Mater Studiorum Università di Bologna; 2008.
[66] Zhang H, Dong M, Zhao S. Which one is more important in chemical flooding for enhanced court heavy oil recovery, lowering interfacial tension or reducing water mobility? Energy Fuel 2010;24:1829–36.
[67] Ma S. Enhanced heavy oil recovery by dilute alkaline flooding. ProQuest; 2007.
[68] Liu Q. Interfacial phenomena in enhanced heavy oil recovery by alkaline flood. ProQuest; 2007.
[69] Ding B, Zhang G, Ge J, Liu X. Research on mechanisms of alkaline flooding for heavy oil. Energy Fuel 2010;24:6346–52.
[70] Pei H, Zhang G, Ge J, Tang M, Zheng Y. Comparative effectiveness of alkaline flooding and alkaline–surfactant flooding for improved heavy-oil recovery. Energy Fuel 2012;26:2911–9.
[71] Pei H, Zhang G, Ge J, Jin L, Ma C. Potential of alkaline flooding to enhance heavy oil recovery through water-in-oil emulsification. Fuel 2013;104:284–93.
[72] Honarvar B, Rahimi A, Safari M, Khajehahmadi S, Karimi M. Smart water effects on a crude oil-brine-carbonate rock (CBR) system: further suggestions on mechanisms and conditions. J Mol Liq 2020;299:112173.
[73] Xie Q, Brady PV, Pooryousefy E, Zhou D, Liu Y, Saeedi A. The low salinity effect at high temperatures. Fuel 2017;200:419–26.
[74] Okasha TM, Alshiwaish A. Effect of brine salinity on interfacial tension in Arab-D carbonate reservoir, Saudi Arabia. In: SPE middle east oil and gas show and conference. Society of Petroleum Engineers; 2009.
[75] Goodrich JH. Review and analysis of past and ongoing carbon dioxide injection field tests. In: SPE/DOE enhanced oil recovery symposium. Society of Petroleum Engineers; 1980.
[76] Emery LW, Mungan N, Nicholson RW. Caustic slug injection in the Singleton field. J Pet Technol 1970;22:1–569.
[77] Olsen DK, Hicks MD, Hurd BG, Sinnokrot AA, Sweigart CN. In: Design of a novel flooding system for an oil-wet Central Texas carbonate reservoir. SPE/DOE enhanced oil recovery symposium. Society of Petroleum Engineers; 1990.
[78] Somerton WH, Radke CJ. Role of clays in the enhanced recovery of petroleum from some California sands. J Pet Techol 1983;35:643–54.
[79] Liu S. Alkaline surfactant polymer enhanced oil recovery process. Rice University; 2008.
[80] Al-Rossies AAS, Al-Anazi BD, Paiaman AM. Effect of pH-values on the contact angle and interfacial tension. Nafta 2010;61:181–6.
[81] Gomari KAR, Hamouda AA. Effect of fatty acids, water composition and pH on the wettability alteration of calcite surface. J Pet Sci Eng 2006;50:140–50.
[82] Rudin J, Wasan DT. Mechanisms for lowering of interfacial tension in alkali/acidic oil systems 1. Experimental studies. Colloids Surf 1992;68:67–79.
[83] Zhang J, Nguyen QP, Flaaten A, Pope GA. Mechanisms of enhanced natural imbibition with novel chemicals. SPE Reserv Eval Eng 2009;12:912–20.
[84] Takahashi S, Kovscek AR. Spontaneous countercurrent imbibition and forced displacement characteristics of low-permeability, siliceous shale rocks. J Pet Sci Eng 2010;71:47–55.
[85] Takahashi S, Kovscek AR. Wettability estimation of low-permeability, siliceous shale using surface forces. J Pet Sci Eng 2010;75:33–43.

[86] Najafabadi NF, Delshad M, Sepehrnoori K, Nguyen QP, Zhang J. Chemical flooding of fractured carbonates using wettability modifiers. In: SPE symposium on improved oil recovery. Society of Petroleum Engineers; 2008.

[87] Leach RO, Wagner OR, Wood HW, Harpke CF. A laboratory and field study of wettability adjustment in water flooding. J Pet Technol 1962;14:206–12.

[88] Rostami P, Mehraban MF, Sharifi M, Dejam M, Ayatollahi S. Effect of water salinity on oil/brine interfacial behaviour during low salinity waterflooding: a mechanistic study. Petroleum 2019;5:367–74.

[89] Behera US, Sangwai JS. Synergistic effect of brine system containing mixed monovalent (NaCl, KCl) and divalent ($MgCl_2$, $MgSO_4$) salts on the interfacial tension of pure hydrocarbon–brine system relevant for low salinity water flooding. Energy Fuel 2020;34:4201–12.

[90] Suleimanov BA, Latifov YA, Veliyev EF, Frampton H. Comparative analysis of the EOR mechanisms by using low salinity and low hardness alkaline water. J Pet Sci Eng 2018;162:35–43.

[91] Li J, Wang W, Gu Y. Dynamic interfacial tension and oil shrinking effect of crude oil in alkaline solutions. In: Canadian international petroleum conference. Petroleum Society of Canada; 2003.

[92] Sarem AM. Secondary and tertiary recovery of oil by MCCF (mobility-controlled caustic flooding) process. In: Paper SPE; 1974. p. 4901.

[93] Cheng KH. Chemical consumption during alkaline flooding: a comparative evaluation. In: SPE enhanced oil recovery symposium. Society of Petroleum Engineers; 1986.

[94] Prosser AJ, Franses EI. Adsorption and surface tension of ionic surfactants at the air–water interface: review and evaluation of equilibrium models. Colloids Surf A Physicochem Eng Asp 2001;178:1–40.

[95] Jiravivitpanya J, Maneeintr K, Boonpramote T. Experiment on measurement of interfacial tension for subsurface conditions of light oil from thailand. In: MATEC web of conferences, vol. 95. EDP Sciences; 2017. p. 18007.

[96] Xi Z. A comprehensive study and mechanism investigation for alkaline-heavy oil recovery process (Dissertation). Faculty of Graduate Studies and Research, University of Regina; 2019.

[97] French TR, Josephson CB. Surfactant-enhanced alkaline flooding with weak alkalis. Bartlesville, OK (USA): National Inst. for Petroleum and Energy Research; 1991.

[98] Maziah binti Abu Mukhtar M. Mechanism of alkaline flooding in low acid number crude (Dissertation). Universiti Teknologi PETRONAS; 2012.

[99] Buckley JS, Fan T. Crude oil/brine interfacial tensions. Petrophysics 2007;48.

[100] Hornof V, Neale GH, Yu A. Effect of flooding sequence on the displacement of acidic oil by alkaline solutions. J Pet Sci Eng 1994;10:291–7.

[101] Hoeiland S, Barth T, Blokhus AM, Skauge A. The effect of crude oil acid fractions on wettability as studied by interfacial tension and contact angles. J Pet Sci Eng 2001;30:91–103.

[102] Liu Q, Dong M, Yue X, Hou J. Synergy of alkali and surfactant in emulsification of heavy oil in brine. Colloids Surf A Physicochem Eng Asp 2006;273:219–28.

[103] Xinheng C, Songbai T. Review and comprehensive analysis of composition and origin of high acidity crude oils. China Pet Process Petrochem Technol 2011;13:6–15.

[104] Dong M, Ma S, Liu Q. Enhanced heavy oil recovery through interfacial instability: a study of chemical flooding for Brintnell heavy oil. Fuel 2009;88:1049–56.

[105] Bortolotti V, Gottardi G, Macini P, Srisuriyachai F. In: Intermittent alkali flooding in vertical carbonate reservoirs. Europec/EAGE conference and exhibition. Society of Petroleum Engineers; 2009.

[106] Sheng JJ. Investigation of alkaline–crude oil reaction. Petroleum 2015;1:31–9.

[107] Li J, Niu L, Lu X. Performance of ASP compound systems and effects on flooding efficiency. J Pet Sci Eng 2019;178:1178–93.

[108] Mohammadi H. Mechanistic modeling, design, and optimization of alkaline/surfactant/polymer flooding. The University of Texas at Austin; 2008.

[109] Arekhov V, Hincapie RE, Clemens T, Tahir M. Variations in wettability and interfacial tension during alkali–polymer application for high and low TAN oils. Polymers 2020;12:2241.

[110] Smith R, Inomata H, Peters C. Introduction to supercritical fluids: a spreadsheet-based approach. Newnes; 2013.

[111] Chen F, Gu J, Jiang H, Yao X, Li Y. Laboratory evaluation and numerical simulation of the alkali–surfactant–polymer synergistic mechanism in chemical flooding. RSC Adv 2018;8:26476–87.
[112] Winsor PA. Hydrotropy, solubilisation and related emulsification processes. Trans Faraday Soc 1948;44:376–98.
[113] de Castro Dantas TN, de Oliveira AC, de Souza TTC, dos Santos Lucas CR, de Andrade Araújo E, Aum PTP. Experimental study of the effects of acid microemulsion flooding to enhancement of oil recovery in carbonate reservoirs. J Pet Explor Prod Technol 2019;1–9.
[114] Mungan N. Alkaline flooding. In: Heavy crude oil recovery. Springer; 1984. p. 317–52.
[115] Seifert WK, Howells WG. Interfacially active acids in a California crude oil. Isolation of carboxylic acids and phenols. Anal Chem 1969;41:554–62.
[116] Farmanian PA, Davis N, Kwan JT, Weinbrandt RM, Yen TF. Participation of selective native petroleum fractions in lowering interfacial tensions of aqueous alkaline systems. ACS Publications; 1978.
[117] Donaldson EC, Crocker ME. Characterization of the crude oil polar compound extract. Bartlesville, OK (USA): Department of Energy, Bartlesville Energy Technology; 1980.
[118] Arihara N, Yoneyama T, Akita Y, XiangGuo L. Oil recovery mechanisms of alkali-surfactant-polymer flooding. In: SPE Asia Pacific oil and gas conference and exhibition. Society of Petroleum Engineers; 1999.
[119] Falls AH, Thigpen DR, Nelson RC, Ciaston JW, Lawson JB, Good PA, et al. Field test of cosurfactant-enhanced alkaline flooding. SPE Reserv Eng 1994;9:217–23.
[120] Krumrine PH, Falcone Jr JS, Campbell TC. Surfactant flooding 1: the effect of alkaline additives on IFT, surfactant adsorption, and recovery efficiency. Soc Pet Eng J 1982;22:503–13.
[121] Shuler PJ, Kuehne DL, Lerner RM. Improving chemical flood efficiency with micellar/alkaline/polymer processes. J Petrol Tech 1989;41:80–8.
[122] Nelson RC, Lawson JB, Thigpen DR, Stegemeier GL. Cosurfactant-enhanced alkaline flooding. In: SPE enhanced oil recovery symposium. Society of Petroleum Engineers; 1984.
[123] Pitts MJ, Dowling P, Wyatt K, Surkalo H, Adams KC. Alkaline-surfactant-polymer flood of the Tanner field. In: SPE/DOE symposium on improved oil recovery. Society of Petroleum Engineers; 2006.
[124] Pratap M, Gauma MS. Field implementation of alkaline-surfactant-polymer (ASP) flooding: a maiden effort in India. In: SPE Asia Pacific oil and gas conference and exhibition. Society of Petroleum Engineers; 2004.
[125] Wang D, Cheng J, Yang Q, Wenchao G, Qun L, Chen F. Viscous-elastic polymer can increase microscale displacement efficiency in cores. In: SPE annual technical conference and exhibition. Society of Petroleum Engineers; 2000.
[126] Demin W, Zhenhua Z, Jiecheng C, Jingchun Y, Shutang G, Li L. Pilot test of alkaline surfactant polymer flooding in Daqing oil field. SPE Reserv Eng 1997;12:229–33.
[127] Qi Q, Gu H, Li D, Dong L. The pilot test of ASP combination flooding in Karamay oil field. In: International oil and gas conference and exhibition in China. Society of Petroleum Engineers; 2000.
[128] Li H, Liao G, Han P, Yang Z, Wu X, Chen G, et al. Alkline/surfactant/polymer (ASP) commercial flooding test in central Xing2 area of daqing oilfield. In: SPE international improved oil recovery conference in Asia Pacific. Society of Petroleum Engineers; 2003.
[129] Yang X, Liao G, Han P, Yang Z, Yao Y. An extended field test study on alkaline-surfactant-polymer flooding in Beiyiduanxi of Daqing oilfield. In: SPE Asia Pacific oil and gas conference and exhibition. Society of Petroleum Engineers; 2003.
[130] Chang HL, Zhang ZQ, Wang QM, Xu ZS, Guo ZD, Sun HQ, et al. Advances in polymer flooding and alkaline/surfactant/polymer processes as developed and applied in the People's Republic of China. J Pet Technol 2006;58:84–9.
[131] Rudin J, Wasan DT. Mechanisms for lowering of interfacial tension in alkali/acidic oil systems: effect of added surfactant. Ind Eng Chem Res 1992;31:1899–906.
[132] Martin FD, Oxley JC, Lim H. Enhanced recovery of a J sand crude oil with a combination of surfactant and alkaline chemicals. In: SPE annual technical conference and exhibition. OnePetro; 1985.
[133] French TR, Burchfield TE. Design and optimization of alkaline flooding formulations. In: SPE/DOE enhanced oil recovery symposium. Society of Petroleum Engineers; 1990.
[134] Baviere M, Glenat P, Plazanet V, Labrid J. Improvement of the efficiency/cost ratio of chemical EOR processes by using surfactants, polymers, and alkalis in combination. SPE Reserv Eng 1995;10:187–93.

[135] Gao S, Li H, Li H. Laboratory investigation of combination of alkaline-surfactant-polymer for Daqing EOR. SPE Reserv Eng 1995;10:194–7.
[136] Al-Hashim HS, Obiora V, Al-Yousef HY, Fernandez F, Nofal W. Alkaline surfactant polymer formulation for Saudi Arabian carbonate reservoirs. In: SPE/DOE improved oil recovery symposium. Society of Petroleum Engineers; 1996.
[137] Vargo J, Turner J, Bob V, Pitts MJ, Wyatt K, Surkalo H, et al. Alkaline-surfactant-polymer flooding of the Cambridge Minnelusa field. SPE Reserv Eval Eng 2000;3:552–8.
[138] Hirasaki G, Zhang DL. Surface chemistry of oil recovery from fractured, oil-wet, carbonate formation. In: International symposium on oilfield chemistry. Society of Petroleum Engineers; 2003.
[139] Zhang D, Liu S, Yan W, Puerto M, Hirasaki GJ, Miller CA. Favorable attributes of alkali-surfactant-polymer flooding. In: SPE/DOE symposium on improved oil recovery. Society of Petroleum Engineers; 2006.
[140] Pope GA, Nelson RC. A chemical flooding compositional simulator. Soc Pet Eng J 1978;18:339–54.
[141] Delshad M, Asakawa K, Pope GA, Sepehrnoori K. Simulations of chemical and microbial enhanced oil recovery methods. In: SPE/DOE improved oil recovery symposium. Society of Petroleum Engineers; 2002.
[142] Zerpa LE, Queipo NV, Pintos S, Salager J-L. An optimization methodology of alkaline–surfactant–polymer flooding processes using field scale numerical simulation and multiple surrogates. J Pet Sci Eng 2005;47:197–208.
[143] Anderson GA, Delshad M, Brown King CL, Mohammadi H, Pope GA. Optimization of chemical flooding in a mixed-wet dolomite reservoir. In: SPE/DOE symposium on improved oil recovery. Society of Petroleum Engineers; 2006.
[144] Arhuoma M, Yang D, Dong M, Li H, Idem R. Numerical simulation of displacement mechanisms for enhancing heavy oil recovery during alkaline flooding. Energy Fuel 2009;23:5995–6002.
[145] Wang J, Dong M. Simulation of O/W emulsion flow in alkaline/surfactant flood for heavy oil recovery. J Can Pet Technol 2010;49:46–52.
[146] Bhuyan D. Development of an alkaline/surfactant/polymer compositional reservoir simulator (Doctoral dissertation). The University of Texas at Austin; 1991.
[147] Delshad M, Pope GA, Sepehrnoori K. A compositional simulator for modeling surfactant enhanced aquifer remediation, 1 formulation. J Contam Hydrol 1996;23:303–27.
[148] Saad N. Field scale studies with a 3-D chemical flooding simulator. Austin, TX: The University of Texas at Austin; 1989.
[149] Delshad M, Han W, Pope GA, Sepehrnoori K, Wu W, Yang R, et al. Alkaline/surfactant/polymer flood predictions for the Karamay oil field. In: SPE/DOE improved oil recovery symposium. Society of Petroleum Engineers; 1998.
[150] Wu W. Optimum design of field-scale chemical flooding using reservoir simulation. The University of Texas at Austin; 1997.
[151] Mohammadi H, Delshad M, Pope GA. Mechanistic modeling of alkaline/surfactant/polymer floods. SPE Reserv Eval Eng 2009;12:518–27.
[152] Hand DB. Dineric distribution. J Phys Chem 2002;34:1961–2000.
[153] Korrani AK, Sepehrnoori K, Delshad M. A mechanistic integrated geochemical and chemical-flooding tool for alkaline/surfactant/polymer floods. SPE J 2016;21:32–54.
[154] Charlton SR, Parkhurst DL. Modules based on the geochemical model PHREEQC for use in scripting and programming languages. Comput Geosci 2011;37:1653–63.
[155] Dang C, Nghiem L, Nguyen N, Yang C, Chen Z, Bae W. Modeling and optimization of alkaline-surfactant-polymer flooding and hybrid enhanced oil recovery processes. J Pet Sci Eng 2018;169:578–601.
[156] Nghiem L, Skoreyko F, Gorucu SE, Dang C, Shrivastava V. A framework for mechanistic modeling of alkali-surfactant-polymer process in an equation-of-state compositional simulator. In: SPE reservoir simulation conference. Society of Petroleum Engineers; 2017.
[157] Dang C, Nghiem L, Nguyen N, Chen Z, Nguyen Q. Mechanistic modeling of low salinity water flooding. J Pet Sci Eng 2016;146:191–209.
[158] Bhuyan D, Lake LW, Pope GA. Mathematical modeling of high-pH chemical flooding. SPE Reserv Eng 1990;5:213–20.

[159] Ertekin T, Sun Q. Artificial intelligence applications in reservoir engineering: a status check. Energies 2019;12:2897.
[160] Wang B, Wang X, Chen Z. A hybrid framework for reservoir characterization using fuzzy ranking and an artificial neural network. Comput Geosci 2013;57:1–10.
[161] Anifowose FA, Labadin J, Abdulraheem A. Hybrid intelligent systems in petroleum reservoir characterization and modeling: the journey so far and the challenges ahead. J Pet Explor Prod Technol 2017;7:251–63.
[162] Lim J-S. Reservoir properties determination using fuzzy logic and neural networks from well data in offshore Korea. J Pet Sci Eng 2005;49:182–92.
[163] Mohaghegh SD. Recent developments in application of artificial intelligence in petroleum engineering. J Pet Technol 2005;57:86–91.
[164] Al-Mudhafer WJM. Coupling programming of an adaptive genetic algorithm with reservoir flow simulation for infill oil well locations optimization. In: SPE Western regional & AAPG Pacific section meeting 2013 joint technical conference. Society of Petroleum Engineers; 2013.
[165] Ani M, Oluyemi G, Petrovski A, Rezaei-Gomari S. Reservoir uncertainty analysis: The trends from probability to algorithms and machine learning. In: SPE intelligent energy international conference and exhibition. Society of Petroleum Engineers; 2016.
[166] Ramgulam A, Ertekin T, Flemings PB. Utilization of artificial neural networks in the optimization of history matching (Dissertation). Pennsylvania State University; 2006.
[167] Sun Q, Ertekin T. Development and application of an artificial-neural-network based expert system for screening and optimization of polymer flooding projects. In: SPE Kingdom of Saudi Arabia annual technical symposium and exhibition. Society of Petroleum Engineers; 2018.
[168] Ramakrishnan TS, Wasan DT. A model for interfacial activity of acidic crude oil/caustic systems for alkaline flooding. Soc Pet Eng J 1983;23:602–12.
[169] Ghosh S, Johns RT. An equation-of-state model to predict surfactant/oil/brine-phase behavior. SPE J 2016;21:1–106.
[170] Ghosh S. A novel equation-of-state to model microemulsion phase behavior for enhanced oil recovery applications (Dissertation). The Pennsylvania State University; 2015.
[171] Ghosh S, Johns RT. A modified HLD-NAC equation of state to predict alkali/surfactant/oil/brine phase behavior. SPE J 2018;23:550–66.
[172] Sun Q. The development of an artificial-neural-network-based toolbox for screening and optimization of enhanced oil recovery projects. The Pennsylvania State University; 2017.
[173] Khazali N, Sharifi M, Ahmadi MA. Application of fuzzy decision tree in EOR screening assessment. J Pet Sci Eng 2019;177:167–80.
[174] Nabil MN. Screening EOR Dengan machine learning berbasis fuzzy logic (Dissertation). Muhammad Naufal Nabil; 2021.
[175] Tarybakhsh MR, Sadeghi MT, Assareh M. Development of a data-driven fuzzy screening model for enhanced oil recovery methods using an adaptive weighting system. Can J Chem Eng 2019;97:3035–51.
[176] Biyanto TR, Rahman AH, Abdillah AI, Bethiana TN, Irawan S. Application of killer whale algorithm in ASP EOR optimization. Procedia Comput Sci 2017;124:158–66.
[177] Tariq Z, Abdulraheem A, Khan MR, Sadeed A. New inflow performance relationship for a horizontal well in a naturally fractured solution gas drive reservoirs using artificial intelligence technique. In: Offshore technology conference Asia. Offshore Technology Conference; 2018.
[178] Dang C, Nghiem L, Fedutenko E, Gorucu E, Yang C, Mirzabozorg A. Application of artificial intelligence for mechanistic modeling and probabilistic forecasting of hybrid low salinity chemical flooding. In: SPE annual technical conference and exhibition. Society of Petroleum Engineers; 2018.

CHAPTER 5

Alkaline-surfactant polymer (ASP)

Meysam Nourani[a] and Saeid Sadeghnejad[b]
[a]Ugelstad Laboratory, Department of Chemical Engineering, Norwegian University of Science and Technology (NTNU), Trondheim, Norway
[b]Department of Petroleum Engineering, Faculty of Chemical Engineering, Tarbiat Modares University, Tehran, Iran

5.1 Introduction

Alkaline surfactant polymer (ASP) flooding is a major area of interest within the field of chemical enhanced oil recovery (CEOR), and is a combination of alkaline flooding, surfactant flooding, and polymer flooding [1]. In the ASP flooding process, alkaline, surfactant, and polymer constituents are added to the same solution slug. The scheme of an ASP flooding process is shown in Fig. 5.1. Among the available CEOR methods, ASP flooding is the most promising technique since it creates synergy between the alkaline, surfactant, and polymer components [4, 5].

ASP flooding benefits from having an integrated chemical phase behavior of the in situ-generated natural surfactant and the injected surfactant. This is one of its main advantages compared to other CEOR methods. Consequently, substantial progress has made ASP flooding a practical approach for field-enhanced oil recovery and more impressive than other CEOR approaches [6]. Other CEOR methods have several disadvantages, such as the long duration of dilute alkaline flooding and adsorptive surfactant loss in surfactant flooding, but the ASP method can alleviate such challenges [7].

A simulation study compared water injection, polymer flooding, surfactant polymer (SP) flooding, and ASP flooding scenarios for a pilot area. As shown in Fig. 5.2, maximum incremental oil recovery was observed within one pore volume (PV) injection of ASP [8]. A series of core flood experiments using heterogeneous sand packs showed that ASP flooding after polymer flooding can lead to the recovery of an additional 32% of the original oil in place (OOIP) and cumulative oil recovery of 82% OOIP [9]. In addition, the ASP process has been recognized as a cost-effective CEOR process that can increase tertiary oil recovery in some oilfields, such as in the Shengli, Henan, and Daqing fields [10–13].

5.2 Synergy of alkaline, surfactant, and polymer constituents

The synergy between an injected surfactant and in situ-produced soap is important in ASP flooding. Usually, the optimal salinity concentration for soap is idealistically low. The injected alkaline concentration must be low to meet the low optimal salinity

Chemical Methods
https://doi.org/10.1016/B978-0-12-821931-7.00011-0

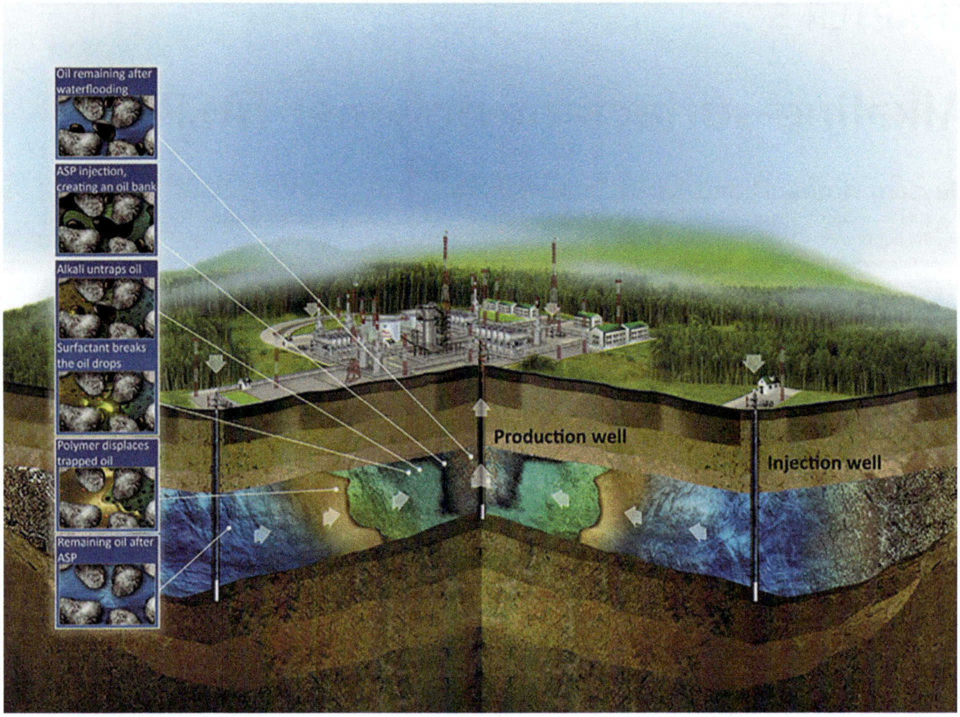

Fig. 5.1 Scheme of an ASP flooding process [2, 3].

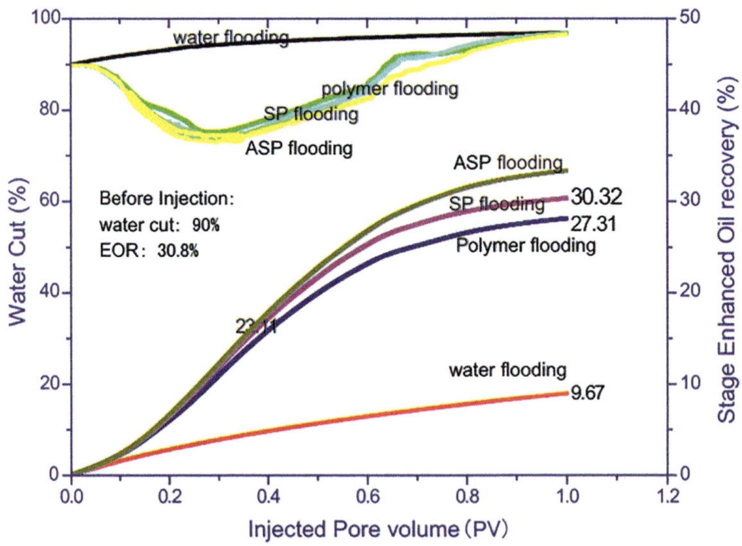

Fig. 5.2 The incremental oil recovery of different scenarios within one PV injection [8].

condition. Therefore, because the injected alkaline level is lower than the amount of consumption, the alkaline does not spread efficiently. Additionally, because the optimum salinity for a surfactant is high, a synthetic surfactant is added. After the surfactant and the soap are combined, the optimum salinity range in which IFT reaches a minimum is extended and increased [14, 15]. The synergistic effect of alkalinities, surfactants, and polymers in ASP flooding can be summarized as follows [16]: (1) the decrease in water production is related to the increase in the viscosity of the displacing fluid as well as scaling and emulsification after the injection of the ASP slug [17]; (2) due to the emulsification of surfactants and alkalis, polymers aid in stabilizing emulsions, which promotes the sweep efficiency; (3) the interaction between the polymer and the surfactant alters the stretched molecular chain and the configuration of polymer aggregation. Additionally, under high-salt conditions, the nonpolar part of the surfactant combines with the molecular chain of the polymer to form an association; these variations lead to an increase in the viscosity of the displacing fluid [18]; (4) the polymer can react with calcium and magnesium cations to avoid the conversion of the surfactant to salt with low activity; (5) the adsorption of surfactant on rock surfaces and the overall cost can be reduced by the addition of polymer due to the adsorption competition between the surfactant and the polymer [15]; (6) surfactant adsorption is reduced under alkaline conditions, which increases the surfactant efficiency; and (7) the soap generated from the reaction between the alkaline component and organic acid benefits the injected surfactant and increases interfacial tension (IFT) reduction and emulsification effects.

5.3 Polymer effect

The polymer flooding method is similar to water flooding, except that a water-soluble polymer is dissolved in water to increase the viscosity of the solution. Therefore, this process is also called thickened water flooding. By increasing the viscosity of the polymer solution, viscous fingering decreases, and the sweep efficiency improves [15]. Eq. (5.1) expresses the corresponding simplified fractional flow equation where the gravity and capillary forces are neglected [19]:

$$f_w = \frac{\frac{K_{rw}}{\mu_w}}{\frac{K_{rw}}{\mu_w} + \frac{K_{ro}}{\mu_o}} \tag{5.1}$$

where μ_o, μ_w, K_{ro}, and K_{rw} are the viscosities and relative permeabilities of oil and water, respectively. The ratio of the effective permeability of a fluid to its viscosity is defined as fluid mobility [20]. Rearranging Eq. (5.1) yields the following expression [15]:

$$f_w = \frac{1}{1 + \frac{\lambda_o}{\lambda_w}} \tag{5.2}$$

where λ_o and λ_w are the mobilities of oil and water, respectively. According to Eqs. (5.1) and (5.2), adding polymer increases the water viscosity and consequently decreases the water mobility and the water cut. Thus, increased oil recovery can be achieved at the same volume of injected water compared with that from water flooding. Crossflow between vertically heterogeneous layers expands the polymer distribution in the vertical layers, which leads to an improved vertical sweep efficiency [21]. In addition to an increase in water viscosity, polymer flooding results in another mechanism called disproportionate permeability reduction. As polymers are dispersed in the water phase, the retention of polymers can reduce the water relative permeability more than the oil permeability. Although the retention of polymer includes adsorption, mechanical entrapment, and precipitation mechanisms, the adsorption of polymer is the main contributor to the diminution of the mean velocity of polymer molecules throughout their distribution in porous media [12]. The combination of decreasing the relative permeability of water and increasing the water viscosity leads to a decrease in water mobility. Therefore, the mobility ratio of water to oil $\left(\frac{\lambda_w}{\lambda_o}\right)$ is decreased. When the mobility ratio becomes less than one, water moves slower than oil, causing more oil to be recovered [22].

As shown in Fig. 5.3, factors such as hydrodynamically trapped polymer molecules in stagnant zones and the mechanical entrapment of polymer macromolecules in a porous medium can lead to the retention of polymers. The entrapment of polymer molecules in porous medium occurs based on the size of polymer molecules, and molecules larger than pore throats are trapped [12].

As the retention of polymers is an irreversible process, permeability reduction can be maintained after polymer flooding, and this result is called the residual permeability effect. The residual permeability effect will lead to increased oil production based on the following water flooding process. The polymer decreases surfactant adsorption in an ASP system as a result of the competition for adsorption sites between the surfactant and polymer [15].

In addition, large polymer molecules can shield some electrically positive locations from covering by anionic surfactants, consequently decreasing surfactant adsorption.

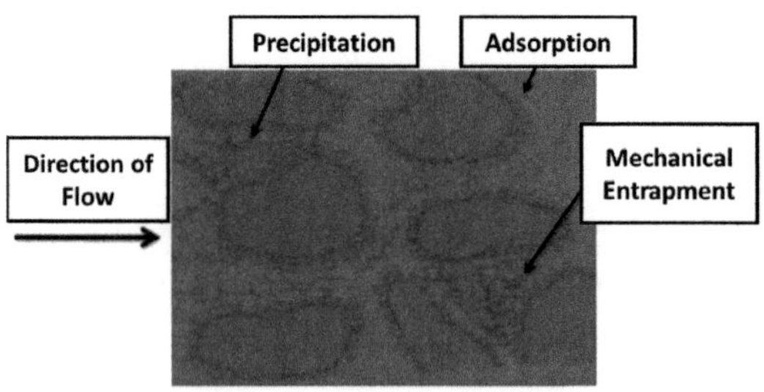

Fig. 5.3 Mechanisms of polymer retention in the porous medium [12].

These mechanisms are vital in polymer flooding in connection with oil recovery [23]. Additionally, a secondary mechanism is linked to the viscoelastic behavior of polymers. There is normal stress between the polymer solution and oil due to the viscoelastic properties of the polymer. Therefore, polymers apply a high pulling force on oil films or oil droplets. Polymers push and pull oil out of dead-end pores. Consequently, ultimate oil recovery is increased [24, 25]. Obviously, the viscosity of the polymer solution increases with the concentration of the polymer. The polymer solution viscosity decreases in the presence of salt due to a reduction in the hydrodynamic volume of the polymer chain. The hydrodynamic diameter of the polymer decreases because of the shielding of ionic groups, which leads to reduced repulsion among the intrachain polymer molecules [26]. To estimate the size of polymer molecules at reservoir temperature and salinity levels, an empirical relationship between the hydraulic diameter of polymer molecules [D (m)], and the salt concentration [C_{Salt} (wt%)], temperature [T (°C)], sulfonation content [$C_{Sulf.}$(wt%)] and molecular weight [M_w (million Daltons)] was developed as [27]

$$\text{Ln}\, D = -2.39 \times 10^{-2} C_{Salt} - 2.6 \times 10^{-4}\, T + 5.06 \times 10^{-3} C_{Sulf.} + 3.40 \times 10^{-2} M_w - 15.0939 \tag{5.3}$$

Estimates of the hydraulic diameter of a polymer molecule with Eq. (5.3) can be used to predict the intrinsic viscosity of polymer solutions under reservoir salinity and temperature conditions using the following equation [27, 28]:

$$[\eta] = \frac{\pi}{600} \frac{N_A D^3}{M_w} \tag{5.4}$$

where N_A and $[\eta]$ are the Avogadro's number ($6.02 \times 10^{23} \text{mol}^{-1}$) and intrinsic viscosity (dl/g), respectively. If the polymer content M_w is high, the hydraulic diameter of the polymer molecule and the solution viscosity will be higher than those for a low-M_w solution at the same concentration. The polymer solution experiences shear thinning under usual reservoir conditions. The viscosity of the polymer solution is a function of the shear rate and can be modeled by a power law equation [29]. Due to the increased salt effect for alkaline solutions, it has been suggested that the polymer solution viscosity decreases with the pH [30]. However, this relation is not always valid, and it has also been reported that the polymer solution viscosity increases with the pH [31]. These contradictory results related to the pH are possibly linked to the early-time hydrolysis effect. The pH relation is complex considering different salt and hydrolysis effects [15].

Water and the treatment of water are expensive in some locations, such as desert areas and offshore environments. Since a polymer can improve the sweep efficiency, less water is produced and injected than in other nonpolymer cases. Therefore, in addition to increasing the viscosity of the solution, polymers generate a positive economic impact by reducing the amounts of injected and produced water compared with those for water flooding [1, 15].

5.4 Emulsion properties and stability

An emulsion is a temporary combination of oil and water produced by distributing one phase into very small droplets. In ASP flooding, emulsification is a vital mechanism for improving oil recovery. Laboratory core flooding and field studies have shown that emulsification increases oil recovery and decreases water cut levels [32, 33].

In alkaline, surfactant, and even in water flooding conditions, stable emulsions can be produced. Polymers aid in stabilizing emulsions and do not produce emulsions with oils. As alkaline components react with crude oil to produce in situ surfactants, stable emulsions are generated in alkaline flooding. Additionally, as a surfactant reduces the IFT in surfactant flooding, stable emulsions can be generated. Due to the existence of natural emulsifiers such as asphaltene in crude oil, stable emulsions can be generated in water injection as well. There are four common categories of emulsions: oil suspended in water (O/W), water suspended in oil (W/O), oil-in-water-in-oil (O/W/O), and water-in-oil-in-water (W/O/W). Normally, W/O emulsions display much higher stability than O/W emulsions. Commonly, O/W and W/O emulsions are produced at high and low water cut levels, respectively. Producing a W/O emulsion is easy in an oil-wet rock. In a continuous heavy oil system, W/O emulsions are much more common than O/W emulsions [1]. When a surfactant solution comes in contact with oil and brine, three types of microemulsions can be produced as the salinity is increased: Winsor type I (O/W), Winsor type II (W/O), and Winsor type III. In Winsor types I and II, the microemulsion phases are in equilibrium with the oil phase in excess and the water phase in excess, respectively. Winsor type III is a bicontinuous middle-phase microemulsion where the microemulsion phase is in equilibrium with both water and oil phases [34]. Winsor types I, II, and III are also called type II(−)\lower-phase microemulsions, type II(+)\upper-phase microemulsions, and middle-phase microemulsions, respectively [35, 36]. Fig. 5.4 shows a schematic representation of Winsor type II(−)/(O/W) and type II(+)/(W/O).

Emulsion sizes range between 0.1 μm and tens of micrometers. Normally, the average diameter of formed emulsions is less than 10 μm. The average diameter of dispersed water droplets in a W/O emulsion and the average diameter of dispersed oil droplets in an O/W emulsion are 5.9 and 8 μm, respectively. A small emulsion size corresponds to high stability. The stability of W/O emulsions increases with increasing surfactant, alkaline, and polymer concentrations. The stability of O/W emulsions increases with surfactant and polymer concentrations, but it decreases with the alkaline concentration [37].

The stability of emulsions increases by decreasing the required energy to create emulsions. This energy can be estimated using the following equation [1]:

$$E = \Delta A \gamma_{ow} \tag{5.5}$$

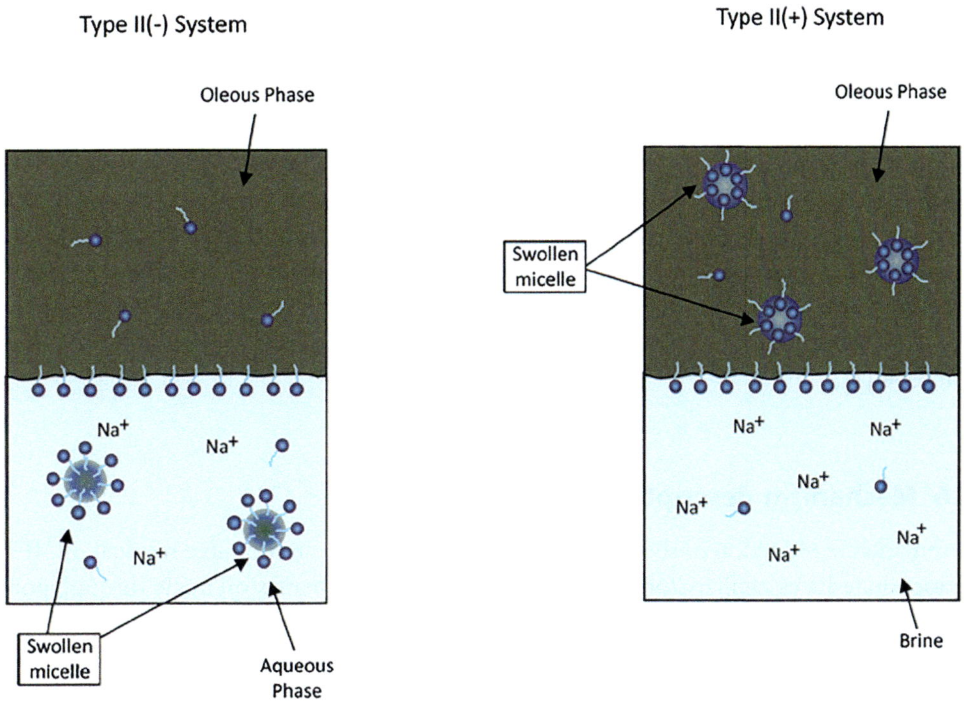

Fig. 5.4 Schematic representation showing Winsor type II(−)/(O/W) and type II(+)/(W/O) [36].

where ΔA and γ_{ow} are the increased interface area and the IFT between oil and water, respectively. The addition of surfactant decreases the IFT and, thus, according to Eq. (5.5), the required energy to create emulsions will be decreased. Consequently, the stability of emulsions increases with surfactant addition and reduction of IFT. Yet despite this, some emulsions were reported unstable even though at very low IFT conditions. Therefore, it should be noted that low IFT is a positive factor but it is not a determining element for emulsion stability [1, 30].

5.5 ASP compatibility

The compatibility of an alkaline-surfactant system can be determined by light absorbance measurements. When light passes through a solution, some of it is absorbed due to interactions with the molecules and atoms in the medium. Light absorbance is the quantity of light absorbed by a solution. Eq. (5.6) expresses the light absorbance definition [38]:

$$A = \log\frac{I_0}{I_1} \tag{5.6}$$

where $\frac{I_0}{I_1}$ is the ratio of the intensity of the incident beam to the intensity of transmitted light. As phase interfaces scatter the light that passes through them, emulsion tends to increase the light absorbance. A low level of light absorbance indicates high compatibility [30]. When light absorbance decreases with the alkaline concentration, the system compatibility improves until reaching the optimum alkaline concentration at which light absorbance is minimized and the compatibility of the alkaline-surfactant system is maximized. When the alkaline concentration increases further, the compatibility decreases. However, polymers promote surfactant transportation by increasing the water viscosity, and weak interactions between the surfactants and polymer exist. By adding polymer to the surfactant solution, the stability of dynamic IFT and solution compatibility slightly improve [1].

5.6 Mechanism descriptions

A remarkable reduction in the IFT between oil and water can mobilize residual oil. IFT reduction reduces capillary forces and facilitates the flow of matrix oil freely through pore throats without renewed trapping. The main mechanism for surfactant in ASP flooding is the reduction in the IFT between brine and trapped oil, which contributes to the development of oil banks and supports mobilization [39, 40]. In addition, emulsion flooding and wettability alteration can be significant under some conditions [15]. A capillary number is a dimensionless number that describes the ratio of viscous to capillary forces and is defined as [41]

$$\mathrm{N_c} = \frac{\mu v}{\gamma_{ow} \cos\theta} \tag{5.7}$$

Many researchers have reported that the residual oil saturation decreases when the capillary number increases [42–45], where μ is the displacing fluid, v is the displacing Darcy velocity and θ is the contact angle. The typical capillary number before CEOR is approximately 10^{-7}. The capillary number must be increased by more than 1000 times to reduce residual oil saturation. This significant increase in the capillary number can occur by adding a surfactant to reduce the IFT from 20 to 30 to 10^{-3} mN/m [15]. One mechanism of alkaline flooding in the ASP process is to reduce the adsorption of the surfactant during displacement by generating an in situ surfactant. This process promotes the efficiency of the surfactant, and consequently, less surfactant will be consumed. The in situ-produced surfactant is also called soap to differentiate it from an injected synthetic surfactant [46, 47]. In addition, the presence of alkaline conditions can change the rock wettability and promote a transition to either more oil-wet or more water-wet conditions [40]. Other alkaline mechanisms comprise bubble entrapment, emulsification, and oil entrainment [32].

5.7 Factors that influence IFT

Crude oil is composed of natural surface-active agents such as organic acids, alcohols and salts. The aggregation of these natural surfactants at the interface of oil and brine produces an adsorbed film that reduces the IFT between the crude oil and brine. The structure of the film can change significantly depending on the properties of the crude oil, such as the sulfur content, oil density, salt and metal content, pour point and viscosity [48].

Hence, the surface properties such as charge, viscosity and elasticity, and molecular packing of the adsorbed film are important factors that control numerous phenomena, such as the oil drop migration in porous medium and coalescence of emulsion droplets [12]. The dynamic IFT between crude oil and ASP is also controlled by the surface-active materials produced from the reaction between the alkaline and heavy components in crude oil [1]. Crude oils with boiling points higher than 350°C are enriched in polar compounds, such as asphaltenes and resins [49]. The interfacial activity increases with an increasing number of reactions between crude oil polar compounds and ASP. The produced emulsion is more stable when the crude oils contain higher molecular weights and molecular polarities than it is under other conditions. Therefore, for heavy crude oil, even with a near-zero acid number, surface-active agents such as asphaltenes and resins interfere with the surfactant in the oil/brine interfacial layer [50]. The accumulation of surfactant molecules at the brine/oil interface is crucial to achieving ultralow IFTs. By increasing the alkaline concentration, the IFT decreases, which indicates that oil and alkaline conditions produce surface-active materials and that surface-active molecules accumulate at the alkaline/oil interface. It is generally assumed that the effect of polymers on the IFT is negligible. By adding polymer, however, slightly additional IFT reduction can be observed, indicating increased interface activity in the presence of polymer [51–53]. There is no agreement on how a polymer affects the IFT between crude oil and alkaline solutions, but some explanations have been given, such as the existence of the maximum electrical density at the interface and low surfactant associations with divalent. A surfactant can be adsorbed at the interface with a polymer to build an adsorption layer of mixtures. This mixed layer has a high electrical density and leads to a reduction in the IFT. Additionally, surfactants can be shielded by polymers from being associated with divalents to produce complexes; therefore, in this case, more surfactant molecules are available to reduce the IFT [1].

5.8 Factors that influence wettability

Alkaline and surfactant solutions can alter wettability toward more water-wet sandstone formations [32, 54, 55], and wettability variations from water-wet conditions to intermediate or oil-wet conditions have also been reported [56–58]. Wettability alteration toward more oil-wet conditions is considered one of the important mechanisms for

CEOR based on alkaline flooding due to adsorption of heavy oil components on the surface of the rock and the blocking of water channels [56]. For sandstone formations, wettability is dependent on optimal salinity. When the salinity is below or above the optimal salinity, wettability is likely to be preferentially water wet or oil wet, respectively [40]. Wettability measurements for different types of carbonate formations with different surfactants and sodium carbonate characteristics reflected wettability alteration toward water wetting with low anionic surfactant concentrations [54, 59, 60]. As the polymer is dissolved in the water phase, the adsorption of the polymer can make a rock surface increasingly water wet. However, the adsorption of polymer on the rock surface can lead to reduced rock polarity and water wettability. Therefore, the presence of polymers can also change rock wettability toward different oil-wet or water-wet states [1, 61].

5.9 Phase separation

Phase separation is the conversion of a single homogeneous phase into two discrete phases [62]. A sharp decrease in light transmittance with the Na^+ concentration is an obvious indication of the SP phase separation. At an equal cation concentration, monovalent anions (Cl^-, OH^-) and trivalent anions (PO_4^{3-}) display the largest and smallest effects on SP phase separation, respectively. The effect of divalent anions (CO_3^{2-} and SO_4^{2-}) on phase separation was reported to be higher than that of trivalent anions and lower than that of monovalent anions. The remaining surfactant concentration in the solution decreases with the divalent concentration. The Mg^{2+} effect on the percentage of remaining surfactant concentration in SP and ASP systems after 35 days at reservoir temperature was reported to be much larger than Ca^{2+}. The Mg^{2+} effect on the remaining surfactant concentration in an ASP system can be substantially larger than that in an AS system. Therefore, Mg^{2+} promotes noncompatibility between surfactants and polymers in ASP systems besides influencing surfactant precipitation in AS systems [1].

5.10 Surfactant polymer adsorption

The loss of the injected surfactants due to adsorption onto reservoir rock surfaces can lead to decreased oil recovery. Therefore, surfactant adsorption is one of the major challenges in determining the feasibility of a surfactant flooding project [63–65]. The rock mineral composition, surface charge, salinity, pore structure, chemical structure of the surfactant, ionic strength, and pH are the main parameters that influence the adsorption of surfactant onto the surface of rock [66]. In addition, the adsorption of surfactant is controlled by a range of ionic and molecular forces between the solid surfaces and solution, involving the interaction between the hydrocarbon chains and hydrophobic sites on the solid, hydrogen bonding, covalent bonding, electrostatic attraction and the association of methyl groups in hydrocarbon chains [67]. Generally, mitigating surfactant adsorption is a considerable challenge in designing a surfactant flooding project, and decreasing surfactant adsorption can be one of the important mechanisms of applying polymers in ASP methods. Anionic surfactants are normally applied for sandstone reservoirs to avoid the high adsorption of surfactants due to the electrostatic repulsion between the anionic surfactants and the negative charge of the rock surface [68]. Due to the adsorption of polymers on the surface of the rock, the permeability of polymer flow paths decreases; therefore, the water flow velocity decreases as well. As polymer adsorption is an irreversible process, the permeability of the subsequent water remains low after polymer flooding, and the polymer cannot be simply flushed away [15]. The experimental results indicate that the application of polymers, either by preflushing or coinjecting, can efficiently decrease the adsorption of surfactant. In a laboratory core flooding study, when only surfactant was injected, the maximum adsorption of surfactant by carbonate rock cores was determined to be 0.203 mg/g-rock. The maximum surfactant adsorption was reduced to 0.0848 and 0.0872 mg/g-rock for preflushed samples with polymer and for surfactant and polymer coinjected samples, respectively. Thus, the presence of polymer in an SP system decreases surfactant adsorption [66]. This outcome is owing to the protection of surfactant molecules through the occupation of electrically positive sites by large polymer molecules and the competition for adsorption sites between the surfactant and polymer constituents [1].

5.11 Modeling and simulations

Several simulation software programs have been developed to assess the ASP process. Simulation studies can minimize the complexity and number of flooding measurements and support reservoir management in recognizing the characteristics of the ASP process [69]. The explanation of a chemical process needs knowledge of numerous parameters. Preferably, these parameters should be measured in the laboratory. However, all these parameters cannot be feasibly measured due to high cost, long time, and lack of access requirements. Occasionally, some of these parameters can be obtained by tuning model variables through history matching by means of simulation. Therefore, a suitable numerical simulation program can history match the results of laboratory core flood experiments to tune ASP flow parameters. The tuned or upscaled parameters can be used in a sector model or at the pilot or field scale to optimize the flooding design and estimate the ASP efficiency. In addition, the calibrated parameters should be applied at both the core flood scale model and in the upscaled model to perform sensitivity studies and engineering analyses [15].

Many simulation studies have been reported for SP and ASP processes using both academic and commercial software, such as CMG-STARS, Eclipse, MoReS, and UTCHEM [15, 36, 70–75]. UTCHEM, which was designed and developed by the University of Texas, is claimed to be the best simulator for capturing ASP mechanisms [74].

The key limitation of commercial simulators is that the three types of microemulsions for surfactant solutions are generally not considered; additionally, the detailed geochemistry reactions linked to alkaline flooding are often not involved in these programs [15]. PHREEQC is a geochemistry software that includes the reactions between an aqueous solution and minerals, gases, exchangers, and sorption surfaces. MoReS, Shell's dynamic reservoir simulator, can be coupled with PHREEQC to simulate ASP processes [75]. An implicit, parallel-based, in-house simulator has been developed to study the SP flooding process. This in-house simulator can work with all two- and three-phase emulsion types based on salinity and includes a shear thinning and thickening rheology model for polymers, although degradation mechanisms are not considered [76]. One of the most complete academic simulators for SP flooding is a 2D mathematical model for a three-phase, six-component system. This academic simulator considers phenomena such as the effects of surfactant-polymer interactions (SPIs) on viscosity, the IFT, and adsorption [77].

5.12 Application of machine learning

The success of ASP flooding depends on numerous factors, such as reservoir and operating conditions, brine and crude oil properties and, most notably, the economic feasibility of the project. However, there are certain concerns, such as the high cost and the large time consumption of extensive core flooding tests and simulations, and these metrics are traditionally used to estimate the CEOR process efficiency prior to being applied at the reservoir and field scales. In addition to considering the inevitably high uncertainties in the flooding process, an injection strategy is essential for optimizing the economic and technical aspects of a project [78–80]. To study proxy models for different ASP flooding situations, an optimization methodology was proposed and used at the full field scale with a slight number of numerical simulations [81]. In addition, a response surface methodology (RSM) was used as an efficient optimization approach for an ASP pilot project with an optimized polymer slug size 69% less than that proposed based on a measurement plan [82]. Artificial neural networks (ANNs) have been applied broadly in many parts of the oil industry, but few studies have considered ANNs in chemical flooding. Additionally, an optimization tool must be applied alongside the network model to fully assess the proficiency of models and estimate the most leading screening values to reach the goal [79]. Enhanced oil recovery (EOR) and inflow performance relationship (IPR) models for polymer flooding have been built based on back-propagation (BP) neural network and orthogonal designs with different oil prices [83]. Additionally, an ANN model was successfully constructed for SP flooding with negligible mean absolute error from a large set of simulation records. The model displayed significant time savings in performance prediction with consistent precision [84]. In another major study, an ANN was effectively created for CEOR based on a large amount of simulation data, and it considered the real conditions of a viscous oil field for a quarter five-spot pattern.

Based on the high level of agreement between the outputs estimated by the ANN and those estimated by the simulation, it was suggested that the model effectively simulated the proficiency of ASP flooding with an extensive range of parameters. The feasibility of this injection strategy considering certain costs and optimal chemical designs verified the effectiveness of ASP flooding for high-viscosity oil reservoirs. In addition, the sensitivity study confirmed the dominant effect of the ASP slug size on the profitability of the flooding scheme. As the considered output of the model was the ultimate recovery factor in place of the net present value (NPV), the model can be submissively adapted based on changes in the market situation induced by variations in the oil price [79].

5.13 Optimization the design of ASP injection

During low-oil-price periods, the economic profit of a chemical flooding process is often low because of complex reservoir conditions, the high cost of chemicals and the uncertainty of the flooding process [85]. However, the chemical cost is the main factor that makes the chemical flooding process unfeasible. It is estimated that the SP process is economically unfeasible when the oil price becomes cheaper than $60/bbl [86]. Therefore, the optimal design of ASP flooding is vital for achieving economically feasible chemical flooding.

In a core flooding study, five experimental Schemes (A, B, C, D, and E) were explored to examine how to maximize incremental oil recovery in ASP flooding while keeping the total amount of each chemical constant. In Scheme A, 0.6-PV ASP was injected. In Scheme B, the alkaline concentration gradient was reduced from 1.4% PV to 1.2% PV and then to 1.0% PV. In Scheme C, the alkaline concentration gradient increased from 1.0% PV to 1.2% PV and then to 1.4% PV. In Scheme D, the surfactant concentration gradient was reduced from 0.3% PV to 0.2% PV and then to 0.1% PV. In Scheme E, the surfactant concentration gradient increased from 0.1% PV to 0.2% PV and 0.3% PV. The incremental oil recovery factors (IORF) in Schemes B and D were clearly higher than those in Scheme A. The surfactant and alkaline concentration gradients from high to low can conquer the negative consequences at the displacement front triggered by alkaline consumption, dilution, and surfactant adsorption. Because alkalis can react with the acid in crude oil to produce in situ surfactants and decrease IFT, a preflushed slug of alkaline polymer (AP) can be injected before an ASP slug to decrease the volume of surfactant in the ASP slug. To investigate the effect of an AP preflushed slag on the IORF, Schemes 1, 2, and 3 were investigated. In Scheme 1, a single ASP slug with a 0.6-PV ASP was injected. In Scheme 2, a preflushed slug with a 0.3-PV AP was injected before a 0.3-PV ASP slug. In Scheme 3, a preflushed slug with a 0.4-PV AP was injected before 0.2-PV ASP flooding. In these three schemes, only the injected surfactant was decreased, whereas the amount of polymer or alkaline was kept constant. The IORFs over water flooding registered almost at the same level for these three schemes. When increasingly low amounts of surfactant were flooded in Schemes 2 and 3, schemes with AP preflushed slags became economically more attractive than Scheme 1 [1, 87].

One study examined a statistical approach called design of experiment (DOE) and used Nelder-Mead optimization method to optimize the design of ASP flooding. This study showed that the application of DOE in ASP flooding is efficient and the effect of polymer on viscosity is the most important factor concerning all synergic effects in the studied reservoir. Moreover, it was shown that the case of no alkaline injection and the great effect of surfactant on water viscosity was the optimal case in the studied reservoir [88].

In another major study, the results of parallel sand pack flooding showed that alternating flooding with a polymer and surfactant-polymer (PASP) approach can increase the EOR efficiency and sweep efficiency compared to those for the simultaneous flooding of the surfactant and polymer. In addition, it was demonstrated that the IORF in the PASP process increases with alternating cycle number [86].

Some studies have applied numerical simulations to optimize the design of SP flooding [89–93]. A hybrid numerical method was suggested to solve a coupled system of nonlinear equations required in modeling two-phase, multicomponent porous media flow during the SP flooding process. The validated numerical method can be applied to estimate the efficiency of different chemical flooding schemes. In addition, the convergence performance of the hybrid numerical method was investigated by considering two-phase, one-component porous media flow [91, 93].

5.14 Chemistry

Soap generation from the reaction between the flooded alkaline and the saponifiable components of crude oil is the key mechanism of oil recovery by alkaline flooding. The saponifiable constituents of a crude oil are defined as naphthenic or petroleum acids. Petroleum acid is a general mixture of cyclohexyl carboxylic acids and several cyclopentyls with molecular weights of 120 to more than 700. The main fractions consist of carboxylic acids, and other fractions can be asphaltene, carboxyphenols, and porphyrins. Naphthenic acid is produced when the naphtha fraction of crude oil raffination is oxidized. The composition depends on the conditions during raffination and oxidation and the crude oil composition [94–100]. The acidic components of crude oil can be determined and denoted by the acid number. The acid number is measured as the milligrams of potassium hydroxide required to neutralize 1 g of crude oil [101]. Nevertheless, the acid number might not be a proper index for the acid constituent available in crude oil that participates in in situ soap production [12]. Therefore, novel procedures for acid number assessments have been developed to reflect the production of soap [102]. Fig. 5.5 shows a schematic representation of the soap generation process. The production of soap

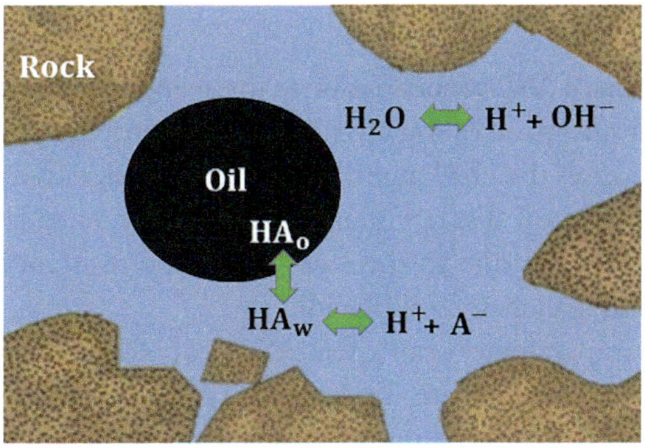

Fig. 5.5 Schematic representation of soap generation process.

can be demonstrated by partitioning the acid in crude oil to water based on the following expression [103]:

$$HA_O \overset{K_D}{\Leftrightarrow} HA_W \tag{5.8}$$

$$K_D = \frac{[HA_W]}{[HA_O]} \tag{5.9}$$

where K_D is the partitioning constant of HA between oil and water. HA_W and HA_O are the concentrations of acid in water and oil, respectively. Then, the acid in water dissociates in the aqueous phase to generate soluble anionic surfactant as follows [103]:

$$\mathrm{HA}_W \overset{K_a}{\Leftrightarrow} H^+ + A^- \tag{5.10}$$

where A^- is the produced soap and K_A is the reaction constant from the following equilibrium equation:

$$K_A = \frac{[H^+][A^-]}{[HA_w]} \tag{5.11}$$

In addition, water dissociates as

$$H_2O \overset{K_w}{\Leftrightarrow} H^+ + OH^- \tag{5.12}$$

$$K_w = [H^+][OH^-] \tag{5.13}$$

By combining Eqs. (5.9), (5.11) and (5.13), the concentration of soap (anionic surfactant) in the aqueous phase can be calculated using the following equation [95]:

$$[A^-] = \frac{K_A K_D [HA_O][OH^-]}{K_w} \tag{5.14}$$

5.15 Screening criteria

The design optimization of ASP processes due to the complex composition of ASP flooding agents requires consideration of the physicochemical properties of brine, oil, and rock formations, which increases the challenge of formulating designs for ASP flooding agents. Generally, ASP composite oil displacement agents are not suitable for carbonate, high saline, and thermal reservoirs due to the lack of alkaline resistance, salt tolerance, and high-temperature resistance characteristics [4].

Although numerous parameters can affect ASP processes in sandstone reservoirs, the most vital parameters are the formation permeability, clay contents, oil viscosity, reservoir temperature, formation water salinity, divalent contents, and oil saturation before the ASP process begins [15, 104–108]. In a major study, the average values of these vital parameters were taken from real ASP projects, as described below and listed in Table 1. The parameters were calculated using the rank and percentile method, and the values at the 50th percentile were considered [15].

Surfactants can increase effective permeability. Therefore, surfactants were also injected in low-permeability reservoirs as a stimulation agent to increase well injectivity [109, 110]. However, as polymer injection requires high permeabilities, the minimum formation permeability required for ASP flooding is reported to be 10 millidarcys [106, 108]. There is agreement in the literature that the clay content for the ASP process should be low [15, 106–108]. The oil density is less important than the effect of viscosity during surfactant flooding. Hence, the oil API gravity was not considered an essential criterion [35]. The maximum and average values for oil viscosity in ASP projects are 35 and 12.9 cP (cP), respectively [15, 105, 106, 108]. The reservoir depth is not a critical factor for oil recovery in ASP and surfactant flooding if the reservoir temperature is below the relevant temperature limit [15, 35]. The maximum and average reservoir temperatures suitable for the ASP process are reported to be 70 degree and 52 degree centigrade, respectively [15, 107]. The harshest and average values for formation water salinity are 20,000 and 7993 ppm (ppm), respectively. The strictest and average values for divalent contents are 500 and 178 ppm, respectively. Even for watered-out reservoirs, surfactant flooding reduces residual oil saturation. Because more oil will be recovered by surfactant flooding in a reservoir with high oil saturation than under other conditions, high oil saturation projects can justify the investment in surfactant flooding [35]. Therefore, the

Table 1 Screening criteria for ASP flooding [15, 35, 105–108].

	K (mD)	μ_o (cP)	*T* (°C)	Salinity (ppm)	Divalent (ppm)	S_o (frac.)
Minimum value	10	35	–	–	–	0.35
Maximum value	–	–	70	20,000	500	–
Ave. from ASP projects	473	12.9	52	7993	178	0.3

minimum and average values for oil saturation before ASP methods are implemented are considered to be 35% and 30%, respectively [15, 106, 108].

5.16 Laboratory tests

Conducting some laboratory experiments is essential for all ASP flooding projects. Adsorption experiments, filtration tests, aqueous stability tests, and intrinsic and apparent viscosity measurements under reservoir conditions, such as typical reservoir temperature and salinity ranges, are required to determine the proper polymer type, molecular weight, and concentration [15, 27, 111]. Aqueous stability tests, phase behavior tests, IFT measurements, surfactant adsorption tests and core flooding tests are used to select a suitable surfactant type and concentration [35, 112–114]. The aqueous stability test is necessary for investigating the compatibility of surfactants with formation water and other chemicals. The phase behavior test is an essential process that uses pipettes. This test is also called the salinity scanning test. The main outcomes of the phase behavior test are:

- Nominating a surfactant or some surfactants, which must possess high solubilization ratios of oil and water volumes to the surfactant volume.
- Determining the optimal salinities, which corresponds to the maximum middle-phase volume [35].

The solubilization parameter reflects the volume of oil or water solubilized in the microemulsion phase per volume of surfactant as follows [115]:

$$SP_i = \frac{V_i}{V_s} \tag{5.15}$$

where SP_i and V_i are the solubilization parameter and volume of water or oil, respectively. The microemulsion phase involves equal amounts of water and oil at an optimal salinity; therefore, the solubilization parameters for brine and oil are equal:

$$SP_{opt} = \frac{V_w}{V_s} = \frac{V_o}{V_s} \tag{5.16}$$

The IFT at optimum salinity can be calculated using Eq. (5.17) as [115, 116]

$$\gamma_{ow} = \frac{C}{SP_{opt}^{\ 2}} \tag{5.17}$$

where C is an empirical constant and characteristic of the surfactant. Therefore, the surfactants associated with high solubilization parameters and the optimal salinity for the selected surfactants can be visually determined by conducting phase behavior tests [35]. For surfactant flooding projects, the water-oil volume ratio is usually considered constant and equal to one during salinity scanning measurements. However, an oil scanning test must also be conducted for an ASP solution. Because alkalis react with the acid components in the crude oil and different saturations of oil in the reservoir in different phases of the project exist, different amounts of in situ soap are produced. Therefore, the phase behavior of the mixed soap and injected surfactant is dynamic and varies with the water-oil ratio. To simulate the dynamic variations in the water-oil ratios inside reservoirs, the water-oil volume ratio is also changed in oil scanning experiments [15]. In addition to salinity and oil scanning measurements, pendant drop, and spinning drop methods can be implemented to measure the IFT as supplementary tests [117, 118]. Additionally, core flooding experiments are required to investigate whether considerable oil recovery can be achieved by surfactant and ASP injections [1, 15, 35].

5.17 Field examples and performance

Most of the pilot ASP flooding projects were at small scales with few injectors used; the injectors were designed with inverted five-spot patterns or five-spot patterns. As of 2014, 22% and 66% of ASP field projects were conducted in the USA and China, respectively. However, only four ASP flooding projects have been performed in India, Canada, and Venezuela. Except for the Lagomar offshore project in Venezuela, all the ASP pilot projects were conducted in onshore reservoir areas [15]. The ASP flooding projects were performed in US fields such as Sho-Vel-Tum, Cambridge Minnelusa, West Kiehl, the Lawrence field in Illinois, and Tanner [119–123]. The majority of ASP flooding in China has occurred in the Daqing Oilfield. The ASP process in the Daqing field was initiated in 1994 but was put into commercial use in 2014. The main reservoir in Daqing is sandstone, with an average permeability of 1445 mD (1.426 μm^2) and a porosity of 26%. The connate water salinity and crude oil viscosity at reservoir temperature are 6800 ppm and 11.5 cP, respectively. Fig. 5.6 shows ASP flooding oil production in Daqing Oilfield. ASP flooding in Daqing covers 22 blocks and 7273 wells, and it accounts for 9% of the total oilfield production, which is approximately 24 million oil barrels (3.5 million tons) in 2015 [124–127]. Strong ASP flooding (SASP) is applied in most ASP field projects in China. SASP refers to an ASP field method that uses NaOH as the alkaline component. When Na_2CO_3 is used, the ASP process is called weak ASP flooding (WASP). Both SASP and WASP have been applied in the Daqing Oilfield. However, ASP flooding is shifting toward WASP dominance, and more SASP flooding tests than WASP tests have been performed so far in the Daqing Oilfield. According to the technical and economical field performance levels, the WASP approach yields better efficiency than the SASP method [15, 125].

In addition to the Daqing ASP field tests, ASP flooding projects have been reported in other Chinese oil fields, such as Shengli, Karamay, Bohai, and Henan [4, 125]. The Shengli pilot test occurred earlier than the first Daqing ASP pilot test. However, experimental ASP flooding research started in the early 1980s, and the first ASP pilot in China started in 1992 in the Gudong Reservoir area. Gudong is an unconsolidated sandstone reservoir

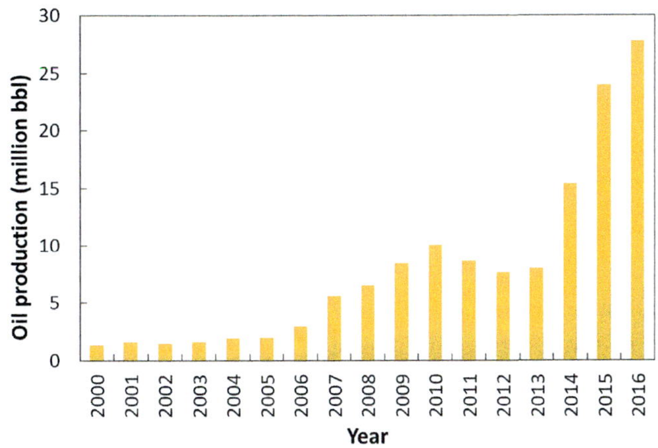

Fig. 5.6 ASP flooding oil production in Daqing Oilfield [124]. *Adapted from Sun C, Guo H, Li Y, Jiang G, Ma R. Alkali effect on alkali-surfactant-polymer (ASP) flooding enhanced oil recovery performance: two large-scale field tests' evidence. J Chem 2020;2020.*

with an average permeability of 2560 mD and a porosity of 35%. The brine salinity, reservoir temperature, and crude oil viscosity in the basin were reported to be 3259 ppm and 68°C and 41.3 cP, respectively [1, 13, 128, 129]. The second ASP pilot test was performed in 1997 in the western part of the Gudao Reservoir over an area of 0.61 km^2. First, a preflushed slug of 0.1-PV polymer solution at a concentration of 0.2% was flooded for 10 months. Then, a 0.3-PV ASP slug containing 12,000 ppm Na_2CO_3, 2000 ppm Surfactant A, 1000 ppm Surfactant B, and 0.17% of the polymer was flooded for 32 months. Finally, a polymer slug of 0.05-PV polymer solution with a 0.15% concentration was injected for 5 months. The injection of chemical slugs was completed in 2002. The total IORF was reported to be 15.5% OOIP [12].

The average IORF and the average decrease in the water cut level achieved by ASP flooding in 21 field tests were reported to be 21.8% OOIP and 18%, respectively. In addition, the maximum IORF of ASP flooding was registered at 33% OOIP [15].

5.18 ASP flooding: Field challenges

Although ASP flooding has been successfully applied globally at the field scale, there are still a number of corresponding difficulties and challenges, such as operational difficulties, formation damage, scaling and precipitation in the process of ASP flooding, issues with the chromatographic separation of the ASP components, difficulties in treating produced emulsions, the need for treatment and disposal of ASP-produced water and the unique challenges associated with the offshore application of ASP methods. These challenges and difficulties have limited industrialized, commercial, and widespread large-scale applications, especially in offshore environments [1, 4, 12, 15, 130].

Polymer degradation, bacterial growth, corrosion, incomplete polymer dissolution, pump failures, the plugging of injection wells, low-volume injection, difficulty separating produced water from oil, issues related to logistics, and handling are common operational issues in ASP projects. Polymer degradation can occur due to bacterial activities, mechanical shear stress, and contamination with oxygen and iron at surface injection facilities. Mechanical shear stress is generated through high-pressure pumps, perforations and pore throats. The existence of alkaline, surfactant, and polymer constituents in a produced fluid can lead to a very tight oil-in-water emulsion and result in complexities in surface water/oil separation. Logistics and handling issues are related to the large volumes of chemicals that must be transported to remote sites and then stored on platforms where space is limited, especially for work in offshore reservoirs [6, 12, 15, 131, 132].

Clays and formation rocks can be eroded by alkaline solutions. However, the permeability can be increased in some cases because of erosion, and it decreases in many cases due to the blockage of pore throats by migrated clays and eroded rocks. The pore throats in low-permeability formations can be blocked easily due to their small size. Therefore, formation damage is much more common in low-permeability formations than in high-permeability rocks. In addition, formation damage can be induced by precipitates and scale associated with the reactions of surfactants, alkaline components, and rock minerals. The permeability reduction that occurs in ASP flooding is lower than that caused by alkaline flooding alone. Post-ASP water flooding can partially reduce formation damage [1].

When seawater is injected into a formation, the sulfate (SO_4^{2-}) concentration increases. In addition, once an alkaline solution is flooded into a formation, the concentration of hydroxide ions (OH^-) increases, and silicate (SiO_3^{2-}) is produced via reactions between alkaline components and formation minerals. The increased hydroxide converts bicarbonate (HCO_3^-) into carbonate (CO_3^{2-}). In the following reactions, the scale can originate from the reactions between the available divalents in the formation, such as Ca^{2+}, Mg^{2+}, and Fe^{2+}, and hydroxide, carbonate, silicate and sulfate constituents; the reaction equations are as follows [1, 12, 15, 133, 134]:

$$M^{2+} + 2OH^- = M(OH)_2 \downarrow \tag{5.18}$$

$$M^{2+} + CO_3^{2-} = MCO_3 \downarrow \tag{5.19}$$

$$M^{2+} + SiO_3^{2-} = MSiO_3 \downarrow \tag{5.20}$$

$$M^{2+} + SO_4^{2-} = MSO_4 \downarrow \tag{5.21}$$

where M^{2+} represents the divalent cations. The reaction of available divalent cations in hard brines with surfactant leads to the generation of a surfactant divalent cation complex through the following equation:

$$2R^- + M^{2+} \rightarrow MR_2 \downarrow \tag{5.22}$$

where R^- and MR_2 are the anionic surfactant and the divalent cation surfactant complex, respectively. The surfactant complex precipitates due to its low solubility in brine [4, 12, 22]. The anionic surfactant reacts with the divalent cation in brine, which has a low hardness level, to generate a monovalent cation. The produced monovalent cation can chemically exchange with cations initially bound to in situ clay as follows [12, 135]:

$$R^- + M^{2+} \rightarrow MR^+ \tag{5.23}$$

$$Na - Clay + MR^+ \rightarrow MR - Clay + Na^+ \tag{5.24}$$

The effluent concentration histories of ASP slug injection exhibit the chromatographic separation of the alkaline, surfactant, and polymer components. The three components should be transported at the same velocity to produce synergy in the ASP flooding process. Normally, breakthrough times and actual effluent concentrations are a function of the individual balance between retention/consumption and the injection concentration. Large polymer molecules cannot access some small pores. This area of a reservoir is called the inaccessible pore volume (IPV). Polymer molecules are transported faster than the solution in the IPV. In addition, the velocity of polymer molecules decreases due to adsorption. Therefore, the consequent transport velocity is the balance between these two phenomena. High retention results in a decreased transport velocity. Normally, the retention levels of surfactant and polymer are the highest and lowest, respectively. However, alkaline adsorption is lower than surfactant adsorption and higher than polymer adsorption. Therefore, the polymer and surfactant have the shortest and longest breakthrough times, respectively. The discrepancy in breakthrough time between the alkaline solution and the polymer is not significant, but the surfactant breaks through far behind the alkaline solution. Because the IPV is intrinsic and retention cannot be substantially modified to solve the separation problem, injection concentrations can be designed to effectively solve the separation issue [1, 15].

Although the sweep efficiency increases in a porous medium with the formation of emulsions, transportation and oil/water separation are difficult. Notably, it was difficult to separate water from oil in a Shengli ASP pilot test, even though WASP flooding occurred. In addition, because the treatment of the produced water was problematic, the produced water had to be reflooded into the formation [1].

Emulsified produced water is oily and has a high suspended solids content. Therefore, this water is different from that produced by other EOR technologies and is problematic to dispose of as a result of the adsorption of the injected chemicals onto the surface of oil droplets. The separation of suspended solids and oil in ASP-produced water is a major challenge, as the treated produced water must meet the reinjection standard; thus, an environmentally friendly produced water disposal method should be developed [12]. Generally, produced water treatment devices, such as that for porous medium separation, hydrocyclones, gravity settling devices, membrane separation equipment, and induced gas floatation and electrodialysis devices, are used to successfully clean produced water. However, due to the reduction in the IFT, increased brine viscosity when the polymer is added, stable O/W emulsions due to the small sizes of oil droplets, and various environmental and economic issues, the abovementioned equipment is not capable of successfully cleaning the produced water in ASP flooding [15].

References

[1] Sheng J. Modern chemical enhance oil recovery: theory and practice. London, UK. Oxford, UK: Gulf Professional; 2011.
[2] Druetta P, Raffa P, Picchioni F. Chemical enhanced oil recovery and the role of chemical product design. Appl Energy 2019;252:113480.
[3] Enhanced Oil Recovery; 2013. Available from: http://salympetroleum.com/technologies/newoil/asp/. 2021.
[4] Yang P, Z-a L, Xia B, Y-j Y, Q-t H, W-l L, et al. Comprehensive review of alkaline–surfactant–polymer (ASP)-enhanced oil recovery (EOR). In: Proceedings of the International field exploration and development conference 2017. Springer; 2019. p. 858–72.
[5] Mohd T, Jaafar MZ, Rasol A, Ali J. A new prospect of streaming potential measurement in alkaline-surfactant-polymer flooding. Chem Eng Trans 2017;56:1183–8.
[6] Bataweel MA, Nasr-El-Din HA. Alternatives to minimize scale precipitation in carbonate cores caused by alkalis in asp flooding in high salinity/high temperature applications. In: SPE European formation damage conference. Society of Petroleum Engineers; 2011.
[7] Stoll M, Al-Harthy S, Van Wunnik J, Faber M. Alkaline-surfactant-polymer Flood–From the Laboratory to the Field. In: IOR 2011—16th European symposium on improved oil recovery. European Association of Geoscientists & Engineers; 2011 [cp-230-00039].
[8] Chen X, Feng Q, Wu X, Zhao G. A pilot numerical simulation case study for chemical EOR feasibility evaluation. J Pet Explor Prod Technol 2016;6(2):297–307.
[9] Aitkulov A, Dao E, Mohanty KK. ASP flood after a polymer flood vs. ASP flood after a water flood. In: SPE improved oil recovery conference. Society of Petroleum Engineers; 2018.
[10] Li Y, Zhang W, Kong B, Puerto M, Bao X, Sha O, et al. Mixtures of anionic/cationic surfactants: a new approach for enhanced oil recovery in low-salinity, high-temperature sandstone reservoir. SPE J 2016;21(04):1164–77.
[11] Li D, Jia Q, Chen X. A study on component retention in reservoirs during ASP flooding field trials at Daqing. Oilfield Chem 2001;18(4):362–6.
[12] Olajire AA. Review of ASP EOR (alkaline surfactant polymer enhanced oil recovery) technology in the petroleum industry: prospects and challenges. Energy 2014;77:963–82.
[13] Zhijian Q, Yigen Z, Xiansong Z, Jialin D. A successful ASP flooding pilot in Gudong oil field. In: SPE/DOE improved oil recovery symposium. Society of Petroleum Engineers; 1998.
[14] Nelson R, Lawson J, Thigpen D, Stegemeier G. Cosurfactant-enhanced alkaline flooding. In: SPE enhanced oil recovery symposium. Society of Petroleum Engineers; 1984.
[15] Sheng JJ. A comprehensive review of alkaline–surfactant–polymer (ASP) flooding. Asia Pac J Chem Eng 2014;9(4):471–89.
[16] Li J, Wang H, Hu J, Jiang H, Zhang R, Tang L. Development performance and pressure field evolution of ASP flooding. Adv Polym Technol 2020;2020.
[17] Yue X, Xia H, Zhang Y, Li J. Calculation of IPR curves of oil wells for polymer flooding reservoirs. In: SPE annual technical conference and exhibition. Society of Petroleum Engineers; 1997.
[18] Yue X, Wang Y, Wang K. Basis for enhanced oil recovery. Beijing, China: Petroleum Industry Press Publishing; 2007.
[19] Ding L, Wu Q, Zhang L, Guérillot D. Application of fractional flow theory for analytical modeling of surfactant flooding, polymer flooding, and surfactant/polymer flooding for chemical enhanced oil recovery. Water 2020;12(8):2195.

[20] Sheng JJ. Mobility control requirement in multiphase displacement processes in porous media. Asia Pac J Chem Eng 2013;8(4):555–66.
[21] Sorbie K. Polymer-improved oil recovery, 115 Glasgow. Scotland: Blackie & Son; 1991. p. 126–63.
[22] Green DW, Willhite GP. Enhanced oil recovery. Henry L. Doherty Memorial Fund of AIME, Society of Petroleum Engineers; 1998.
[23] Sheng JJ. Modern chemical enhanced oil recovery: theory and practice. Gulf Professional Publishing; 2010.
[24] Wang D, Cheng J, Xia H, Li Q, Shi J. Viscous-elastic fluids can mobilize oil remaining after waterflood by force parallel to the oil-water interface. In: SPE Asia Pacific improved oil recovery conference. Society of Petroleum Engineers; 2001.
[25] Wang D, Cheng J, Yang Q, Wenchao G, Qun L, Chen F. Viscous-elastic polymer can increase microscale displacement efficiency in cores. In: SPE annual technical conference and exhibition. Society of Petroleum Engineers; 2000.
[26] Ramazani SAA, Nourani M, Emadi MA. Rheological behaviour and hydrodynamic diameter of high MolecularWeight, partially hydrolyzed poly (acrylamide) in high salinity and temperature conditions; 2010.
[27] Ramazani SAA, Nourani M, Emadi MA, Jafariesfad N. Analytical and experimental study to predict the residual resistance factor on polymer flooding process in fractured medium. Transp Porous Media 2010;85(3):825–40.
[28] Grubisic Z, Rempp P, Benoit H. A universal calibration for gel permeation chromatography. J Polym Sci Part B: Polym Lett 1967;5(9):753–9.
[29] Bird RB, Stewart WE, Lightfoot EN. Transport phenomena. 413. New York: John Wiley & Sons; 1960.
[30] Kang W. Study of chemical interactions and drive mechanisms in Daqing ASP flooding. Beijing: Petroleum Industry Press; 2001.
[31] Szabo MT. An evaluation of water-soluble polymers for secondary oil recovery-parts 1 and 2. J Pet Technol 1979;31(05):553–70.
[32] Johnson Jr C. Status of caustic and emulsion methods. J Pet Technol 1976;28(01):85–92.
[33] Cheng J, Liao G, Yang Z, Li Q, Yao Y, Xu D. Overview of Daqing ASP pilots. Pet Geol Oilfield Dev Daqing 2001;20(2):46–9.
[34] Friberg SE, Bothorel P. Microemulsions: structure and dynamics; 1987.
[35] Sheng JJ. Status of surfactant EOR technology. Petroleum 2015;1(2):97–105.
[36] Druetta P, Picchioni F. Surfactant-polymer interactions in a combined enhanced oil recovery flooding. Energies 2020;13(24):6520.
[37] Li H. Advances in alkaline-surfactant-polymer flooding and pilot tests. China: Science Press; 2007.
[38] Fleming I, Williams DH. Spectroscopic methods in organic chemistry. Springer; 1966.
[39] Liu S, Zhang D, Yan W, Puerto M, Hirasaki G, Miller C. Favorable attributes of alkaline-surfactant-polymer flooding. SPE J 2008;13:5–16 [SPE-99744-PA].
[40] Hirasaki GJ, Miller CA, Puerto M. Recent advances in surfactant EOR. In: SPE annual technical conference and exhibition. Society of Petroleum Engineers; 2008.
[41] Moore T, Slobod R. The effect of viscosity and capillarity on the displacement of oil by water. Prod Mon 1956;20(10):20–30.
[42] Koh HS. Experimental investigation of the effect of polymers on residual oil saturation; 2015.
[43] Morrow NR, Chatzis I, Taber J. Entrapment and mobilization of residual oil in bead packs. SPE Reserv Eng 1988;3(03):927–34.
[44] Jamaloei BY, Asghari K, Kharrat R. The investigation of suitability of different capillary number definitions for flow behavior characterization of surfactant-based chemical flooding in heavy oil reservoirs. J Pet Sci Eng 2012;90:48–55.
[45] Lake LW. Enhanced oil recovery; 1989.
[46] Mohammadi H, Delshad M, Pope GA. Mechanistic modeling of alkaline/surfactant/polymer floods. SPE Reserv Eval Eng 2009;12(04):518–27.
[47] Lu J, Liyanage PJ, Solairaj S, Adkins S, Arachchilage GP, Kim DH, et al. New surfactant developments for chemical enhanced oil recovery. J Pet Sci Eng 2014;120:94–101.
[48] Gary JH, Handwerk GE. Petroleum refining: technology and economics. New York: M. Dekker; 1994.

[49] Merdrignac I, Espinat D. Physicochemical characterization of petroleum fractions: the state of the art. Oil Gas Sci Technol-Rev IFP 2007;62(1):7–32.
[50] Buijse MA, Prelicz RM, Barnes JR, Cosmo C. Application of internal olefin sulfonates and other surfactants to EOR. Part 2: the design and execution of an ASP field test. In: SPE improved oil recovery symposium. Society of Petroleum Engineers; 2010.
[51] Li KW. The interfacial tension between Daqing crude oil fractions and asp and as flooding solutions and the stability of their emulsions. Oilfield Chem 1999;4.
[52] Wu X, Han M, Zahrani BH, Guo L. Effect of surfactant-polymer interaction on the interfacial properties for chemical EOR. In: SPE middle east oil & gas show and conference. Society of Petroleum Engineers; 2015.
[53] Ma BD, Gao BY, Zhang L, Gong QT, Jin Z, Zhang L, et al. Influence of polymer on dynamic interfacial tensions of EOR surfactant solutions. J Appl Polym Sci 2014;131(15).
[54] Hirasaki G, Zhang DL. Surface chemistry of oil recovery from fractured, oil-wet, carbonate formations. SPE J 2004;9(02):151–62.
[55] Standnes DC, Austad T. Wettability alteration in chalk: 2. Mechanism for wettability alteration from oil-wet to water-wet using surfactants. J Pet Sci Eng 2000;28(3):123–43.
[56] Gong H, Li Y, Dong M, Ma S, Liu W. Effect of wettability alteration on enhanced heavy oil recovery by alkaline flooding. Colloids Surf A Physicochem Eng Asp 2016;488:28–35.
[57] Ding M, Wang Y, Yuan F, Zhao H, Li Z. A comparative study of the mechanism and performance of surfactant-and alkali-polymer flooding in heavy-oil recovery. Chem Eng Sci 2020;115603.
[58] Nourani M, Tichelkamp T, Khanamiri HH, Johansen T, Hov IK, Gaweł B, et al. Dynamic wettability alteration for combined low salinity brine injection and surfactant flooding on silica surface. SN Appl Sci 2020;2(7):1–13.
[59] Adibhatla B, Mohanty KK. Oil recovery from fractured carbonates by surfactant-aided gravity drainage: laboratory experiments and mechanistic simulations. In: SPE/DOE symposium on improved oil recovery. Society of Petroleum Engineers; 2006.
[60] Seethepalli A, Adibhatla B, Mohanty KK. Physicochemical interactions during surfactant flooding of fractured carbonate reservoirs. SPE J 2004;9(04):411–8.
[61] Li Z, Ayirala S, Mariath R, AlSofi A, Xu Z, Yousef A. Microscale effects of polymer on wettability alteration in carbonates. SPE J 2020.
[62] Khabibullaev PK, Saidov A. Phase separation in soft matter physics: micellar solutions, microemulsions, critical phenomena. Springer Science & Business Media; 2013.
[63] ShamsiJazeyi H, Hirasaki G, Verduzco R. Sacrificial agent for reducing adsorption of anionic surfactants. In: SPE International symposium on oilfield chemistry. Society of Petroleum Engineers; 2013.
[64] Nandwani SK, Chakraborty M, Gupta S. Adsorption of surface active ionic liquids on different rock types under high salinity conditions. Sci Rep 2019;9(1):1–16.
[65] Ahmadall T, Gonzalez MV, Harwell JH, Scamehorn JF. Reducing surfactant adsorption in carbonate reservoirs. SPE Reserv Eng 1993;8(02):117–22.
[66] Wang J, Han M, Fuseni AB, Cao D. Surfactant adsorption in surfactant-polymer flooding for carbonate reservoirs. In: SPE middle east oil & gas show and conference. Society of Petroleum Engineers; 2015.
[67] Zhang R, Somasundaran P. Advances in adsorption of surfactants and their mixtures at solid/solution interfaces. Adv Colloid Interf Sci 2006;123:213–29.
[68] Saxena N, Kumar A, Mandal A. Adsorption analysis of natural anionic surfactant for enhanced oil recovery: the role of mineralogy, salinity, alkalinity and nanoparticles. J Pet Sci Eng 2019;173:1264–83.
[69] Kudaibergenov S. Alkaline/surfactant/polymer (ASP) flooding. Int J Biol Chem 2015.
[70] Sinha AK, Bera A, Raipuria V, Kumar A, Mandal A, Kumar T. Numerical simulation of enhanced oil recovery by alkali-surfactant-polymer floodings. Pet Sci Technol 2015;33(11):1229–37.
[71] Atthawutthisin N. Numerical simulation of low salinity water flooding assisted with chemical flooding for enhanced oil recovery. Institutt for petroleumsteknologi og anvendt geofysikk; 2012.
[72] Abadli F. Simulation study of enhanced oil recovery by ASP (alkaline, surfactant and polymer) flooding for Norne field C-segment. Institutt for petroleumsteknologi og anvendt geofysikk; 2012.

[73] Hosseini-Nasab SM, Padalkar C, Battistutta E, Zitha P. Mechanistic modeling of the alkaline/surfactant/polymer flooding process under sub-optimum salinity conditions for enhanced oil recovery. Ind Eng Chem Res 2016;55(24):6875–88.
[74] Delshad M, Han C, Veedu FK, Pope GA. A simplified model for simulations of alkaline–surfactant–polymer floods. J Pet Sci Eng 2013;108:1–9.
[75] Parkhurst DL, Appelo C. User's guide to PHREEQC (version 2): a computer program for speciation, batch-reaction, one-dimensional transport, and inverse geochemical calculations. Water Resour Invest Rep 1999;99(4259):312.
[76] Yuan C. Commercial scale simulations of surfactant/polymer flooding; 2012.
[77] Jin B, Jiang H, Zhang X, Wang J, Yang J, Zheng W. Numerical simulation of surfactant-polymer flooding. Chem Technol Fuels Oils 2014;50(1):55–70.
[78] Dang C, Nghiem L, Nguyen N, Yang C, Chen Z, Bae W. Modeling and optimization of alkaline-surfactant-polymer flooding and hybrid enhanced oil recovery processes. J Pet Sci Eng 2018;169:578–601.
[79] Le Van S, Chon BH. Artificial neural network model for alkali-surfactant-polymer flooding in viscous oil reservoirs: generation and application. Energies 2016;9(12):1081.
[80] Elkamel A. An artificial neural network for predicting and optimizing immiscible flood performance in heterogeneous reservoirs. Comput Chem Eng 1998;22(11):1699–709.
[81] Zerpa LE, Queipo NV, Pintos S, Salager J-L. An optimization methodology of alkaline–surfactant–polymer flooding processes using field scale numerical simulation and multiple surrogates. J Pet Sci Eng 2005;47(3–4):197–208.
[82] Zerpa LE, Queipo NV, Pintos S, Tillero E, Alter D. An efficient response surface approach for the optimization of ASP flooding processes: ASP pilot project LL-03 reservoir. In: Latin American & Caribbean petroleum engineering conference. Society of Petroleum Engineers; 2007.
[83] Wang Y, Hou J, Cao X, Wu G, Wang C. Evaluation of polymer flooding potential based on orthogonal design and BP artificial neural network. In: International petroleum technology conference; 2014.
[84] Al-Dousari MM, Garrouch AA. An artificial neural network model for predicting the recovery performance of surfactant polymer floods. J Pet Sci Eng 2013;109:51–62.
[85] Alkhatib A, Babaei M. Applying the multilevel Monte Carlo method for heterogeneity-induced uncertainty quantification of surfactant/polymer flooding. SPE J 2016;21(04):1,192–203.
[86] He H, Chen Y, Yu Q, Wen X, Liu H. Optimization design of injection strategy for surfactant-polymer flooding process in heterogeneous reservoir under low oil prices. Energies 2019;12(19):3789.
[87] Yang P-H, Wang H, Li Z, Yan H. Fundamental and advances in combined chemical flooding (in Chinese). China Petroleum Press; 2002.
[88] Azamifard A, Bashiri G, Gerami S, Hemmati-Sarapardeh A. On the evaluation of alkaline-surfactant-polymer flooding in a field scale: screening, modelling, and optimization. Can J Chem Eng 2017;95(8):1615–25.
[89] Druetta P, Picchioni F. Surfactant–polymer flooding: influence of the injection scheme. Energy Fuel 2018;32(12):12231–46.
[90] Kamari A, Gharagheizi F, Shokrollahi A, Arabloo M, Mohammadi AH. Integrating a robust model for predicting surfactant–polymer flooding performance. J Pet Sci Eng 2016;137:87–96.
[91] Daripa P, Dutta S. On the convergence analysis of a hybrid numerical method for multicomponent transport in porous media. Appl Numer Math 2019;146:199–220.
[92] Zhou K, Hou J, Ren X, Yu B. Economic design of surfactant-polymer flooding under low oil price. In: SPE oil and gas india conference and exhibition. Society of Petroleum Engineers; 2017.
[93] Daripa P, Dutta S. Modeling and simulation of surfactant–polymer flooding using a new hybrid method. J Comput Phys 2017;335:249–82.
[94] Seifert W. Carboxylic acids in petroleum and sediments. In: Fortschritte der chemie organischer naturstoffe/progress in the chemistry of organic natural products. Springer; 1975. p. 1–49.
[95] Sheng J. Investigation of alkaline–crude oil reaction. Petroleum 2015;1(1):31–9.
[96] Shuler P, Kuehne D, Lerner R. Improving chemical flood efficiency with micellar/alkaline/polymer processes. J Pet Technol 1989;41(01):80–8.

[97] Pasquarelli C, Wasan D. The effect of film-forming materials on the dynamic interfacial properties in crude oil-aqueous systems. In: Surface Phenomena in Enhanced Oil Recovery. Springer; 1981. p. 237–48.
[98] Rudzinski WE, Oehlers L, Zhang Y, Najera B. Tandem mass spectrometric characterization of commercial naphthenic acids and a Maya crude oil. Energy Fuel 2002;16(5):1178–85.
[99] Dunning H, Moore J, Denekas M. Interfacial activities and porphyrin contents of petroleum extracts. Ind Eng Chem 1953;45(8):1759–65.
[100] Mayer E, Berg R, Carmichael J, Weinbrandt R. Alkaline injection for enhanced oil recovery-a status report. J Pet Technol 1983;35(01):209–21.
[101] Mahajan S, Konar SK. Boocock DGb. Determining the acid number of biodiesel. J Am Oil Chem Soc 2006;83(6):567–70.
[102] Fan T, Buckley JS. Acid number measurements revisited. SPE J 2007;12(04):496–500.
[103] DeZabala E, Vislocky J, Rubin E, Radke C. A chemical theory for linear alkaline flooding. Soc Pet Eng J 1982;22(02):245–58.
[104] Lake LW, Schmidt RL, Venuto PB. A niche for enhanced oil recovery in the 1990s. Oilfield Rev 1992;4(1):55–61.
[105] Dickson JL, Leahy-Dios A, Wylie PL. Development of improved hydrocarbon recovery screening methodologies. In: SPE improved oil recovery symposium. Society of Petroleum Engineers; 2010.
[106] Taber J, Martin F, Seright R. EOR screening criteria revisited—part 2: applications and impact of oil prices. SPE Reserv Eng 1997;12(03):199–206.
[107] Al-Bahar MA, Merrill R, Peake W, Jumaa M, Oskui R. Evaluation of IOR potential within Kuwait. In: Abu Dhabi international conference and exhibition. Society of Petroleum Engineers; 2004.
[108] Taber JJ, Martin F, Seright R. EOR screening criteria revisited-part 1: introduction to screening criteria and enhanced recovery field projects. SPE Reserv Eng 1997;12(03):189–98.
[109] Rilian NA, Sumestry M, Wahyuningsih W. Surfactant stimulation to increase reserves in carbonate reservoir "a case study in Semoga field". In: SPE EUROPEC/EAGE annual conference and exhibition. Society of Petroleum Engineers; 2010.
[110] Miller BJ, Pitts MJ, Dowling P, Wilson D. Single well alkaline-surfactant injectivity improvement test in the big sinking field. In: SPE/DOE symposium on improved oil recovery. Society of Petroleum Engineers; 2004.
[111] Al-Hajri S, Mahmood SM, Akbari S, Abdulelah H, Yekeen N, Saraih N. Experimental investigation and development of correlation for static and dynamic polymer adsorption in porous media. J Pet Sci Eng 2020;189:106864.
[112] Tichelkamp T, Teigen E, Nourani M, Øye G. Systematic study of the effect of electrolyte composition on interfacial tensions between surfactant solutions and crude oils. Chem Eng Sci 2015;132:244–9.
[113] Hosseinzade Khanamiri H, Nourani M, Tichelkamp T, Stensen JÅ, Øye G, Torsæter O. Low-salinity-surfactant enhanced oil recovery (EOR) with a new surfactant blend: effect of calcium cations. Energy Fuel 2016;30(2):984–91.
[114] Hosseinzade Khanamiri H, Baltzersen Enge I, Nourani M, Stensen JÅ, Torsæter O, Hadia N. EOR by low salinity water and surfactant at low concentration: impact of injection and in situ brine composition. Energy Fuel 2016;30(4):2705–13.
[115] Spildo K, Sun L, Djurhuus K, Skauge A. A strategy for low cost, effective surfactant injection. J Pet Sci Eng 2014;117:8–14.
[116] Huh C. Interfacial tensions and solubilizing ability of a microemulsion phase that coexists with oil and brine. J Colloid Interface Sci 1979;71(2):408–26.
[117] Taylor KC, Schramm LL. Measurement of short-term low dynamic interfacial tensions: application to surfactant enhanced alkaline flooding in enhanced oil recovery. Colloids Surf 1990;47:245–53.
[118] Liu S. Alkaline surfactant polymer enhanced oil recovery process; 2008.
[119] Vargo J, Turner J, Vergnani B, Pitts MJ, Wyatt K, Surkalo H, et al. Alkaline-surfactant-polymer flooding of the Cambridge Minnelusa field. In: SPE Rocky mountain regional meeting. Society of Petroleum Engineers; 1999.
[120] Meyers J, Pitts MJ, Wyatt K. Alkaline-surfactant-polymer flood of the West Kiehl, Minnelusa Unit. In: SPE/DOE enhanced oil recovery symposium. Society of Petroleum Engineers; 1992.

[121] French T. Evaluation of the Sho-Vel-tum alkali-surfactant-polymer (ASP) oil recovery project-Stephens County, OK. Tulsa, OK: National Petroleum Technology Office (NPTO); 1999.
[122] Pitts MJ, Dowling P, Wyatt K, Surkalo H, Adams KC. Alkaline-surfactant-polymer flood of the Tanner field. In: SPE/DOE symposium on improved oil recovery. Society of Petroleum Engineers; 2006.
[123] Seyler B, Grube J, Huff B, Webb N, Damico J, Blakley C, et al. Reservoir characterization of Bridgeport and Cypress sandstones in Lawrence field Illinois to improve petroleum recovery by alkaline-surfactant-polymer flood. University Of Illinois; 2012.
[124] Guo H, Li Y, Wang F, Gu Y. Comparison of strong-alkali and weak-alkali ASP-flooding field tests in Daqing oil field. SPE Prod Oper 2018;33(02):353–62.
[125] Sun C, Guo H, Li Y, Jiang G, Ma R. Alkali effect on alkali-surfactant-polymer (ASP) flooding enhanced oil recovery performance: two large-scale field tests' evidence. J Chem 2020;2020.
[126] Shutang G, Huabin L, Zhenyu Y, Pitts M, Surkalo H, Wyatt K. Alkaline/surfactant/polymer pilot performance of the West Central Saertu, Daqing Oil Field. SPE Reserv Eng 1996;11(03):181–8.
[127] Pu H. An update and perspective on field-scale chemical floods in Daqing oilfield, China. In: SPE middle east oil and gas show and conference. Society of Petroleum Engineers; 2009.
[128] Song W, Yang C, Han D, Qu Z, Wang B, Jia W. Alkaline-surfactant-polymer combination flooding for improving recovery of the oil with high acid value. In: International meeting on petroleum engineering. Society of Petroleum Engineers; 1995.
[129] Wang C, Wang B, Cao X, Li H. Application and design of alkaline-surfactant-polymer system to close well spacing pilot Gudong oilfield. In: SPE Western regional meeting. Society of Petroleum Engineers; 1997.
[130] Raney KH, Ayirala SC, Chin RW, Verbeek P. Surface and subsurface requirements for successful implementation of offshore chemical enhanced oil recovery. In: Offshore technology conference; 2011.
[131] Weatherill A. Surface development aspects of alkali-surfactant-polymer (ASP) flooding. In: International petroleum technology conference; 2009.
[132] Li J, Liu Y, Wu D, Meng X, Zhao F. The synergistic effects of alkaline, surfactant, and polymer on the emulsification and destabilization of oil-in-water crude oil emulsion produced by alkaline-surfactant-polymer flooding. Pet Sci Technol 2013;31(4):399–407.
[133] He L, Gang C, Guochen S. Technical breakthrough in PCPs' scaling issue of ASP flooding in Daqing oilfield. In: SPE annual technical conference and exhibition. Society of Petroleum Engineers; 2007.
[134] Arensdorf JJ, DB MD, Yuan M. Static and dynamic testing of silicate scale inhibitors. In: Hoster D. International oil and gas conference and exhibition in China. Society of Petroleum Engineers; 2010.
[135] Hill H. Cation exchange in chemical flooding: part 3-experimental. Soc Pet Eng J 1978;18(06): 445–56.

CHAPTER 6

Improved oil recovery by gel technology: Water shutoff and conformance control

Saeid Sadeghnejad[a], Marjan Ashrafizadeh[b], and Meysam Nourani[c]
[a]Department of Petroleum Engineering, Faculty of Chemical Engineering, Tarbiat Modares University, Tehran, Iran
[b]Department of Chemical Engineering, College of Engineering, University of Tehran, Tehran, Iran
[c]Ugelstad Laboratory, Department of Chemical Engineering, Norwegian University of Science and Technology (NTNU), Trondheim, Norway

6.1 Introduction

Water is usually coproduced along with oil. Excessive water production is the cause of many technical and operational problems during oil production that may result in early abandonment of the well and leave a large amount of oil unrecovered in mature oil fields. The source of produced water can be from either aquifers or water breakthrough during waterflooding. Phenomena such as water conning (due to high production rates), moving water/oil contact (WOC), completion problems (e.g., flow behind pipe), and mechanical problems (casing, tubing, packer leaks) can be the source of unwanted water production. Moreover, water naturally tends to flow through high-permeable strata (e.g., fractures, faults, high-permeable layers with and without crossflow) toward production wells. The injected water can break through high-conductive layers into production wells, reduce the oil recovery factor, and increase excessive water production [1–3].

Reservoir heterogeneity is responsible for excessive water production and poor hydrocarbon sweep efficiency (or reduction in recovery) [4]. This heterogeneity can consist of high-permeable strikes or thief zones including fracture networks, channels, faults, joints, permeability contrast in production layers, wormholes, vugs, or caves. In general, severe vertical and lateral formation heterogeneity can result in the rapid water-cut increase in production wells, and they are the main reason for low sweep efficiency [5]. Prolonged water injection into mature reservoirs can also cause the erosion of high-permeability layers, creating channels (because of mineral dissolution and/or sand production) and increasing the permeability contrast in reservoir rocks. As a result, the injected water can flow rapidly through these high-permeable thief zones and bypasses large amounts of hydrocarbons in unswept low-permeability zones (Fig. 6.1). This problem, in which the flooding front is not uniform, is considered as a conformance problem. Moreover, water from aquifers can also be coproduced from these high-permeable

https://doi.org/10.1016/B978-0-12-821931-7.00001-8

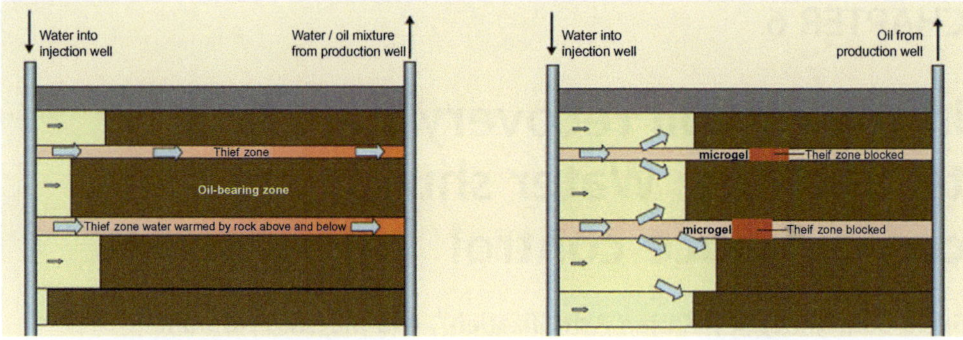

Fig. 6.1 The injected water breaks through the thief zones and increases the water cut in the production well. Water can be diverted into low-permeable areas by plugging the thief zones [6].

heterogeneities, if the production rate is not controlled below the critical rate, which results in water conning or rising WOC.

In such reservoirs, the production water-cut can increase even more than 90%, influencing the field production economy [7]. Excessive water production increases costs (e.g., operational, separation, and disposal costs) and causes environmental issues (disposal of water containing mercury, arsenic, and other salts) [8, 9]. Therefore, one effective mechanism is reducing the permeability contrast in heterogeneous reservoirs by implementing improved oil recovery (IOR) methods.

Several IOR methods are used to address water shutoff (WSO) problems, which are divided into various categories: physical (or temporary) (e.g., installation of plug set, inflatable sleeves), chemical plugging agents (e.g., cement squeezing, resin, gel injection), and well technologies [10, 11]. Implementing well completion methods such as perforating water legs using dual completions and well technologies such as drilling horizontal and multilateral wells and sidetracking can also be used to reduce excessive water production.

The chemical WSO treatment is one of the water control processes that has shown feasibility and practicality in oil fields for decades [12]. Among chemical treatments, implementing polymer gels, to reduce excessive water production, has many advantages, including more flexibility, no need for additional equipment (pills are usually bullhead into the formation or placed at the determined depth by coiled tubing), and the ability to penetrate deep into reservoir far from injection wells (i.e., conformance control). Polymer gel treatments have many applications in the petroleum industry including, WSO, cementing, well stimulation, and sand control [13] and they have been developed and improved for WSO operations since the 1970s [14,15]. By effectively blocking the high-permeable thief zones, they divert the injected water toward adjacent low-permeable zones and increase the productive life of reservoirs [9].

Historically, in situ bulk gels (BGs) were used to block excessive water flow in areas next to production or injection wells. Other polymer gel systems such as colloidal dispersion gels [16,17], preformed particle gels [18,19], pH-sensitive microgels [20–22], temperature-sensitive gels [23,24], salinity-triggered gels [25], or other smart polymers [26–28] can be implemented to reduce the excessive water production problem in both injection and production wells. Some polymer gels can even reduce water permeability without affecting oil permeability [29–31]. These polymer gels are called disproportionate permeability reducers (DPR) or relative permeability modifiers. The polymer gel treatment techniques for conformance purposes are categorized as IOR methods. However, they have a very similar concept of enhanced oil recovery (EOR) methods to increase the reservoir sweep efficiency. These treatments are widely used during secondary and tertiary oil recovery (EOR) in both production and injection wells.

Polymer gels are primarily compatible with all well equipment and materials and are not harmful to the environment. They can also be removed by water injection or ordinary acids, and thus the reservoir can be cleaned easily. An appropriate mechanical resistance to reservoir pressure drop during production and good stability against temperature, salinity, and pH, is another advantage of polymer gels. Therefore, using gel systems as diverting or blocking agents to improve oil or gas production processes is growing. Each of these gel types has its advantages and disadvantages. This chapter first discusses the excessive water production problem and then clarifies the role of gel conformance control. A common polymer gel classification used in conformance control treatments is then discussed. Subsequently, the characteristics, gel treatment mechanisms, gel transport in porous media, gel modeling approaches, and field applications of various gel treatment methods are explained. The advantages and disadvantages of these gel treatment methods are also highlighted.

6.2 Excessive water control

Excessive water production has become one of the world's significant problems, which causes early abandonment of production wells, reducing their economic production life, and leaving unrecovered hydrocarbons in reservoirs. It is estimated that three water barrels are coproduced for each oil barrel produced in the world [32]. Handling, lifting, separation, pumping, and disposal of a large amount of water increases the production costs [9,32]. Excessive water production also causes other problems such as corrosion, scale formation in surface facilities, and environmental problems. Thus, managing excessive water production is an essential task during production [33].

Conformance problems occur during waterflooding when water breakthrough happens and bypasses a large amount of oil. Thus, excessive water production is a result of this conformance problem. Excessive water production can either be treated by WSO in production wells (i.e., near-wellbore treatments or WSO from producers) or be reduced by

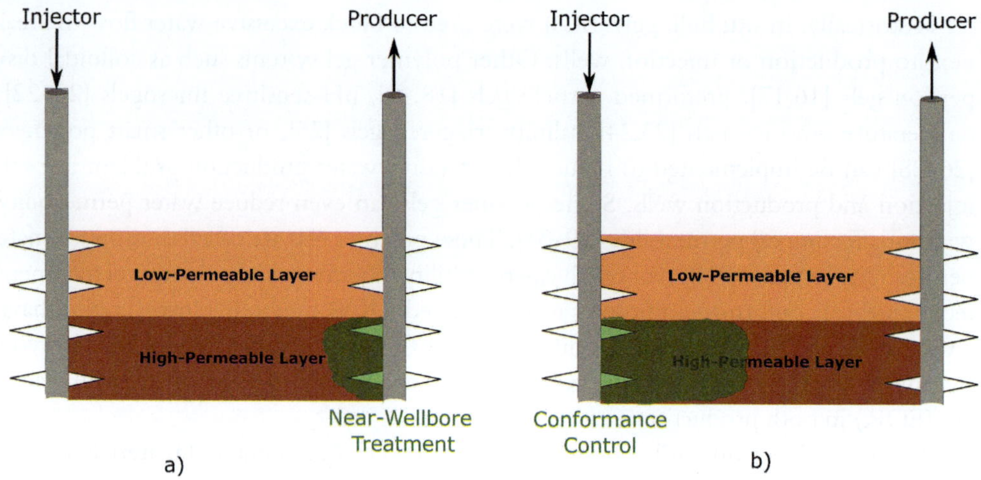

Fig. 6.2 Comparison of excessive water treatment in (A) production well (near-wellbore treatment) and (B) injection well (conformance control).

controlling the injected fluid front far from injection wells (i.e., profile control from injectors) (see Fig. 6.2). The latter method is called conformance control, which is a key IOR method associated with the sweep efficiency improvement during oil recovery [34]. Conformance control is defined as any technique that encourages the movement of normally unswept hydrocarbon zones in reservoirs and brings the flooding drive closer to perfect conforming conditions [35]. Conformance control should only be implemented in reservoirs with an economically feasible volume of recoverable oil [7].

Different solutions have been proposed to control or minimize unwanted water production [32]. These solutions are divided into two general categories:

- Chemical and physical plugging agents (also called permeability blockers or sealants)
- Mechanical and well technologies

Chemical agents can be injected in either injection or production wells (Fig. 6.2). The most applicable chemicals for excessive water control treatments include (ultrafine) cement, sand, calcium carbonate, silicate, solid slurry (clay), resin, polymer, gels, and a combination of these materials (e.g., a mixture of gel with cement or sand) [32]. Upon reaching the location, the cement slurry isolates the target zone; however, this method has some inherent problems regarding the cement hardening [36]. The use of cement is suitable for flow problems behind casings/tubings and leakage from casings, while gels are usually used for excessive water blocking operations near wellbores. Gels are more flexible than cements and can be removed from wells using various ways (e.g., acid wash) [37]. Resins and elastomers are strong enough to close channels and leakage points. Resins and elastomers include phenolic, epoxy, and furfuryl alcohol [38]. One of the advantages of epoxy is that it is stronger than phenolic and furfuryl alcohol resins.

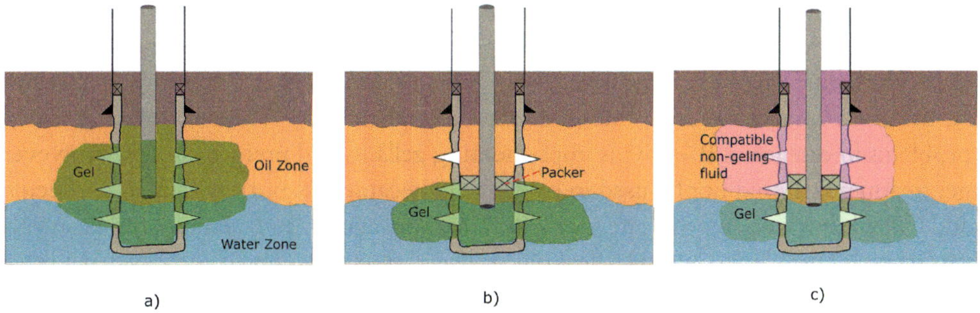

Fig. 6.3 (A) Bullhead placement technique, (B) mechanical-packer placement technique, and (C) dual-injection placement technique. *(Modified after reference Jaripatke OA, Dalrymple ED. Water-control management technologies: a review of successful chemical technologies in the last two decades. SPE international symposium and exhibition on formation damage control. Society of Petroleum Engineers; 2010.)*

The simplest and most economical form of placing gel chemicals in wells is bullheading (Fig. 6.3A). In BGs, polymer and cross-linker are mixed in an aqueous phase (i.e., called gellants). Oil with an appropriate emulsifier is then added to the solution to create a suitable emulsion at the well sides [31]. The solution is then bullheaded into the formation. After well shut-in, the emulsion is separated into the oil and water phases. The oil phase protects upper oil flowing zones, and gellants change to a gel state at proper deeper depths. With this method, there is no need for formation isolation; however, it increases the damage risk to productive zones if not applied correctly or in complicated WSO scenarios. Mechanical packers (bridge plugs) can isolate the target zone and guide the treatment chemical to desired zones (Fig. 6.3B). Moreover, the dual-injection placement method (Fig. 6.3C) can be implemented. The treatment and another compatible fluid are simultaneously pumped through the tubing and annulus for better control of the treatment fluid toward the target interval [39]. The main advantage of this method is it does not require workover rigs.

The physical technologies such as installing packers, bridge plugs, casing replacement, and inflatable sleeves can also prevent excessive water production, which are categorized as mechanical near-well treatments. For example, packers are used when water flow blocking in wellbores is required. However, when there is crossflow between production layers, these mechanical treatments fail to control the WSO problem. The well techniques are also implemented to control excessive water production. The simplest form of this category is the abandonment of wells with high water cuts. Moreover, more sophisticated technologies such as infill drilling, horizontal wells, and implementing well completion methods such as perforating water legs using dual completion, drilling multilateral wells, and sidetracking can also be used for WSO purposes.

The water control solutions (e.g., gel properties, gel volume, and placement procedures) for reservoirs with a linear water flow source (e.g., fractures, faults) is fundamentally different from those reservoirs with a radial water flow [40]. The problems of excessive water production and their proper treatments were discussed by Seright et al. [32]. The treatments were categorized from A to D based on their difficulties. In category A, the conventional treatment methods can be implemented. The best solution for the problems in categories B and C is implementing gellants and preformed gels. However, gel treatments should not be used in category D because it is not easy and cost effective and will seldom work for this category [32]. Chemical agents (e.g., cement, resins, or gels) can treat the problems caused by leaking or flow behind casings. If the excessive water production is due to fractures or fracture-like features during linear flow toward production wells, gel treatments can be the right candidate. However, if the excessive water production is due to crossflow between reservoir layers or 3D coning, gel treatments should not be implemented. Even by injecting the gellant into a single zone, it is a risk for crossflowing and damaging the adjacent productive zones far from wellbores. Therefore, gel treatment is a solution for cases in which there is no crossflowing between zones and proper cement sealing behind casing exists [32].

Suppose that the water and hydrocarbon areas are entirely separated from each other, or the problem of excessive water production is near wellbores. In this case, the installation of mechanical and chemical barriers (e.g., cement or resin) in water production layers can also be effective. However, if the oil and water areas are not separable, the use of flow blockers may be risky. A more attractive way to control the flow profile in these cases is an in-depth fluid diversion. This process increases the overall sweep efficiency of flooding significantly more than the near-well treatments, especially when there is a crossflow and good connection between different production layers. Unlike the production well treatments, there is no need for a production shut-in in this method. Polymer gel injection acts as the most-effective deep conformance control treatments [10]. In this case, reducing the permeability of high-permeable strata and improving the sweeping efficiency of the injected fluid in heterogeneous reservoirs using in-depth conformance control methods (e.g., DPRs, pH-sensitive microgels, PPGs) is one of the IOR applications of polymer gel treatments.

6.3 Polymer gels

Polymer gels are composed of cross-linked polymer networks which are swollen in solvents [41]. The main property of cross-linked polymer gels is that they swell in a well-matched fluid [42]. Those polymer gels compatible with water are called hydrogels that can absorb large amount of water without dissolving [43]. The swelling capacity of some polymer gels changes based on environmental variations (e.g., pH, salt concentration, temperature). These gels sometimes are categorized as intelligent materials [44].

A gelation process has three phases: induction, reaction acceleration, and equilibrium stage [34]. The induction phase is mainly dedicated to the mobility of polymer chains as the temperature and pressure increase. In this phase, only a small amount of cross-linking bonds forms. In the reaction acceleration phase, the reaction rate of cross-linking rises suddenly. As a result, a 3D network is formed, and the viscosity of the gel system increases. Finally, the reaction completes in the equilibrium stage. Although minor cross-linking reactions continue to harden the internal gel structure, the viscosity maintains steadily in this phase. The polymer gel viscosity, gelation time, and gel strength are a function of environmental conditions (e.g., temperature, fluid salinity, presence of divalent cations, pH), gel composition (e.g., polymer concentration, polymer to cross-linker ratio, and the presence of additives and retarders), and applied shear forces.

The base polymers used in polymer gels are of two types [45,46]:

- ***Synthetic polymers***: They include polyacrylamide (PAM) or partially hydrolyzed polyacrylamide (HPAM), polyacrylonitrile (PAN), polyvinyl alcohol (PVA), polyvinylamine (PVAm), and copolymer of AMs and 2-acrylamido-2-methylpropanesulfonic acid (AMPSA), terpolymer of AM, AMPSA, and *N*,*N*-dimethyl acrylamide (N, N-DMA). Fig. 6.4 depicts the chemical structure of these polymers, which can be components of homo-, co-, or terpolymers.
- ***Biopolymers***: They include xanthan gum, guar gum, hydroxyethyl cellulose, scleroglucan, lignin, and tannin.

Cross-linkers can also be catagorized into [45]:

- ***Inorganic*** *cross-linkers*: These cross-linkers are composed of multivalent metal cations (chromium, zirconium, aluminum, or titanium). These cations form ionic bonds with negatively charged groups of the polymer structure. These groups can be acetate, citrate, lactate, malonate, propionate, and so on. High temperatures can make inorganic bonds unstable. So they have low thermal stability.
- ***Organic*** *cross-linkers*: The functional groups on polymer structures are cross-linked by covalent bonding with an organic cross-linker (e.g., glyoxal, PEI, phenol, dihydroxy-naphthalene, phenylacetate, hydroquinone, or formaldehyde (paraformaldehyde, hexamethylenetetramine, terephthalaldehyde). Compared to inorganic cross-linkers, these cross-linkers have prolonged gelation time, higher thermal stability, and higher salinity resistance.
- ***Natural*** *cross-linkers:* These cross-linkers are environment-friendly cross-linkers. This type of cross-linkers includes cross-linking reaction of chitosan with acrylamide-based polymers such as PHPA, PAtBA, and AMPS/acrylic acid copolymer [47]. Chitosan is a nontoxic and biodegradable linear polysaccharide. It can be cross-linked through covalent bonds with acrylamide-based polymers. However, it has a limited application in the oil industry because of its precipitation over a pH of 6.

Both organic [such as polyethylenimine (PEI) and phenol/formaldehyde] and inorganic [mainly metal ions such as zirconium (IV), aluminum (III), and chromium (III)] cross-linkers may be used in polymer gel structures [46]. The type of cross-linker controls the physical and thermal stability of the gel. An essential factor in the selection of polymer gel is

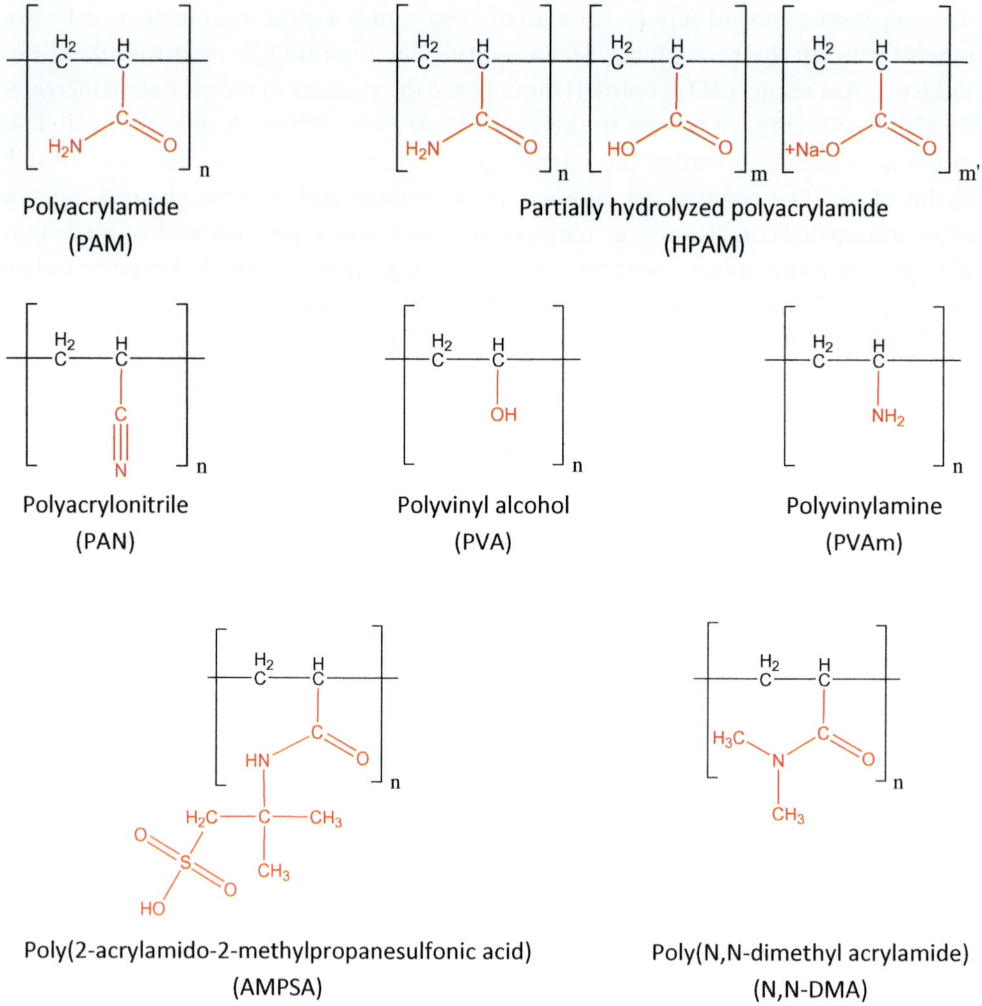

Fig. 6.4 Chemical structures of synthetic polymers.

the reservoir temperature. The reservoir temperature of 60–80°C is considered high, and above 120°C, the reservoir temperature is considered ultrahigh-temperature [46]. Most gels can withstand temperatures less than 60°C. Conventional polymers such as PAM and xanthan can degrade over 80°C [13,46,48]. High temperature makes the cross-linking process too fast; therefore, inorganically cross-linked polymer gels have rapid gelation time, which causes gel formation in wellbores and its pumping too difficult. At temperatures over 120°C, these inorganic cross-linkers cannot be implemented because of decrease in their gelation time and thermal instability due to their fast cross-linking reactions [46]. Furthermore, acrylamide groups on the polymer chains are hydrolyzed and produce some cross-linking sites for interaction with Ca^{2+} and Mg^{2+} (available in formation brines), resulting in syneresis of gels [49]. Generally, organically cross-linked polymer gels are more stable than inorganic ones at higher reservoir temperatures (>80°C) [13].

An appropriate polymer gel selection for a specific reservoir is challenging because of the complexity of an underlying conformance problem [9]. The significant challenges of WSO treatments by gels include cost optimization, treatment performance, gel stability at reservoir conditions, and minimization of environmental effects [50]. The compatibility of gel systems directly influences the efficiency of WSO treatments at reservoir conditions (i.e., reservoir rock mineralogy, temperature, pressure, salinity, pH, and H_2S content). A proper decision between polymer stability and operational costs must be made to have an optimum gel performance at reservoir conditions [51]. The performance of gels is influenced by their chemistry and concentration of cross-linker and polymer. The long-term thermal, mechanical, and electrolyte stability and the gelation time and the gellant stability (in not preformed gels) are among factors that should be enhanced in a gel polymer for a specific application. This improvement can be achieved using additives [13,52]. The addition of these materials (e.g., nanomaterials, particularly metal oxide nanoparticles and 2D materials like graphene) could increase the applicability range of polymer gels at operating conditions such as high temperature or saline formations [51]. In some cases, retarders can be added to gel polymer systems to delay the gelation time (e.g., ammonium chloride in PAM/PEI systems) [11]. Nanocomposite gels also have strong gel networks that are highly desirable for isolating thief zones [36].

6.3.1 Polymer gel classification

Today, two types of polymer gel systems (Fig. 6.6) are used for conformance control purposes [17,35]:

- In situ gels (sometimes called immature gels)
- Preformed gels

The main difference between these two gels is in the gelation mechanism. In the first category (in situ gels), a polymer/monomer and cross-linker solution is first prepared on the surface (Fig. 6.5A). The resulting solution is called gellant. Gellant is then injected into wells at desired locations by coiled tubing or bullheading. Because of the low viscosity of gellant solutions, they can penetrate far from injection wells through fractures or fractured-like features in a reservoir. The gellant components react with each other inside the reservoir and form a BG in place, acting as a flow diverting or blocking agent. The disadvantage of this process is difficulty in predicting the exact gelation time and in situ physical properties of the formed gels. Moreover, continuous injection of gellants into a porous rock might cause chromatographic fractionation of gellant components. Besides, most BGs are not stable under high reservoir temperature conditions (e.g., >80°C).

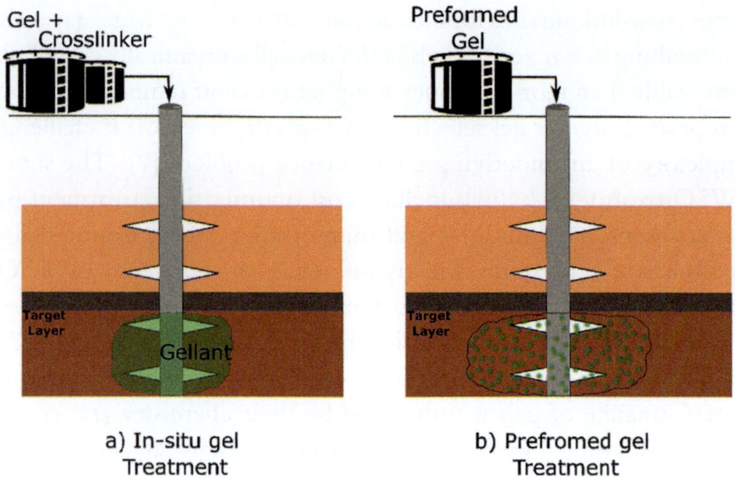

Fig. 6.5 Comparison of (A) in situ and (B) preformed gel during treatments.

Some in situ microgels (CDGs) are made from lower concentrations of both polymer and cross-linker than BGs. Under this condition, polymer microgels are formed instead of continuous BG, which can be quickly injected deep into formations. CDGs involve low polymer and cross-linker concentrations [$(100–1200) \times 10^{-6}$ and $(10–25) \times 10^{-6}$, respectively] that make an injection of a large volume of these gels cost-effective [35].

Although in situ gel systems are cost-effective, they have some intrinsic drawbacks, including the lack of control over gelling time, uncertainty in gelling due to shear degradation, changes in gelling ratio components (chromatographic fractionation), and dilution by formation water [35]. The gelation time of in situ gels is highly sensitive to reservoir conditions such as temperature, pH, salinity, H_2S content, and the type of materials in reservoirs, which affect the gel stability [11]. Therefore, a new conformance control agent called preformed gels has been developed to control fluid profiles at reservoir conditions. Unlike in situ gels, preformed gels are prepared before being injected. All constituents are reacted on the surface, and then gel particle suspensions are injected into wells (Fig. 6.5B). Preformed gels improve waterflooding processes by in-depth plugging of high-permeable channels/strata and thief zones, thereby preventing the flow channeling through them [53]. In other words, they divert the injected fluid toward areas with low sweep efficiency (i.e., deep conformance control). If (micro) gels even penetrate less permeable areas, they can produce oil from those areas without damaging formation rocks.

Preformed gels are usually classified into the following categories (Fig. 6.6) [6,35,54–56]. Their main differences are particle sizes (from millimeter to submicron or nanometer), swelling mechanisms, and operational conditions.

- preformed particle gels (PPGs)
- temperature-sensitive or temperature-activated polymer (TAP) gels
- pH-sensitive microgels

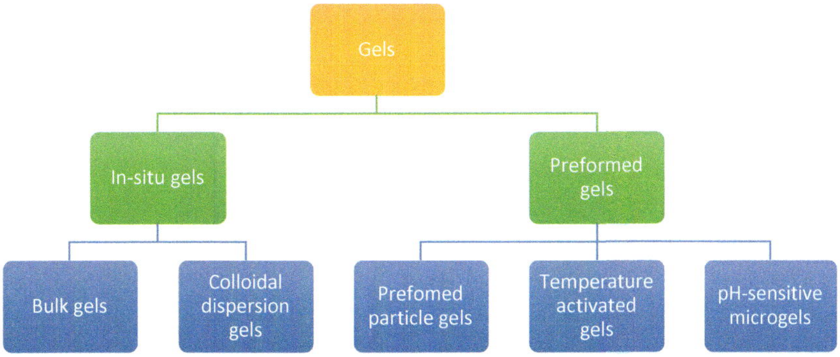

Fig. 6.6 Classification of polymer gels used in conformance control.

The preformed gels used in conformance control treatments have polymer chains with 3D networks formed by cross-linker agents and can absorb water or aqueous solutions more than several times their weight [35]. These particles' adsorption power mainly depends on the concentration of networking agents (cross-linkers). Polymers with lower cross-linking densities can absorb more significant amounts of aqueous solutions, but their gel strength will be relatively weaker [35].

TAPs and pH-sensitive microgels can be categorized as intelligent polymers for deep conformance control. These materials are injected in the colloid form of an unswollen gel particle. Their swelling occurs during flowing through reservoir rocks due to changes in environmental pH or temperature. Thus, they can penetrate far from injection wells before their swelling and plug thief zones. Accordingly, they can provide in-depth conformance control. pH-sensitive (or pH-responsive) microgels are injected in acidic conditions. These microgels are in an unswollen state and have low viscosity at a low pH environment. Therefore, they can be injected at substantially reduced pressure and penetrate in thief zones and high-permeable strata far from injection wells. The TAP swelling is triggered by the rising of the solution temperature (i.e., thermo-responsive). The labile cross-linking of the gel breaks at a higher temperature and allows gels to swell. They are injected with cold water into formations. By flowing through the formation and increasing temperature, TAPs start to swell inside the formation, and an in-depth flow diversion can be achieved. Tables 6.1 and 6.2 summarize the advantages, disadvantages, and operational conditions of each of the introduced gels.

6.3.2 Resistance factor and residual resistance factor

Two parameters, the resistance factor (*RF*) and the residual resistance factor (*RRF*), can be used to examine the retention behavior of polymer gels. The *RF* is used to measure mobility reduction because of a gel polymer flood and is defined as the ratio of the pressure drop during gel injection (or the water mobility) to the pressure drop during the waterflooding (or gel mobility) at the same flow rates (Eq. 6.1). The *RRF* is another

Table 6.1 Comparison of in-situ gel systems and preformed gels. Modified after Bai, et al. [35].

Gels	Description	Advantage	Disadvantage	Typical examples	Field applications
In situ gels	• Gellant is injected into formations. • It has close viscosity to polymer solution. • Gellant will form gel under reservoir conditions.	• High injectivity • In-depth-conformance control.	• Gelation is a function of injection rate (shear stress), Chromatographic adsorption, reservoir pH, temperature, salinity, etc., so the gelation time is difficult to control. • Can penetrate and plug low permeability zones.	• Strong BG • Weak BG • CDG	It is widely applied across the world.
Preformed gels	• Preformed gels are injected into formations.	• Gelation is less affected by surface facilities, wellbore conditions, and formation conditions. • Can treat reservoirs with high-permeability contracts	• Cannot penetrate low permeability reservoirs	• PPG • TAP • pH-sensitive microgels	PPG has been widely used; TAP has also been used globally; pH-sensitive has not been applied yet.

Table 6.2 Comparison of different microgels [57].

Name	Gelation	At surface	Trigger	In situ	Size before	Size after
BG	In-Situ	Polymer and cross-linker	–	Swollen BG	–	–
CDG	In-Situ	Polymer and cross-linker	Transition pressure	Swollen microgel	nm to μm-sized particles	μm-sized particles
PPG	Preformed	Particle gel	Transition pressure	Swollen particle gel	μm to cm-sized particles	20–200 times larger
BrightWater® (TAP)	Preformed	Microgel	Temperature	Swollen microgel	0.1–1 μm	1–10 μm
pH-sensitive	Preformed	Microgel	pH	Swollen microgel	μm-sized particles	Up to 1000 times larger

(Modified after Abdulbaki MR. Simulation study of polymer microgel conformance treatments; 2012.)

critical parameter related to the retention of gels and evaluates their stability effectiveness in porous media. This parameter is defined as the pressure drop ratio during the postwaterflooding (or chase water) to the pressure drop of the primary waterflooding before the gel injection at the same flow rate (Eq. 6.2). The *RRF* is a measure of permeability reduction insurance during postwaterflooding.

$$RF = \frac{\Delta P_{gel}}{\Delta P_{Waterflooding}} = \frac{k_w/\mu_w}{k_{gel}/\mu_{gel}} \tag{6.1}$$

$$RRF = \frac{\Delta P_{Postwaterflooding}}{\Delta P_{Waterflooding}} = \frac{k_w/\mu_w}{k_{w2}/\mu_{w2}} \tag{6.2}$$

where k_w and μ_w are the permeability and viscosity of the water phase, respectively, and k_{gel} and μ_{gel} are the permeability and viscosity of the gel solution, respectively. The properties of the postwaterflooding are shown with "2" subscript.

6.4 In situ gel

6.4.1 Bulk gel (BG)

In the past, in situ BGs were mostly used to control water flow in areas close to production or injection wells. BGs are squeezed into high-permeable layers and/or fractures in wells and can plug them. By blocking high-permeable thief zones, they are typically used in production wells to control excessive water production. These gels have a continuous semisolid 3D network. They mainly consist of organic or inorganic cross-linkers (3000–12,000 mg/L) with high-molecular-weight PAM (8–13 × 10^6 Da) [46]. Moreover, the other most used and cost-effective BGs are sodium silicate gels [58]. For this purpose, the polymer with its cross-linker (i.e., gellant) are injected into wells, and BGs are then formed in formations to fully or partially plug high-permeable strata. Polymer and cross-linker can either be injected together or sequentially in one slug [54,59]. In the latter injection method, the controlling of when and where polymer and cross-linker meet each other in porous media is a complex task. The first implementation of BGs as conformance control agents was tested by Philips Co. (now ConocoPhilips) in the 1970s [35]. They used HPAM and aluminum citrate as a conformance control agent. Treatments using BGs usually have higher efficiency when excess water is broken through channels or fractured systems [60].

Gellants are liquid-like solutions with low viscosity (e.g., water-like viscosity). Before the onset of the gelation time, the gellant should be injected into treatment wells and placed into the desired intervals. A very low flow rate should be used for pumping to effectively penetrate high-permeable zones in formations and reduce gel invasion in adjacent rock matrixes [59]. A sufficient shut-in time is necessary for the gellant to be converted to a solid gel state (i.e., BG) and gain its full gel strength. This time depends on the gellant recipe and reservoir conditions. BGs should also maintain their rigidness for an extended period, at a wide range of formation temperatures, salinity concentration, H_2S content, and pH [61]. The next important step is the well flowing after the gel treatment, which should be gentle enough before returning to the initial production rate. The gel treatment design is still heavily based on conformance engineers' experience because of the lack of efficient designing tools [62].

The conventional BGs usually plug the immediate vicinity of the wellbore [63] because of their quick cross-linking reactions. To overcome this difficulty, some researchers proposed a method to decrease polymer and cross-linker concentration in a gel, called weak BGs. The cross-linking time increases and the gellant of weak gels has enough time to reach and plug targeted zones far from injection wells. This can improve in-depth profile control of the injected chase water. Because of low gelation kinetics, such gels can be mixed in surface facilities and injected as a single phase without the risk of premature gelation in surface facilities and wellbores [64]. These partially cross-linked gellants absorb much water in formations [35] and still be cross-linked at reservoir conditions [65]. However, they have relatively low strength. Weak BGs are formed by intramolecular (and few intermolecular) cross-linking reactions because the polymer and cross-linker concentrations are relatively low [66].

Sun et al. [63] prepared a moderate delayed-HPAM hydrogel with a mature delay time of approximately 10 days. Thus, the prepared gellant with low viscosity had sufficient time for penetrating deep into formations. Because of the dense structure of the prepared BGs, they had a promising elastic property to plug large-width channels. Moreover, redox agents (e.g., thiourea, sodium bisulfite, H_2S, and sodium thiosulfate [67]) can be mixed with gellants to delay the gelation time [35]. For prolonging the gelation time, thiourea, which slowly reduces Cr^{+6} to Cr^{+3}, is widely used [67].

There have been many successful applications of HPAM gel systems, cross-linked with organic and inorganic complexes, in fractured reservoirs with low-to-moderate-high temperatures [68]. During the injection, the partial penetration of gellant into adjacent zones or loss of cross-linker into the matrix can influence the final strength of the final BG. By varying polymer/cross-linker content, it was possible to have enough stability of HPAM-Cr^{+3} BGs at temperatures up to 124°C [58]. Chromium cross-linked PAM system has some inherent weaknesses, lack of gelation time predictions, and gel instability in the presence of some chemical species or high temperature; therefore, some extra chemical agents should be used [59].

HPAM gels, cross-linked by inorganic cross-linkers (e.g., zirconium and chromium cross-linker), become inefficient at reservoir temperatures higher than 80°C because of the occurrence of ionic bonds between inorganic ions cross-linkers and the $-COO^-$ group of HPAM at higher temperatures [12,14,60,69]. To overcome this issue, the polymeric gels with organic cross-linkers (e.g., PEI and phenol/formaldehyde) are considered and preferred [12]. AM-based copolymers and terpolymers usually have higher thermal stability [69]. For example, BGs formed of nonionic PAM and phenolic resin are stable at reservoir temperatures lower than 143°C [70]. Zhu et al. [70] characterized in situ terpolymer BGs with a controllable gelation time (i.e., from 7 to 14h) composed of resorcinol-hexamethylenetetramine for the WSO purposes at reservoir temperatures above 150°C.

BGs can also be formed by adding acidic solutions (activators) to sodium or potassium silicate. For example, sodium silicate gels are usually formed by mixing sodium silicate solutions with HCl, called *acidic silicate gels* [58]. They are rigid and opalescent materials that have low-to-moderate cost [67]. Sodium silicate gels proved to be stable at high-temperature conditions (300°C) and up to a pressure of 1250psi for 12weeks in laboratory conditions [71]. The gelation time can be influenced by ions (especially divalent cations) during the preparation phase [72]. Weak acids can also be used to make *alkaline silica gel systems* with longer gelation times [67]. The precipitates of calcium and magnesium silicates can be observed when gellants contact formation brines. Moreover, gellant leaking can affect the final gel stability unless gellants can quickly be converted to gels at the fracture walls and prevent leaking [58]. Moreover, these BGs can be implemented for in-depth conformance control. By changing sodium silicate concentrations (10–13wt%) and NaCl (activator) concentrations (0.6–1.74wt%) above 40°C, a variable gelation time (1 − 100h) was achieved [58,73].

Among cross-linked polymer gels, PEI gels have received more attention because of their environment-friendly characteristics and enough stability at high temperatures [45]. PEI is a branched polymer with low-molecular weight, enhancing its biodegradability and thermal stability [74]. It is categorized as an organically cross-linked polymer. The reason for higher thermal resistance in comparison to inorganic cross-linkers is the stable covalent bonds between amide groups of the polymer structure and organic cross-linkers [34]. Moreover, it has a low viscosity, can be propagated deep into formations, and has less sensitivity to available fluids formation and rock lithology [34]. Thus, at high temperatures (up to 190°C), PEIs are more favorable than gels composed of metallic cross-linkers. PEI was usually successfully copolymerized by acrylamide-based polymers (e.g., PAtBA, PAM, PHPAM, AM/AMPS/N, and N-DMA AM/AMPS), which can provide a technical advantage for field applications (e.g., from 4.4°C to 205°C) [45]. PEI gels have been applied for WSO purposes [75–80] and the lower gas-oil ratio [78] in the Middle East, North Sea, and the Gulf of Mexico.

Homo-, co-, and ter-acrylamide polymers with phenol/formaldehyde cross-linkers are also well known in the field [59]. In contrast to PEIs, these gels are toxic because of their cross-linking agent containing phenol and the carcinogenic character of formaldehyde. However, these BGs are thermally stable, depending on the polymer composition, and have a controllable gelation time over a wide range of temperatures [47]. The activation temperature of these gels is in the range of 70–80°C; therefore, at higher temperatures (e.g., 140°C), the gel formulation should be enhanced by adding other monomers into the polymer structure [34].

Aldhaheri et al. [62] reviewed the designs of injection-well BG treatments conducted between 1985 and 2014, including 61 field projects. The gel treatments typically started with a polymer concentration in a range of 3000–7600 ppm, a constant polymer-cross-linker ratio of 40:1, and molecular weights of 10–11 MM Da in the reviewed fields. The designing criteria for gel strength were designed in those case studies based on each specific problem, required volume of gels, and gel breakthrough. Interestingly, a more significant polymer concentration (with smaller treatment volume) was used in the unsuccessful gel treatments than that of the successful ones. Moreover, in cases with long-flooding histories, with higher driving fluid channeling, gels with higher strength were implemented.

The BG treatments, in general, have some shortcomings, including [29,68,81–83]:

- the lack of gelation time control,
- uncertainties of gelation time because of shear degradation in the surface facility and porous media,
- chromatographic fractionation of gellant components (polymer, cross-linker, and additives) in porous media,
- dilution of gellant with formation water,
- potential damage to low-permeability hydrocarbon zones,
- limited permeability range of applicability,
- reduction of permeability to oil in productive layers,
- different results for treatments of two wells under the same condition (or even non-repeated results in a single well for several trials),
- back production and washout, and
- syneresis, precipitation, and gel breakage can also influence the effectiveness of gel treatments.

6.4.1.1 Disproportionate permeability reduction (DPR)

Disproportionate permeability reduction (DPR) or relative permeability modifying (RPM) is a phenomenon in which polymer gel reduces the relative permeability of the water phase higher than the hydrocarbon phase. The ultimate effect of DPR treatment will be a reduction in the producing water/oil ratio (known as the positive DPR effect) [84,85]. It should be noted that there is always a reduction in the oil permeability; however, this amount is much lower than the water phase. Some studies reserved the term "DPR" and "RPM" for relatively strong and weak gels that result in water permeability reduction, respectively [29].

It should be noted that DRP gels can be implemented only in production wells for WSO purposes. Many hydrogels such as linear polymers including PAMs and polysaccharides are the most well-known examples. Liang et al. [86] treated a Berea sandstone with various BGs and showed that HPAM/Cr^{+3} reduced the water and oil permeability of 2–100 and 2–15 folds, respectively.

Several mechanisms have been introduced for oil permeability development and water permeability reduction in DRP gels. Because of polymer retention, an adsorbed layer of polymer on walls of porous media is created (i.e., wall effect), which generally reduces the permeability to the water phase [59,87]. In general, this adsorbed polymer layer can also change the wettability of grain surfaces [88]. The oil phase tends to finger and pass through the gel at the center of pores by the gel dehydration mechanism, whereas water diffuses through the swelled gel [89–91]. The dehydration of cross-linked-gel structure in the treated porous rock by injection of oil was also confirmed by Ganguly et al. [92]. However, the segregation of oil and water pathways may have an important role in DPR [40,93]. Theoretically, DPR can also be a function of the amount of cross-sectional area changes of both oil and water phases because of a tendency toward more water-wet conditions after the polymer adsorption [87,94]. Nilsson et al. [95] showed that the segregated pathways are governed by wettability and, to some extent, the pore size. Moreover, placing gellant at the lower oil saturations results in stronger permeability reduction toward both phases. Grattoni et al. [96] proposed that the pore size changes and the redistribution of oil and water phases can be considered as the primary mechanisms for the DPR. Al-shajalee et al. [88] focused on the adsorption and lubrication effect. So far, various DPR mechanisms were introduced, including wall effect, fluid segregation, residual oil blocking, restricted pore throat model; however, none of these mechanisms proved to be the primary reason for DRP [40]. Table 6.3 summarizes the mechanisms, the conflict opinions, and weak points [97,98].

DRP systems are attractive for the petroleum industry as no mechanical isolation tools are required during the treatment and can be easily placed into the desired zones by bull-heading into production wells without any costly workover operations [29,32,61,97]. However, temperature, shear, and availability of divalent cations can affect their effectiveness [59]. DRP treatments can be effectively applied for a long period to reservoirs having water and oil-producing strata with no crossflow [29]. Moreover, hydraulic-fractured wells and natural fracture reservoirs in which fractures unintentionally break into water zones can be treated using this method. The gel treatment in these wells relies on reducing permeability to the water phase in matrixes around fractures [29]. Sydansk and Seright [29] mentioned that this method could be the right choice for systems in which water and oil are being produced from matrix to fractured networks; however, if water flows through fractures from sources far from the wellbore, this cannot be an appropriate treatment. During fracture treatments, the amount of permeability reduction to oil and the distance of the gellant treated radius around the wellbore should be considered. For cases with large gellant penetrations far from the wellbore, the pressure gradient is not enough to initiate oil flow through the gel [40].

The DRP gel treatment is ineffective when [29] the following scenarios occur:

- ***Production from a single oil zone with high permeability.*** After the treatment, water can be blocked from entering the gel-treated zone because of forming a relative-

Table 6.3 The proposed DRP mechanisms. Modified after Alfarge, et al. [97].

DPR mechanisms	Proposal	Investigated by
1. Wall effect	1. Wall effect can explain DPR when the gellant is prepared from or match the wetting phase of the rock 2. Gel droplet model explains DPR when the gellant is prepared from or match the nonwetting phase of rock	[99]
2. Gravity effect	The density of water-soluble gel (usually 99% water) = density of brine. Therefore, Gels would go to water rather than oil. Then, gel would reduce K_{rw} more than K_{ro}	[93]
3. Lubrication effect	The interface between oil and adsorbed polymer would lubricate the path of oil rather than water	[30,93,100]
4. Rock wettability change and water/oil pathways constriction	DPR is due to polymer adsorption on water-wet rock walls	[30,32,93]
5. Segregated pathways mechanism	The water-based gel would flow through most parts of pores which are available to brine	[95,101]
6. Capillary forces and gel elastic effect	DPR resulted from the balance between capillary forces and gel elasticity	[91,102]
7. Polymer leaching from gel and reducing brine mobility mechanism	DPR due to polymer leaching from gel during water injection and not leaching through oil injection	[102]
8. Gel swelling in water and shrink in oil	DPR due to water-based gel is shrinking in oil and swelling in water	[93,101]
9. Polymer adsorption entanglement	A polymer layer would be formed on the crevices between grains and handers only water	[103,104]
10. Gel deformation or dehydration	Oil would deform and dehydrate the gels while water would not	[105,106]

permeability block at the outer radius of the treatment area. By passing the time, water saturation rises beyond the treated zone and blocks oil flowing.

- ***Crossflow exists between reservoir zones.*** Because the water is blocked by gels in a high-permeable layer, it will flow through nontreated less-permeable layers (crossflow) toward production wells later.
- ***Water coning through unfractured matrix block.*** RPM/DPRs treatments only prolong the water coning for a short period.

6.4.2 Colloidal dispersion gel (CDG)

Polymer concentrations in CDGs are usually very low (approximately 100–1000 ppm) [107]. Thus, they form microgel colloidal dispersions (separate microgels) instead of continuous networks in BGs. Fig. 6.7 visualizes the difference between CDGs and BGs. These spherical microgel colloids usually have nanometer diameters (e.g., approximately 1–150 nm), which are suspended in solutions [16,107,108]. The mechanism of this gel treatment method is an increase in microscopic diversion because of the blocking of pore throats far from wellbore areas by the aggregates of CDGs. This performance results in deep conformance control in reservoirs without plugging near-wellbore areas [16]. Moreover, CDGs can increase both microscopic and macroscopic sweep efficiencies during IOR/EOR [107].

CDGs are commonly produced by combining HPAM as a polymer and polyvalent cations (e.g., Cr^{+3}, Fe^{+3}, Al^{+3}, and Zr^{+4}) as cross-linkers at low concentrations [11,46,83,109,110]. Chromium triacetate was also used as a cross-linker to form CDGs in many applications [17]. CDGs show physiochemical properties such as colloidal aqueous solutions [107] and have a controllable gelation time.

CDG solutions, contrary to polymer solutions, express approximately Newtonian behavior with lower viscosity and lower electrophoretic mobility [108]. For example, similar viscosity to an aqueous solution was found for a 400–600 ppm CDG system with a polymer-to-aluminum ratio of 40:1 during 13 days [111]. The resistance of CDGs to flow depends on the amount of *transition pressure*. When the applied pressure on these particles is greater than the transition pressure, microgels move quickly through porous media. In contrast, at pressures lower than the transition pressure, microgels tend to agglomerate and block porous paths [110,112]. In other words, they create a resistance to flow.

The viscosity of a CDG solution changes progressively during the injection phase into a porous medium and finally forms gels at later times [113]. There is a critical overlap

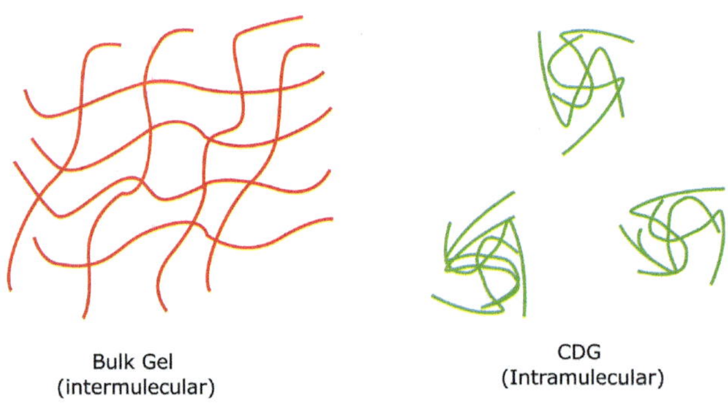

Fig. 6.7 Gel structure difference between BGs and CDGs. *(Adapted from Diaz D, Somaruga C, Norman C, Romero JL. Colloidal dispersion gels improve oil recovery in a heterogeneous argentina waterflood. SPE symposium on improved oil recovery. Tulsa, Oklahoma, USA: Society of Petroleum Engineers; 2008.)*

concentration [108] above which intermolecular cross-links among polymer particles occur. Most of the cross-link bonds below the critical concentration are intramolecular links [108]. By continuing these cross-linking, a high viscous gel will eventually be created inside the formation that blocks the porous media. As reviewed by Spildo et al. [16], a CDG solution is composed of both intra- and intermolecular cross-linking. The dominant type of cross-linking is a function of cross-linkers, polymers, molecular weight, concentration, and polymer-to-cross-linker ratio, salinity, reaction time, and temperature [16]. Bjørsvik et al. [108] compared the viscosity of HPAMs to that of CDG solutions (polymer-to-aluminum ratio of 10:1). The CDG viscosity at the same concentration (600 ppm) was lower than that of the HPAM solution. Salinity rising also decreased the viscosity difference between these two solutions because salt forms coiled polymers, which is like intramolecularly cross-linked polymer units. This viscosity reduction was lower at higher salinities.

In CDGs, the gelation rate depends on the polymer concentration, water salinity, temperature, and polymer to cross-linker ratio. The gelation time should be long enough for polymers to change to a gel state far away from injection wells in field applications [110]. However, the reaction between the polyvalent cross-linker cations and HPAM is usually so fast that it may form pregels, which prohibits having a stable CDG system [108]. Among cross-linkers, aluminum citrate proved to have a slow reaction, forming largely intramolecular bonds. Sheng [112] showed that the resistance of this type of microgels is dependent on the molecular weight of polymers, the degree of hydrolysis, and salinity concentrations (i.e., decreases with increasing salinity). However, it was found that CDGs may form stable solutions even up to 5 wt% NaCl solutions [108].

The linear core flood experiments of Lu et al. [114] showed that the CDG treatment could recover 40% of the remaining oil after waterflooding. These experiments also demonstrated that gel formation occurs at the front end of the core samples. This result indicated that in field applications gel can be formed in the areas around the wellbore and not in-depth of the reservoir [10,114,115]. Al-Assi et al. [111] studied the CDG system composed of HPAM with aluminum citrate and its propagation through sandpacks. They found that CDGs treated approximately 12 ft in depth of a 10-darcy porous medium with an injection velocity of 5 ft./D. The main reason for permeability reduction was gel aggregate retention in porous media. Moreover, other researchers questioned the deep penetration of CDGs into reservoirs [114–116]. All laboratory core flood experiments reported a massive retention of CDGs at preliminary lengths of cores with no indication of in-depth conformance control of CDGs [107]. However, some other literature proved deep conformance control of CDGs [107,108]. These conflicting results can be because of implementation of different experimental conditions (e.g., polymer type, polymer concentration, polymer-cross-linker ratio, CDG mechanism, salinity, and temperature) [113].

Moreover, a new in situ polymeric microsphere (ISPM) agent was introduced with application in low-permeable reservoirs [117]. ISPMs are injected into a target formation, then after a certain period (8.4 h), the emulsion could polymerize into CDG microspheres with a diameter of 1.88–30.60 μm that can plug water channels. The polymerization reaction time is a function of salinity, while pH has negligible effect on it. The pressure experiments and SEM results indicate that these emulsions had proper injectivity in low-permeable samples [117].

CDGs have been implemented in various field applications. They were first implemented as a conformance control agent in the region of Rocky Mountains [110]. The incremental oil recoveries ranged from 1% to 18% in 29 field projects [110]. After that, several successful applications were reported [16,17,109,118–120]. These polymer gels have also been implemented in the low-salinity sandstone formation of the Daqing oil field in China [118]. The CDG flooding in this field showed a 19.8% reduction in water cut and a 10.5% rise in the incremental oil recovery. Moreover, an injection of CDGs in a multilayer heterogeneous sandstone reservoir in Argentina showed a 2.9% incremental oil recovery [17].

In general, CDGs cannot work well in field applications under the following conditions [10]:

- Fields with high salinity (>30,000 ppm) [83,110].
- High-temperature fields (>90°C) [83].
- Fields flooded with water for a long time (mature waterflooded reservoirs) [110].
- Fields that have wells with high positive skin factors (damaged wells) or very low permeability around the wellbore [110].

6.4.2.1 CDG transport modeling

Shi et al. [121] introduced a model for the CDG transport in porous media, incorporated in UTGEL, a reservoir simulator developed at the University of Texas at Austin. They defined the jamming ratio (*JR*) that is the ratio between the pore throat diameters and the layer thickness of adsorbed microgels.

$$JR = \frac{2r_h}{\varepsilon_h} \tag{6.3}$$

where ε_h and r_h are the layer thickness of adsorbed microgels and the mean pore radius, respectively. The pore throat radius can be predicted from the porosity, ϕ, and the permeability, K, of the rock sample by:

$$r_h = 1.15\sqrt{\frac{8K}{\phi}} \tag{6.4}$$

In that study, pores were classified into three categories based on *JR*: inaccessible, plugged, and adsorbed pores. When the pore throat diameter is less than or equal to the adsorbed layer thickness, ($JR \leq 1$), microgels cannot enter into inaccessible pores.

By JR increasing, pores become partially or wholly plugged. By further JR rising, pores are considered as adsorbed ones. The transition JR value of 4 was considered for switching from plugged to adsorbed pores.

The subsequent steps were used to assess the permeability reduction factor after CDG flooding in reservoir grid blocks. The combination of Hagen-Poiseuille equation and Darcy's law was used to calculate the flow rate of porous media. Gels cannot enter into those grid blocks during the gel treatment whose pore throat radii is less than the adsorbed gel thickness. Thus, the flow rate of these blocks is not modified. On the other hand, flow rates in grid blocks that their pore throat radii are more significant than adsorbed gel thickness are affected due to the decreasing of the pore radii (approximated by ε_h). The permeability of these grid blocks is updated based on the new radius (r_h–ε_h). The adsorbed layer thickness, ε_h, can be calculated through data matching with laboratory results from:

$$\varepsilon_h = (aC_M + b)\dot{\gamma}_{eq}^{c} \tag{6.5}$$

in which, a, b, and c are empirical constants, $\dot{\gamma}_{eq}$ is the equivalent shear rate and C_M is the microgel concentration.

A log-normal distribution was considered for permeability reduction with μ and σ as the mean and standard deviation of the variable's logarithm, respectively. Having ε_h and the JR, the dimensionless permeability reduction factor (R_k) can be calculated from the following capillary model relationship:

$$R_k = 1.5 + \left[\frac{g}{(JR-1)\sqrt{2\pi\sigma^2}}\exp\left(-\frac{(\ln(JR-1)-\mu)^2}{2\sigma^2}\right)\right]^f \tag{6.6}$$

where g and f are two matching parameters.

It should be noted that if microgels are formed at the surface and before the injection, this treatment fails [113]. Shi et al. [113] proposed a viscosity model for CDG as a function of applied shear rates and microgel concentrations. The model predictions were experimentally validated.

$$\mu_M^0 = \mu_s\left(1 + [\eta]C_M + [\eta]^2 K_H C_M^2 + \frac{[\eta]^3 K_H^2}{2}C_M^3\right) \tag{6.7}$$

where $[\eta]$ is the zero-shear intrinsic viscosity, an indicator of colloid internal density, and K_H is the Huggins constant. This constant is an indicator of microgel colloid interactions in the solution, $\mu\mu_M^0$ is the viscosity at zero-shear rate, and μ_s is the solvent viscosity. The variation of viscosity (μ_M) as a function of shear rate for a shear-thinning component was defined from the following equations:

$$\frac{\mu_M - \mu_s}{\mu_M^0 - \mu_s}\mu_M^0 = \left[1 + \left(\lambda \dot{\gamma}_{eq}\right)^2\right]^{(n-1)/2}$$
$$\lambda = de^{eC_M} \tag{6.8}$$

where $\dot{\gamma}_{eq}$ is the equivalent shear rate, and d, e, and n are three empirical constants. d and e are obtained by regression of the viscosity of polymer solutions versus shear rates at various concentrations. The shear-thinning behavior of CDG can assist the selective penetration into the high-permeability layers only [67]. The comparison between the experimental data and the viscosity model is illustrated in Fig. 6.8. Several simulation runs were used to validate the simulator's accuracy and good agreement was obtained [121].

The CDG transport can also be modeled by considering the gel reaction kinetics and considering the *RF* and *RRF* in porous media. The following equation can be implemented for a simplified gel formation model [67,122].

$$Polymer(C_1) + Crosslinker(C_2) \rightarrow Gel(C_3) \tag{6.9}$$

The reaction kinetics between cross-linkers and polymers is given [122]:

$$r_k = r_{rk} e^{-\frac{Ea}{RT}} \prod_{i=1}^{n} C_i^{e_k} \tag{6.10}$$

where r_k is the reaction rate, E_a is the activation energy, R is the modal gas constant, T is the temperature, e_k is the order of reaction for the ith component, and C_i is the concentration of reactant i.

The permeability reduction factor after gel treatment can be defined as [122]:

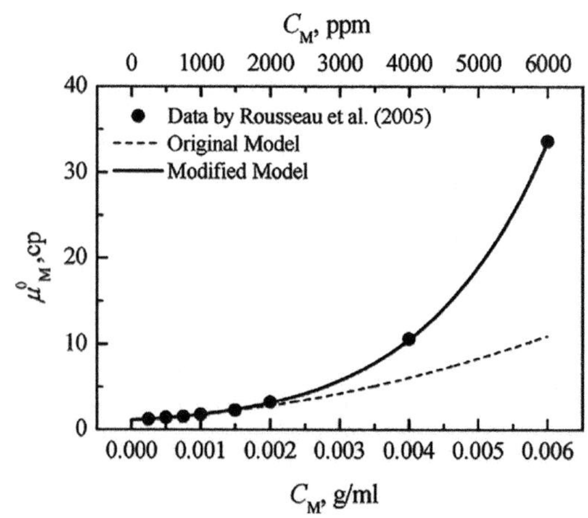

Fig. 6.8 Comparison between the experimental viscosity at zero-shear rate with predictions of the viscosity model (Eq. 6.8). *(Reprinted with permission from reference Shi J, Varavei A, Huh C, Delshad M, Sepehrnoori K, Li X. Viscosity model of preformed microgels for conformance and mobility control. Energy Fuel 2011;25(11):5033–7. Copyright (2011) American Chemical Society.)*

$$PRF = 1 + (RF_{\max} - 1)\left(\frac{\theta_i}{\theta_{\max}}\right) \tag{6.11}$$

in which, $RF_{\max}$ is the maximum resistance factor and $\theta_{\max}$ is the maximum adsorption level at maximum concentration.

6.5 Preformed particle gel (PPG)

PPG is a cross-linked superabsorbent polymer particle that can swell up to 200 times its original size [5]. This type of microgels is prepared by drying BGs, grinding, and then sieving them to desired sizes (e.g., from nanometer to millimeter). They are injected into reservoirs in a suspension form, and they then swell by water absorption to several times their original size [35,54,56]. Fig. 6.9 depicts a typical form of PPG particles after swelling. These gels have less sensitivity to reservoir conditions (e.g., pH, salinity, hydrogen sulfide, and temperature) [123]. They are designed to overcome the inherent problems of other gel technologies including gellant composition alteration during injection and gel treatment controlling problems. This treatment method has also been combined with numerous other EOR techniques, including low-salinity water injection [123,124] and surfactant flooding [125] to improve macroscopic and microscopic efficiencies.

The development of PPGs was first introduced in 1996 by PetroChina [35]. Since then, it has been implemented in more than several thousand wells to treat excessive

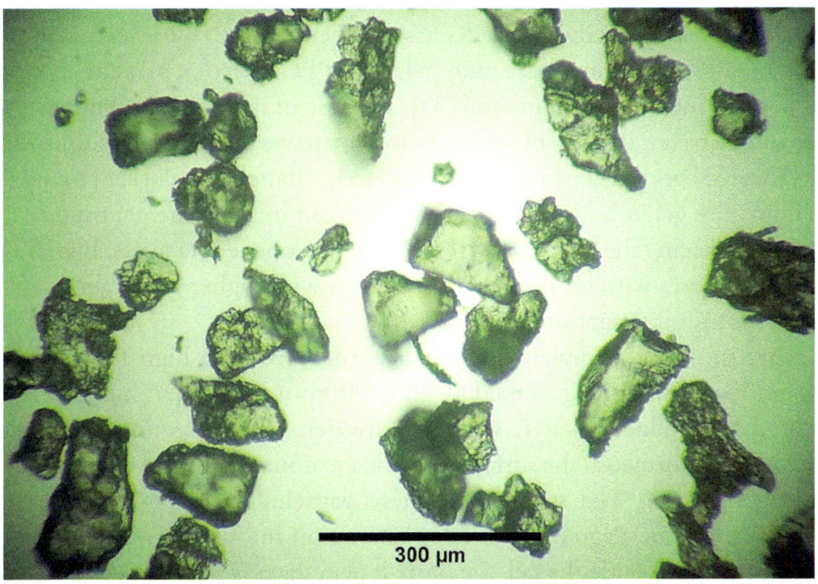

Fig. 6.9 The microscopic photo of PPG particles with a particle diameter of 40–50 μm [4].

water production problems. There are many successful case studies of implementing PPGs in fields, most of which are related to fractured and nonfractured mature oil fields in China (e.g., Daqing, Lamadian, Shengli, Xingbei, Liaohe, Jidong, Zhongyuan, Dagang, Tuha, Shengli); at temperatures from 20°C to 110°C; and salinity between 2000 and 280,000 mg/L [82,124,126].

The ability of PPGs for correcting the reservoir heterogeneity is a function of PPG properties and pore morphology; therefore, PPGs properties (e.g., elastic properties) at near-wellbore areas can influence the design of conformance control treatments [4]. The PPG blocking efficiency can be optimized by controlling the gel strength and its particle size for each specific case study. These two properties are functions of PPG composition, temperature, salinity, and pH [82]. The low swelling rate of PPG particles to reach equilibrium that can take even up to weeks is so attractive for conformance control purposes [127].

PPG treatments showed an excellent result in excessive water plugging control in reservoirs containing fractures and fracture-like features, which cannot be treated with other gel treatment methods. PPGs, because of their size, do not usually penetrate into a low-permeability matrix. After the PPG treatment of fractures, the fracture conductivity decreases, and the pressure gradients across adjacent matrix blocks increase during subsequent chase floods [128]. Nevertheless, PPGs cannot be implemented in very severe open channels or superhigh-permeability fractures because they may be washed out from porous media [11]. Different factors can reduce the washout of PPGs from fractures, including the increase of gel concentration, the decrease of fracture width, and gel injection pressure [32]. Flooding experiments in fractures showed that PPG fronts have a piston-like movement [129,130]. PPG propagation investigations in fractured rocks revealed that the elastic modulus of PPG has a higher effect on the injection pressure gradient than the PPG-to-channel size ratio [54,131]. PPG injection pressure increases as the conduit became more heterogeneous [131]. Because of this pressure difference between fractures and matrixes, the gel solvent leaks into matrixes and the gel volume in fractures reduces [132,133] or damage the gel structure [134]. Brattekås et al. [135] found that placing the immature gel in both matrix and fracture can increase the gel rupture resistance more than just placing the gel in fractures. In partially opened fractures, higher RRF was observed in fractures with greater length, matrixes with higher permeability, or samples with low residual oil saturation [136,137].

PPGs are suitable for implementation in formations with high-temperature (up to 120°C) and high salinity (e.g., 200,000–300,000 ppm) [11,138]. Thus, PPGs can even be prepared by produced water, saving freshwater, and protecting the environment [5]. Since they are formed at the surface, there is a reasonable control over the gel strength and time of gelation. The strength of these particles depends on the cross-linker-to-polymer ratio. The degree of particle penetration into a formation is a function of its gel strength. The weaker the gel, the more it penetrates [139]. The conventional PPGs can also be reinforced by nanomaterials (e.g., clays, nanosilica, montmorillonite, cellulose) [12,63,140] to increase their strength; starch [141] and coal fly ash [142] to improve their viscoelasticity behavior; or curable resins [143] to prevent their washout. These combinations can improve the conformance control of PPGs in mature reservoirs. For example, incorporating nanoparticles can solve the conventional problem of PPGs including, inadequate ability to swell, gel elasticity, gel strength, and thermal resistance [144].

Because of the PPG particle size, they cannot be implemented in low-permeability mature reservoirs. A novel micron-size branched PPG (B-PPG) was developed to solve this issue: a PPG with a cross-linked network and some linear branched chains [145]. The branches in the molecular structure of B-PPGs make the gel system more viscous than traditional PPG suspensions in a way that they combine the excellent high viscosity characteristic of linear polymers with the outstanding profile control ability of PPGs [19,146]. It was shown that B-PPGs could significantly enhance oil recovery in low-permeability reservoirs by improving the permeability contrast [147]. A combination of B-PPG, HPAM, and surfactant was injected in the Shengli oil field, which showed a remarkable incremental oil recovery [145].

Another class of PPGs used for water shutoff and improving sweep efficiency in reservoirs containing fractures or fractured-like features is recross-linkable PPGs (RPPGs). RPPGs, classified as self-healing materials, can recross-link to form rubber-like bulky material and increase the plugging efficiency [148]. RPPGs are placed at the desired location in a reservoir, just similar to PPGs. After the desired aging time, the assembly groups in the structure of RPPGs release at the reservoir condition and recross-link the placed PPG particles together to make an RPPG structure. The result is a network of RPPGs with a very dense structure in which no particle edge can be identified (Fig. 6.10) [149]. RPPGs show excellent mechanical robustness and thermal stability to reduce water permeability in heavily fractured media [148,150]. For example, dual cross-linking chemistry can be implemented to design such materials. The first cross-linker is responsible for forming

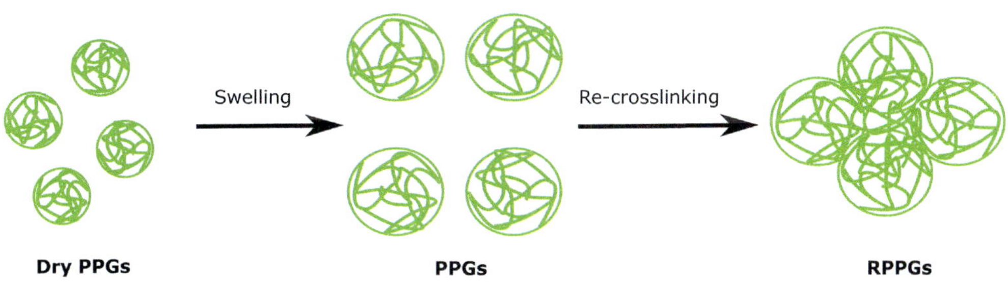

Fig. 6.10 RPPG recross-linking procedure. *(Adapted from reference Pu J, Bai B, Alhuraishawy A, Schuman T, Chen Y, Sun X. A novel re-crosslinkable preformed particle gel for conformance control in extreme heterogeneous reservoirs. SPE annual technical conference and exhibition. Society of Petroleum Engineers; 2018.)*

the 3D structure of gel particles during the surface preparation. The second cross-linker is responsible for merging various PPG particles after the injection. In this situation, RPPGs swell after the injection. The swollen contacted particles can recross-link with each other to form a bulky gel after a predetermined time by the second cross-linker [148]. Thus, RPPGs can withstand higher pressure gradients than conventional PPGs [148].

6.5.1 PPG preparation

The primary PPG preparation method has different steps (Fig. 6.11). First, monomer, cross-linker, and other additives are mixed. After adding the initiator, the cross-linking (or polymerization) process starts, and a BG is finally formed. The BG is mechanically cut into small pieces and dried to form solid particles in the next step. The prepared solid particles are ground and sieved to desired sizes.

PPGs can also be prepared by applying high shearing rates in a colloid mill [46]. In the conventional PPG preparation method, specific concentrations of the polymer, initiator, cross-linking agent, and other additives are mixed; after that, the PPGs are formed by drying and pulverization [152]. In the latter method, the formed gels are called dispersed particle gels (DPGs). Polymer and cross-linkers are first mixed (Fig. 6.12A), and the cross-linking reaction begins forming a BG with a 3D network structure (Fig. 6.12B). Next, the BG is added to the colloid mill. The high shear forces break and convert the formed BG into discontinuous particle dispersions (Fig. 6.12C). By applying continuous shear forces, the particle distribution becomes uniform (Fig. 6.12D) and the final DPG product is obtained (Fig. 6.12E).

The initial polymer concentration, salinity, and applied shear rates can adjust the DPG sizes from nanometer to micrometer [153]. DPGs can significantly block channels and increase sweep efficiency and, consequently, increase oil recovery. DPG particles have viscoelastic behavior and will deform during propagating through pore throats, contributing to in-depth conformance control [154]. Nanosilica particles were also implemented to reinforce the inner framework of DPGs and increase their gel strength [155].

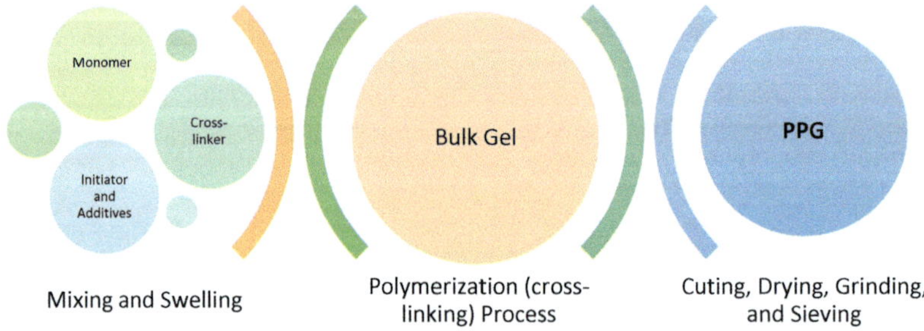

Fig. 6.11 PPG preparations steps [151]. *(Adapted from Ref. Bai B, Li L, Liu Y, Wang Z, Liu H. Preformed particle gel for conformance control: factors affecting its properties and applications. SPE/DOE symposium on improved oil recovery. Society of Petroleum Engineers; 2004.)*

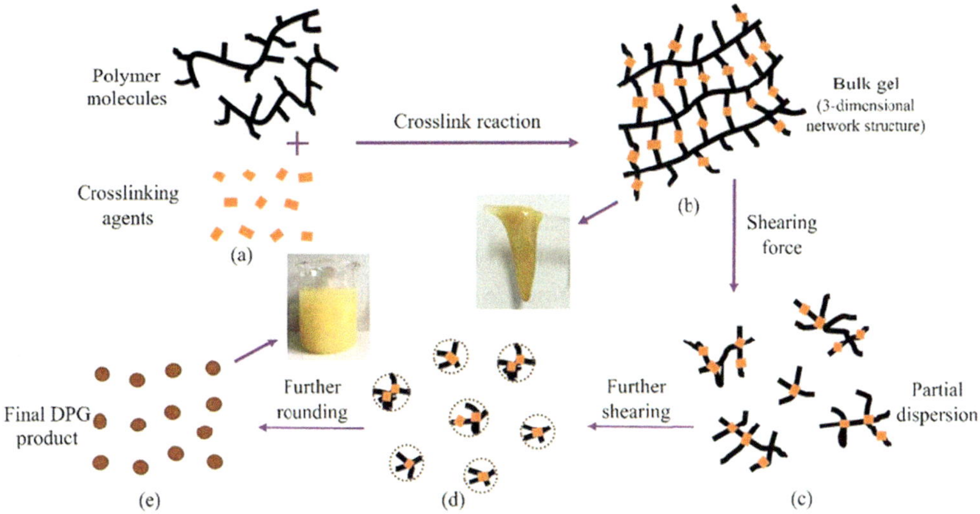

Fig. 6.12 The preparation procedure of DPGs. *(Reprinted with permission from reference Liu Y, Dai C, Wang K, Zhao M, Gao M, Yang Z, et al. Investigation on preparation and profile control mechanisms of the dispersed particle gels (DPG) formed from phenol–formaldehyde cross-linked polymer gel. Ind Eng Chem Res 2016;55(22):6284–92. Copyright (2016) American Chemical Society.)*

6.5.2 Swelling characteristic of PPGs

Swelling and water absorption capacity are the main characteristic of polymeric gels [156]. Swelled gels have higher volume and different properties than dry gels in formations, influencing the effectiveness of a gel treatment [157]. PPGs are insoluble in water; however, they absorb and swell several times (e.g., 20–200 folds) their original mass [128]. The evaluation of swelling ratio performance can reveal valuable insights about the mass-related behavior of PPGs at reservoir conditions [4].

In many previous studies, the PPG swelling ratio was considered only a function of salinity [54,128,158–161]. For example, PPG swelling in a range of $CaCl_2$ salinity from 0 to 30,000 ppm and pH from 3 to 11 was investigated [162]. These particle gels were composed of chromium (III) acetate, 2-acrylamide-2-methyl-propane sulfonic acid, sodium salt, and AM. The results reveal that the PPG swelling ratio was a function of more water salinity than pH. Increasing gel solvent salinity reduces swelling ratio and increases swollen particle strength [158,160]. On the contrary, low-salinity postwater flooding can improve the gel-blocking capacity [128]. Lenji et al. [163] showed that the PPG swelling ratio decreases when both polymer and cross-linker concentrations and salinity increase.

However, there are some evidences that temperature variations can also influence the PPG swelling ratio. Molloy et al. [164] examined the simultaneous effect of salinity from 5000 to 35,000 ppm and temperature from 5°C to 40°C on the PPG swelling ratio composed of acrylamide and methylene bisacrylamide. At each constant salinity, the effect of temperature was studied. The result showed that the PAM gels shrunk in all saline solutions, which was more severe at higher salinity levels. The temperature effect was not as much as the salinity effect. However, with increasing temperature, the slope of swelling volume decreases. This result was also confirmed by other studies [163].

The swelling properties of composite PPGs (PAM, chromium acetate, and bentonite clay) were also investigated [82]. It was revealed that the swelling capacity depends on the gel composition, temperature, water salinity, and pH. With an increase in the amount of cross-linking agent and clay, the gel swelling decreased and their elastic modulus increased. On the other hand, the presence of salt reduced the amount of gel swelling. Increasing the temperature caused the swelling of the studied gels to increase and their elastic modulus to decrease. In acidic environments, with increasing pH, the swelling capacity of particles increased and remained constant from 6 onward. Farasat et al. [43] coupled a three-level full factorial experimental design method with laboratory analyses to examine the functionality of swelling ratio of a new class of composite PPGs (AM, *N,N*-dimethyl AM, *N*-vinylpyrrolidone, and 2-AM-2-methylpropane sulfonic sodium as monomers with the cross-linker of *N,N*-methylenebis, and montmorillonite nano-clay). The results illustrated that the PPG swelling ratio was a function of water salinity in low water salinity regions (<10,000 ppm) and temperature in high water salinity regions (between 10,000 and 225,000 ppm). Moreover, the swelling ratio showed a logarithmic decrease when salinity increases. It means that the swelling ratio decreased sharply when the salinity increased (<10,000 ppm). However, further increase of salinity (between 10,000 and 225,000 ppm) had less impact on the swelling ratio reduction.

6.5.3 PPG injection in porous media

The PPG performance in a porous medium depends on the transfer mechanisms of particle gels and their interactions in pore spaces [138,165,166]. Therefore, it is essential to study the behavior and suspension performance of PPGs inside pore spaces.

6.5.3.1 PPG transport mechanisms in porous media

The success of a PPG treatment rests on the gel ability to reduce the permeability of high-conductive features in reservoirs; therefore, understanding the gel mechanisms to block the high-conductive features during a conformance control treatment is essential [54]. PPGs move through porous media under several mechanisms. During flowing, gel particles smaller than pore throats can pass easily. Observations show that some of these particles may be adsorbed on pore walls because of particle-rock interactions. Larger gel particles can plug pore throats or even break into smaller particles because of flow shear stress and then pass through pore throat openings. Therefore, the gel particle transport through a porous medium is categorized into six mechanisms [5]. Fig. 6.13 illustrates the schematic representation of these transport mechanisms.

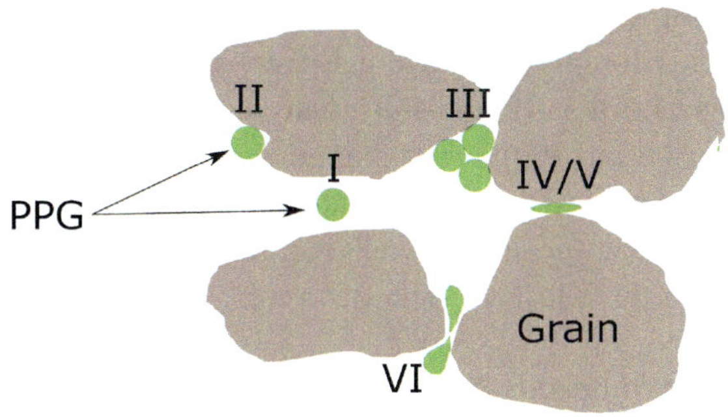

Fig. 6.13 Schematic illustration of PPG transport mechanisms through porous media. (I) direct pass, (II) adsorption and retention, (III) trap, (IV) deform and pass, (V) shrink and pass, (VI) snap-off and pass.

I. **Direct pass:** When the particle size is smaller than a pore throat, it can move through it by the water displacement.
II. **Adsorption and retention:** Because of attraction forces between rock surfaces and particles, some of them can be adsorbed on rock surfaces. When the particle size is larger than the pore throat size, the retention mechanism is dominant.
III. **Trap:** Particles block pore throats and stop flowing forward.
IV. **Deform and pass:** The shape of a deformable particle can be changed because of applied displacement forces by the water flow and then it passes through pore throats. After passing, it can be reshaped to its original form.
V. **Shrink and pass:** Because of displacement forces applied on a swollen particle, it shrinks and losses a portion of its water. Thus, the particle diameter reduces and can pass through the pore throat. After entering the downstream pore, the particle can reabsorb the lost water and be reshaped to its original size.
VI. **Snap-off and pass:** A particle can be broken into smaller size when passing through a pore throat.

The last four mechanisms occur when particles are larger than pore throat diameters. Determining the main patterns responsible for particle transports depends on the relative sizes of swollen particles, pore throat openings, particle mechanical strength, and the driving force (shear stress) of the injected fluid. It is challenging to determine the dominant mechanism, and always several mechanisms coexist simultaneously.

The injection experiments in sandpacks [81] during the deform and pass mechanism showed that the pressure drop across the system increaseed linearly with a continued injection. During this PPG flow mechanism, the size distribution of effluent particles remained similar to that of the primary injected particles. This mechanism had a minimum effect on the *RF* and *RRF* values. The broke (or snap-off) and pass mechanism also showed as the same pressure drop behavior as the deform and pass mechanism. The only difference was the increase in the *RF* and *RRF* in the initial sections of the sandpack, which indicates that the particles are being trapped there. The pressure drop rised rapidly in the trap mechanism, while the middle and end sections of the sandpack did not show any pressure changes. Moreover, particle gels were usually not observed at the outlet flow. In this mechanism, the *RF* and *RRF* had very large values at the initial length of the sandpack (ca., 600 and 100), while the remaining lengths showed a number equal to unity [81]. A similar process can occur inside injection wells. If large-sized PPGs are used, the entrapment inside and around wells increases, and the penetration depth of the particles far away from the injection wells decreases.

6.5.3.2 PPG retention in porous media

The interactions between PPGs and rock surfaces in pores and throats control the amount of retention (or trapping) at reservoir conditions and directly influence the gel particle capability to reduce the rock permeability. The penetration depth of microgels also depends on the degree of retention [138,165,166]. Sandpack experiments [167] showed that nano-PPGs had good injectivity into cores with high permeabilities (e.g., above 143.4 mD), but the injectivity was reasonably low for rocks with low permeability (e.g., 41.2 mD). The *RF* and *RRF* were inversely proportional to fluid velocity, while they were directly related to PPG concentration [167,168].

The retention amount is a function of several parameters: injection rate, particle gel elasticity, concentration, particle-to-pore throat size ratio (DPT), and load pressure. When retention increases, both *RF* and *RRF* increases as well. The PPG retention increases when the particle size, cross-linker concentration, and salinity increase; however, it decreases with temperature, flow rate, and gel particle elasticity [43,159,169]. Adsorption on the surface of rocks or mechanical entrapment is responsible for the retention mechanisms. Mechanical entrapment happens when the polymer gel radius is larger than the radius of pore throats [107]. PPGs also have elastic behavior, so they are deformable, which complicates the prediction of retention behavior.

The nature of rigid particle retention (nonelastic particles) in a porous medium was investigated by Donaldson et al. [170]. A suspension of solid rigid particles was injected into a sandpack with a known average pore throat size distribution. The results showed that when the particle-to-pore throat diameter ratio was less than 0.08, no significant retention occurred, and particles moved through a sandpack until being produced from the outlet. However, the particles could be superficially adsorbed on the rock surface. At ratios of 0.17–0.25, adsorption and blockage occurred because of particle accumulation in pore throats. At ratios above 0.39, the particles did not penetrate the sandpack at the inlet surface by creating a filter cake.

Unlike rigid particles, PPGs (as deformable particles) can penetrate pore throats smaller than their size due to their viscoelastic behavior [171,172]. Fig. 6.14 illustrates the difference between the rigid and deformable particles passing through a pore throat. Particle gels with a DPT greater than one can block pore throats or even deform and pass through them when a threshold pressure gradient is exceeded. The mechanical

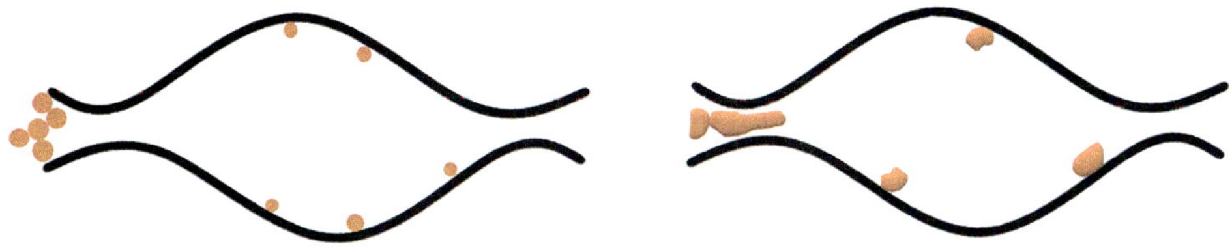

Fig. 6.14 Difference between rigid and deformable particles passing through pore throat. *(Adapted from reference Al-Ibadi A, Civan F. Experimental study of gel particles transport through porous media. SPE Latin America and caribbean petroleum engineering conference. Society of Petroleum Engineers; 2012.)*

entrapment mechanism of PPGs has a more significant effect on permeability reduction than the adsorption, leading to a long-term persistence of PPG against water flow in high-permeable zones [169]. Bai et al. [5] showed that swollen week and strong PPG particles could penetrate through pore throats whose diameter are only 1/5.7 and 1/1.3 times of swollen particles.

PPG particles generally show less adsorption and a high remigration behavior; thus, they migrate deep into formations instead of blocking near-wellbore areas [173]. The pressure gradient for remigration of trapped PPGs exponentially increases by the DPT [173]. Therefore, the retention amount of these particles decreases with increasing shear rate (or fluid velocity) and rises with increasing particle concentration or diameter. By increasing fluid velocity, they penetrate deep into formations, act as in-depth injection profile controllers, and increase the oil recovery factor. The previous investigations showed that these particle gels act more effectively in more heterogeneous porous media [167,173–175].

PPG gels have a low-permeability value and show compressible behavior. The placed PPG permeability (bulk permeability) is a function of the particle size. For example, large-sized particles have mostly a lower bulk permeability than smaller particles [54]. The PPG pack permeability is also a function of load pressure. The bulk permeability of placed PPGs is lower at reservoir conditions than at surface conditions because of the gel compressibility [54]. This gel compressibility varies with salinity and particle sizes. If subjected to overburden pressure, the bulk permeability decreases and become almost constant because of compression of gel particles [54].

Farasat et al. [4] investigated the mechanical entrapment of PPGs in porous media. They found that the amount of entrapment (or retention) is a function of reservoir temperature, gel particle-pore size ratio, and injection fluid velocity (Fig. 6.15). The PPG entrapment declines as the fluid velocity (Fig. 6.15A, B) and reservoir temperature (Fig. 6.15C, D) increase. With the increased fluid velocity and higher shear stress on particle gels, more particles deform, and gel retention decreases [4,167,169]. At different temperature levels, entrapment behaves differently with velocity changes [4]. The entrapment reduction rate is more evident at lower temperatures (Fig. 6.15A). The

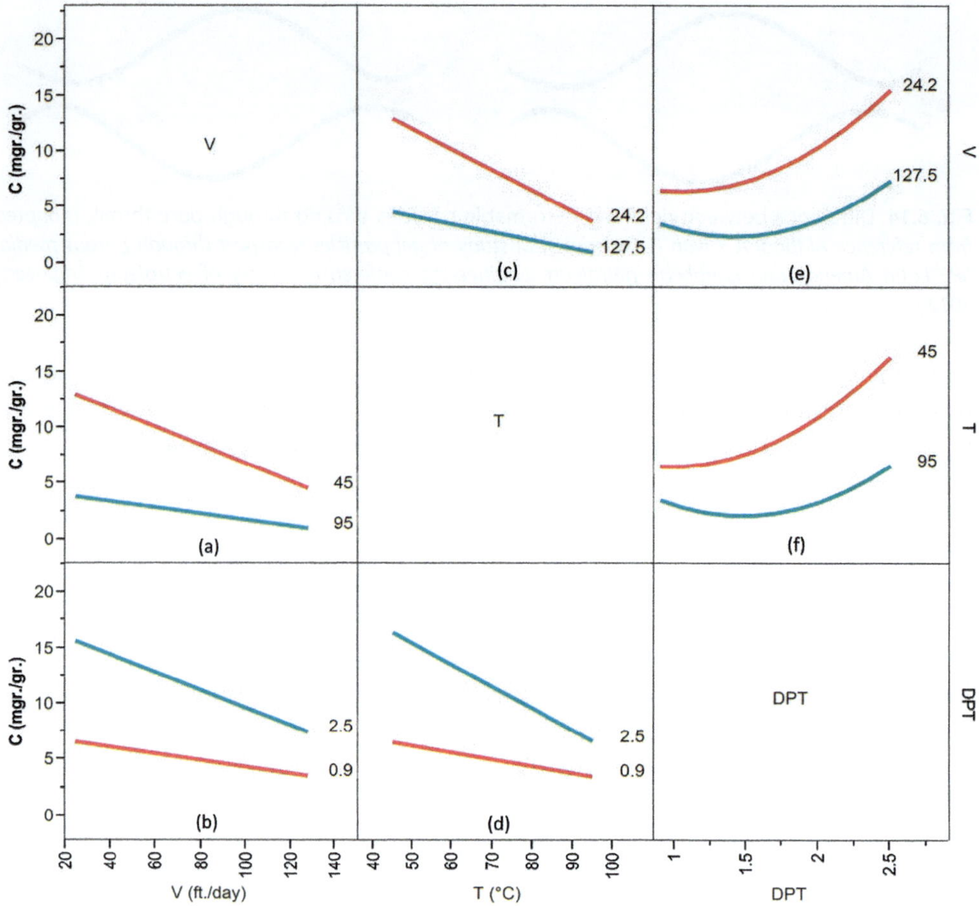

Fig. 6.15 The interaction plots of the amount of the mechanical entrapment (C). Each line (*red* and *blue*) shows the lowest and highest level of parameters during the experiment [4].

entrapment can be significantly affected by temperature variations (Fig. 6.15C, D) if the fluid velocity will be at the lowest level or the DPT will be at the highest level. In both cases, the temperature increment can soften the particles and reduces the differential pressure needed to pass the throats. However, the gel particle DPT displays a complex behavior because of the elastic nature of PPGs (Fig. 6.15E, F). The PPG entrapment amount decreases slightly by increasing the DPT and then increases dramatically. The minimum entrapment (Fig. 6.15F) is observed at the maximum studied temperature of 95°C with an injection velocity of 127.5 ft./day and a DPT of 1.6 [4].

The elasticity (or strength) of gel particles also plays an important role in the amount of retention and the performance of PPGs in porous media. The gel strength usually rises when the molecular weight of polymer increases; nevertheless, those gels

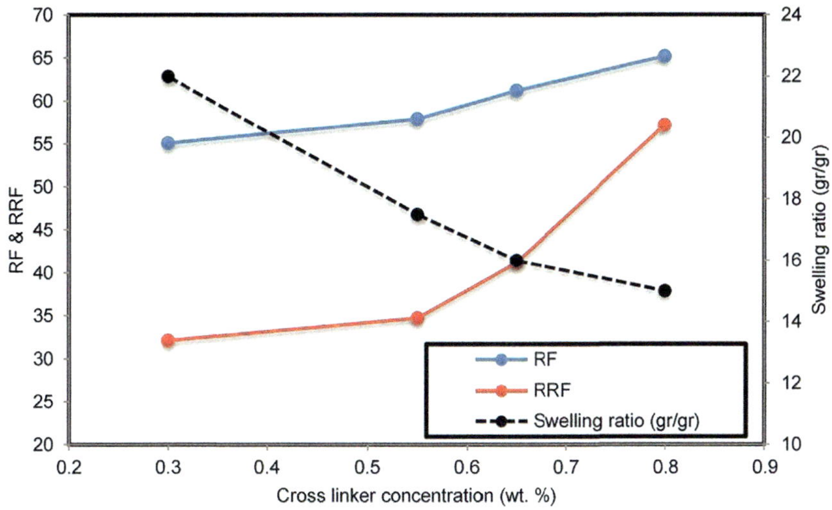

Fig. 6.16 Effect of cross-linker concentration on the *RF*, *RRF*, and swelling ratio [169].

prepared with high-molecular-weight polymers show more elastic behavior [62]. The effect of particle gel elasticity on injectivity (or increase in the *RF* and *RRF*) sometimes can be much greater than the DPT ratio. Soft (or weak) gel particles with a higher swelling capacity can pass through smaller pore throat openings than hard (or strong) gel particles with a lower swelling capacity. As the PPG concentration increases, the amount of gel retention in porous media increases to some extent, after which it exhibits a stable behavior [81]. Weak gels can easily break into smaller particles or lose their water (dehydration) more than strong gels; thus, they will have smaller injection pressure than the strong gels [159]. Moreover, increasing the water salinity increases the particle gel strength and increases the *RF* and *RRF* [159,169]. However, changes in PPG injection pressure, *RF*, and *RFF* can become negligible at high brine concentrations because of the lack of change in PPGs swelling ratio beyond that amount of salinity [169]. Correspondingly, increasing the cross-linker concentration (see Fig. 6.16) raises the gel strength and increases the *RF* and *RRF* values and gel retention in porous media [169].

6.5.4 PPG transport modeling

As discussed, PPGs can be deformed during flowing in porous media. Their primary transportation mechanism includes pass, broken and pass, and plug (or trap) [5]. The threshold pressure, P_{Gmax}, controls the displacement mechanism [176]. During the PPG migration, they can plug those pore throats with a smaller pore diameter. When the injection pressure is higher than P_{Gmax}, particles can break or deform and pass.

The rate of PPG plugging, $\frac{\partial \delta}{\partial t}$, is defined as the difference between the rate of plugging, r_p, and the rate of restarting (remigrating), r_r, of previously plugged particles [177].

$$\frac{\partial \delta}{\partial t} = r_p - r_r \tag{6.12}$$

A kinematic filtration model can be considered for the PPG deposition in porous media [81,176–179]. Iwasaki showed that the particle concentration through a filter at low flow rates is described by the following equation [180]:

$$\frac{\partial C_{ppg}}{\partial r} = -\lambda_{ppg} C_{ppg} \quad P_G \leq P_{G\max} \tag{6.13}$$

where C_{ppg} is the volume fraction of PPG, λ_{ppg} is the filtration rate coefficient, and $P_{G\max}$ is the threshold value. λ_{ppg} is a measure of the filtration efficiency. The model of Maroudas and Eisenklam [181] were used in different studies for this parameter [81,176,178].

$$\lambda_{ppg} = \lambda_0 \left(1 - \frac{\delta}{\delta_{\max}}\right) \tag{6.14}$$

where λ_0 is the initial filter coefficient (or plugging rate coefficient), δ is the reduced porosity (the deposited volume per unit rock volume), and $\delta_{\max}$ is a dimensionless number showing the saturation value of the deposited particles.

The initial retention coefficient can easily be obtained from laboratory experiments while $\delta_{\max}$ is hard to measure. The permeability decline during the PPG injection is implemented to find the maximum retention [176]. The permeability decline from k_0 to $k_{\min}$ is determined by the plugging ratio.

$$\eta = \frac{k_0 - k_{\min}}{k_0} \tag{6.15}$$

By replacing the permeability terms in the above equation with the well-known Carmen-Kozeny equation ($K = \frac{\phi r^2}{8\tau^2}$), we have [176]:

$$\eta_{\max} = \frac{k_0 - k_{\min}}{k_0} = \frac{\frac{\phi_0 r^2}{8\tau^2} - \frac{(\phi_0 - \delta_{\max}) r^2}{8\tau^2}}{\frac{\phi_0 r^2}{8\tau^2}} = \frac{\delta_{\max}}{\phi_0} \tag{6.16}$$

in which, ϕ_0 and k_0 are the initial porosity and permeability, τ is the tortuosity, and $k_{\min}$ is the postflooding permeability of the medium.

The plugging rate coefficient, λ_0, was considered fixed in some modeling approaches [81,176,178]. However, PPG diameter and the applied pressure gradients can affect this parameter [179]. To account for this functionality, Wang et al. [177] and Jing et al. [179] implemented an approach in which two size distributions were considered for pore throats and PPG particles. A log-normal and normal distribution were considered for throat diameters and PPG particle sizes, respectively. The critical pressure gradient connects pore throats and PPG diameters together. A relationship between the initial plugging rate coefficient and the pore size distributions was then obtained.

The applied pressure on gel particles (∇p) needs to overcome the threshold pressure (∇p_{Gmax}) gradient (Eq. 6.18). The restarting rate (r_r, in Eq. 6.12) shows that the remigrated particles are a function of pressure gradient, Darcy velocity, time, and specific deposit parameter [178]. The rate of particle remigration can be described by the Heaviside function [81,176–178].

$$r_r = \frac{\partial \delta_r}{\partial t} = \chi \delta v_{ppg} \frac{\nabla p - \nabla p_{\mathrm{Gmax}}}{\nabla p} \Theta(\nabla p - \nabla p_{\mathrm{Gmax}}) \tag{6.17}$$

$$\Theta(\nabla p - \nabla p_{\mathrm{Gmax}}) = \begin{cases} 0 & \nabla p < \nabla p_{\mathrm{Gmax}} \\ 1 & \nabla p \geq \nabla p_{\mathrm{Gmax}} \end{cases} \tag{6.18}$$

This equation has a kinematic form describing the migration of trapped particles because of deformation. In this equation, χ is the removal coefficient of plugged particles (or remigration rate coefficient), when the restarting condition is met; $\Theta(x)$ is the Heaviside function; ∇p is the instantaneous pressure gradient; v_{ppg} is PPG fluid velocity (or filtration rate).

In reality, both captured PPGs and pore throats have noncircular diameters, there is a chance that a part of captured PPGs may still remigrate even when the local pressure gradient is smaller than P_{Gmax}. Moreover, the remigration rate does not linearly depend on P_{Gmax} changes. Jing et al. [179] introduced a general model to explain the relationship between the migration coefficient and local pressure gradient to account for these behaviors.

$$r_r = \frac{\partial \delta_r}{\partial t} = \chi \delta v_{ppg} \left(\frac{\nabla p - \alpha \nabla p_{\mathrm{Gmax}}}{\nabla p} \right)^{\beta} \Theta\left(\nabla p - \alpha \nabla p_{\mathrm{Gmax}}\right) \tag{6.19}$$

$$\Theta(\nabla p - \alpha \nabla p_{\mathrm{Gmax}}) = \begin{cases} 0 & \nabla p < \alpha \nabla p_{\mathrm{Gmax}} \\ 1 & \nabla p \geq \alpha \nabla p_{\mathrm{Gmax}} \end{cases} \tag{6.20}$$

where α and β are fitting parameters. $\alpha < 1$ shows that trapped PPGs can remigrate more easily. β incorporates the nonlinear behavior between the remigration rate coefficient and the local pressure gradient.

A similar viscous model of polymers can be used for gel particles [176,177]. According to experimental results (Fig. 6.17), the dependency of viscosity on concentration is [177,179].

$$\mu_{ppg} = \mu_w \left(a_1 c_{ppg}^2 + a_2 c_{ppg} + a_3 \right) \tag{6.21}$$

where a_1, a_2, and a_3 are some constants, which are obtained by experiments.

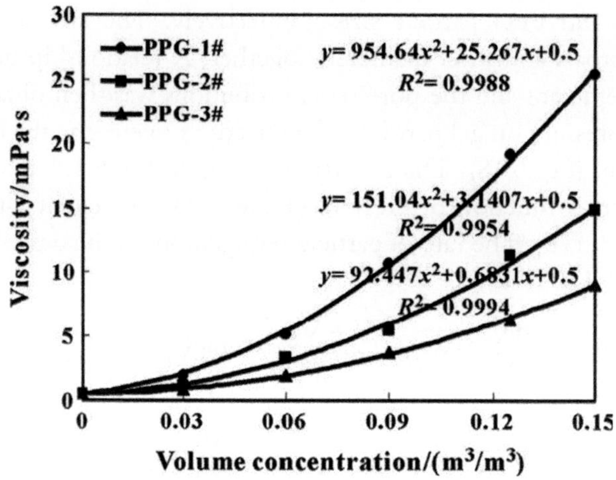

Fig. 6.17 Relationship between viscosity and volume concentration of PPG [177].

Particle retention changes the porosity and permeability of a porous medium. The plugged PPG porosity and permeability of the porous medium can be computed from.

$$\phi = \phi_0 - \delta \tag{6.22}$$

$$k(\delta) = k_0 \, e^{\frac{\chi\delta}{\phi_0}} \tag{6.23}$$

The permeability is reduced exponentially as plugged PPGs increase [173,179]. Other forms of permeability modification equations have also been used in the literature [176,178].

The coupled model for PPG injection in a three-phase system of the black oil model can be described by continuity equation of oil, water, and gas:

$$\nabla \cdot \left[\frac{kk_{ro}}{\mu_o B_o} (\nabla P_o - \gamma_o \nabla z) \right] + q_o = \frac{\partial}{\partial t} \left[\frac{\phi S_o}{B_o} \right] \tag{6.24}$$

$$\nabla \cdot \left[\frac{kk_{rw}}{\mu_w B_w} (\nabla P_w - \gamma_w \nabla z) \right] + q_w = \frac{\partial}{\partial t} \left[\frac{\phi S_w}{B_w} \right] \tag{6.25}$$

$$\nabla \cdot \left[R_s \frac{kk_{ro}}{\mu_o B_o} (\nabla P_o - \gamma_o \nabla z) + \frac{kk_{rg}}{\mu_g B_g} \left(\nabla P_g - \gamma_g \nabla z \right) \right] + R_s q_o + q_g = \frac{\partial}{\partial t} \left[\frac{\phi S_o R_s}{B_o} + \frac{\phi S_g}{B_g} \right] \tag{6.26}$$

in which, k_{ri} are the relative permeability of phases ($i =$ oil, water, and gas); P_i and S_i are the pressure and the saturation of the available phases. B_i is the formation volume factor of phases; γ_i is the specific gravity of phases, R_s is the solution gas-oil ratio, q_i is the source or sink term of phases, and z is the depth of each grid block.

The continuity equation for gel particles can be written as:

$$\nabla\left(v_w c_{ppg}\right) + q_w c_{ppg} = \frac{\partial}{\partial t}\left(\varphi s_w c_{ppg}\right) + \frac{\partial \delta}{\partial t} \tag{6.27}$$

The mathematical description of multiphase flow with PPG propagation (Eqs. 6.24–6.27), is completed by writing Darcy's law for all three phases, fluid saturation equation ($\sum_i S_i = 1$), and capillary pressure relations, $P_c = P_{nw} - P_w$, in which P_{nw} and P_w are the pressure of nonwetting and wetting phase, respectively. It is worth mentioning that the gel particles are considered to suspend in the aqueous phase uniformly. The effect on the density, formation volume factor, interfacial tensions, and relative permeabilities are ignored. Moreover, gel particles are supposed not to interact with the oil and gas phases and moved with an equal velocity to the water phase.

Another approach for numerical modeling of PPG blocking in porous media was implemented using the *RF* and *RRF* factors [6,33,182]. In each grid block, they calculate the pore throat size using Eq. (6.4).

As discussed earlier in this chapter, PPGs movement through a grid block is a function of the particle diameter-to-pore throat size ratio of that grid. The passing conditions of both weak and strong PPG particles through pore throats were considered as follows [182]:

- Weak particles: particle diameter $< 5.7 \times d_p$
- Strong particles: PPG particle diameter $< 1.3 \times d_p$

PPGs can pass through the pore throats of a grid block if the above conditions apply. In this situation, a new viscosity for the aqueous phase is calculated based on the *RF* and *RRF* during PPG flooding and postwaterflooding, respectively [182].

$$\mu_{effecfive} = RF \times \mu_w \rightarrow \text{During PPG injection} \tag{6.28}$$

$$\mu_{effecfive} = RRF \times \mu_w \rightarrow \text{During post water flooding} \tag{6.29}$$

They accounted for the *RF* and *RRF* with salinity effect and injection rate by conducting various core flooding experiments. The following correlations were considered for these parameters [182]:

$$RF = a_1\left(C_{salinity}\right)^{b_1}\left(q_w\right)^{c_1} \tag{6.30}$$

$$RRF = a_2\left(C_{salinity}\right)^{b_2}\left(q_w\right)^{c_2} \tag{6.31}$$

where a_l, b_l, and c_l ($l = 1, 2$) are model parameters determined from fitting to laboratory experiments of the *RF* and *RRF*; $C_{salinity}$ is the salinity of the water phase.

Farasat et al. [183] implemented the same methodology [6,33]. They developed a model for a carbonate sample by considering PPG swelling ratio and entrapment amount. The PPG entrapment was a function of PPG-to-pore/throat size ratio, fluid velocity, temperature, salinity, PPG concentration, and swelling ratio. They found the formulations of the *RF* and *RRF* (Eqs. 6.31 and 6.30) from 1D flooding experiment at different injection velocities and PPG concentrations. The model was coded in the MATLAB Reservoir Simulation Toolbox (MRST) and then validated on laboratory results of PPG injection experiments at various flow rates. They used their validated model to predict the behavior of PPG injection in various 3D field-scale scenarios. The sensitivity results reveal a higher recovery factor for reservoirs with higher permeability contrast (more heterogenous). Moreover, an optimum PPG size to avoid PPG washing out during postwaterflooding was also reported.

6.6 Temperature-activated polymer gel (TAP)

TAPs are submicron-sized particle gels that can swell multiple times their original volume as a result of their environment temperature increasing. As a flow diverting agent, these gels were introduced through a collaborative effort among a consortium of BP, Chevron, Texaco, and Nalko companies in 1997 and named BrightWater® [35]. The main aim was to provide delayed and swellable materials to improve the sweeping efficiency. These submicron-sized particle gels (also called kernels) are injected into reservoirs by cold water at a minimum viscosity [23,184]. For the ease of injection, the injected viscosity must be as close as possible to the water viscosity [59]. Therefore, a thermal front can also be created by preinjection of cold water into a high-temperature reservoir before the TAP injection [11]. During the TAP flooding process, particles move through high-permeable zones (thief zones), while slowly absorbing heat from surrounding rocks. They swell rapidly and irreversibly at a critical temperature up to about 10 times their original volume, which increases their viscosity, blocks thief zones, and diverts the injected water to other unswept areas (i.e., conformance control) [10,11,185]. After activation, the viscosity of the injected fluid rises several folds (e.g., 17-folds after 45 days of aging at 99°C [186]). This increases the sweep efficiency during chase water injection and reduces excessive water production. Therefore, these gels show a proper conformance control behavior. As most of the polymer would flow through the high-permeable rocks (thief zones), the permeability reduction is more severe (stronger and over a larger distance) in these areas [24]. The main reason for this resistance to flow is due to polymer viscosity rising and its adsorption/retention in porous media [23].

The TAP injection phases are given as follow [185]:

1. ***Injection***: Particles easily enter the formation at low temperatures (typically 50–120 °F). They show a low viscosity at this temperature.
2. ***Propagation phase***: Particles are transferred deep into the formation by a chase waterflood. The particle velocity decreases and their temperature increases by adsorbing heat from the formation.
3. ***Swelling***: It occurs when the particles reach the critical temperature (typically between 120°F and 170°F). With time and temperature, internal cross-links break,

and the particles absorb water and swell. They can then plug the pore spaces of thief zones.

TAPs are commonly composed of AMPS as the main polymer, diacrylamides, and methacrylamides of diamines as the labile cross-linker (reversible), and MBA as the non-labile cross-linker (permanent) [46]. For example, the liable and nonliable concentrations are usually between 20,000–60,000 ppm and 0–100 ppm, respectively [187]. The stimuli-responsive behavior of these particles owes to the presence of the labile cross-linker in their structure. This cross-linker degrades by adsorbing heat and passing the time; thus, particles can expand in size by adsorbing water [46]. Fig. 6.18 illustrates the TAP activation mechanism. Both labile and nonlabile cross-linkers keep TAP particles tight in the original state. However, the labile cross-linkers break by hydrolysis in the presence of high temperature, which results in the swelling of gel particles. While nonlabile cross-linker remains responsible for keeping the 3D structure integrated.

BrightWater®'s particles are primarily spherical with a size range of 0.1–1 μm [24,185]. These particles are dispersed in an emulsion of light mineral oil with a 30% concentration [11,24]. Having small particles and low viscosity makes these particles easy to inject and allows them to migrate in-depth of reservoirs before their complete swelling. The density of TAPs is very close to water density; therefore, they have minimum gravity segregation in the reservoir during the flooding process [188]. The other advantages of this method include no need for production shut down, low environmental risks, and not affected by chromatographic separation because of having only one component [184,188].

Salehi et al. [189] investigated the capability of TAPs to create low resistance in porous media with high permeability and the effect of solution concentration. Their slim-tube experiments showed that these microgels could change the flow path even in a high-permeable porous medium, and their efficiency increases with increasing microgels

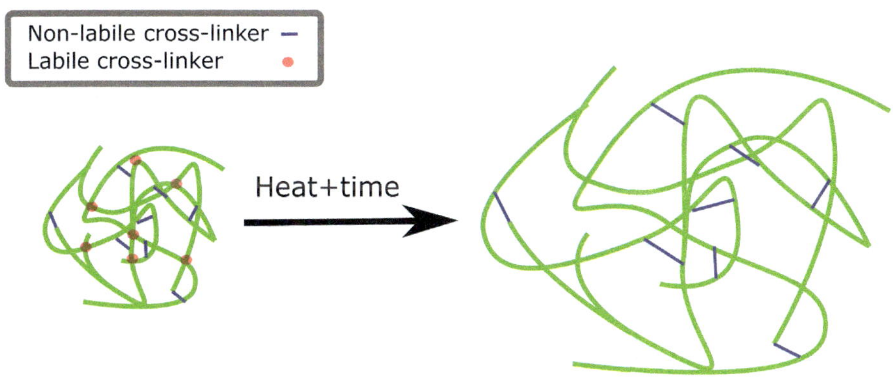

Fig. 6.18 As TAP particles flow deep into a reservoir, the particle temperature increases and the labile cross-linkers degrade, and TAPS then adsorb water and swell. The nonlabile cross-linkers keep the integrity of particles after swelling.

concentration. Fabbri et al. [190] studied the effect of TAPs on permeability reduction in high-permeable unconsolidated sandstones. They observed a high amount of inactivated TAP adsorption (130 μg/g rock) on the surface of the porous medium grains. This result could be due to the nature of the simulated porous medium (i.e., type and heterogeneity distribution of the implemented sands) and the absence of oil. Pritchett et al. [188] indicated that BrightWater® kernels with a mean diameter of one-tenth of the pore throat sizes could easily be transported through porous media.

The optimal positioning of the microgels within reservoir thief zones requires sufficient information about the pore size of thief zones, reservoir temperature, and degree of particle gel penetration [10,33,187,189]. Particle sizes should be specifically designed for a target porous medium by adjusting the main polymer as well as the types and ratio of labile to nonlabile cross-linkers [10]. For a given TAP particle grade, the plugging point will be deeper in the cases of a lower temperature reservoir or of the faster injection rate [185]. Thief zones close to injection wells can be treated by choosing a faster-reacting TAP [185]. For example, Ohms et al. [185] used slow-reacting TAPs (4 months to reach their final equilibrium) to treat thief zones far from injection wells.

The first BrightWater® field application was conducted at the Minas field on the Sumatra island in Indonesia [35,186,188]. BP and Chevron have confirmed the positive results of using this technology in more than 60 wells there. Particle gels with a 1500–6000 ppm concentration were injected with surfactants. Surfactants prevent the agglomeration of submicron-sized microparticles. The simultaneous use of TAPs and surfactants makes it challenging to distinguish the effect of each component in the production enhancement result. This field trial showed that BrightWater® could be propagated far from injection wells (e.g., gelation at 125 ft. from the injection well) [188]. The commercial potential of BrightWater® was demonstrated in other case studies in Alaska [185,191], Argentina [192,193], Brazil [24], Tunisia [194], and Equatorial Guinea [195]. This technology was also implemented offshore of the North Sea [196]. The use of BrightWater® in these case studies led to a decrease in water cuts and an increase in the incremental oil production. The primary screening for the applicability of TAPs in a reservoir includes [23]:

1. availability of unswept zones and moveable oil
2. early water breakthrough in formations
3. moderate to high permeability contrast in reservoirs
4. $T < 150°C$
5. sandstone formations are preferred
6. pH ≥ 6
7. water salinity $<150{,}000$ ppm
8. TAP transit time >30 days
9. no existence of natural fractures

This gel system (TAPs) is summarized as follows [186–188,193]:

- TAP is not sensitive to the shear rate and salinity of a formation. It is applicable in a varied range of salinity and pH.
- This gel system has two types of labile and nonlabile cross-linkers. Breaking the labile cross-linker at the critical temperature causes the microgel to swell by water absorption. In contrast, the nonlabile cross-linker prevents the microgel 3-D structure disintegration.
- At the beginning of the swelling process, the critical temperature is related to the type of cross-linkers and their de-cross-linking mechanism.
- The plugging effect of TAPs depends on TAP concentrations.
- Different TAP grades can be used at different reservoir temperatures.
- TAP treatments (unlike PPGs) are not proper for fractured systems.

6.6.1 TAP transport modeling

Garmeh et al. [184] described the details of TAPs simulation approaches. Combined thermal and compositional simulations should be used at the Darcy scale to model a TAP treatment. A thermal simulator predicts the reservoir temperature distribution after cold water injection to determine the trigger temperature for the TAP swelling. Temperature gradients can be obtained from temperature logs. TAPs have low adsorption near injection points. As TAPs swell, their viscosity increases, and they start to stick to rock surfaces. This adsorption gradually increases with heat and time. A compositional simulator is implemented to track the pressure and saturation of TAPs and the distribution of available phases (water, polymer, cross-linker, oil). Temperature changes at activation zones trigger the viscosity and adsorption/retention. The rock and fluid conductivities and heat capacity are the primary inputs of this simulation model. Field tracer-tests or their equivalent simulations can be used to determine the required transit time of TAPs to reach the target location [184].

Two approaches can be used for the TAP modeling [184]:

- ***Single-component approach:*** A single component is considered for the injected components (polymer, cross-linker, and TAP) during the simulation. The physical viscosity and adsorption of TAPs change with heat and time.
- ***Chemical-reaction approach:*** Three separate components of polymer, cross-linker, and TAP are considered during the simulation. TAP is formed when polymer and cross-linker are reacted together because of the temperature activation. The produced TAP has a higher viscosity and adsorption than the other two components.

The irreversible adsorption level in simulations depends on fluid concentration, temperature, and rock minerals. The Langmuir adsorption equation can be used for determining the adsorption value (θ_i) of component i at the concentration of C_i. The adsorption rises as the solution concentration increases. In Eq. (6.32), ($\frac{a}{b}$) is the slope of the Langmuir type curve and shows the adsorption sensitivity [184].

$$\theta_i = \left(\frac{a}{b}\right)\left(\frac{bC_i}{1+bC_i}\right) \tag{6.32}$$

The permeability reduction factor (*PRF*) is related to the amount of adsorption, retention, and filtration. The TAP viscosity and static adsorption (determined from bottle tests) should be determined using laboratory experiments at the reservoir temperatures to have the *RF* and *RRF* [184].

$$PRF = 1 + (RF_{\max} - 1)\left(\frac{\theta_i}{\theta_{\max}}\right) \tag{6.33}$$

in which, $RF_{\max}$ is the maximum resistance factor and $\theta_{\max}$ is the maximum adsorption level at the maximum concentration. Slim-tube and core-flooding tests can be implemented at reservoir conditions to measure the *RF* development versus time at different temperatures and polymer concentrations [24]. Moreover, the shear-thinning viscosity model of Eqs. (6.7) and (6.8) can be used for TAPs in porous media.

The kinetics of the gel formation reaction can also be modeled by the gelation equation (Eq. 6.34), which allows for a temperature dependency [24]. The parameters of this equation can also be determined from the laboratory experiments and will vary for different polymers.

$$R_g = R_k C_M \exp\left(B_k\left(\frac{1}{T} - \frac{1}{T_{ref}}\right)\right) \tag{6.34}$$

in which, R_g is the rate of gelation reaction, R_k is a constant, C_M is the polymer concentration, B_k is the temperature exponent, T is the grid block temperature, and T_{ref} is the reference temperature.

6.7 pH-sensitive microgel

One of the water profile control methods, which has many advantages over other similar methods, is pH-sensitive microgel treatment. The main effective mechanisms of these gels are deep conformance controlling of injection profile and increasing reservoir sweep efficiency [10]. These polymer gels have a low viscosity in acidic environments and their viscosity increases with environment pH rising. Therefore, they can be easily injected at a low pH. pH rising during the flooding process causes the swelling of these microgels and consequently plugging of the high-permeable zones. An acid preflushing is sometimes recommended to reduce the pH around injection points as much as possible, enhancing the effectiveness of pH-sensitive microgels [197].

The idea of pH-sensitive microgels using for conformance control purposes was first introduced by Al-Anazi and Sharma [20,198]. Some gels made up of specific polyelectrolytes, like poly(acrylic acids), are sensitive to pH and can swell even up to 1000 times

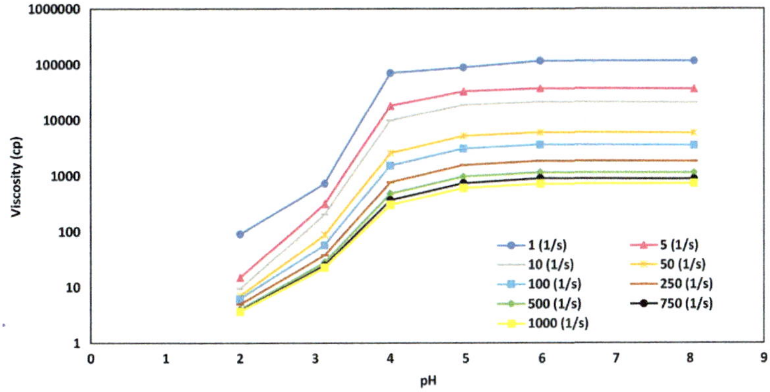

Fig. 6.19 The impact of shear rate and pH on the apparent viscosity of Carbopol 934 (0.5 wt%) at the ambient condition [200].

their original volume by absorbing water [199]. This massive increase in volume increases their viscosity significantly [33]. According to the observations, increasing the pH from 3.5 to 5 increases the apparent viscosity of the solutions by several folds (Fig. 6.19). Recently, new dual responsive cross-linked amphiphilic microgels were synthesized that their viscosity increases at pH greater than 5 [21,22]. Thus, they are well-suited for carbonate reservoirs.

This behavior is due to carboxylic functional groups (i.e., COOH) in the chains of these polymer gels. The following figure shows the behavior of these polymer gels at different pH values. At low pH (Fig. 6.20A), these polymer gels are electrically neutralized and show a low viscosity [201]. At high pH, the repulsive forces between ionized

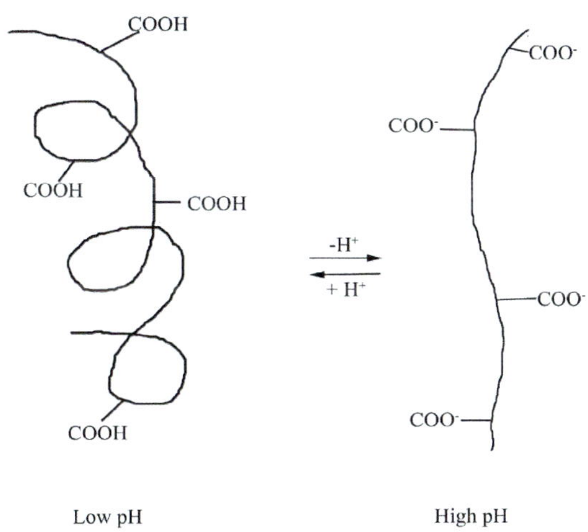

Fig. 6.20 A pH-sensitive polymer chain changes from a coiled shape in a low pH solution (left) to an uncoiled form at higher pH values (right) [197].

carboxylic groups (COO^-) cause polymer chains to uncoil, resulting in a dramatic increase in viscosity [197,198]. (Fig. 6.20B).

Such observations led to implementing pH-sensitive microgels as a method for deep conformance control of injection profiles. The low pH environment causes polymer gel molecules to be in a coiled form that results in a very low viscosity solution; therefore, the injection pressure of these microgels significantly reduces, and they can migrate far from injection points [35]. An injection of polymer gels at low viscosity significantly saves the electrical power of injection pumps. On the other hand, lower differential injection pressure reduces mechanical damage risks. Besides, this method prevents the forming of unwanted fractures near injection wells. As these pH-sensitive materials are produced in large volumes for other applications [e.g., poly(acrylic acid) for cosmetic], they have a low price [199]. The reversibility (de-gelling) of the swelling process by post-acid injection is also possible in sandstones [202]. This reversibility is a significant advantage of these microgels over BrightWater® (TAPs). There is no need for an expensive de-gelling chemical to recover damaged areas or unsuccessful WSO operations; however, some studies showed some unsuccessful post-acid flooding for the de-gelling process in carbonates [203]. Being environmentally friendly can be added to the advantages of these polymer gels as well [10].

Fig. 6.21 explains how a pH-sensitive microgel treatment works [197]. Fig. 6.21A represents the early water breakthrough problem during a conventional waterflooding project. Because of this early breakthrough, a considerable amount of oil in low permeable zones remains unattached (Fig. 6.21). The acid preflushing into the formation (Fig. 6.21B) can decrease the system pH. Fig. 6.21C indicates the pH-sensitive microgel injection step in which the injected polymer flows at a low viscosity through the porous media. The injected fluid pH gradually increases by spontaneous reactions of the acidic injected fluid with rock minerals far away from injection wells (Fig. 6.21D). This increase in pH results in the uncoiling of the polymer gel chains and a considerable increase in viscosity by absorbing water into the polymer gel network. Swollen polymers reduce

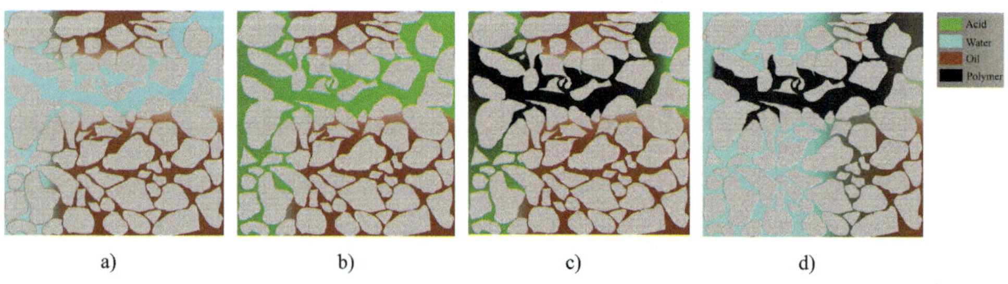

Fig. 6.21 Steps of pH-sensitive injection. (A) Water breakthrough problem through high-permeable zones; (B) acid preflushing to reduce the system pH; (C) pH-sensitive polymer injection at low viscosity in an acidic condition; and (D) postwaterflooding [197].

water mobility and improve injected water profile. On the other hand, they block the penetrated high-permeable areas. By postwaterflooding (Fig. 6.21D), water flows through the adjacent low-permeable zones (i.e., deep conformance control). Both of these phenomena increase the recovery factor of oil trapped in low-permeability and unswept areas [197,204].

It should be noted that the presence of divalent cations (e.g., Ca^{2+}) in reservoir fluids decreases the critical swelling pH [205–208]. Lalehrokh et al. [205] reported that these microgels could penetrate to a depth of 1000 ft in fractured sandstone rocks with an initial permeability of 230 md, but their penetration ability, before polymer gelation, reduced to about 150 ft. in carbonates [207]. This can be attributed to the faster increase in pH in carbonates containing large amounts of highly reactive minerals. Therefore, acid pre-flushing operations in carbonate reservoirs are usually required before utilizing pH-sensitive microgels. Lowering the pH of porous media results in increasing microgel mobility before the final swelling and gelation.

Potential leak of the hydrocarbons or carbon dioxide (CO_2) through wells with fractured cement or deboned microannuli is a common problem during oil and gas production [209–211]. The presence of fractures (Fig. 6.22) with a size of 10–300 μm that have an effective permeability in a range of 0.1–1 md can increase the risk of fluid leakage [210]. These kinds of leaks from cement fractures can usually be repaired by cement squeezing. However, cement squeezing is not so effective in reservoirs with small fractures (e.g., < 400 μm) because of the cement slurry screening problem [211]. Therefore, a treating agent with low viscosity and no solid particles, which can quickly enter these small fractures, is required [209,210]. pH-sensitive microgels can be used effectively to close such small fractures in cement behind casings and tubing.

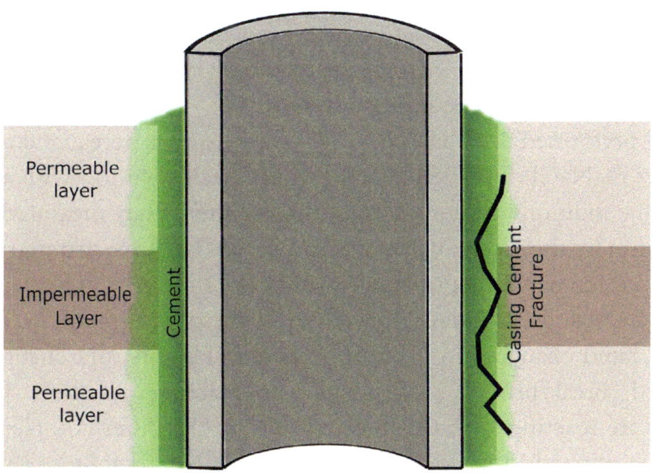

Fig. 6.22 Fractured cement behind casing causes fluid transport among different layers.

Casing cement such as Portland contains calcium hydroxide ($Ca(OH)_2$), and its hydrated form is very alkaline and has a pH of about 13 [212]. The basis of this method is the injection and transfer of pH-sensitive microgels into cement fractures and reactions between low-pH polymer gel and the highly alkaline cement. Upon neutralization with cement, pH-sensitive microgels swell and block the fluid flow [210]. Ho et al. [209] investigated the application of pH-sensitive microgel in closing very small fractures in cement by conducting various core flooding experiments. They injected pH-sensitive microgels into microfractures and calculated the gel resistance against flow. The gel formation tests inside cement fractures were promising; however, the main problem of this method was the reaction between polymer gels and calcium cations inside of cement, which results in the formation of calcium deposits. This reaction causes the liquid to leave the gel structure, which reduces the gel strength. Therefore, before the main polymer solution injection, another solution is needed to be flushed to remove calcium cations. After testing various materials, they found that sodium triphosphate solution ($Na_5P_3O_{10}$) solves the deposition problem, improves the injection process, and increases gel strength. The cement core rinsed with this solution represents a swollen gel withstand a pressure drop of up to 70 psi/ft.

6.7.1 Acid preflushing before pH-sensitive microgel injection

The typical pH of sandstone reservoirs is between 8 and 9, which causes a rapid increase in the viscosity of pH-sensitive microgel solutions around injection wells. To create a low pH environment and improve the penetration of microgel solutions deep into formations, preflushing with acidic solutions before the main microgel injection is recommended [197,206,207,213]. Injection of polymer gel solutions at a low pH improves the solution flow efficiency in porous media. It also reduces the required pump power and injection pressure at surface facilities. However, under these conditions, the polymer gel adsorption on rock surfaces increases, influencing operational costs. Therefore, it is essential to find an optimal injection condition [214].

Choi [206] performed three injection experiments on a Berea sandstone core with a brine solution (3% NaCl) followed by a 0.1 M HCl acid solution with a pH of 1 at different injection conditions (flow rates and shut-in time). They measured the percentage of effluent cations by the inductively coupled plasma (ICP) spectrometry method at the core outlet. All three experiments showed similar pH changes. The equilibrium pH of 9.5 during the brine injection was achieved across the porous medium. During the HCl injection, two rapid changes in pH were observed. The first pH change was observed after the acid breakthrough, in which fast-reacting minerals [e.g., anorthite ($CaAl_2Si_2O_8$)] are reacting with the injected acid (HCl). After complete dissolution of the fast-reacting minerals, the slow-reacting minerals [e.g., kaolinite ($Al_2Si_2O_5(OH)_4$)] enter into the reactions. However, unlike the fast-reacting minerals, the slow-reacting minerals are not entirely consumed during the injection process. They also concluded that lower injection rates, due to the longer residence time of acid, increase the reactions. This reduces the required volume of acids during the preflushing process.

Implementation of weak acids (instead of strong acids like HCl) in lowering the pH of rocks can be more efficient than strong acids during the acid preflushing phase. For example, citric acid (as a weak acid) can keep the porous media pH low because of its lower reactivity compared to HCl (as a strong acid) [197,204,213,214]. This condition causes the injected gel viscosity to remain low enough in the areas close to injection points, which results in deeper migration of microgel into formations.

In bottle tests, crushed Brea sandstone and Texas Cream limestone rocks (e.g., 100 g) were submerged into the solutions of hydrochloric acid (strong acid), citric acid (weak acid), and acetic acid (which is a pH-sensitive microgel monomer) with an initial pH of 2 [207]. The pH change through time showed that in sandstones, both citric and acetic acid solutions maintained a low pH value during the tests (i.e., pH of 3.2 after 5 days). In the limestone carbonate samples, the pH of hydrochloric acid, citric acid, and acetic acid solutions after 5 days were 7.1, 6.5, and 4.9, respectively, which indicates the higher reactivity of limestones in comparison to sandstones. A suitable acid for a preflushing treatment in carbonate reservoirs was also investigated in detail [213]. They evaluated various weak acids (acetic, formic, and citric acids) and HCl (strong acid) by both batch (static) and injection (dynamic) experiments (Fig. 6.23). It was concluded that acetic acid could retain the carbonate reservoir pH low enough, which is preferred for acid preflushing treatment prior to the pH-sensitive microgel injection. Moreover, a compositional reactive flow model (convection-diffusion-reaction) was implemented to predict acid preflushing dynamics. This model could precisely predict the equilibrium pH and

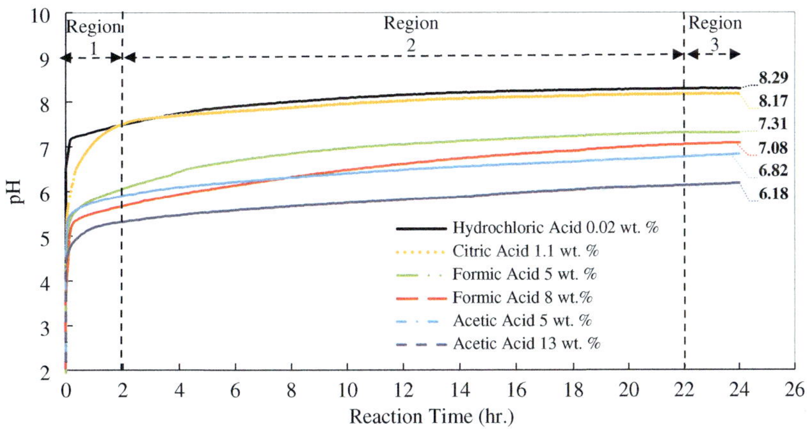

Fig. 6.23 Batch experiments showing pH changes of various acids. All figures have three distinct regions, i.e., sharp pH increasing at early time (region 1), slow pH changes (region 2), a constant pH value (region 3) [213].

calcium concentration at different injection conditions. Moreover, it indicated that after establishing an equilibrium state, the remaining unionized weak acids could effectively contribute toward maintaining a low-pH environment that guarantees deep conformance control of pH-sensitive microgels [213].

Acid preflushing tests (HCL solution with pH of 2) in fractures sandstones showed that acid reactions had little effect on the injection efficiency of the main pH-sensitive microgels [207]. Thus, this acid preflushing step can be ignored in these formations. However, acid preflushing is highly recommended in fractured carbonate rocks due to the higher reaction rate between polymer solutions and rock minerals.

6.7.2 pH-sensitive microgel injection in sandstones and carbonates

So far, various pH-sensitive microgels have been used to control injection profiles. In most cases, commercial polyacrylic acid microgels under the Carbopol brand were used for conformance control purposes in sandstone reservoirs [20,198,206]. Core flow experiments showed that the swelled microgels inside the porous space could reduce the permeability of reservoir rocks several times.

Due to the penetration problem of microgels during profile control in sandstone reservoirs, the use of pH-sensitive HPAM has been proposed [204,214]. The investigation of this polymer in the control of anisotropic permeability profile in heterogeneous reservoirs showed a fourfold decrease in permeability [215]. Moreover, hydrogels composed of these polymers with higher cross-linking density have a more remarkable ability to control anisotropic permeability in inhomogeneous reservoirs.

Al-Wahabi et al. investigated different types of Carbopol microgels for conformance control purposes [216,217]. The efficiency of polyacrylic acid microgels to reduce water production in fractured reservoirs was also investigated [218]. Observations showed that increasing the standing time after the polymer gel injection (shut-in time between the polymer gel solution injection and the postwaterflooding) to 40 h increased the percentage of permeability reduction to 97%, which indicates the higher flow resistance of these solutions in porous media. Injecting the Carbopol 934 solution into a carbonate sandpack system showed a 92% permeability reduction percentage [203] that shows these gel polymers have an excellent performance in permeability reduction in carbonate formations. Furthermore, the constant pressure injection experiments in sandstone and carbonate samples showed that the microgels inside fractures were very stable against postwaterflooding and could withstand pressure gradients up to about 90 psi/ft. [207] and 98 psi/ft. [219] in sandstones and carbonates, respectively. Under this condition, the *RRF* could grow up to 51 times [207].

The application of pH-sensitive microgels in plugging high-permeability fractures and improving the sweep efficiency of fractured carbonate reservoirs has also been evaluated [205,207,208]. In these studies, core samples with a length of 1 ft. from different

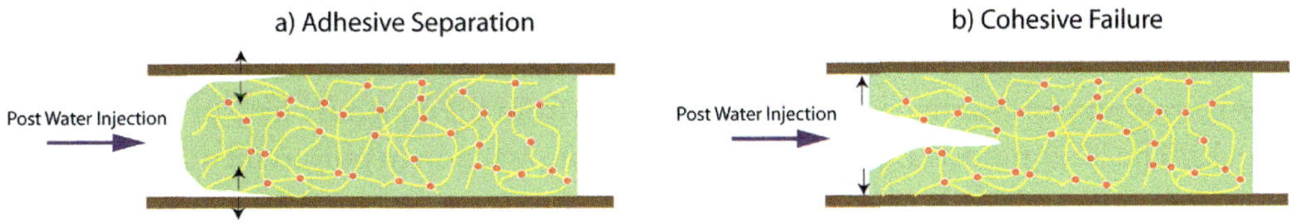

Fig.6.24 (A) Gel failure by the adhesive separation and (B) cohesive failure of gels [219].

carbonate and sandstone rocks have been analyzed. The polymer solution used included 3% Carbopol 934, 3% NaCl, and 94% water, and its initial pH was between 2 and 2.2. The experiments [207] showed that the *RRF* is highly dependent on the shut-in time. Moreover, the constant pressure experiments revealed that polymer microgels blocked the fracture and prevented water channeling inside the fracture against 90 psi/ft. of injection pressure gradient.

The effect of temperature on the rheological properties of the microgels up to 80°C is not very significant [20,198]. However, as the salinity of pH-sensitive microgel dispersion increases, the apparent viscosity decreases [206,207]. The reason is that the negative charges in pH-sensitive microgel chains (carboxyl groups) by the salt cations resulted in the reduction of the repulsive forces among these ionized groups. It should also be noted that increasing the salt concentration to more than 3% will result in difficulties in microgel dispersion into brine [207]. Microgel swelling occurs at low pH by reducing salt concentrations, and dispersion viscosity reaches a higher plateau value [206]. The polyvalent cations such as Ca^{2+}, Mg^{2+}, Al^{3+}, and Fe^{2+} can result in gel precipitation at high pH values. $CaCl_2$ does not cause precipitation at pH values less than 3. Preflushing of NaCl, NH_4Cl, and KCl can minimize the negative effect of polyvalent cations on the viscosity of pH-sensitive microgel solutions [20,198].

The gel failure mechanisms (e.g., adhesive separation, cohesive failure) during post-waterflooding in fractures plugged by pH-sensitive microgels were investigated in detail [219]. A visual cell was designed to monitor carbonate rock-gel reactions during a pH-sensitive microgel injection. Two main gel failure mechanisms were presented; adhesive separation and cohesive failure. In the former mechanism (Fig. 6.24A), the gel lost its adhesive strength from the fracture wall. In contrast, in the later failure mechanism (Fig. 6.24B), the BG breaks [219]. This study also showed an optimum concentration of gels at which the failure mechanism changes from cohesive failure to adhesive separation. It means that higher gel concentrations are not always the base choices.

6.7.3 pH-sensitive microgel transport modeling

Reactive flow models were widely investigated in the literature [197]. These models can be categorized into the pore-scale [220–224], Darcy-scale [225–227], and hybrid models [228–230]. The dominant reaction in acidizing carbonate reservoirs is the reaction between hydrogen ions and calcite. The two-scale continuum model (TSC) has attracted much attention during the past decade [179,197,231,232]. This model considers a first-order reaction rate for transferring hydrogen ions from the aqueous phase to the surface of rocks. The TSC models are applicable for only one reaction in porous media (e.g., strong acid reaction with carbonate rocks); nevertheless, a more sophisticated model that can consider both equilibrium and kinetic reactions should be implemented to react to organic acids with several rocks minerals [197].

For the first time, the KGEOFLOW simulator, a 1D geochemical simulator based on partial local equilibrium theory, was used to simulate the behavior of an acid preflushing before pH-sensitive microgel injection in Berea sandstones [206]. The basis of this model was geochemical reactions in a porous medium and is also dependent on dissolution kinetics. The kinetics considered for all acid-rock reactions were considered primarily. They used a three-mineral model (i.e., quartz, feldspar, and kaolinite clay) to represent the Berea sandstone. The reactions were first-order concerning the reactants. In this research, an attempt was made to predict the results of laboratory injection by changing the reaction constants of acid with sandstone reservoir rock minerals. The history match results of the injection process showed a relatively acceptable prediction for the effluent pH.

The efficiency of the GEM-GHG simulator was also tested to estimate geochemical reactions in a Berea core sample [202]. A model of four minerals (quartz, calcite, potassium feldspar, and kaolinite) was implemented in this study. The simulator could model the low pH fluid propagation deep into a reservoir. The Geochemist workbench professional simulator was also used to match the laboratory experiments of Choi [206]. The model used the specific reactive surface area as a matching parameter instead of the reaction rate constants in the previous study [206], which provided better predictions.

The technical feasibility of both TAP and pH-sensitive polymers was examined by running sensitivity analysis on 3D field-scale models by the UTGEL simulator [233]. An attempt was made to predict the acid preflushing behavior in porous media by considering the pH-dependent equilibrium constants during the simulations. Sensitivity analysis on oil-to-water viscosity ratio, vertical-to-horizontal permeability ratio, slug size, and polymer concentration was performed to improve the conformance control efficiency. The results showed that increased polymer concentration, rising of the injected slug size, and high viscosity ratios (oil-to-water) increase the recovery factor.

The applicability of acid preflushing before pH-sensitive polymer injection was also investigated in detail [197,234]. The introduced mathematical model considered both homogeneous and heterogeneous geochemical reactions between acid and rock minerals in these studies. A global implicit approach was used to solve the three nonlinear conservation equations of mass balance, Darcy, and species concentration balance simultaneously. The model was examined on strong and weak acid injection into carbonate rock. They found that weak acids can be a better choice for acid preflushing in carbonates, resulting in a better conformance control during the pH-sensitive injection.

The mass balance for a slightly compressible fluid can be written as:

$$\frac{\partial(\rho u_x)}{\partial x} = \rho\phi(c_l + c_r)\frac{\partial P}{\partial t} \tag{6.35}$$

where P is the pressure, u_x is the fluid velocity in the x-direction, ρ is the fluid density, ϕ is the porosity, and c_l and c_p represents the compressibility of fluid and rock, respectively.

The local equilibrium assumption can be considered which divides the reactions into two categories to increase the computational speed [197]. The first reaction category includes those high-speed reactions (also called homogenous reactions) in the aqueous phase. The remaining slow reactions (heterogeneous reactions) are considered in the second category.

The equilibrium among N_C primary (A_j) and N_X secondary (A_i) species are [197,226,235–237]:

$$A_i \rightleftarrows \sum_{j=1}^{N_c} v_{ij} A_j \quad i = 1, \ldots, N_x \tag{6.36}$$

where v_{ij} is the stoichiometric coefficient of primary species j. Two mass balance equations can be written for both primary (C_j) and secondary species (X_i) [197].

$$\frac{\partial(uC_j)}{\partial x} - D\frac{\partial^2 C_j}{\partial x^2} + \frac{\partial(\phi C_j)}{\partial t} + R_j^{\min} + R_i^{aq} = 0 \quad j = 1, \ldots, N_C \tag{6.37}$$

$$\frac{\partial(uX_i)}{\partial x} - D\frac{\partial^2 X_i}{\partial x^2} + \frac{\partial(\phi X_i)}{\partial t} + R_i^{\min} + R_i^{aq} = 0 \quad i = 1, \ldots, N_X \tag{6.38}$$

in which $R^{\min}$ and R^{aq} represents the primary and secondary reaction rates, respectively. The above mass balance equations can be simplified into the following form by a simple mathematical substitution, which considers the effect of all species [197],

$$\frac{\partial\left(u\left(C_j + \sum_{i=1}^{N_X} v_{ij}X_i\right)\right)}{\partial x} - D\frac{\partial^2\left(C_j + \sum_{i=1}^{N_X} v_{ij}X_i\right)}{\partial x^2} + \frac{\partial\left(\phi\left(C_j + \sum_{i=1}^{N_X} v_{ij}X_i\right)\right)}{\partial t} + R_j^{\min} + \sum_{i=1}^{N_X} v_{ij}R_i^{aq} = 0 \tag{6.39}$$

The term of $U_j = C_j + \sum_{i=1}^{N_X} v_{ij}X_i$ in the above equation is called the total concentration [197].

The reaction rate of the pth mineral can be calculated from [197]:

$$R_p^{\min} = \left\{ k_f A_p \prod_{i=1}^{N_C + N_X} C_i^{v_{ip}} \right\} \left(1 - \left(\frac{Q_s}{K_{eq}}\right)^n\right)^m \tag{6.40}$$

where k_f is the forward reaction rate, v_{ip} is the stoichiometric coefficient of i in the pth reaction, n and m are the reaction orders, A_p is the reactive surface area, K_{eq} is the equilibrium constant, and Q_s is defined as [197]:

$$Q_s = \prod_{i=1}^{N_C + N_X} C_i^{\nu_{ip}} \tag{6.41}$$

For nonideal solutions (e.g., high salinity solutions), where there is a deviation from Raoult's law, the concentrations should be modified to $C_i\gamma_i$. The Truesdell-Jones equation can be used for activity coefficients of species [238]:

$$\log\gamma = \frac{-Az^2\sqrt{I}}{1 + Ba\sqrt{I}} + bI \tag{6.42}$$

where γ is the activity coefficient, z is the ion electrical charge, and I is the solution ionic strength. A and B are connected to the solvent type. The ionic strength is defined as:

$$I = \frac{1}{2}\sum_{i=1}^{N_T} m_i z_i^2 \tag{6.43}$$

where m_i characterizes the molarity and N_T is the total number of species.

References

[1] Javaheri P, Sadeghnejad S. Effect of injection pattern arrangements on formation connectivity during water flooding. In: SPE Europec featured at 79th EAGE conference and exhibition. Society of Petroleum Engineers; 2017.

[2] Sadeghnejad S, Manteghian M, Ruzsaz H. Simulation optimization of water-alternating-gas process under operational constraints: a case study in the Persian Gulf. Sci Iran 2019;26(6):3431–46.

[3] Sadeghnejad S, Masihi M. Water flooding performance evaluation using percolation theory. J Petrol Sci Technol 2011;1(2):19–23.

[4] Farasat A, Sefti MV, Sadeghnejad S, Saghafi HR. Mechanical entrapment analysis of enhanced preformed particle gels (PPGs) in mature reservoirs. J Petrol Sci Eng 2017;157:441–50.

[5] Bai B, Liu Y, Coste J-P, Li L. Preformed particle gel for conformance control: transport mechanism through porous media. SPE Reserv Eval Eng 2007;10(02):176–84.

[6] Goudarzi A, Almohsin A, Varavei A, Taksaudom P, Hosseini SA, Delshad M, et al. New laboratory study and transport model implementation of microgels for conformance and mobility control purposes. Fuel 2017;192:158–68.

[7] Sydansk RD, Southwell GP. More than 12 years of experience with a successful conformance-control polymer gel technology. In: SPE/AAPG Western regional meeting. Society of Petroleum Engineers: Long Beach, California; 2000.

[8] Hibbeler JC, Rae PJ. The environmental benefits of reducing unwanted water production. SPE Asia Pacific Health, Safety and Environment Conference and Exhibition. Society of Petroleum Engineers; 2005.

[9] Kazemi S. Review of polymer gels for conformance control in oil reservoirs. In: Major papers, vol. 97; 2019. p. 1–102.

[10] Abdulbaki M, Huh C, Sepehrnoori K, Delshad M, Varavei A. A critical review on use of polymer microgels for conformance control purposes. J Petrol Sci Eng 2014;122:741–53.

[11] El-Karsani KSM, Al-Muntasheri GA, Hussein IA. Polymer systems for water shutoff and profile modification: a review over the last decade. SPE J 2014;19(01):135–49.

[12] Kumar A, Mahto V, Choubey AK. Synthesis and characterization of cross-linked hydrogels using polyvinyl alcohol and polyvinyl pyrrolidone and their blend for water shut-off treatments. J Mol Liq 2020;301:112472.

[13] Hamza A, Shamlooh M, Hussein IA, Nasser M, Salehi S. Polymeric formulations used for loss circulation materials and wellbore strengthening applications in oil and gas wells: a review. J Petrol Sci Eng 2019;180:197–214.
[14] Sydansk RD, Romero-Zerón L. Reservoir conformance improvement. Society of Petroleum Engineers; 2011.
[15] Sydansk RD. A new conformance-improvement-treatment chromium (III) gel technology. In: SPE Enhanced Oil Recovery Symposium. OnePetro 1988.
[16] Spildo K, Skauge A, Aarra MG, Tweheyo MT. A new polymer application for North Sea reservoirs. SPE Reserv Eval Eng 2009;12(03):427–32.
[17] Diaz D, Somaruga C, Norman C, Romero JL. Colloidal dispersion gels improve oil recovery in a heterogeneous argentina waterflood. In: SPE symposium on improved oil recovery. Tulsa, Oklahoma, USA: Society of Petroleum Engineers; 2008.
[18] Bai B, Wei M, Liu Y. Injecting large volumes of preformed particle gel for water conformance control. Oil Gas Sci Technol 2012;67(6):941–52.
[19] Yu L, Sang Q, Dong M. Enhanced oil recovery ability of branched preformed particle gel in heterogeneous reservoirs. Oil Gas Sci Technol 2018;73(35):65.
[20] Al-Anazi HA, Sharma MM. n.d.Evaluation of a pH-sensitive polymer for gravel-packing operations. SPE-76813-PA 2002.
[21] Ashrafizadeh M, Tam KC, Javadi A, Abdollahi M, Sadeghnejad S, Bahramian A. Dual physically and chemically cross-linked polyelectrolyte nanohydrogels: compositional and pH-dependent behavior studies. Eur Polym J 2020;122:109398.
[22] Ashrafizadeh M, Tam KC, Javadi A, Abdollahi M, Sadeghnejad S, Bahramian A. Synthesis and physicochemical properties of dual-responsive acrylic acid/butyl acrylate cross-linked nanogel systems. J Colloid Interface Sci 2019;556:313–23.
[23] Garmeh G, Izadi M, Salehi M, Romero JL, Thomas CP, Manrique EJ. Thermally active polymer to improve sweep efficiency of water floods: simulation and pilot design approaches. In: SPE enhanced oil recovery conference. Kuala Lumpur, Malaysia: Society of Petroleum Engineers; 2011.
[24] Roussennac BD, Toschi C. Brightwater trial in Salema field (Campos Basin, Brazil). In: SPE EUROPEC/EAGE annual conference and exhibition. Society of Petroleum Engineers; 2010.
[25] Gussenov I, Zhappasbayev B, Kudabergenov S, Akhmedzhanov T. Permeability reduction of heterogeneous oil reservoirs by brine-triggered gellan gel. J Nanosci Nanotechnol 2017;17(12):9198–202.
[26] Ashrafizadeh M, Ahmad RS, Sadeghnejad S. Improvement of polymer flooding using in-situ releasing of smart nano-scale coated polymer particles in porous media. Energy Explor Exploit 2012;30(6):915–39.
[27] Ashrafizadeh M, Ahmad Ramazani SA, Sadeghnejad S. Enhanced polymer flooding using a novel nano-scale smart polymer: experimental investigation. Can J Chem Eng 2017;95(11):2168–75.
[28] Pu J, Zhou J, Chen Y, Bai B. Development of thermotransformable controlled hydrogel for enhancing oil recovery. Energy Fuel 2017;31(12):13600–9.
[29] Sydansk RD, Seright RS. When and where relative permeability modification water-shutoff treatments can be successfully applied. In: SPE/DOE symposium on improved oil recovery. Society of Petroleum Engineers; 2006.
[30] Zaitoun A, Kohler N. Two-phase flow through porous media: Effect of an adsorbed polymer layer. In: SPE annual technical conference and exhibition. Society of Petroleum Engineers; 1988.
[31] Stavland A, Andersen KI, Sandoey B, Tjomsland T, Mebratu AA. How to apply a blocking gel system for bullhead selective water shutoff: from laboratory to field. In: SPE/DOE symposium on improved oil recovery. Society of Petroleum Engineers; 2006.
[32] Seright RS, Lane RH, Sydansk RD. A strategy for attacking excess water production. SPE Prod Facil 2003;8(03):158–69.
[33] Goudarzi A. Modeling conformance control and chemical EOR processes using different reservoir simulators (Doctoral dissertation). The University of Texas at Austin; 2015.
[34] Amir Z, Said IM, Jan BM. In situ organically cross-linked polymer gel for high-temperature reservoir conformance control: a review. Polym Adv Technol 2019;30(1):13–39.
[35] Bai B, Zhou J, Yin M. A comprehensive review of polyacrylamide polymer gels for conformance control. Pet Explor Dev 2015;42(4):525–32.

[36] Dijvejin ZA, Ghaffarkhah A, Sadeghnejad S, Sefti MV. Effect of silica nanoparticle size on the mechanical strength and wellbore plugging performance of SPAM/chromium (III) acetate nanocomposite gels. Polym J 2019;51(7):693–707.
[37] Perez D, Fragachan F, Barrera AR, Feraud J. Applications of polymer gel for establishing zonal isolations and water shutoff in carbonate formations. SPE Drill Complet 2001;16(03):182–9. SPE-76813-PA.
[38] Takahashi T, Okura M, Takahashi S, Kobayashi T, Kobayashi F, Sato H. Temporary plugging agent for well drilling. Google Patents; 2016.
[39] Jaripatke OA, Dalrymple ED. Water-control management technologies: a review of successful chemical technologies in the last two decades. In: SPE international symposium and exhibition on formation damage control. Society of Petroleum Engineers; 2010.
[40] Liang B, Jiang H, Li J, Chen F, Miao W, Yang H, et al. Mechanism study of disproportionate permeability reduction using nuclear magnetic resonance T2. Energy Fuel 2018;32(4):4959–68.
[41] Masuda T, Akimoto AM, Yoshida R. Self-oscillating polymer materials. In: Ebara M, editor. Biomaterials nanoarchitectonics. William Andrew Publishing; 2016. p. 219–36.
[42] Pal K, Paulson AT, Rousseau D. Biopolymers in controlled-release delivery systems. In: Ebnesajjad S, editor. Handbook of biopolymers and biodegradable plastics. Boston: William Andrew Publishing; 2013. p. 329–63.
[43] Farasat A, Sefti MV, Sadeghnejad S, Saghafi HR. Effects of reservoir temperature and water salinity on the swelling ratio performance of enhanced preformed particle gels. Korean J Chem Eng 2017;34 (5):1509–16.
[44] Özeroglu C, Birdal A. Swelling properties of acrylamide-N,N′-methylene bis(acrylamide) hydrogels synthesized by using meso-2,3-dimercaptosuccinic acid-cerium(IV) redox couple. Express Polym Lett 2009;(3):168–76.
[45] Ghriga MA, Grassl B, Gareche M, Khodja M, Lebouachera SEI, Andreu N, et al. Review of recent advances in polyethylenimine crosslinked polymer gels used for conformance control applications. Polym Bull 2019;1–29.
[46] Zhu D, Bai B, Hou J. Polymer gel systems for water management in high-temperature petroleum reservoirs: a chemical review. Energy Fuel 2017;31(12):13063–87.
[47] Reddy B, Eoff L, Dalrymple ED, Black K, Brown D, Rietjens M. A natural polymer-based crosslinker system for conformance gel systems. SPE J 2003;8(02):99–106.
[48] Caulfield MJ, Qiao GG, Solomon DH. Some aspects of the properties and degradation of polyacrylamides. Chem Rev 2002;102(9):3067–84.
[49] Moradi-Araghi A. A review of thermally stable gels for fluid diversion in petroleum production. J Petrol Sci Eng 2000;26(1–4):1–10.
[50] Kelland MA. Production chemicals for the oil and gas industry. CRC press; 2014.
[51] Fathima A, Almohsin A, Michael FM, Bataweel M, Alsharaeh EH. Polymer nanocomposites for water shutoff application-a review. Mater Res Express 2018;6(3), 032001.
[52] He H, Wang Y, Sun X, Zhang P, Li D. Development and evaluation of organic/inorganic combined gel for conformance control in high temperature and high salinity reservoirs. J Pet Explor Prod Technol 2015;5(2):211–7.
[53] Rousseau D, Chauveteau G, Renard M, Tabary R, Zaitoun A, Mallo P, et al. Rheology and transport in porous media of new Water shutoff/conformance control microgels. In: SPE international symposium on oilfield chemistry. The Woodlands, Texas: Society of Petroleum Engineers; 2005.
[54] Imqam A, Bai B. Optimizing the strength and size of preformed particle gels for better conformance control treatment. Fuel 2015;148:178–85.
[55] Imqam A, Aldalfag A, Wang Y, Bai B. Evaluation of preformed particle gels penetration into matrix for a conformance control treatment in partially open conduits. In: SPE annual technical conference and exhibition. Society of Petroleum Engineers; 2016.
[56] Long Y. Study and development of robust preformed particle gels for conformance control. Master of Science, Missouri University of Science And Technology; 2016.
[57] Abdulbaki MR. Simulation study of polymer microgel conformance treatments. OnePetro 2012. https://doi.org/10.2118/17329-MS.
[58] Gussenov I, Nuraje N, Kudaibergenov S. Bulk gels for permeability reduction in fractured and matrix reservoirs. Energy Rep 2019;5:733–46.

[59] Chung T, Bae W, Nguyen N, Dang C, Lee W, Jung B. A review of polymer conformance treatment: a successful guideline for water control in mature fields. Energy Sources, Part A 2011;34(2):122–33.
[60] Seright R, Zhang G, Akanni O, Wang D. A comparison of polymer flooding with in-depth profile modification. J Can Pet Technol 2012;51(05):393–402.
[61] Sachdeva JS. Evaluation of silicate and polymer systems for disproportionate permeability reduction in oil reservoirs. Norway: University of Stavanger; 2015.
[62] Aldhaheri M, Wei M, Zhang N, Bai B. A novel survey of bulk gel treatment designs in injection wells-part I: gel strength. In: Offshore technology conference; 2018.
[63] Sun F, Lin M, Dong Z, Zhang G. Delayed-crosslink hydrogel for improving oil recovery in differential heterogeneous reservoirs. ACS Omega 2019;5(1):228–35.
[64] Kohler N, Rahbari R, Han M, Zaitoun A. Weak gel formulations for selective control of water production in high-permeability and high-temperature production wells. In: SPE international symposium on oilfield chemistry. Society of Petroleum Engineers; 1993.
[65] Wang W, Gu Y, Liu Y. Applications of weak gel for in-depth profile modification and oil displacement. J Can Pet Technol 2003;42(06):PETSOC-03-06-04.
[66] Xie K, Cao W, Lu X, Song K, Liu Y, Zhang Y, et al. Influence of water dilution on performance of chromium polymer weak gel in porous medium. J Dispers Sci Technol 2019;1549–58.
[67] Khamees TK. Evaluation of the performance of different polymer gel systems as an in-depth gel treatment using numerical simulation. Missouri University of Science and Technology; 2020.
[68] Kabir A. Chemical water & gas shutoff technology-an overview. In: SPE Asia Pacific improved oil recovery conference. Society of Petroleum Engineers; 2001.
[69] Zhu D, Hou J, Meng X, Zheng Z, Wei Q, Chen Y, et al. Effect of different phenolic compounds on performance of organically cross-linked terpolymer gel systems at extremely high temperatures. Energy Fuel 2017;31(8):8120–30.
[70] Zhu D, Hou J, Wei Q, Wu X, Bai B. Terpolymer gel system formed by resorcinol–hexamethylenetetramine for water management in extremely high-temperature reservoirs. Energy Fuel 2017;31(2):1519–28.
[71] Bauer S, Gronewald P, Hamilton J, LaPlant D, Mansure A. High-temperature plug formation with silicates. In: SPE international symposium on oilfield chemistry. Society of Petroleum Engineers; 2005.
[72] Fleury M, Sissmann O, Brosse E, Chardin M. A silicate based process for plugging the near well bore formation. Energy Procedia 2017;114:4172–87.
[73] Pham LT, Hatzignatiou DG. Rheological evaluation of a sodium silicate gel system for water management in mature, naturally-fractured oilfields. J Petrol Sci Eng 2016;138:218–33.
[74] Vasquez J, Civan F, Shaw TM, Dalrymple ED, Eoff L, Reddy B, et al. Laboratory evaluation of high-temperature conformance olymer systems. In: SPE production and operations symposium. Society of Petroleum Engineers; 2003.
[75] Hardy M, Botermans W, Hamouda A, Valdal J, Warren J. The first carbonate field application of a new organically crosslinked water shutoff polymer system. Society of Petroleum Engineers: SPE international symposium on oilfield chemistry; 1999.
[76] Al-Muntasheri GA, Sierra L, Garzon FO, Lynn JD, Izquierdo GA. Water shut-off with polymer gels in a high temperature horizontal gas well: a success story. In: SPE improved oil recovery symposium. Society of Petroleum Engineers; 2010.
[77] Uddin S, Dolan JD, Chona RA, Gazi NH, Monteiro K, Al-Rubaiyea JA, et al. Lessons learned from the first openhole horizontal well water shutoff job using two new polymer systems-a case history from Wafra Ratawi Field, Kuwait. Middle East oil show. Society of Petroleum Engineers; 2003.
[78] Beltagy AE, Taha I, Shawky A, Gaber NS, Elwan MA. Offshore coiled tubing interventions maximized the value of watered out mature fields using selective water shutoff treatments with an organic crosslinked polymer in a high temperature sour well. In: SPE/IATMI Asia Pacific oil & gas conference and exhibition. Society of Petroleum Engineers; 2015.
[79] Ortiz Polo RP, Monroy RR, Toledo N, Dalrymple ED, Eoff L, Everett D. Field applications of low molecular-weight polymer activated with an organic crosslinker for water conformance in south mexico. In: SPE annual technical conference and exhibition. Society of Petroleum Engineers; 2004.
[80] Vasquez JE, Jurado I, Santillan A, Hernandez R. Organically crosslinked polymer system for water reduction treatments in Mexico. In: International oil conference and exhibition in Mexico. Society of Petroleum Engineers; 2006.

[81] Wu Y-S, Bai B. Modeling particle gel propagation in porous media. In: SPE annual technical conference and exhibition. Society of Petroleum Engineers; 2008.
[82] Bai B, Li L, Liu Y, Liu H, Wang Z, You C. Conformance control by preformed particle gel: factors affecting its properties and applications. SPE Reserv Eval Eng 2007;10(4):415–21.
[83] Coste J-P, Liu Y, Bai B, Li Y, Shen P, Wang Z, et al. In: In-depth fluid diversion by pre-gelled particles. laboratory study and pilot testing. SPE/DOE improved oil recovery symposium. Society of Petroleum Engineers; 2000.
[84] Askarinezhad R, Hatzignatiou DG. Improved disproportionate permeability reduction of silicate-gelant treatments: A laboratory scale approach. In: SPE international conference on oilfield chemistry. Society of Petroleum Engineers; 2017.
[85] Askarinezhad R, Hatzignatiou D. Water-soluble silicate gelants for disproportionate permeability reduction: importance of formation wetting and treatment conditions. In: International symposium of the society of core analysts held in snowmass; 2016. p. 21–6.
[86] Liang J, Sun H, Seright R. Reduction of oil and water permeabilities using gels. In: SPE/DOE enhanced oil recovery symposium. Society of Petroleum Engineers; 1992.
[87] Grattoni C, Luckham P, Jing X, Norman L, Zimmerman RW. Polymers as relative permeability modifiers: adsorption and the dynamic formation of thick polyacrylamide layers. J Petrol Sci Eng 2004;45(3–4):233–45.
[88] Al-shajalee F, Wood C, Xie Q, Saeedi A. Effective mechanisms to relate initial rock permeability to outcome of relative permeability modification. Energies 2019;12(24):4688.
[89] Dawe RA, Zhang Y. Mechanistic study of the selective action of oil and water penetrating into a gel emplaced in a porous medium. J Petrol Sci Eng 1994;12(2):113–25.
[90] Nguyen TQ, Green DW, Willhite GP, McCool CS. Effects of gelant composition and pressure gradients of water and oil on disproportionate permeability reduction of sandpacks treated with polyacrylamide-chromium acetate gels. SPE J 2006;11(02):145–57.
[91] Seright RS, Prodanovic M, Lindquist WB. X-ray computed microtomography studies of fluid partitioning in drainage and imbibition before and after gel placement: disproportionate permeability reduction. SPE J 2006;11(02):159–70.
[92] Ganguly S, Willhite G, Green D, McCool C. Effect of flow rate on disproportionate permeability reduction. In: International symposium on oilfield chemistry. Society of Petroleum Engineers; 2003.
[93] Liang J-T, Sun H, Seright R. Why do gels reduce water permeability more than oil permeability? SPE Reserv Eng 1995;10(04):282–6.
[94] Song Z, Hou J, Zhang L, Chen Z, Li M. Experimental study on disproportionate permeability reduction caused by non-recovered fracturing fluids in tight oil reservoirs. Fuel 2018;226:627–34.
[95] Nilsson S, Stavland A, Jonsbraten H. Mechanistic study of disproportionate permeability reduction. In: SPE/DOE improved oil recovery symposium. Society of Petroleum Engineers; 1998.
[96] Grattoni C, Jing X, Zimmerman R. Disproportionate permeability reduction when a silicate gel is formed in-situ to control water production. In: SPE Latin American and Caribbean petroleum engineering conference. Society of Petroleum Engineers; 2001.
[97] Alfarge D, Bai B, Almansour A. Numerical simulation study to understand the performance of RPM gels in water-shutoff treatments. J Petrol Sci Eng 2018;171:818–34.
[98] Alfarge DK, Wei M, Bai B. Numerical simulation study of factors affecting relative permeability modification for water-shutoff treatments. Fuel 2017;207:226–39.
[99] Liang J, Seright R. Wall-effect/gel-droplet model of disproportionate permeability reduction. In: SPE/DOE improved oil recovery symposium. Society of Petroleum Engineers; 2000.
[100] Prado Paez M, Rauseo O, Reyna M, Ferreira I. Evaluation of the effect of oil viscosity on the disproportionate permeability reduction of a polymeric gel used for controlling excess water production. In: Latin American and Caribbean petroleum engineering conference. Society of Petroleum Engineers; 2009.
[101] Al-Sharji HH, Grattoni CA, Dawe RA, Zimmerman RW. Pore-scale study of the flow of oil and water through polymer gels. In: SPE annual technical conference and exhibition. Society of Petroleum Engineers; 1999.
[102] Liang J, Seright R. Further investigations of why gels reduce kw more than ko. In: International symposium on oilfield chemistry: Houston TX, 18–21 February 1997; 1997. p. 393–400.

[103] Al-Sharji H, Grattoni C, Dawe R, Zimmerman R. Disproportionate permeability reduction due to polymer adsorption entanglement. In: SPE European formation damage conference. Society of Petroleum Engineers; 2001.

[104] Zitha PL, Vermolen FJ, Bruining H. Modification of two phase flow properties by adsorbed polymers and gels. In: SPE European formation damage conference. Society of Petroleum Engineers; 1999.

[105] Krishnan P, Asghari K, Willhite G, McCool C, Green D, Vossoughi S. Dehydration and permeability of gels used in in-situ permeability modification treatments. In: SPE/DOE improved oil recovery symposium. Society of Petroleum Engineers; 2000.

[106] Willhite GP, Zhu H, Natarajan D, McCool C, Green D. Mechanisms causing disproportionate permeability in porous media treated with chromium acetate/HPAAM gels. In: SPE/DOE improved oil recovery symposium. Society of Petroleum Engineers; 2000.

[107] Spildo K, Skauge A, Skauge T. Propagation of colloidal dispersion gels (CDG) in laboratory Corefloods. In: SPE improved oil recovery symposium. Society of Petroleum Engineers; 2010.

[108] Bjørsvik M, Høiland H, Skauge A. Formation of colloidal dispersion gels from aqueous polyacrylamide solutions. Colloids Surf A Physicochem Eng Asp 2008;317(1–3):504–11.

[109] Fielding Jr RC, Gibbons D, Legrand F. In-depth drive fluid diversion using an evolution of colloidal dispersion gels and new bulk gels: an operational case history of north rainbow ranch unit. In: SPE/DOE improved oil recovery symposium. Society of Petroleum Engineers; 1994.

[110] Mack J, Smith J. In-depth colloidal dispersion gels improve oil recovery efficiency. In: SPE/DOE improved oil recovery symposium. Society of Petroleum Engineers; 1994.

[111] Al-Assi AA, Willhite GP, Green DW, McCool CS. Formation and propagation of gel aggregates using partially hydrolyzed polyacrylamide and aluminum citrate. SPE J 2009;14(03):450–61.

[112] Sheng J. Modern chemical enhanced oil recovery: theory and practice. Gulf Professional Publishing; 2010.

[113] Shi J, Varavei A, Huh C, Delshad M, Sepehrnoori K, Li X. Viscosity model of preformed microgels for conformance and mobility control. Energy Fuel 2011;25(11):5033–7.

[114] Lu X, Song K, Niu J, Chen F. Performance and evaluation methods of colloidal dispersion gels in the Daqing oil field. In: SPE Asia Pacific conference on integrated modelling for asset management. Society of Petroleum Engineers; 2000.

[115] Wang D, Han P, Shao Z, Hou W, Seright RS. Sweep-improvement options for the Daqing oil field. SPE Reserv Eval Eng 2008;11(01):18–26.

[116] Seright RS. Propagation of an aluminum citrate-HPAM colloidal dispersion gel through Berea sandstone. In: Report No PRRC 1994:94-29; 1994.

[117] Zou J, Yue X, Dong J, Shao M, Wang L, Gu J, et al. Novel in-depth profile control agent based on in-situ polymeric microspheres in low permeability reservoir. J Dispers Sci Technol 2020;41 (8):1254–64.

[118] Chang HL, Sui X, Xiao L, Guo Z, Yao Y, Yiao Y, et al. Successful field pilot of in-depth colloidal dispersion gel (CDG) technology in Daqing oilfield. SPE Reserv Eval Eng 2006;9(06):664–73.

[119] Chang H, Zhang Z, Wang Q, Xu Z, Guo Z, Sun H, et al. Advances in polymer flooding and alkaline/surfactant/polymer processes as developed and applied in the People's Republic of China. J Petrol Tech 2006;58(02):84–9.

[120] Muruaga E, Flores MV, Norman C, Romero JL. Combining bulk gels and colloidal dispersion gels for improved volumetric sweep efficiency in a mature waterflood. In: SPE symposium on improved oil recovery. Society of Petroleum Engineers; 2008.

[121] Shi J, Varavei A, Huh C, Delshad M, Sepehrnoori K, Li X. Transport model implementation and simulation of microgel processes for conformance and mobility control purposes. Energy Fuel 2011;25(11):5063–75.

[122] Anon. Guide C-SUs. Calgary, AB, Canada: Computer Modeling Group Ltd.; 2012.

[123] Alhuraishawy AK, Sun X, Bai B, Wei M, Imqam A. Areal sweep efficiency improvement by integrating preformed particle gel and low salinity water flooding in fractured reservoirs. Fuel 2018;221:380–92.

[124] Sun X, Alhuraishawy AK, Bai B, Wei M. Combining preformed particle gel and low salinity waterflooding to improve conformance control in fractured reservoirs. Fuel 2018;221:501–12.

[125] Chen X, Feng Q, Liu W, Sepehrnoori K. Modeling preformed particle gel surfactant combined flooding for enhanced oil recovery after polymer flooding. Fuel 2017;194:42–9.

[126] Liu Y, Bai B, Shuler PJ. Application and development of chemical-based conformance control treatments in China oilfields. In: SPE/DOE symposium on improved oil recovery. Tulsa, Oklahoma, USA: Society of Petroleum Engineers; 2006.
[127] Sun L, Han Q, Li D, Zhang X, Pu W, Tang X, et al. Water plugging performance of preformed particle gel in partially filled fractures. Ind Eng Chem Res 2019;58(16):6778–84.
[128] Brattekås B, Graue A, Seright R. Low-salinity chase Waterfloods improve performance of Cr(III)-acetate hydrolyzed polyacrylamide gel in fractured cores. SPE Reserv Eval Eng 2016;19(02):331–9.
[129] Zhang H, Bai B. Preformed particle gel transport through open fractures and its effect on water flow. In: SPE improved oil recovery symposium. Society of Petroleum Engineers; 2010.
[130] Song Z, Bai B, Zhang H. Preformed particle gel propagation and dehydration through semi-transparent fractures and their effect on water flow. J Petrol Sci Eng 2018;167:549–58.
[131] Imqam A, Wang Z, Bai B. Preformed-particle-gel transport through heterogeneous void-space conduits. SPE J 2017;22(05):1437–47.
[132] Seright R. n.d.Mechanism for gel propagation through fractures. Paper SPE 1999;55628:15-9.
[133] Brattekas B, Haugen A, Graue A, Seright RS. Gel dehydration by spontaneous imbibition of brine from aged polymer gel. In: SPE improved oil recovery symposium. Society of Petroleum Engineers; 2012.
[134] Sun X, Bai B. Dehydration of polyacrylamide-based super-absorbent polymer swollen in different concentrations of brine under CO2 conditions. Fuel 2017;210:32–40.
[135] Brattekås B, Pedersen SG, Nistov HT, Haugen A, Graue A, Liang J-T, et al. Washout of Cr (III)-acetate-HPAM gels from fractures: effect of gel state during placement. SPE Prod Oper 2015;30 (02):99–109.
[136] Wang Z, Bai B, Sun X, Wang J. Effect of multiple factors on preformed particle gel placement, dehydration, and plugging performance in partially open fractures. Fuel 2019;251:73–81.
[137] Wang Z, Bai B. Preformed-particle-gel placement and plugging performance in fractures with tips. SPE J 2018;23(06):2316–26.
[138] Cozic C, Rousseau D, Tabary R. Broadening the application range of water shutoff/conformance-control microgels: an investigation of their chemical robustness. In: SPE annual technical conference and exhibition. Society of Petroleum Engineers; 2008.
[139] Bai B, Huang F, Liu Y, Seright RS, Wang Y. Case study on preformed particle gel for in-depth fluid diversion. In: SPE symposium on improved oil recovery. Society of Petroleum Engineers; 2008.
[140] Aqcheli F, Salehi MB, Pahlevani H, Taghikhani V. Rheological properties and the micromodel investigation of nanosilica gel-reinforced preformed particle gels developed for improved oil recovery. J Petrol Sci Eng 2020;192:107258.
[141] Long Y, Wang Z, Ding H, Geng J, Bai B. Investigation and characterization of a robust nanocomposite preformed particle gel for enhanced oil recovery. Energy Fuel 2019;33(6):5055–66.
[142] Pereira KA, Pereira KA, Oliveira PF, Mansur CR. Behavior of partially hydrolyzed polyacrylamide/polyethyleneimine reinforced with coal fly ash for preformed particle hydrogels. J Appl Polym Sci 2020;137(46):49423.
[143] Sun L, Li D, Pu W, Li L, Bai B, Han Q, et al. Combining preformed particle gel and curable resin-coated particles to control water production from high-temperature and high-salinity fractured producers. SPE J 2020;25(02):938–50.
[144] Tongwa P, Baojun B. A more superior preformed particle gel with potential application for conformance control in mature oilfields. J Pet Explor Prod Technol 2015;5(2):201–10.
[145] Li Z, Wang Y, He H, Yuan F, Liu H, Chen Y. Insights into the effects of salinity on the transport behavior of polymer-enhanced branched-preformed particle gel suspension in porous media. Energy Fuels 2021;35(2):1104–12.
[146] Li J, Jiang Z, Wang Y, Zheng J, Huang G. Stability, seepage and displacement characteristics of heterogeneous branched-preformed particle gels for enhanced oil recovery. RSC Adv 2018;8(9):4881–9.
[147] Long Y, Yajun L, Houjian G, Qian S, Kangning L, Mingzhe D. EOR of branched-preformed particle gel (B-PPG) flooding after polymer flooding in heterogeneous reservoirs. Pet Geol Recovery Effic 2016;23(1):113–8.
[148] Pu J, Bai B, Alhuraishawy A, Schuman T, Chen Y, Sun X. A recrosslinkable preformed particle gel for conformance control in heterogeneous reservoirs containing linear-flow features. SPE J 2019;24 (04):1714–25.

[149] Pu J, Bai B, Alhuraishawy A, Schuman T, Chen Y, Sun X. A novel re-crosslinkable preformed particle gel for conformance control in extreme heterogeneous reservoirs. In: SPE annual technical conference and exhibition. Society of Petroleum Engineers; 2018.
[150] Wang L, Long Y, Ding H, Geng J, Bai B. Mechanically robust re-crosslinkable polymeric hydrogels for water management of void space conduits containing reservoirs. Chem Eng J 2017;317:952–60.
[151] Bai B, Li L, Liu Y, Wang Z, Liu H. Preformed particle gel for conformance control: factors affecting its properties and applications. SPE/DOE symposium on improved oil recovery. Society of Petroleum Engineers; 2004.
[152] Wu D, Zhou K, Hou J, An Z, Zhai M, Liu W. Review of experimental and simulation studies of enhanced oil recovery using viscoelastic particles. J Dispers Sci Technol 2021;42(7):956–69.
[153] Zhao G, Dai C, Zhao M. Investigation of the profile control mechanisms of dispersed particle gel. PLoS ONE 2014;9(6), e100471.
[154] Liu Y, Dai C, Wang K, Zhao M, Gao M, Yang Z, et al. Investigation on preparation and profile control mechanisms of the dispersed particle gels (DPG) formed from phenol–formaldehyde cross-linked polymer gel. Ind Eng Chem Res 2016;55(22):6284–92.
[155] Dai C, Chen W, You Q, Wang H, Yang Z, He L, et al. A novel strengthened dispersed particle gel for enhanced oil recovery application. J Ind Eng Chem 2016;41:175–82.
[156] Okay O. General properties of hydrogels. In: Gerlach G, Arndt K-F, editors. Hydrogel sensors and actuators: engineering and technology. Berlin, Heidelberg: Springer Berlin Heidelberg; 2010. p. 1–14.
[157] Young TS, Hunt JA, Green DW, Willhite GP. Study of equilibrium properties of Cr(III)/polyacrylamide gels by swelling measurement and equilibrium Dialysis. SPE Reserv Eng 1989;04.
[158] Elsharafi MO, Bai B. Effect of strong preformed particle gel on Unswept oil zones/areas during conformance control treatments. In: EAGE annual conference & exhibition incorporating SPE Europec. Society of Petroleum Engineers; 2013.
[159] Imqam A, Bai B, Al Ramadan M, Wei M, Delshad M, Sepehrnoori K. Preformed-particle-gel extrusion through open conduits during conformance-control treatments. SPE J 2015;20(5):1083–93.
[160] Muhammed FA, Bai B, Al Brahim A. A simple technique to determine the strength of millimeter-sized particle gel. Society of Petroleum Engineers; 2014.
[161] Tu TN, Wisup B. Investigating the effect of polymer gels swelling phenomenon under reservoir conditions on polymer conformance control process. International petroleum technology conference.
[162] Mousavi Moghadam A, Vafaie Sefti M, Baghban Salehi M, Dadvand KA. Preformed particle gel: evaluation and optimization of salinity and pH on equilibrium swelling ratio. J Pet Explor Prod Technol 2012;2(2):85–91.
[163] Lenji MA, Haghshenasfard M, Sefti MV, Salehi MB. Experimental study of swelling and rheological behavior of preformed particle gel used in water shutoff treatment. J Petrol Sci Eng 2018;169:739–47.
[164] Molloy PJ, Smith MJ, Cowling MJ. The effects of salinity and temperature on the behaviour of polyacrylamide gels. Mater Des 2000;21(3):169–74.
[165] Chauveteau G, Tabary R, Blin N, Renard M, Rousseau D, Faber R. Disproportionate permeability reduction by soft preformed microgels. In: SPE/DOE symposium on improved oil recovery. Society of Petroleum Engineers; 2004.
[166] Dupuis G, Lesuffleur T, Desbois M, Bouillot J, Zaitoun A. Water conformance treatment using SMG microgels: a successful field case. In: SPE EOR conference at oil and gas west asia. Society of Petroleum Engineers; 2016.
[167] Almohsin AM, Bai B, Imqam AH, Wei M, Kang W, Delshad M, et al. Transport of Nanogel through porous media and its resistance to water flow. In: SPE improved oil recovery symposium. Society of Petroleum Engineers; 2014.
[168] Al-Ibadi A, Civan F. Experimental study of gel particles transport through porous media. In: SPE Latin America and caribbean petroleum engineering conference. Society of Petroleum Engineers; 2012.
[169] Saghafi HR, Emadi MA, Farasat A, Arabloo M, Naderifar A. Performance evaluation of optimized preformed particle gel (PPG) in porous media. Chem Eng Res Des 2016;112:175–89.
[170] Donaldson EC, Chilingarian GV, Yen TF. Microbial enhanced oil recovery. Newnes; 1989.
[171] Cohen Y, Christ FR. Polymer retention and adsorption in the flow of polymer solutions through porous media. SPE Reserv Eng 1986;1(02):113–8.
[172] Sorbie KS. Polymer-improved oil recovery. Springer Science & Business Media; 2013.

[173] Wang J, Liu H-Q, Wang Z-L, Hou P-C. experimental investigation on the filtering flow law of pre-gelled particle in porous media. Transp Porous Media 2012;94(1):69–86.
[174] Sang Q, Li Y, Yu L, Li Z, Dong M. Enhanced oil recovery by branched-preformed particle gel injection in parallel-sandpack models. Fuel 2014;136:295–306.
[175] Zhang H. Preformed particle gel transport through fractures. Doctoral dissertations, Missouri University of Science and Technology; 2014.
[176] Feng Q, Chen X, Zhang G. Experimental and numerical study of gel particles movement and deposition in porous media after polymer flooding. Transp Porous Media 2013;97(1):67–85.
[177] Wang J, Liu H, Wang Z, Xu J, Yuan D. Numerical simulation of preformed particle gel flooding for enhancing oil recovery. J Petrol Sci Eng 2013;112:248–57.
[178] Feng Q, Zhang G, Yin X, Luan Z. Numerical simulation of the blocking process of gelled particles in porous media with remaining polymers. Pet Sci 2009;6(3):284–8.
[179] Jing W, Hui-Qing L, Hong-ling Z, Kamy S. Simulation of deformable preformed particle gel propagation in porous media. AICHE J 2017;63(10):4628–41.
[180] Iwasaki T, Slade J, Stanley WE. Some notes on sand filtration [with discussion]. J Am Water Works Ass 1937;29(10):1591–602.
[181] Maroudas A, Eisenklam P. Clarification of suspensions: a study of particle deposition in granular media: part I—some observations on particle deposition. Chem Eng Sci 1965;20(10):867–73.
[182] Goudarzi A, Zhang H, Varavei A, Taksaudom P, Hu Y, Delshad M, et al. A laboratory and simulation study of preformed particle gels for water conformance control. Fuel 2015;140:502–13.
[183] Farasat A, Younesian-Farid H, Sadeghnejad S. Conformance control study of preformed particle gels (PPGs) in mature waterflooded reservoirs: numerical and experimental investigations. J Petrol Sci Eng 2021;203:108575.
[184] Garmeh G, Izadi M, Salehi M, Romero JL, Thomas C, Manrique EJ. Thermally active polymer to improve sweep efficiency of waterfloods: simulation and pilot design approaches. SPE Reserv Eval Eng 2012;15(01):86–97.
[185] Ohms DS, JD ML, Graff CJ, Frampton H, Morgan J, Cheung SK, et al. Incremental oil success from waterflood sweep improvement in Alaska. In: SPE International symposium on oilfield chemistry. Society of Petroleum Engineers; 2009.
[186] Frampton H, Morgan JC, Cheung SK, Munson L, Chang KT, Williams D. Development of a novel Waterflood conformance control system. In: SPE/DOE symposium on improved oil recovery. Tulsa, Oklahoma: Society of Petroleum Engineers; 2004.
[187] Chang K-T, Frampton H, Morgan JC. Composition and method for recovering hydrocarbon fluids from a subterranean reservoir. Google Patents; 2002.
[188] Pritchett J, Frampton H, Brinkman J, Cheung S, Morgan J, Chang K, et al. In: Field application of a new in-depth waterflood conformance improvement tool. SPE international improved oil recovery conference in Asia Pacific. Society of Petroleum Engineers; 2003.
[189] Salehi M, Thomas CP, Kevwitch R, Garmeh G, Manrique EJ, Izadi M. Performance evaluation of thermally-activated polymers for conformance correction Applications. In: SPE improved oil recovery symposium. Society of Petroleum Engineers; 2012.
[190] Fabbri C, Klimenko A, Jouenne S, Cordelier P, Morel D. Laboratory and simulation investigation of the effect of thermally activated polymer on permeability reduction in highly permeable unconsolidated sand. In: SPE Asia Pacific enhanced oil recovery conference. Society of Petroleum Engineers; 2015.
[191] Husband ME, Ohms DS, Frampton H, Carhart SR, Carlson BH, Chang K-T, et al. Results of a three-well waterflood sweep improvement trial in the Prudhoe Bay field using a thermally activated particle system. In: SPE improved oil recovery symposium. Society of Petroleum Engineers; 2010.
[192] Paez Yanez PA, Mustoni JL, Frampton H, Relling MF, Chang K-T, Hopkinson PC. New attempt in improving sweep efficiency at the Mature Koluel Kaike and Piedra Clavada waterflooding projects of the S. Jorge Basin in Argentina. In: Latin American & Caribbean petroleum engineering conference. Society of Petroleum Engineers; 2007.
[193] Mustoni JL, Denyer P, Norman C. Deep conformance control by a novel thermally activated particle system to improve sweep efficiency in mature waterfloods of the san Jorge Basin. In: SPE improved oil recovery symposium. Society of Petroleum Engineers; 2010.

[194] Ghaddab F, Kaddour K, Tesconi M, Brancolini A, Carniani C, Galli G, Borma E. BrightWater: a tertiary method for enhanced oil recovery for a mature field. In: Paper SPE 136140 presented at the SPE production and operations conference and exhibition, Tunis, Tunisia, 8–10 June; 2010.
[195] Choudhary M, Parekh B, Solis H, Meyer B, Shepstone K, Dezabala E, et al. Reservoir in-depth waterflood conformance: an offshore pilot implementation. In: SPE improved oil recovery symposium. Society of Petroleum Engineers; 2014.
[196] Frampton H, Denyer P, Ohms D, Husband M, Mustoni J. Sweep Improvement from the Lab to the Field. In: IOR 2009-15th European symposium on improved oil recovery. European Association of Geoscientists & Engineers; 2009. cp-124-00041.
[197] Younesian-Farid H, Sadeghnejad S. Geochemical performance evaluation of pre-flushing of weak and strong acids during pH-triggered polymer flooding. J Petrol Sci Eng 2019;174:1022–33.
[198] Al-Anazi HA, Sharma MM. Use of a pH sensitive polymer for conformance control. In: International symposium and exhibition on formation damage control. Society of Petroleum Engineers; 2002.
[199] Huh C, Choi SK, Sharma MM. A rheological model for pH-sensitive ionic polymer solutions for optimal mobility control applications. In: SPE annual technical conference and exhibition. Dallas, Texas: Society of Petroleum Engineers; 2005.
[200] Koochakzadeh A. Laboratory investigation of preflushing acid during EOR using pH-sensitive polymers. Master of Science, Iran: Department of Petroleum Engineering. Tarbiat Modares University; 2019.
[201] Kontturi K, Mafe S, Manzanares J, Svarfvar B, Viinikka P. Modeling of the salt and pH effects on the permeability of grafted porous membranes. Macromolecules 1996;29(17):5740–6.
[202] Benson IP, Nghiem LX, Bryant SL, Huh C. Development and use of a simulation model for mobility/conformance control using a pH-sensitive polymer. In: SPE annual technical conference and exhibition. Society of Petroleum Engineers; 2007.
[203] Koochakzadeh A, Ashrafizadeh M, Sadeghnejad S. Experimental Evaluation of Enhanced Oil Recovery by pH-sensitive Microgels in Carbonate Formations. In: 82nd EAGE annual conference & exhibition. European Association of Geoscientists & Engineers; 2020. p. 1–5.
[204] Choi SK. pH sensitive polymers for novel conformance control and polymer flooding applications. The University of Texas at Austin; 2008.
[205] Lalehrokh F, Bryant SL, Huh C, Sharma MM. Application of pH-triggered polymers in fractured reservoirs to increase sweep efficiency. In: SPE symposium on improved oil recovery. Society of Petroleum Engineers: Tulsa, Oklahoma, USA; 2008.
[206] Choi SK. A study of a pH-sensitive polymer for novel conformance control applications. M.Sc, The University of Texas at Austin; 2005.
[207] Lalehrokh F. Applicability of pH-triggered polymers to increase sweep efficiency in fractured reservoirs. The University of Texas at Austin; 2010.
[208] Lalehrokh F, Bryant SL. Application of pH-triggered polymers for deep conformance control in fractured reservoirs. In: SPE annual technical conference and exhibition. Society of Petroleum Engineers; 2009.
[209] Ho JF, Patterson JW, Tavassoli S, Shafiei M, Huh C, Bommer PM, et al. The use of a pH-triggered polymer gelant to seal cement fractures in wells. SPE-76813-PA; 2016.
[210] Ho M. The use of a pH-triggered polymer gelant to seal cement fractures in wells. M. Sc. The University of Texas at Austin; 2015.
[211] Tavassoli S, Ho JF, Shafiei M, Huh C, Bommer P, Bryant S, et al. An experimental and numerical study of wellbore leakage mitigation using pH-triggered polymer gelant. Fuel 2018;217:444–57.
[212] Patterson JW. Placement and performance of pH-triggered polyacrylic acid in cement fractures. The University of Texas at Austin; 2014.
[213] Koochakzadeh A, Younesian-Farid H, Sadeghnejad S. Acid pre-flushing evaluation before pH-sensitive microgel treatment in carbonate reservoirs: experimental and numerical approach. Fuel 2021;297.
[214] Choi SK, Sharma MM, Bryant S, Huh C. pH-sensitive polymers for novel conformance-control and polymer-flood applications. SPE Reserv Eval Eng 2010;13(6):926–39.
[215] Garcia Cordeiro Tessarolli F, Gomes De Carvalho Queirós Y, Regina Elias Mansur C. Evaluation of pH-sensitive hydrogels to control the permeability anisotropy of oil reservoirs. J Appl Polym Sci 2014;131(17):40665.

[216] Al-Wahaibi Y, Al-Wahaibi T, Abdelgoad M. Characterization of pH-sensitive polymer microgel transport in porous media for improving oil recovery. Energy Sources, Part A 2011;33(11):1048–57.
[217] Al-Wahaibi T, Al-Wahaibi Y, Abdel-Goad M. Geochemical characterization of Omani sandstone core-analysis of change in pH and rock mineral constituents concentrations. Pet Coal 2012;54 (3):260–70.
[218] Zhang P, Huang W, Jia Z, Zhou C, Wang Y, Li D, et al. Synthesis and application of a pH-triggered microgel in reservoir mobility-control. Int J Oil Gas Coal Technol 2015;9(2):180–91.
[219] Teimouri A, Sadeghnejad S, Dehaghani AHS. Investigation of acid pre-flushing and pH-sensitive microgel injection in fractured carbonate rocks for conformance control purposes. Oil Gas Sci Technol 2020;75:52.
[220] Liu M, Mostaghimi P. High-resolution pore-scale simulation of dissolution in porous media. Chem Eng Sci 2017;161:360–9.
[221] Sadeghi MA, Aghighi M, Barralet J, Gostick JT. Pore network modeling of reaction-diffusion in hierarchical porous particles: the effects of microstructure. Chem Eng J 2017;330:1002–11.
[222] Soulaine C, Roman S, Kovscek A, Tchelepi HA. Mineral dissolution and wormholing from a pore-scale perspective. J Fluid Mech 2017;827:457–83.
[223] Tian Z, Wang J. Lattice Boltzmann simulation of dissolution-induced changes in permeability and porosity in 3D CO2 reactive transport. J Hydrol 2018;557:276–90.
[224] Sadeghnejad S, Enzmann F, Kersten M. Digital rock physics, chemistry, and biology: challenges and prospects of pore-scale modelling approaches. Appl Geochem 2021;131:105028.
[225] Langergraber G, Šimůnek J. Modeling variably saturated water flow and multicomponent reactive transport in constructed wetlands. Vadose Zone J 2005;4(4):924–38.
[226] Steefel C, Appelo C, Arora B, Jacques D, Kalbacher T, Kolditz O, et al. Reactive transport codes for subsurface environmental simulation. Comput Geosci 2015;19(3):445–78.
[227] Altree-Williams A, Brugger J, Pring A, Bedrikovetsky P. Exact solution for coupled reactive flow and dissolution with porosity changes. Transp Porous Media 2018;124(2):655–79.
[228] Battiato I, Tartakovsky DM, Tartakovsky AM, Scheibe TD. Hybrid models of reactive transport in porous and fractured media. Adv Water Resour 2011;34(9):1140–50.
[229] Xiong Y, Fakcharoenphol P, Winterfeld P, Zhang R, Wu Y-S. Coupled geomechanical and reactive geochemical model for fluid and heat flow: application for enhanced geothermal reservoir. In: SPE reservoir characterization and simulation conference and exhibition. Society of Petroleum Engineers; 2013.
[230] Siuliukina N, Tartakovsky DM. A hybrid multiscale model of miscible reactive fronts. Water Resources Research 2018;54(1):61–71.
[231] Panga MK, Ziauddin M, Balakotaiah V. Two-scale continuum model for simulation of wormholes in carbonate acidization. AICHE J 2005;51(12):3231–48.
[232] Liu N, Liu M. Simulation and analysis of wormhole propagation by VES acid in carbonate acidizing. J Petrol Sci Eng 2016;138:57–65.
[233] Onbergenov U. Simulation of thermally active and pH-sensitive polymers for conformance control. The University of Texas at Austin; 2012.
[234] Younesian-Farid H, Sadeghnejad S. Modeling geochemical reactions of citric acid with ankerite cement during sandstone matrix acidizing. J Petrol Sci Eng 2020;185:106650.
[235] Mayer KU, Frind EO, Blowes DW. Multicomponent reactive transport modeling in variably saturated porous media using a generalized formulation for kinetically controlled reactions. Water Resour Res 2002;38(9):13–1–13–21.
[236] Balashov VN, Guthrie GD, Hakala JA, Lopano CL, Rimstidt JD, Brantley SL. Predictive modeling of CO2 sequestration in deep saline sandstone reservoirs: impacts of geochemical kinetics. Appl Geochem 2013;30:41–56.
[237] Mahrous M, Sultan A, Sonnenthal E. Towards geochemically accurate modeling of carbonate acidizing with HCl acid. In: SPE annual technical conference and exhibition. Society of Petroleum Engineers; 2017.
[238] Kielland J. Individual activity coefficients of ions in aqueous solutions. J Am Chem Soc 1937;59 (9):1675–8.

CHAPTER 7

Smart water injection

Arastoo Abdi, Zahra Bahmani, Behnam Ranjbar, and Masoud Riazi
Enhanced Oil Recovery (EOR) Research Centre, IOR/EOR Research Institute, Shiraz University, Shiraz, Iran

7.1 Basic concepts

In the 20th century, most operators began to study the feasibility of water injection after natural depletion. The use of water in the oil industry has a long history and dates back to more than a century. The use of water as an injection method was considered a simple technique for displacing oil in hydrocarbon reservoirs due to its advantages such as low costs, easy access, environmental compatibility, the possibility of combining with other methods, and the possibility to use in different conditions.

Smart water is an EOR method that changes the composition of injected water to increase recovery. This method was recognized by large oil companies due to its benefits. The study of smart water was proposed by various companies with different terms such as LoSal by British Petroleum [1], Designer WaterFlood by Shell [2], Advanced Ion Management by ExxonMobil [3], and Smart Water Flood by Saudi Aramco [4], however, all these terms represent the same concept.

The classification of smart water can be classified as follows.

- Low salinity water (salinity less than 5000 ppm).
- A special type of high salinity water with changes in its main composition (seawater is in the second category due to its important role in natural wettability modification) [5].

7.2 Condition for smart water injection in sandstone reservoirs

Austad et al. [6] defined the conditions for effective injection of smart water in sandstone reservoirs based on studies by Morrow [7] and researchers at British Petroleum [8, 9] using laboratory studies as shown in Table 7.1.

Chemical Methods
https://doi.org/10.1016/B978-0-12-821931-7.00008-0

Table 7.1 Conditions for smart water injection in sandstone reservoirs [6].

Porous medium
• The presence of clay minerals in sandstone reservoirs is vital
Oil
• The presence of polar components is vital
Formation water
• Existence of formation water is absolutely necessary and must be included divalent cations
Low salinity injection fluid
• The salinity less than 5000 ppm
Permeability
• Changing permeability curves by switching from high salinity water to smart water
Temperature
• No limitation

7.3 Condition for smart water injection in carbonate reservoirs

The presence of clay in sandstone reservoirs is an important factor for the effectiveness of smart water injection. The absence of clay in carbonate reservoirs dismissed the idea that smart water injection did not have the potential to improve recovery based on studies by Lager et al. [9]. RezaeiDoust et al. [10] showed that the binding of oil components to the surface of calcite with a positive charge is higher than that of quartz with a negative charge. Creating an electrostatic repulsion between the polar components of oil and the rock surface during smart water injection in carbonates is also possible by interaction of smart water with the rock surface and the polar components of oil.

Various studies are aimed at investigating the impact of different ions on the effectiveness of smart water injection. Accordingly, Awolayo et al. [11] used studies performed by Fathi et al. [12, 13], Gupta et al. [3], Yi and Sarma [14], Yousef et al. [15], Zahid et al. [16], Zhang et al. [17], and Austad et al. [18] to describe conditions in which smart water in carbonate reservoirs is efficient, these findings are shown in Table 7.2.

Table 7.2 Conditions for smart water injection in carbonate reservoirs [11].

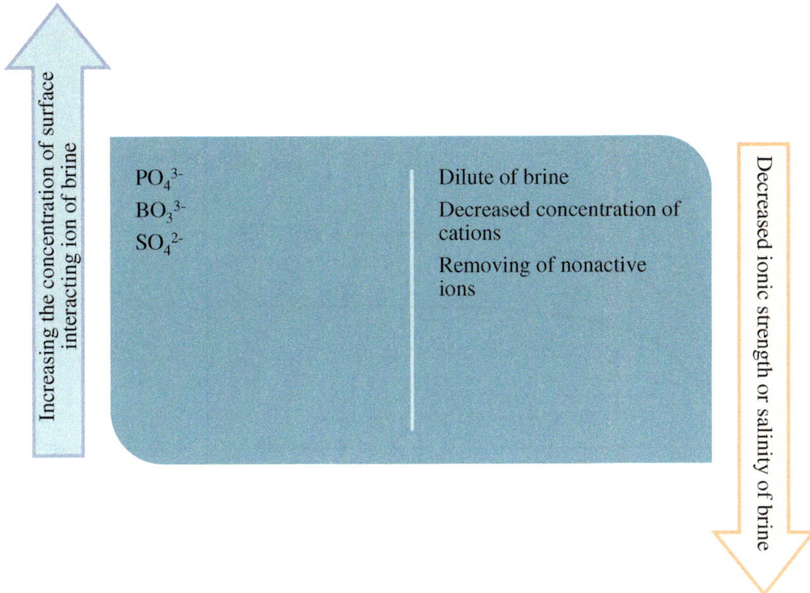

7.4 Factors influencing smart water

7.4.1 Effect of potential determining ions

Ions can alter wettability when a rock surface with surface charge comes in contact with the brine. In the smart water injection method, the presence of SO_4^{2-} ions in injected water has a considerable effect on wettability alteration. The importance of SO_4^{2-} ions in changing the wettability of rock has been investigated by various researchers. It is generally agreed that SO_4^{2-} ions have a strong tendency to adsorb on the surface of carbonate minerals. As the SO_4^{2-} ions get close and adsorb on to the carbonate rock surface, the positive charge density on the surface decreases, and as the concentration of SO_4^{2-} ions near the surface increases, the negative charge density on the surface increases. The created negative charge leads to electrostatic repulsion between the rock surface and the negative oil groups, and as a result, facilitates the separation of the adsorbed oil from the carbonate surface and causes wettability alteration of the rock to more water-wet. In most cases, Mg^{2+} and Ca^{2+} cations get close to the surface by reducing the positive charge of the rock and form complex compounds as the separation of negative oil groups occur [19]. The results of studies by Zhang et al. (Fig. 7.1) and Yi and Sarma (Figs. 7.2 and 7.3) showed that decreasing salinity of seawater and increasing the concentration of SO_4^{2-} in spontaneous imbibition and flooding processes caused a considerable increase in oil recovery [14,20].

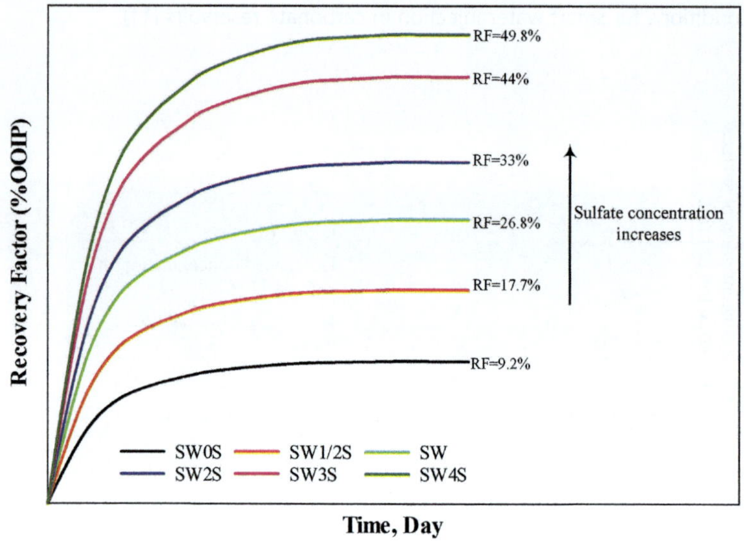

Fig. 7.1 The effect of modification in the salinity of seawater (increasing SO_4^{2-} concentration) on oil recovery in spontaneous imbibition [20].

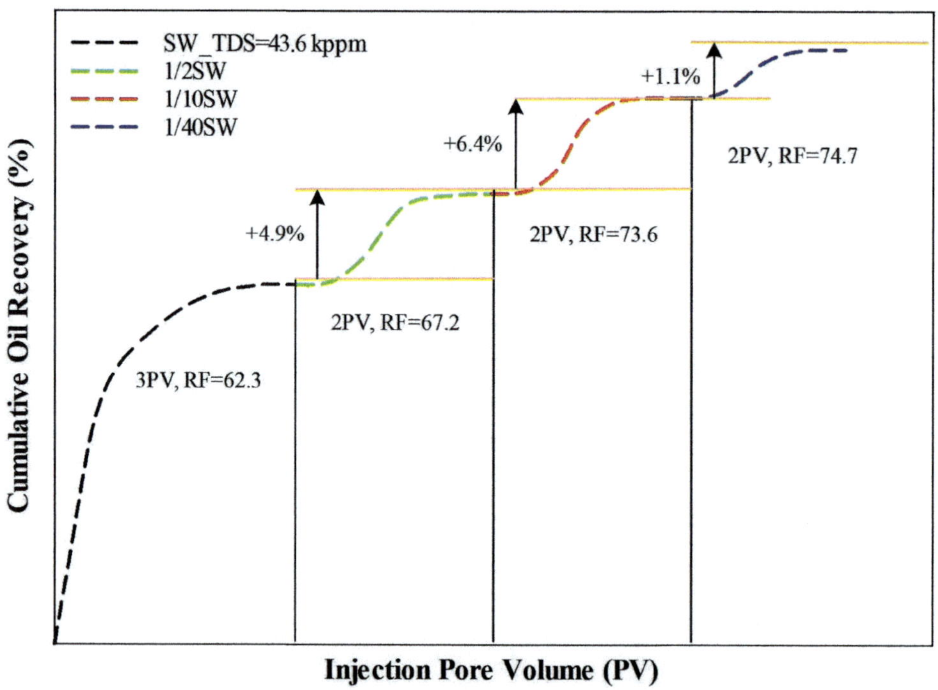

Fig. 7.2 The effect of seawater dilution on oil recovery in coreflooding [14].

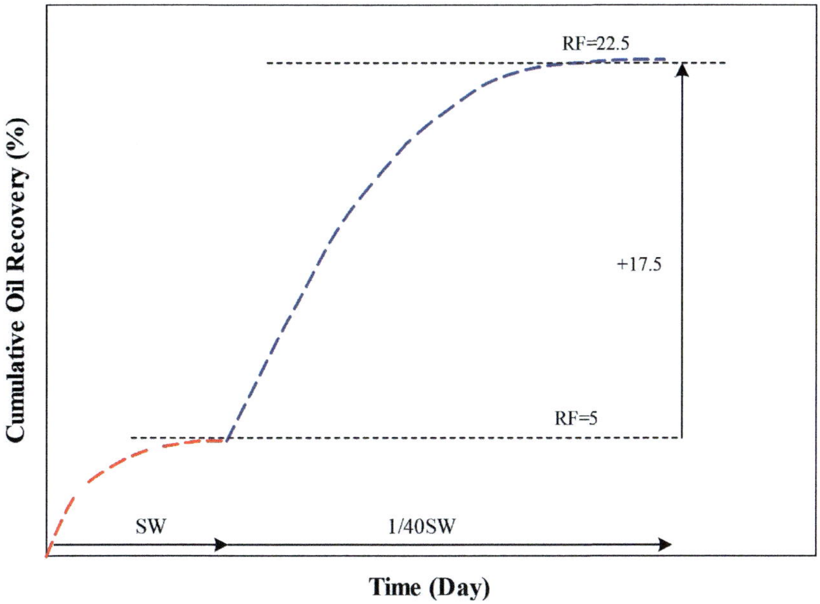

Fig. 7.3 The effect of seawater dilution on oil recovery in spontaneous imbibition [14].

The activity and interaction of Ca^{2+} and Mg^{2+} ions in the electrostatic layer along with SO_4^{2-} ions will change the wettability in the system toward water-wet and separate the oil from the surface of the reservoir rock and increase oil recovery. Doubling the concentration of Ca^{2+} (SW2C), Mg^{2+} (SW2M), and SO_4^{2-} (SW2S) ions in seawater composition while TDS was constant was done in a study by Purswani and Karpyn [21] which showed that modifying the composition of seawater by increasing the concentration of active ions Ca^{2+}, Mg^{2+}, and SO_4^{2-} will significantly increase oil recovery as illustrated in Fig. 7.4.

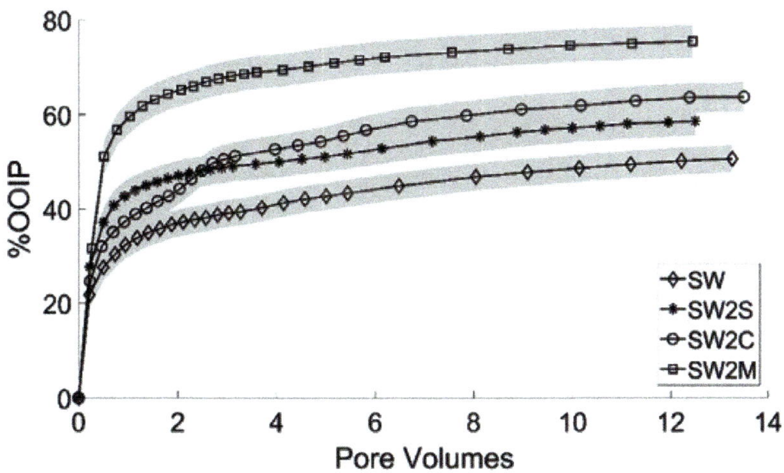

Fig. 7.4 The effect of modification in the salinity of seawater (increasing SO_4^{2-}, Mg^{2+}, and Ca^{2+} concentration) on oil recovery in secondary recovery [21].

7.4.2 Effect of nonpotential determining ions

Some ions do not belong to the electrostatic layer. One of these ions is sodium, which has no effect on recovery and, in addition, reduces the activity of effective ions. Studies have shown that the activity of calcium, magnesium, and sulfate ions is strongly dependent on their chemical activity near the electrostatic layer, but monovalent ions such as Na^+, which are not active ions in the electrostatic layer, decrease the activity of Mg^{2+}, Ca^{2+}, and SO_4^{2-} ions and also reduce oil recovery.

Fathi et al. [13] investigated the effect of ion composition on the process of smart water EOR in carbonate reservoirs and showed that the concentration of Mg^{2+}, Ca^{2+}, and SO_4^{2-} ions is insignificant in changing the wettability of carbonate reservoirs, but the concentration of NaCl salt as non-active ions is very important. The results of this study showed (Fig. 7.6) that there is an increase of about 10% in oil recovery for synthetic seawater without NaCl (SW0NaCl) when compared to ordinary seawater (SW). The study also showed (Figs. 7.5 and 7.6) that 4 times increase in the concentration of SO_4^{2-} ions in seawater without NaCl (SW0NaCl-4SO4) increased the ultimate oil recovery by 5%–18%.

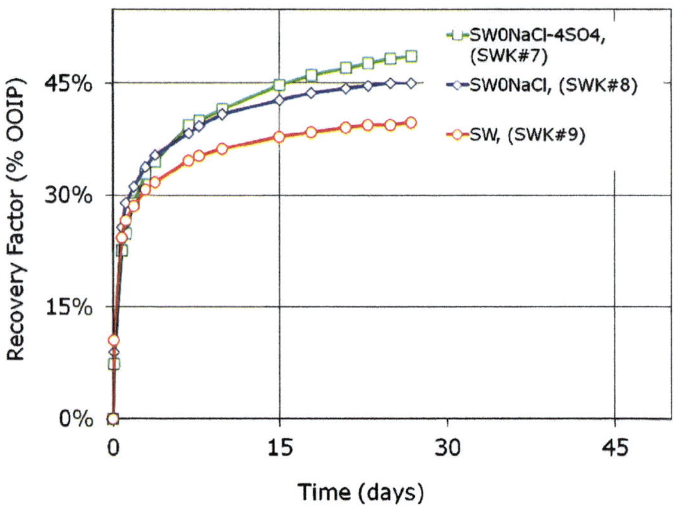

Fig. 7.5 The effect of seawater modification (removal of non-active (Na^+ Cl^- ions)) and the increase of SO_4^{2-} ion concentration on spontaneous imbibition at 70°C [13].

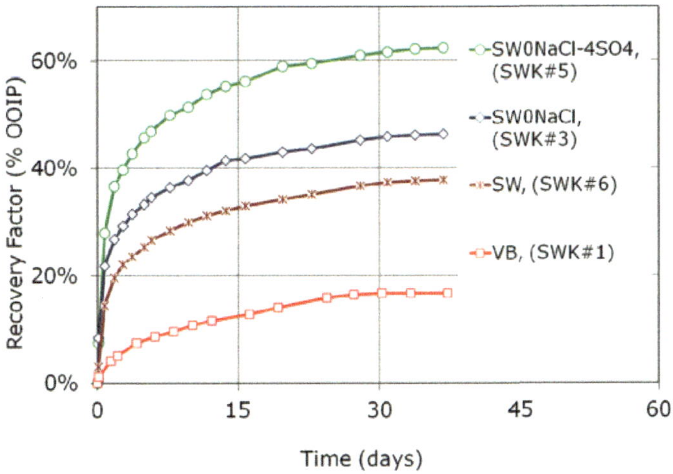

Fig. 7.6 The effect of seawater modification (removal of non-active (Na^+ Cl^- ions)) and the increase of SO_4^{2-} ion concentration on spontaneous imbibition at 90°C [13].

7.4.3 Effect of rock type

Rock type (sandstone and carbonate) impacts wettability alteration and consequently influences the amount of oil that is recovered [22]. Therefore, it is important to analyze each rock type separately for smart water injection; some influencing conditions for this type of injection are shown in Table 7.3.

Table 7.3 Influencing conditions in smart water injection based on rock type.

Parameter	Sandstone	Carbonate
Temperature	–	High temperature is required to activate wettability alteration [23–25]
Carboxylic material	Weak bond	Strong bond
pH	In a low pH environment more attachment of acidic and basic components occur [26]	In a basic environment the rock will become water-wet [27]
Salinity	Low salinity injection	High salinity/low salinity injection

Studies for sandstone and carbonates indicate that injected water is usually diluted (low salinity water injection) and tuned (engineered water injection) in these reservoirs, respectively. Smart water injection has been studied more extensively for sandstone reservoirs because clays are unique due to cation exchange, and their negative charge is the main reason for increase in pH and wettability alteration [28]

Carbonate rocks with positive surface charge have stronger chemical bonds with negative polar components of crude oil (carboxylic material) rather than sandstones which have a negative surface charge, therefore the wettability alteration mechanism differs in each rock type. In addition, the weaker bond between organic material and clay or silica surface in sandstones will enhance the probability of salting-in effect in comparison to carbonate rocks [29].

Seawater can activate wettability alteration in carbonate rocks, however, in a sandstone reservoir the injected water must have salinity less than 5000 ppm [28, 30].

As low salinity water (with decreased inactive ions) is injected into a carbonate rock containing anhydrite, dissolution of anhydrite occurs, which increases sulfate concentration. As Na^+ (inactive ion) decreases and sulfate increases wettability alteration is positively impacted [31].

7.4.4 Effect of temperature

Higher temperatures in carbonates lead to better smart water flooding recovery [23–25]. Rao [32] believes that carbonate reservoirs with higher temperatures will be more water-wet. It seems that temperature and acid number are related in such a way that decarboxylation occurs by an increase in temperature and as a consequence the crude oil acidity decreases, thus the wettability of water also increases [33]. On the other hand, the temperature is directly related to SO_4^{2-} adsorption, which increases production [20,34]. Accordingly, a research focused on calcite shows a direct relationship between temperature and water-wetness [35].

Mahani et al. did not find a significant and integrated relationship between temperature increase and contact angle in high salinity water injection. Calculation of Amott wettability indexes on carbonate and sandstone cores indicate wettability alteration from weak-medium to strongly water-wet with increasing temperature [36]. In reservoirs containing anhydrite, the significant point is to investigate the solubility of anhydrite in injected brine as temperature increases because if the solubility decreases by growth of temperature, the concentration of Ca^{2+} and SO_4^{2-} ions (which cause wettability alteration) decreases and this reduces oil recovery [37].

7.5 Physical and chemical mechanisms of recovery

Sandstone

Due to complex interactions in an oil/brine/rock system the fundamental mechanism in smart water injection is yet to be discovered. However, extensive studies have been performed in this field and mechanisms such as: (1) wettability alteration [7, 38–40] in sandstones with low clay content, (2) formation damage by clay migration, a mechanism for sandstones with high clay contents [41] in which sand particles and some fine minerals are released due to smart water injection and this results in fluid flow diversion which can improve displacement and sweep efficiency [42], (3) pH increase [6, 9, 43, 44], (4) multiion exchange (MIE) or cation exchange [9, 10, 44–46], (5) salting in [2, 9, 29], and (6) double-layer expansion have been introduced for sandstone reservoirs [47, 48].

Carbonates

In order to activate smart water and achieve water-wetness in carbonate rocks, high temperature and Ca^{2+}, Mg^{2+}, and SO_4^{2-} ions are needed [10]. Multiple mechanisms have been introduced as underlying mechanisms in carbonate rocks such as: (1) microemulsions [49, 50], (2) wettability alteration [17, 51–55], (3) MIE can also occur in carbonates, Ca^{2+}, Mg^{2+}, and SO_4^{2-} ions influence the performance of MIE [10, 34, 56], (4) mineral dissolution [16, 57, 58], (5) fine migration [16, 59], and (6) pH increase [60, 61].

7.5.1 Fines migration

Unlike fresh water, low-salt water prevents swelling and stability of clay. As clay and silt become unstable and disperse in water, small pores are gradually closed by them which reduces permeability, so the injected water selects new paths which increase sweeping efficiency and improve recovery [62–65]. In fact, trapping of particles in pore throats and consequently permeability reduction leads to tortuosity flow, this helps particle separation (Fig. 7.7) [66]. Tang and Morrow [65] believe that oil recovery increase in smart water injection is most likely linked to particle release. They assumed that the separation of clay particles from the pore walls are related to how the mechanical and colloidal forces were balanced. With smart water injection, an electrical double-layer mechanism is created, which is one of the causes of clay particle separation (Fig. 7.8) [64,67]. However, this mechanism is not very desirable now due to formation damage, and consequently

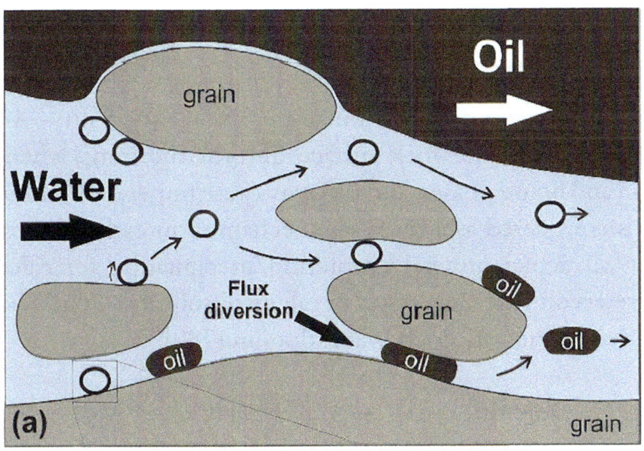

Fig. 7.7 Trapping of fines in pores and the formation of tortuosity flow [66].

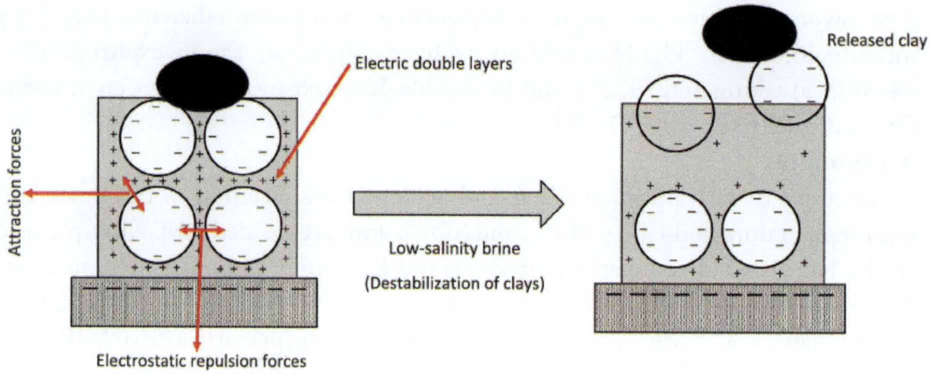

Fig. 7.8 Schematic representation of how forces between particles and the effect of smart water injection affect the separation of clay particles from the walls and surface (A) Colloidal forces between clay particles and cations in electrical double layer (EDL) (B) smart water injection and resulting separation of clay particles by mechanism EDL [76]

damage and loss in production [68–70]. In general, smart water injection causes formation damage under three mechanisms; fines migration, swelling due to migration, and particle swelling [71]. In some studies, smart water injection has been shown to reduce S_{or} despite permeability damage [72–74]. It is not clear to what extent oil recovery is influenced by the benefits of fines migration or its damaging effects [71]. However, Chequer et al. [66] stated that flux diversion from the permeable layer to the low permeability layer (which leads to increased sweep efficiency) occurs when the permeability reduction due to formation damage is relatively low. Valdya and Fogler [75] consider gradual reduction of injected water salinity as a solution to minimize formation damage. In this case, concentration of particles in the water flow remains low.

7.5.2 Mineral dissolution

As smart water—a brine with different ion composition and salinity—is injected into the reservoir, its reaction with the rock surface disturbs the initial chemical equilibrium between the solid and liquid phases, and this may cause mineral dissolution [77, 78]. Mineral dissolution is considered as a recovery mechanism in carbonate rocks [57, 79, 80]. Eqs. (7.1) and (7.2) depict mineral dissolution/precipitation for calcite and dolomite in a carbonated reservoir [81]. As a result of calcite dissolution, salinity is increased which will have a positive impact on smart water flooding [80].

$$\text{Dolomite} + 2H^+ \leftrightarrow Ca^{2+} + 2HCO_3^- + Mg^{2+} \tag{7.1}$$

$$\text{Calcite} + H^+ \leftrightarrow Ca^{2+} + HCO_3^- \tag{7.2}$$

Studies have shown that sandstone cores containing anhydrite become more water-wet after anhydrite dissolution [82].

Experimental tests can help us determine mineral dissolution in reservoir rocks, furthermore, geochemical modeling programs such as PHREEQC can be used as a tool to simulate and predict chemical reactions between rock and water in order to reveal the occurrence of mineral dissolution [83].

As a result of chemical reactions between carbonate reservoir rock and injected water, the pore surface is also dissolved, subsequently, the oil is detached from the rock surface and wettability is altered toward more water-wet [77] (Fig. 7.9).

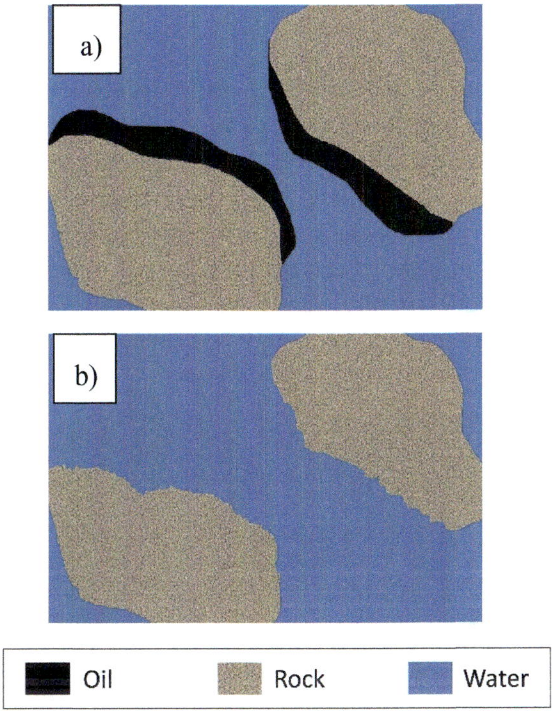

Fig. 7.9 Wettability alteration caused by surface dissolution. (A) oil-wet rock before dissolution, (B) water-wet rock after dissolution.

7.5.3 Emulsion formation

As a result of water/oil interaction in a reservoir, emulsions are formed. These emulsions are categorized as: oil-in-water (o/w) and water-in-oil (w/o) [84].

Microemulsions are optically isotropic emulsions with a droplet diameter of 1–100 nm [85–87]. Initially, the formation of microemulsions was considered to occur spontaneously as brine/crude oil interacted, this was observed visually in an experiment [50, 88].

7.5.3.1 Oil-in-water emulsions

During smart water injection, insoluble fines such as sulfate and clay react with oil and produce oil-in-water emulsions. Results show that an increase in temperature seems to have a demulsifying effect and brine salinity impacts the emulsion formation, but the trend is not clear. Although how the formation of emulsions occur is not completely clear, researchers believe the presence of fines in the oil-water system is the main reason [89–92]. This type of emulsion will flow through more permeable regions leading water to pass through regions with low permeability which will improve sweep efficiency [84]. Findings for this type of emulsion indicate that oil recovery improves as droplet size increases slightly bigger than the pore throat size [93, 94].

7.5.3.2 Water-in-oil emulsions

In a high salinity environment, accumulation of surface-active agents of crude oil at the water/oil interface occurs and this leads to a limited number of surface-active agents in the bulk of the oil, thus, an exceedingly small number of microemulsions might appear. Conversely, in a low salinity environment, the ionic strength of water decreases and active agents of oil move toward the bulk of oil, and as the number of polar components increase in the bulk, an increasing number of microemulsions are formed and stabilized [88], subsequently as polar components of oil abandon the water/oil interface and the rock surface to form w/o emulsions (Fig. 7.10), wettability will alter toward water-wet [50].

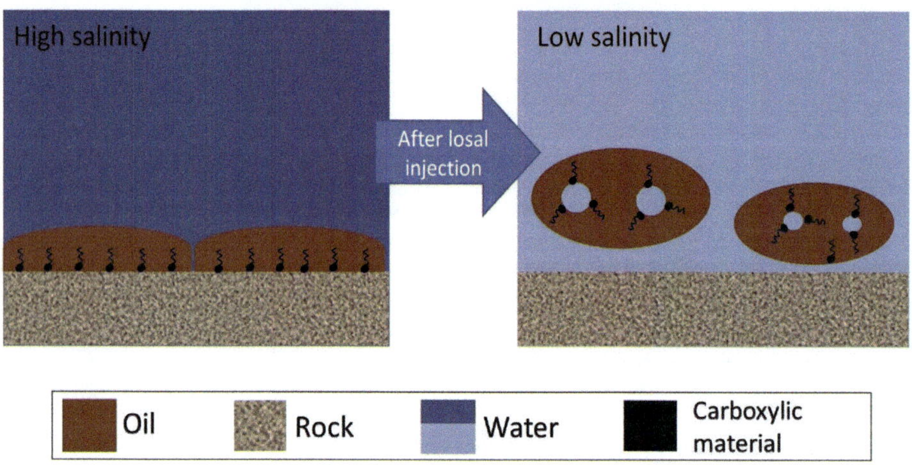

Fig. 7.10 Water in oil (w/o) emulsions formed in a low salinity environment.

Emulsions are proven to have positive effects on chemical EOR processes [95–98], in addition, for smart water injection, the amount of oil recovery is directly dependent on the crude oils ability to form emulsions [49].

7.5.4 pH increase

As illustrated in Fig. 7.11, the initial situation acidic and basic materials are adsorbed onto clay alongside a cation (Ca^{2+}). After low salinity flooding, cations detach from the clay surface as a result of brine-rock equilibrium disturbance. To compensate for this loss, protons adsorb onto clay and finally as H^+ replaces Ca^{2+}, pH increases [6]. pH increase is one of the main recovery mechanisms for sandstone rocks, which enables the desorption of carboxylic material from the clay surface [10]. Based on experimental observations, this process requires the presence of clay, polar components in oil, and active cations in the formation water [6].

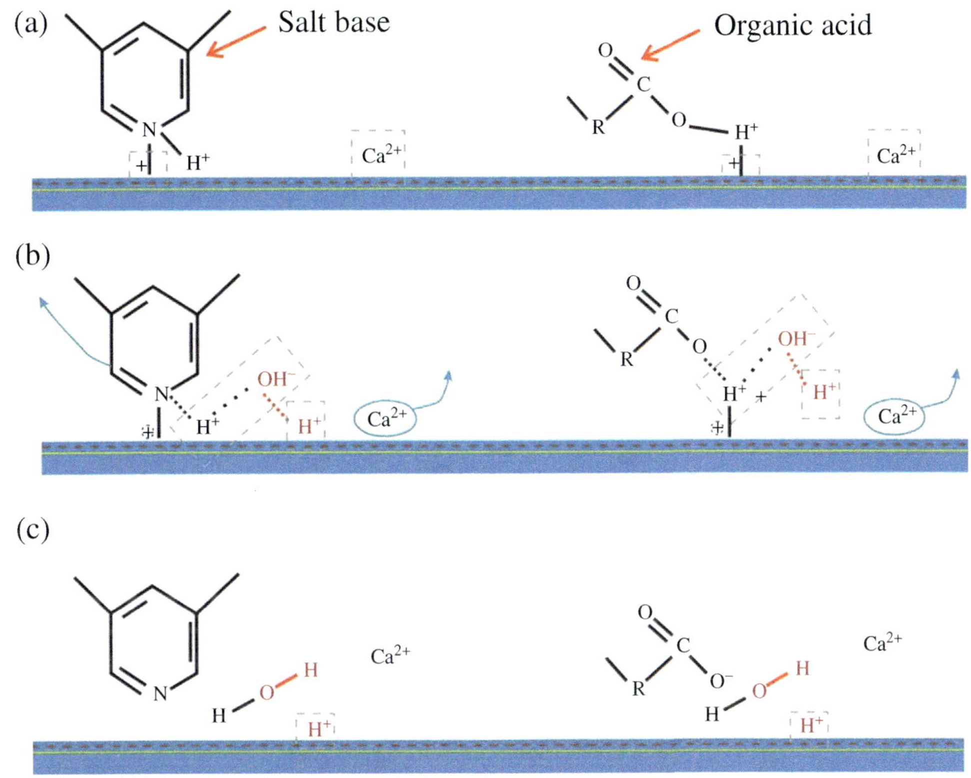

Fig. 7.11 Desorption of acidic and basic organic material from clay surface. (A) Initial state; (B) during smart water injection; (C) final state [99].

In a carbonate reservoir, the oil carboxylic group can form a complex with cations such as Ca^{2+}. As a result of this complex Ca^{2+} ion reduces and calcite dissolution occurs. Calcite dissolution in the reservoir causes hydrogen removal and this can possibly increase pH [100]. In carbonates, when the pH of the solution pH is higher than the (PZC),[a]

[a] Point of zero charge.

a change will occur in the rock surface charge, consequently, expansion of double-layer occurs and eventually it will lead to wettability alteration [26]. On the other hand in a low pH system (pH < 5.5) which consists of crude oils that have high base numbers, dilution of water results in an oil-wet system [60].

7.5.5 Double-layer expansion

EDL is used to describe electrostatic reactions and has a unique application in smart water injection [60]. EDL expands due to a decrease in ionic strength and enhances water-wetness by creating and increasing desorption between oil and calcite. Consequently, a charged colloidal particle forms two charged layers around its surface called EDL. According to Fig. 7.12, the inner layer is called the stern layer and the outer layer, which consists of loosely bond ions, is called the diffuse layer [8, 61]. If the surface and the oil charge are unlike, it will lead to the oil-wetness of the rock surface, as explained in the Wettability alteration section. Fig. 7.15A shows the wettability condition of carbonate rock in a natural state, which demonstrates that when the rock surface and oil molecule charges are unlike, the rock is oil-wet. As smart water is injected, the polarity of the brine-rock and oil-brine interface becomes equal, resulting in increased repulsion between rock and oil (Fig. 7.15C). This condition is accompanied by EDL expansion and a rise in the width of the water film. As a result, the wettability of the rock surface alters from oil-wet to water-wet piecemeal [8, 22, 63, 64]. lowering the salinity of waterflooding is a very effective way in developing the aforementioned conditions [8].

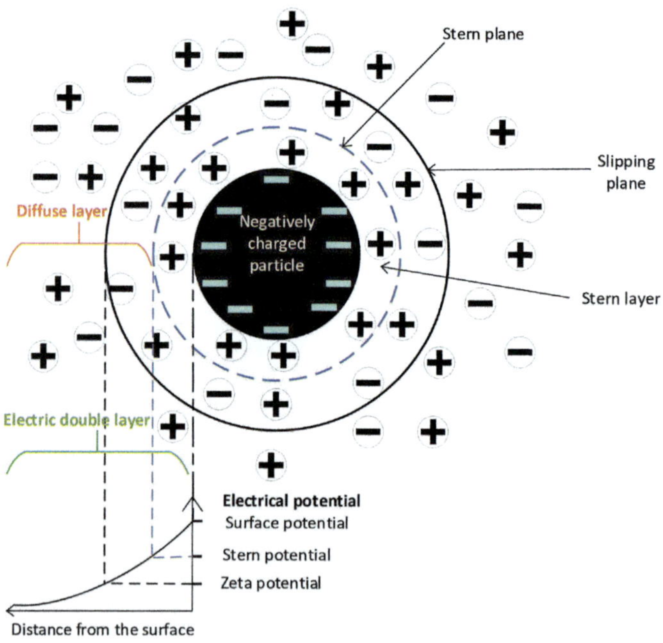

Fig. 7.12 The structure of EDL of a colloid surface, including stern and diffuse layers [101].

7.5.6 Multiple-ion exchange

Ion exchange occurs between surface active components in liquid and clay minerals [102, 103]. Ca^{2+} and Mg^{2+} as the examples of divalent cations create a bridge between the carboxylic material and the clay surface to separate them from each other to expand the water-wetness of the rock surface (Fig. 7.13). The gradual separation of the oil from the clay surface caused by their increasing and continuous repulsion leads to improve oil recovery. Before smart water injection, the polar components in the crude oil and the solid-surface multivalent cations form an organometallic complex, which often causes the rock surface to become more oil-wet. Oil-wetness is increased by the direct adsorption of polar organic material of oil on the clay surface and the replacement of unstable cations. During smart water injection, some organic polar components are eliminated from the surface and uncomplex cations present in the injected water replace the organometallic complex [103]. Another method that has led to oil-wetness is the direct adsorption on the clay mineral surface by the replacement of mobile cations on the clay surface. Lager et al. [9] showed that when multivalent cations are present in connate brine, smart water flooding improves recovery.

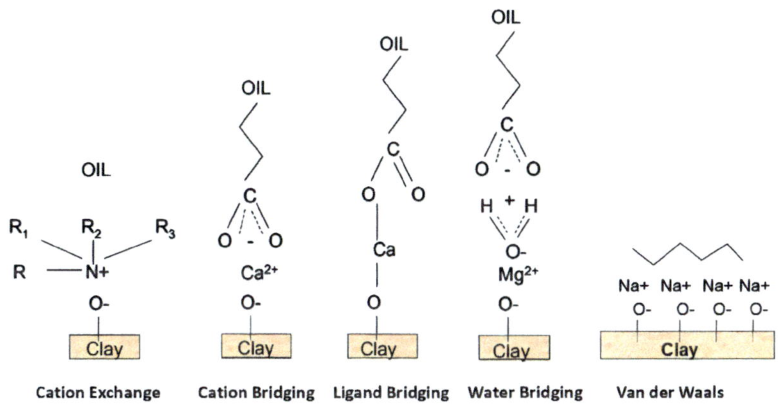

Fig. 7.13 The MIE mechanism: the placement of divalent cations such as Mg^{2+} and Ca^{2+} between oil molecules and surfaces which are both negatively charged that move the oil bonds due to the ion exchange equilibrium [56].

Another ion that has an important effect on the MIE mechanism is the sulfate ion, which is also known as the potential ion. This ion changes wettability by loosening the calcium bonds between the oil and the rock surface. Water-wet condition finally improves in the mechanism by removing/ weakening the bonds between oil and the rock surface [56, 104].

7.5.7 Salting effect

Salting-in and salting-out effect for salts (e.g., NaCl, KCl, $CaCl_2$, and $MgCl_2$) can cause wettability alteration. As the salinity in the aqueous phase increases, salt ions are more likely to migrate to the bulk of water, which increases the salt concentration in the oil/water interface (Fig. 7.14A), Consequently, the organic components tend to move toward the oil phase as a result of solubility reduction in the aqueous phase and this process is called the salting-out effect [105].

On the other hand, a reverse effect called salting in (Fig. 7.14B) occurs as salt is removed from the aqueous phase, which enhances the solubility of organic materials. The organic materials have a strong bond with the surface of carbonate rocks. On the other hand, in sandstones, this bond is weaker with the clay surface, therefore salting-in effect is not functional for carbonates as appose to sandstones where it is found to be effective. In sandstones, as salt is removed from the system cations which act as a bridge between the clay surface and organic material are desorbed which leads to the desorption of organic materials. Eventually, a positive effect on wettability alteration occurs [10].

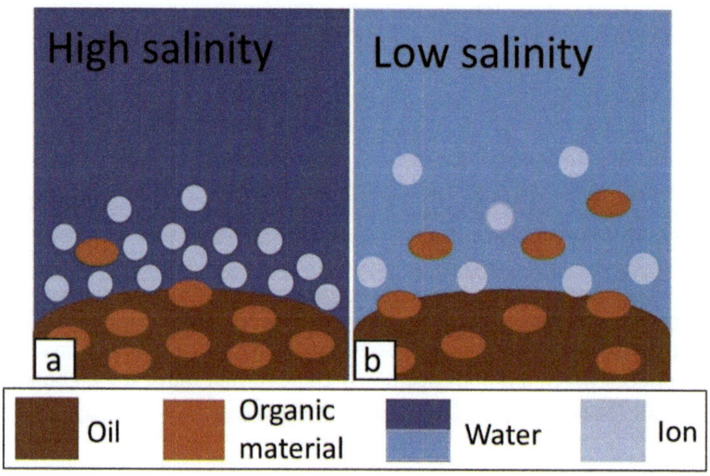

Fig. 7.14 Salting effect at the oil/brine interface (A) salting-out effect and (B) salting-in effect.

7.5.8 Wettability alteration

Wettability alteration is usually not evaluated as an independent mechanism but is evaluated by a set of mechanisms and factors that lead to wettability alteration. In fact, it can be said that the change in wettability is the end result of other mechanisms [62]. Wettability alteration is generally dependent on the salinity of the injected fluid, oil, and brine composition, and type of rock minerals.

Carbonate reservoirs mainly have a positive surface charge and formation brine with pH less than 8–9, which is why these reservoirs are considered oil-wet [23]. The oil-wetness of these reservoirs causes an average recovery of less than 30% [33]. As shown in Fig. 7.15, oil-wet alters to water-wet at pH > 9, while water-wet to intermediate-wet changes at pH < 9. In smart water flooding, the pH is mostly below 9 [106]. Generally, in a carbonate reservoir, charges of crude oil and the surface of the rock will attract, and the surface will become oil-wet. Smart water injection changes the wettability to water-wet by increasing the negative surface charge that causes a gradual repulsion and by creating films of residual water (Fig. 7.15) [103, 107]. Components of oil such as carboxylic material, which have a negative charge and are also the most determining factors in the wettability of carbonate systems, make a strong bond with the positive surface of the rock and form an oil-wet surface by forming an electrostatic force [23, 33]. Acid number is the main parameter in determining the wettability of rocks [108]. One of the issues with carboxylic material is the effect of their acid number (AN) on wettability. For example, Fig. 7.16 indicates an inverse relationship between AN and the oil recovery on chalk. As it is known, increasing the acidity of oil components has caused them to be absorbed more to the positively charged surface of carbonate rocks and has led to further

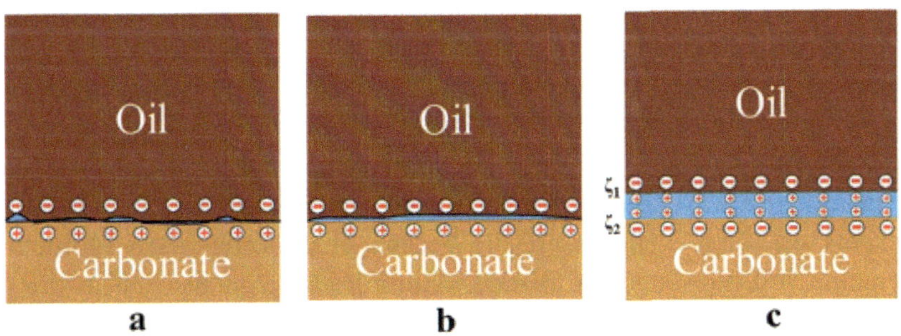

Fig. 7.15 The process of wettability alteration by smart water injection [110].

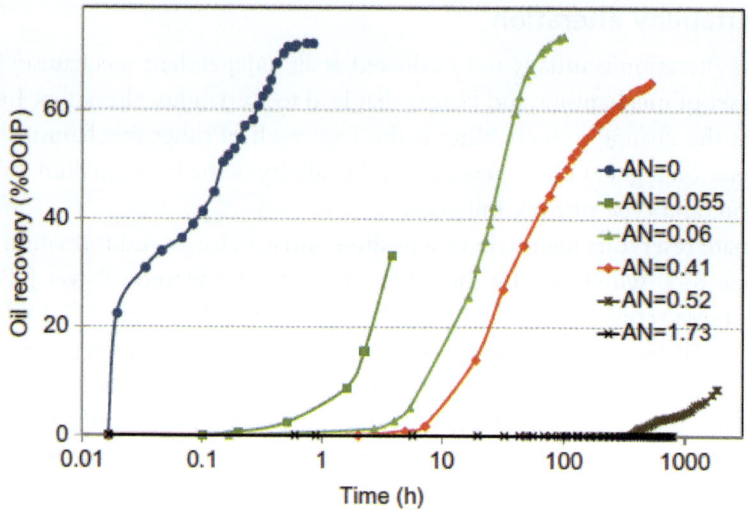

Fig. 7.16 Spontaneous imbibition on chalk cores saturated with oils with different AN [109].

oil-wet surface and consequently reduced oil recovery [33, 37, 51, 109]. In smart water injection, an attempt is made to move the wettability to lower oil-wet or more water-wet. For instance, in research by McGuire et al. [43], injecting smart water (1500 ppm) into a sandstone reservoir reduced the IFT between oil and water and increased the pH to more than 9. These changes led to wettability alteration from oil-wet to water-wet and increased the oil recovery. Sharma and Filoco [72] found that cores with lower calcium concentrations were more water-wet. In cores with higher initial salinity, due to more water-wetness, smart water injection is more effective and oil recovery is higher [62].

Formation water injection may reduce or increase oil recovery depending on the stability of the Crude oil/Brine/Rock (CBR)system, while LSW injection can significantly increase the recovery of carbonate and sandstone reservoirs by wettability alteration (Fig. 7.17) [33]. When the rock is in the mixed-wet state, in addition to water filling events, oil filling events also occur, because at the same time with the formation of water films, oil films are also formed, which leads to an increase in the oil phase communication [68]. Therefore, in the injection of smart water, the goal is to alter wettability toward mixed-wet so less oil remains on the rock surface, compared to a strongly water-wet condition, and finally, oil recovery increases [33, 62, 111]. In some studies, the oil-wet condition alters to mixed-wet due to LSWF in the carbonate reservoirs [112, 113].

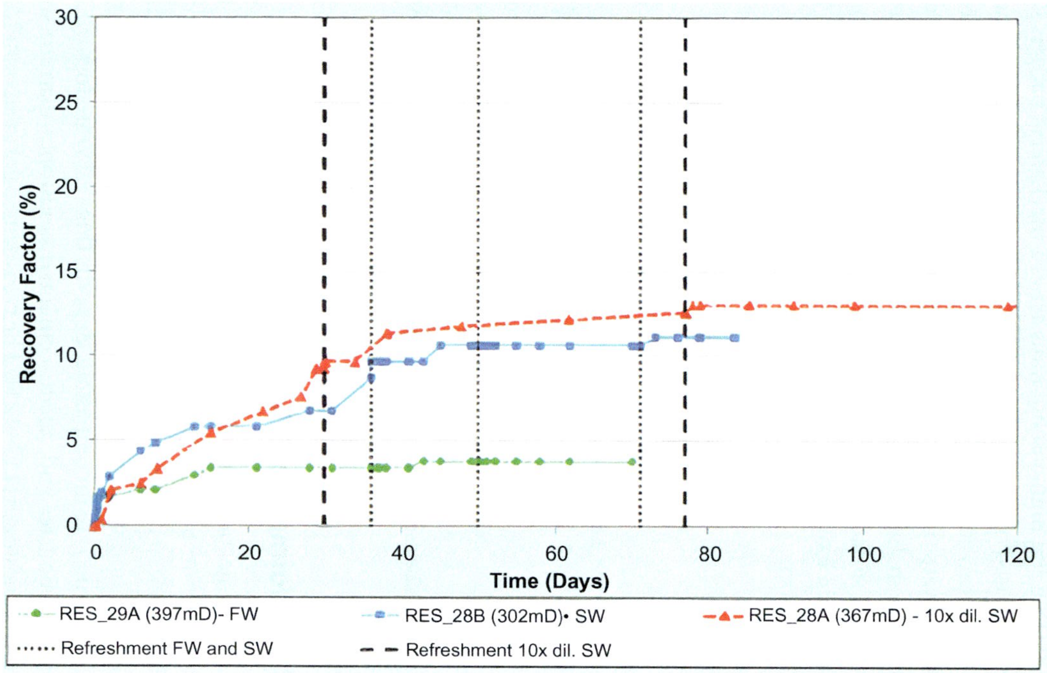

Fig. 7.17 Significant difference between spontaneous imbibition formation water and smart water (seawater) on limestone core recovery [114].

7.5.9 Mobilization of oil on solid

One of the most important effects of smart water injection is the creation of "oil mobilization based on coalescence" mechanism. The results of some studies show that ion-determining potentials including Ca^{2+}, Mg^{2+}, and SO_4^{2-} contribute to oil mobilization by reducing the stability of the oil-water contact surface, especially in carbonate rocks [115,116]. Some stable and high-viscose films act as a mechanical barrier, slowing down the coalescence of released oil droplets. Smart water injection allows penetration and thus the reaction of water ions at the rock/fluid interface by weakening the stability of interfacial films at the oil/water interface. Therefore, oil droplets attached to the rock surface are separated due to wettability alteration. On the other hand, these ions reduce the viscosity of water and oil surface, which increases the probability of coalescence of released oil droplets, resulting in easy mobilization of oil and increased recovery [115]. Fig. 7.18 clearly shows this process.

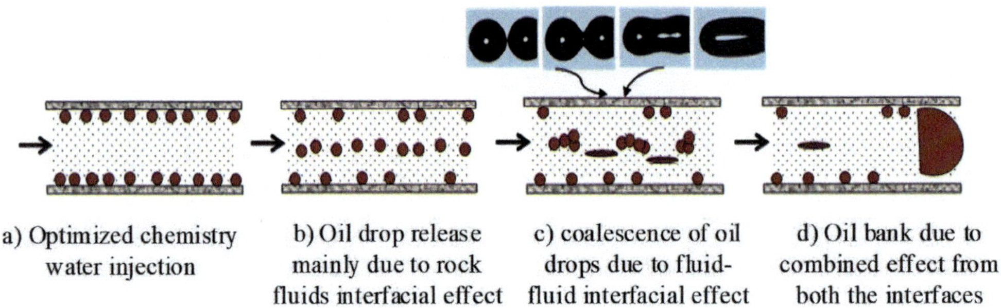

Fig. 7.18 The process of coalescence of released oil droplets by smart water injection [115].

7.6 Injected and formation brine interaction

During a smart water injection, formation and injected brine with different compositions are mixed in the reservoir. Seawater is generally high in sulfate ions, carbonate formation water has a high amount of divalent and trivalent ions such as Ba^{2+}, Sr^{2+}, Ca^{2+}, and Fe^{3+} while sandstone formation water is rich in Ba^{2+} and Sr^{2+} cations. If the two waters (formation and injected brine) are incompatible, deposits (Fig. 7.19) are formed. Calcium sulfate precipitates in carbonates and calcite-cemented sandstones, barium sulfate, and/or strontium deposit in sandstones [117–121]. The two main scales in an oilfield are carbonate scales which are due to variation of pressure and pH and sulfate scales which are due to mixing of incompatible brine [122–126]. These deposits reduce permeability and can damage the rock quality. Therefore to achieve a successful smart water injection it is important to test the compatibility between the two waters [127]. A reliable method to do so is the saturation index (SI) which calculates the solubility of the minerals and predicts if each mineral in the solution (consisting of formation an injected brine) is going to deposit or not [128]. SI is expressed as

$$SI = \log \frac{[Me][An]}{K_{sp}} \tag{7.3}$$

where [Me] is the concentration of metal (e.g., Ca^{2+}), [An] the concentration of anion (e.g., SO_4^{2-}), and K_{sp} is the solubility product.

The solution and the solid scale are at equilibrium when $SI = 0$ and nonscaling condition is met when $SI < 0$. In the above equation k_{sp} is a function of pressure, temperature, and ionic strength [118].

Recent studies have investigated how active ions (Mg^{2+}, Ca^{2+}, SO_4^{2-}) and salinity of injected brine impact calcium sulfate precipitation, results show that magnesium in injected water can decrease the formation of calcium sulfate, on the other hand, the presence of Ca^{2+} can increase precipitation of $CaSO_4$, in addition, the reduction of salinity in the injected water can decrease precipitation [130–133].

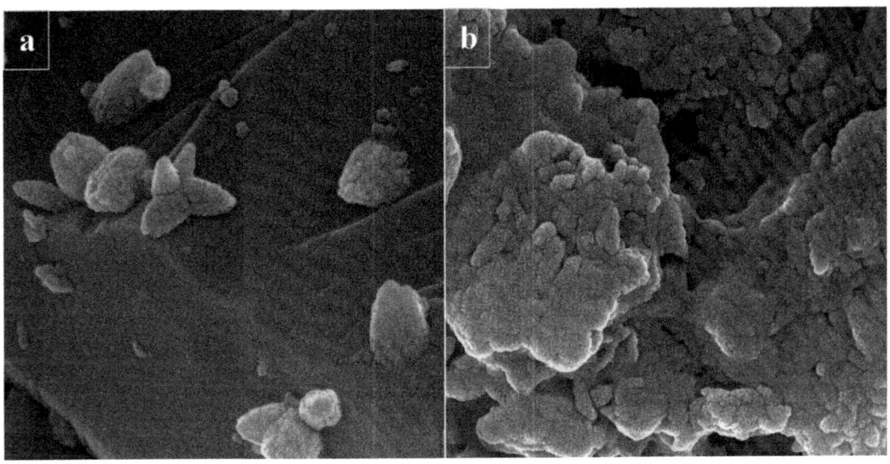

Fig. 7.19 FESEM image (A) before scale deposition and (B) after scale deposition on rock surface [129].

7.7 Optimum salinity

A water salinity concentration in which the lowest IFT between water/oil is reached, and wettability of rock alters to a range between weak water-wet and neutral-wet is called the optimum salinity [105]. This salinity of smart water injection will help obtain maximum oil recovery [134]. Therefore researchers have attempted to find an optimal salt concentration for monovalent (NaCl and KCl) and divalent ($CaCl_2$, $MgCl_2$, and Na_2SO_4) ions in which the mentioned criteria are met [27]. This optimum salinity is achieved by experimental studies using different salt concentrations. For example, Al-Shalabi et al. [135]

and Yousef et al. [59] have concluded that 10 times diluted seawater is considered an optimum level of salinity for smart water injection in carbonate reservoirs.

As illustrated in Fig. 7.20 by reducing salt concentration in any solution, IFT is lowered to a minimum point and then increases again, therefore the salt concentration in the IFT minimum point is the optimum salt concentration [27].

An optimum salinity is also introduced for scaling problems during smart water injection, generally by dilution of seawater, minimum scale is deposited. In order to design an optimum concentration for smart water, ions with higher and faster deposition rates must be recognized [129,131].

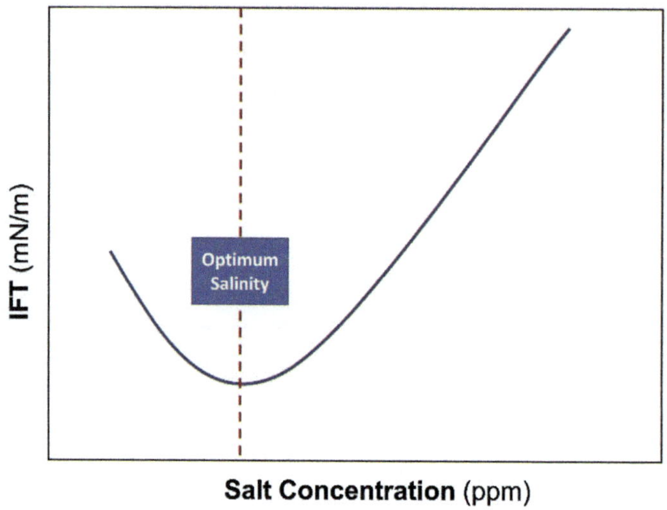

Fig. 7.20 IFT changes by increasing salinity.

7.8 Zeta potential

In various studies, wettability alteration has been expressed as the most common mechanism of smart water [2,6,7,136–138]. Zeta potential is one of the effective parameters in wettability alteration as well as fluid/-fluid interaction in smart water injection [139–144]. Molecular interaction between dissimilar surfaces in contact with each other creates an electric potential known as the zeta potential due to surface charges.

Jackson et al. showed that in order to achieve the optimal composition as well as the effectiveness of smart water injection in line with wettability alteration and increased oil recovery in carbonate reservoirs, the zeta potential for both oil-water and rock-water interfaces must have the same polarity. This means that when the oil-water surface potential is negative, a composition of injected water that creates a negative potential at the rock surface will further improve oil recovery. Jackson et al. results showed that the stability of the thin water film on the rock surface can cause more water-wetness and increase oil recovery. The stability of a wetting film is affected by the result of Van der Waals forces, structural forces, and electrostatic forces, which is expressed as disjoining pressure [140].

Van der Waals force is always attractive that causes destabilization of the water film due to the negative effect on disjoining pressure, but structural force always is repulsion that has a positive effect on the disjoining pressure due to stabilize the water film. The electrostatic force can be attractive or repulsive depending on the zeta potential of the oil-brine and rock-brine surfaces and has a different effect on the disjoining pressure. Therefore, when the electrostatic force is attractive, it stabilizes the water film due to the same polarity of the oil-brine and rock-brine surfaces [140] (Fig. 7.21).

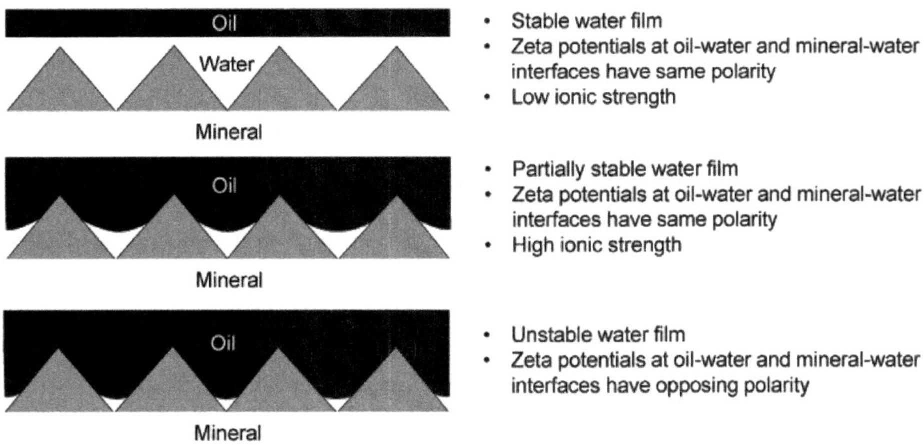

Fig. 7.21 The effect of zeta potentials on the stability of water film [140].

7.9 Dynamic investigation of contact angle and interfacial tension

Crude oil/brine/rock interactions lead to the development of a set of mechanisms that result in the separation of oil from the rock surface minerals [64]. Honarvar et al. [27] investigated the impact of various ion concentrations on the CBR system using contact angle (CA) and IFT. The results showed the highest decrease in IFT for $CaCl_2$ salt among the five salts studied. However, as mentioned in Section 7.7, increasing the concentration of all solutions to reach an optimal point reduces IFT and then increases. The effect of temperature on IFT was more than the effect of pressure, so that increasing pressure had a small negative effect on IFT while decreasing temperature had a significant negative effect on smart water injection performance. The effect of time on IFT was also investigated. Fig. 7.22 shows the IFT decrease during the time for Na_2SO_4, and finally, it remains at the steady condition after achieving 1000 ppm concentration of it.

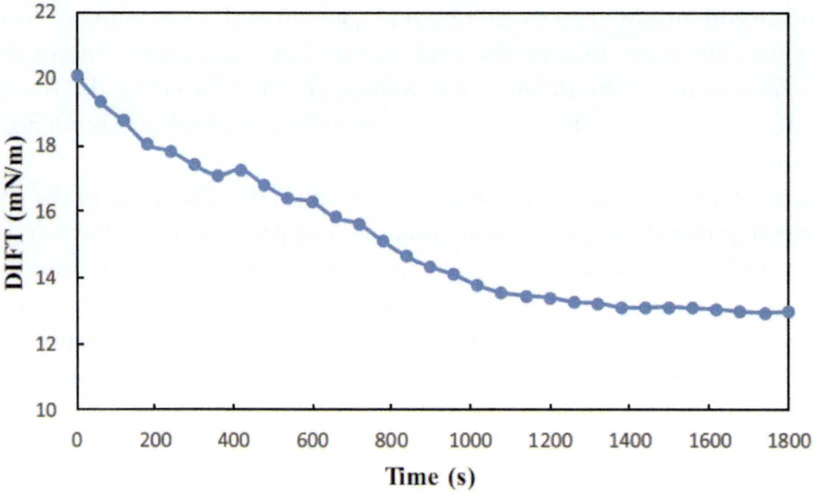

Fig. 7.22 The trend of dynamic interfacial tension (DIFT) changes during the time for Na_2SO_4 [27].

In another study, Safari et al. [145] tried to alter the wettability of a reservoir rock with shale sandstone lithology. Safari et al. investigated NaCl and $MgCl_2$ by measuring IFT and CA. Decreasing concentration of multivalent ions reduces wettability of shale from neutral to slightly water-wet (CA = 63) and decreasing divalent ions in an oil-wet shale CBR system first increases and then decreases the CA. They expressed optimum concentrations of NaCl and $MgCl_2$ for the surface of shale at 1000 ppm and 8000 ppm, respectively. The effect of decreasing the concentration of these ions on IFT results was as expected to be a decreasing trend. As shown in Fig. 7.23, increasing aging time of NaCl solution with shale surface also reduced CA and consequently wettability alteration.

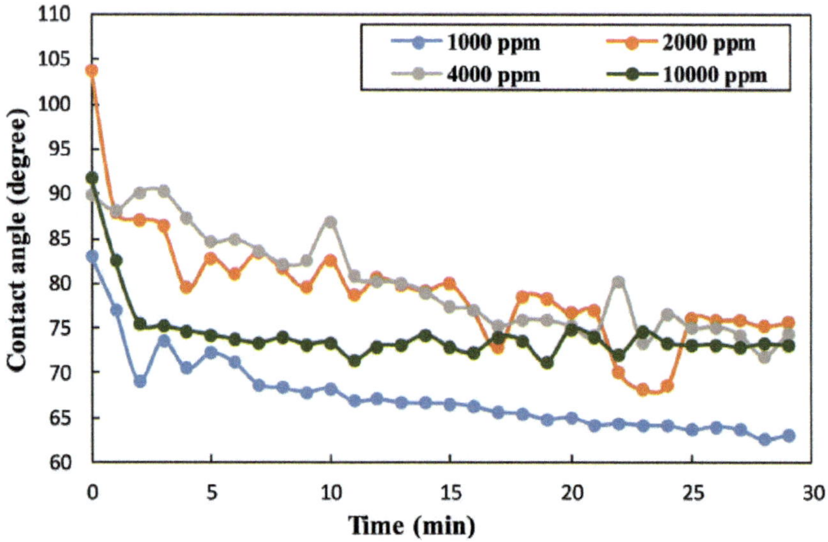

Fig. 7.23 Effect of time on the CA in the different concentration of NaCl solution [145].

7.10 Heterogeneity and fluid diversion

Core-scale smart water injection is usually successful, but field-scale risk increases when spatial distribution of petrophysical characteristics is unclear. Therefore, heterogeneity has a great impact on flooding and fluid behavior in the reservoir and eliminates the possibility of proper interpretation [146–151]. Many studies report an increase in oil recovery due to smart water injection, however, others have not shown a change in recovery [152,153] due to differences in mineralogy [48], and oil and water contents and components [154] in different fields.

Al-Ibadi et al. [146] examined the most important factors influencing heterogeneity. Using modeling, they showed that spatial distribution of porosity and permeability affects fluid performance, and the larger the permeability CV,[b] the weaker the effectiveness of smart water injection. Another factor that highlights the impact of heterogeneity is the effective salinity. Al-Ibadi et al. pointed out that the heterogeneity effect is reduced by a wide effective salinity range. Consequently, since this range is less for sandstone cores, there are more challenges in terms of adverse heterogeneous effects in such reservoirs (sandstone). Hao and Shapiro [155], evaluated the effect of compaction on several chalk cores with varying degrees of heterogeneity during smart water injection. For outcrop cores, oil production from heterogeneous cores was 2.5 times higher than from homogeneous cores; the reason for this increase in production, in addition to compaction, was fluid diversion caused by fines migration and micro-pores closure. For reservoir cores, an additional 1.1% recovery was observed for the heterogeneous core while no additional oil was produced for the homogeneous core. A great connection was observed between these cores (reservoir cores) and the fluid diversion mechanism, such that certain degrees of heterogeneity led to the occurrence of the mentioned mechanism and increased recovery.

7.11 Effect on relative permeability curve

To ensure the occurrence of wettability alteration due to smart water injection, we can conduct relative permeability measurements. Brooks and Corey model which is an analytical widely used model for a water-oil system [156] can be used for further estimations. If the relative permeability curve shifts in a direction that can be interpreted as more water-wet (toward the right) then wettability alteration occurs [21]. This alteration is observed as a decrease in the end point water relative permeability [81] and an increase in end point oil relative permeability [157] (Fig. 7.24).

In a smart water injection for heavy oil, by reducing water relative permeability, the mobility ratio is positively impacted and as a result, viscous fingering effects are reduced [81].

[b] Coefficient of variation.

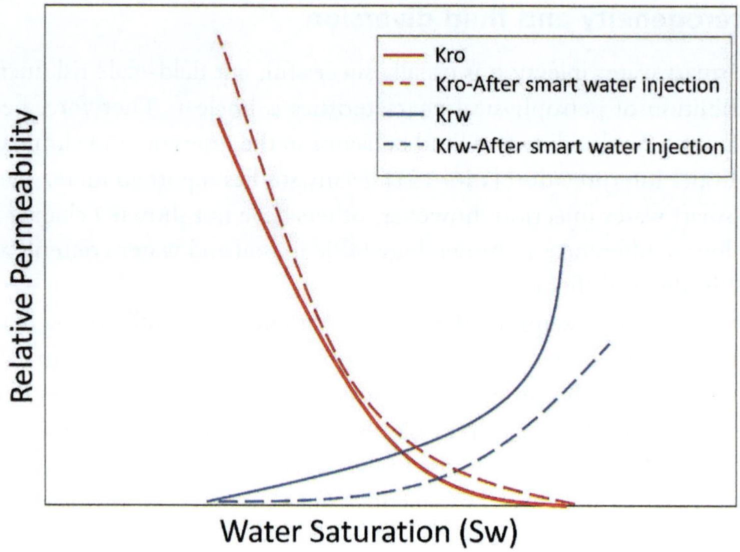

Fig. 7.24 Relative permeability curve in conventional waterflooding vs smart water injection.

7.12 Simulation

Smart water displacement through porous media can be simulated to better understand the mechanisms obtained in experimental studies and predict performance for years ahead. The first modeling of smart water injection using simulation based on wettability alteration was performed in 2006 [158]. It is a general agreement that the main mechanism behind smart water is wettability alteration. To better understand this mechanism through simulation, changes in the curvature of the relative permeability curve can be investigated [159]. Other parameters such as ion exchange solubility and geochemistry can be used for modeling [160,161]. In addition, molecular dynamic simulation can be used to thoroughly observe the interaction between rock/brine for example the effect of salinity concentration on a calcite rock can be investigated [162].

Eclipse, UT-Chem [159], DL_Poly [162], CMG_GEM [163], PHREEQC [77] are simulation programs that can be used in the topic of low salinity water injection.

7.13 Machine learning

Modifying the ionic composition of the injected fluid in an EOR process requires the knowledge of physical and chemical properties at the atomic scale [164]. Machine learning can be utilized to correlate these properties [165]. In addition, on a laboratory scale, machine learning can be used to investigate the relationship between experimental variables and incremental oil recovery in smart water injection [166]. Other applications for this method are sensitivity analysis and optimization [100]. Furthermore, there is a wide range of applications for this method in smart water injection such as forecasting permeability damage caused by scale deposition during water injection [167].

7.14 Upscaling

Simulation and investigation of various processes in field scale are always difficult and impossible due to implementation constraints, so the use of upscaling techniques is of particular importance in the development of EOR studies. The use of upscaling methods depend on solving the two main challenges: (1) using a model in terms of dimension that reflects the characteristics of the reservoir and (2) minimizing numerical dispersion, it is faced with [67].

Upscaling studies for smart water injection have not been developed and extensively researched. Al-Attar [67] used Hearn's method to study the upscaling for smart water flooding after high salinity flooding. Hearn [168] showed that pseudo-relative permeabilities for ordinary flooding assuming a constant pressure boundary condition can be generated for a two-dimensional multilayer reservoir model and used in a homogeneous one-dimensional model. Hearn's method was initially developed for high salinity flooding. Al-Attar studied the upscaling study for smart water flooding by modifying Hearn's method by Buckley Leverett. The upscaling steps in Al-Attar's study in a multilayer reservoir in which initially high salinity flooding and then smart water flooding was performed can be described as follows [67]:

Step 1: First, we calculate the water saturation profile in each PV injected using the analytical Buckley Leverett solution for each layer.

$$x_{S_w} = \frac{tv_i \dfrac{df_w}{dS_{w_{S_w}}}}{\phi} \tag{7.4}$$

where x_{sw} is the water saturation position, L, t the saturation profile calculation time, t, v_i the velocity of front, L/t, ϕ the porosity, (dimensionless), and $\frac{df_w}{dS_{w_{S_w}}}$ is the slope of the tangent line at S_w.

Step 2: In any layer, the saturation profile of water is segregated into several regions based on salinity. Fig. 7.25 shows the water saturation profile for smart water flooding for a two-layer reservoir.

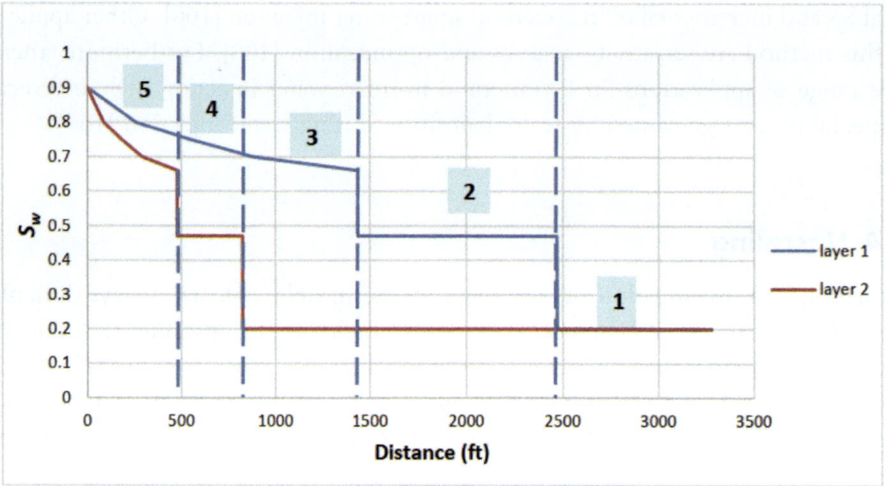

Fig. 7.25 Regions of saturation profile of water in each layer based on the salinities for low salinity water injection [67].

Step 3: Calculating the average saturation and average salinity of water using Eqs. (7.5), (7.6) for the regions determined in the second step.

$$\tilde{S_w} = \frac{\sum_{i=1}^{N} h_i \phi_i S_{wi}}{\sum_{i=1}^{N} h_i \phi_i} \tag{7.5}$$

$$\tilde{C} = \frac{\sum_{i=1}^{N} h_i \phi_i S_{wi} C_i}{\sum_{i=1}^{N} h_i \phi_i S_{wi}} \tag{7.6}$$

where $\tilde{C}$ is the average salinity of water, ppm, I the layer number, and N is the total number of layers.

Step 4: Generating pseudo-relative permeability of water and oil using Eqs. (7.7), (7.8) for the regions defined in the second step.

$$\hat{k_{rw}} = \frac{\sum_{i=1}^{N} h_i K_i k_{rwi}}{\sum_{i=1}^{N} h_i K_i} \tag{7.7}$$

$$\hat{k_{ro}} = \frac{\sum_{i=1}^{N} h_i K_i k_{roi}}{\sum_{i=1}^{N} h_i K_i} \quad (7.8)$$

where $\hat{k_{rw}}$ is the pseudo-water relative permeability and $\hat{k_{ro}}$ is the pseudo-oil relative permeability.

Flooding was performed with two fluids that had different salinities, but the pseudo-relative permeability of water and oil was calculated regardless of salinity. In the pseudo-relative permeability curve for smart water flooding, the following items are considered based on the regions specified in Fig. 7.25.

- In regions 1 and 2, $\hat{k_{rw}}$ and $\hat{k_{ro}}$ calculated by Eqs. (7.7), (7.8) will be considered only for high salinity water because in these regions there is only high salinity water.
- In regions 5, $\hat{k_{rw}}$ and $\hat{k_{ro}}$ calculated by Eqs. (7.7), (7.8) will be considered only for smart water because in this region there is only smart water.
- Regions 3 and 4 have a medium salinity that pseudo-relative permeability is calculated using the following equations.

$$\hat{k_{rw}} = \hat{k_{rwH}}\left(1 - \tilde{C}\right) + \tilde{C}\,\hat{k_{rwL}} \quad (7.9)$$

$$\hat{k_{ro}} = \hat{k_{roH}}\left(1 - \tilde{C}\right) + \tilde{C}\,\hat{k_{roL}} \quad (7.10)$$

7.15 Screening criteria

One of the most important issues before the development of an enhanced oil recovery method is screening. Numerous laboratory studies on smart water have shown that this method has good potential for increasing oil recovery in sandstone and carbonate reservoirs, but due to the complexity in the functional mechanism of smart water, there are still uncertainties. Various studies have shown that the rock-injected water-reservoir fluid interaction is a factor in changing wettability that will increase oil recovery.

Lack of implementation of smart water in field scale has caused insufficient data to be available for smart water screening. Therefore, based on various laboratory and field pilot studies in different reservoirs, important parameters for the effectiveness of smart water injection can be expressed as shown in Table 7.4.

Table 7.4 Smart water injection screening parameters [169,170].

Reservoir
• Sandstones • Carbonates
Clay minerals
• Presence of sufficient clay in the reservoir
Reservoir minerals
• Calcite • Dolomite
Crude oil
• Presence of polar components (asphaltenes, resins, acids, bases) in crude oil • Acid-base interactions for systems oil/brine/rock
Reservoir depth and reservoir temperature
• Not limited
S_{or} after high salinity waterflooding
• S_{or} should be high enough
Primary wettability of reservoir rock
• Oil-wet • Mixed-wet
Formation water
• Must be included divalent cations (Ca^{2+}, Mg^{2+})
Injected water
• Salinity concentration is below certain amount (<5000 ppm) • Salinity concentration is above certain amount (>critical flocculation concentration) • Must be included divalent ions • Injected water must increase the adsorbtion of divalent ions

7.16 Field study

Continuous production of an oil reservoir requires knowledge of reservoir characteristics and its conditions. So far, according to laboratory data, smart water injection has been successful in increasing oil recovery. The good potential of the smart water injection method in laboratory studies has led most of the world's major oil companies to plan smart water injection field studies. Studies of smart water injection by conducting pilot tests shown in Table 7.5 illustrate that smart water injection will have good potential in increasing oil recovery, especially in sandstone reservoirs.

In 2012, the department of UK Energy reported the status of research records on low-salinity water. One of the most important results of this report is that low salinity water injection has not yet become a mature technology in the world and no field experience has been reported, but all major companies in the world have a strong desire to develop it. Evidence shows that numerous laboratory studies on the effect of injected water composition on improving oil recovery from distant years, as well as unpublished studies

Table 7.5 Summary of field studies of smart water injection.

References	Year	Location	Type of rock	Incremental oil recovery (%OOIP)
Webb et al. [171]	2004	–	Sandstone	20–50
McGuire et al. [43]	2005	Alaska's North slope (Ivishak, Kuparuk, and Kekiktuk formation)	Sandstone	8–19
Robertson [1]	2007	Powder River basin of Wyoming	Limestone and sandstone	Recovery tends to decrease as the salinity ratio increases
Lager et al. [172]	2008	Alaskan oil field	Sandstone	10
Seccombe et al. [173]	2010	Endicott field (Alaska)	Sandstone	13
Veledder et al. [174]	2010	Omar field (Syria)	Sandstone	10–15
Skrettingland et al. [152]	2011	Snorre oil field (North Sea, Norwegian)	Sandstone	Ineffective
Yousef et al. [175]	2012	Tow Saudi Arabian reservoirs	Carbonate	Reduction of 7 units S_{or}
Abdulla et al. [176]	2013	Greater Burgan (Kuwait)	Sandstone	23.7
Callegaro et al. [177]	2014	West Africa	Sandstone	Ineffective
Akhmetgareev and Khisamov [41]	2015	Pervomaiskoye oil field (Russian)	Sandstone	5–9
Bedrikovetsky et al. [178]	2015	Bastrykskoye and Zichebashskoe oil field (Russian)	Sandstone	Low incremental recovery
Al-Qattan et al. [179]	2018	Greater Burgan (Kuwait) (Wara reservoir)	Sandstone	11.3

by British Petroleum from the late 20th century, led to the first field pilot test of this method in a single well in 2004. Ongoing studies and assurance of the efficiency of this method led to the testing of a multiwell field pilot in 2010, and all of these studies confirmed that reducing the salinity of injected water could increase the oil recovery factor and lead to increased oil production. In October 2011, British Petroleum approved the first EOR project by low-salinity water injection in the Claire Ridge field in the North Sea. According to the available records, to approve the implementation of this project, more than 10 years of continuous fundamental research on various aspects of this method, conducting and analyzing tests of special properties of rock, conducting and reviewing of core flood tests, extensive simulation studies and prediction of different conditions, sensitivity analysis, and economic studies have been performed [180].

Most field studies shown in Table 7.5 for smart water injection have shown significant results in increasing recovery, but there are some cases where smart water injection has not made a significant difference. The main reason for the failure of smart water injection in the Snorre oil field and West Africa is the lack of necessary conditions for wettability alteration.

Evaluation of smart water injection in Snorre oil field showed that the reasons for the failure of smart water injection were the lack of sufficient clay and that the reservoir was weakly water-wet. The presence of clay in the reservoir is considered an effective parameter for the effectiveness of smart water injection. Jerauld et al. [158] showed that for smart water flooding and pilot tests, the increase in oil recovery is directly related to the amount of clay in the system. Most studies mentioned the change of reservoir rock wettability to water-wet as a mechanism for smart water injection and increased oil recovery. The weakly water-wet rock in the Snorre oil field has been mentioned as another factor in the failure of smart water injection. In smart water injection, one of the parameters that affect efficiency is initial wettability before injection. Various studies have shown that in smart water injection, the oil-wetness of the reservoir rock can significantly impact the recovery increasing mechanisms and cause oil recovery due to smart water injection to increase significantly [152,158,170].

Callegaro et al. stated that the two ineffectiveness factors of smart water injection in West Africa are the mineralogy of the reservoir rock and the low S_{or} after high salinity water flooding [177].

7.17 Success rate

There have been multiple field studies that resulted in the successful performance of smart water injection (Table 7.5). This accomplishment is believed to be the result of mechanisms (e.g., multicomponent ion exchange and fine migration) that alter wettability from mixed-wet conditions to water-wet [181]. Moreover, the positive impact of smart water depends on oil and reservoir rock quality, consequently, the initial wetting condition of the reservoir is highly important [152]. Generally, compared to water injection, smart water has increased oil recoveries up to 30% and has decreased residual oil saturation from 2% to 50% in field trials [182].

7.18 Field challenges

Low salinity injection might cause formation damage, specifically for clay, the injection design should be in a way to avoid this damage. Usually, in a smart water field project, incremental oil is estimated based on restricted data which could be a challenge.

Particles present in the formation water can lead to formation plugging and reduction of injectivity. Reverse osmosis units (RO) can be used as a treatment for this issue [183]. Another reason for loss of injectivity is clay swelling, coreflooding can be utilized to examine the effects of swelling by evaluating brine reduction and permeability changes. Produced water management can also be a challenge in offshore and onshore fields [184].

Field trial data are an essential part of decision-making regarding an EOR method. For smart water injection, these data may be incurred due to the challenging environment of a reservoir, particularly in offshore fields where these trials might be costly or challenging to perform.

7.19 Operation problems

Generally, compared to other chemical EOR processes, smart water injection can be operated in harsh conditions as it does not need significant amounts of imported chemicals and complex mixing operations [43]. However, the produced water will need treatment and this requires some challenges related to system design in order to remove specific components [185]. A reverse osmosis (RO) which removes sodium chloride and sulfate can be used for the desalination process. This unit has low efficiency and requires large amounts of intake water which leads to challenges on how to dispose of the rejected stream of water. In addition, this process is highly energy consuming [183].

Based on field trials that have been performed for smart water it can be concluded that challenges associated with operation and technology are manageable [184].

7.20 Economic and environmental feasibility

As mentioned in the previous section, costly or toxic chemicals are not required for smart water injection. This is economically and environmentally appealing [174,185]. Moreover, capital and operational costs are lower compared to other methods. Additional costs can be optimized by considering factors such as accessibility of water sources and designing a produced water reinjection system [186].

Disposal of produced water should be planned to meet the environmental requirements, depending on the environmental conditions, reinjection or other treatment methods are utilized [184]. For example, in Russia, it is prohibited to inject produced water to avoid freshwater pollution [41].

References

[1] Robertson EP. Low-salinity waterflooding to improve oil recovery-historical field evidence. In: SPE annual technical conference and exhibition [Internet], Anaheim, California, U.S.A; 2007. https://doi.org/10.2118/109965-MS.

[2] Ligthelm DJ, Gronsveld J, Hofman J, Brussee N, Marcelis F, van der Linde H. Novel waterflooding strategy by manipulation of injection brine composition. In: EUROPEC/EAGE conference and exhibition [Internet]. Amsterdam, The Netherlands: Society of Petroleum Engineers; 2009. https://doi.org/10.2118/119835-MS.

[3] Gupta R, Smith GG, Hu L, Willingham T, Lo Cascio M, Shyeh JJ, et al. Enhanced waterflood for carbonate reservoirs—impact of injection water composition. In: SPE Middle East oil and gas show and conference, [Internet]. Society of Petroleum Engineers; 2011. https://doi.org/10.2118/142668-ms.

[4] Yousef AA, Al-Saleh S, Al-Jawfi MS. Improved/enhanced oil recovery from carbonate reservoirs by tuning injection water salinity and ionic content. In: SPE improved oil recovery symposium [Internet]. Tulsa, Oklahoma, USA: Society of Petroleum Engineers; 2012. https://doi.org/10.2118/154076-ms.

[5] Gachuz MH. Investigation of waterflood improvement for heavy oil carbonate reservoirs-crude oil/brine/rock interactions. Heriot-Watt University; 2016.

[6] Austad T, RezaeiDoust A, Puntervold T. Chemical mechanism of low salinity water flooding in sandstone reservoirs. In: SPE improv oil recover symp [Internet]. Tulsa, Oklahoma, USA: Society of Petroleum Engineers; 2010. https://doi.org/10.2118/129767-MS.

[7] Tang GQ, Morrow NR. Salinity, temperature, oil composition, and oil recovery by waterflooding. SPE Reserv Eng 1997;12:269–76 [Society of Petroleum Engineers (SPE)].

[8] Lager A, Webb KJ, Black CJJ. Impact of brine chemistry on oil recovery. In: IOR 2007—14th Eur symp improv oil recover [Internet]. European Association of Geoscientists & Engineers; 2007. https://doi.org/10.3997/2214-4609-pdb.24.a24.

[9] Lager A, Webb K, Black C, Singleton M, Sorbie K. Low salinity oil recovery—an experimental investigation1. Petrophysics 2008;49.

[10] RezaeiDoust A, Puntervold T, Strand S, Austad T. Smart water as wettability modifier in carbonate and sandstone: a discussion of similarities/differences in the chemical mechanisms. Energy Fuels 2009;23:4479–85. https://doi.org/10.1021/ef900185q [Internet. American Chemical Society].

[11] Awolayo A, Sarma H, AlSumaiti AM. A laboratory study of ionic effect of smart water for enhancing oil recovery in carbonate reservoirs. In: SPE EOR conf oil gas West Asia [Internet]. Muscat, Oman: Society of Petroleum Engineers; 2014. https://doi.org/10.2118/169662-ms.

[12] Fathi SJ, Austad T, Strand S. "Smart water" as a wettability modifier in chalk: the effect of salinity and ionic composition. Energy Fuels 2010;24:2514–9. https://doi.org/10.1021/ef901304m [Internet. American Chemical Society (ACS)].

[13] Fathi SJ, Austad T, Strand S. Water-based enhanced oil recovery (EOR) by "smart water": optimal ionic composition for EOR in carbonates. Energy Fuels 2011;25:5173–9. https://doi.org/10.1021/ef201019k [Internet. American Chemical Society (ACS)].

[14] Yi Z, Sarma HK. Improving waterflood recovery efficiency in carbonate reservoirs through salinity variations and ionic exchanges: A promising low-cost "smart-waterflood" approach. In: Abu Dhabi int pet conf exhib [Internet]. Society of Petroleum Engineers; 2012. https://doi.org/10.2118/161631-ms.

[15] Yousef AA, Al-Saleh S, Al-Kaabi AU, Al-Jawfi MS. Laboratory investigation of novel oil recovery method for carbonate reservoirs. In: Can unconv resour int pet conf [Internet]. Society of Petroleum Engineers; 2010. https://doi.org/10.2118/137634-ms.

[16] Zahid A, Shapiro A, Skauge A. Experimental studies of low salinity water flooding in carbonate reservoirs: a new promising approach. In: SPE EOR conf oil gas West Asia [Internet]; 2012. https://doi.org/10.2118/155625-MS.

[17] Zhang P, Tweheyo MT, Austad T. Wettability alteration and improved oil recovery in chalk: the effect of calcium in the presence of sulfate. Energy Fuels 2006;20:2056–62. https://doi.org/10.1021/ef0600816 [Internet. American Chemical Society].

[18] Austad T, Shariatpanahi SF, Strand S, Black CJJ, Webb KJ. Conditions for a low-salinity enhanced oil recovery (EOR) effect in carbonate oil reservoirs. Energy Fuels 2011;26:569–75. https://doi.org/10.1021/ef201435g [Internet. American Chemical Society (ACS)].

[19] Puntervold T, Strand S, Austad T. New method to prepare outcrop chalk cores for wettability and oil recovery studies at low initial water saturation. Energy Fuels 2007;21:3425–30. https://doi.org/10.1021/ef700323c [Internet. American Chemical Society (ACS)].

[20] Zhang P, Austad T. Wettability and oil recovery from carbonates: effects of temperature and potential determining ions. Colloids Surf A Physicochem Eng Asp 2006;279:179–87. https://doi.org/10.1016/j.colsurfa.2006.01.009 [Internet. Elsevier BV].

[21] Purswani P, Karpyn ZT. Laboratory investigation of chemical mechanisms driving oil recovery from oil-wet carbonate rocks. Fuel 2019;235:406–15. https://doi.org/10.1016/j.fuel.2018.07.078 [Internet. Elsevier BV].

[22] Al-Shalabi PEE. Modeling the effect of injecting low salinity water on oil recovery from carbonate reservoirs. The University of Texas at Austin; 2014.

[23] Tetteh JT, Brady PV, Barati GR. Review of low salinity waterflooding in carbonate rocks: mechanisms, investigation techniques, and future directions. Adv Colloid Interface Sci 2020;284:102253. https://doi.org/10.1016/j.cis.2020.102253 [Internet. Elsevier BV].

[24] Sohal MA, Thyne G, Søgaard EG. Review of recovery mechanisms of ionically modified waterflood in carbonate reservoirs. Energy Fuels 2016;30:1904–14. https://doi.org/10.1021/acs.energyfuels.5b02749 [Internet. American Chemical Society (ACS)].

[25] Zaeri MR, Hashemi R, Shahverdi H, Sadeghi M. Enhanced oil recovery from carbonate reservoirs by spontaneous imbibition of low salinity water. Pet Sci 2018;15:564–76. https://doi.org/10.1007/s12182-018-0234-1 [Internet. Springer Science and Business Media LLC].

[26] Al-Shalabi EW, Sepehrnoori K. A comprehensive review of low salinity/engineered water injections and their applications in sandstone and carbonate rocks. J Pet Sci Eng 2016;139:137–61. [Internet]. Available from: http://www.sciencedirect.com/science/article/pii/S0920410515301935.

[27] Honarvar B, Rahimi A, Safari M, Khajehahmadi S, Karimi M. Smart water effects on a crude oil-brine-carbonate rock (CBR) system: further suggestions on mechanisms and conditions. J Mol Liq 2020;299, 112173. [Internet]. Available from: http://www.sciencedirect.com/science/article/pii/S0167732219338383.

[28] Al-Shalabi EW, Sepehrnoori K, Pope G. Geochemical interpretation of low-salinity-water injection in carbonate oil reservoirs. SPE J 2015;20:1212–26. https://doi.org/10.2118/169101-PA [Internet].

[29] Doust AR, Puntervold T, Strand S, Austad T. "Smart Water" as wettability modifier in carbonate and sandstone. In: 15th Eur symp improv oil recover. Paris, France; 2009. p. 17–29.

[30] Ayirala SC, Yousef AA. Injection water chemistry requirement guidelines for IOR/EOR. In: SPE improv oil recover symp. Society of Petroleum Engineers; 2014.

[31] Austad T, Shariatpanahi SF, Strand S, Aksulu H, Puntervold T. Low salinity EOR effects in limestone reservoir cores containing anhydrite: a discussion of the chemical mechanism. Energy Fuels 2015;29:6903–11. https://doi.org/10.1021/acs.energyfuels.5b01099 [Internet. American Chemical Society].

[32] Rao DN. Wettability effects in thermal recovery operations. SPE Reserv Eval Eng 1999;2:420–30. https://doi.org/10.2118/57897-pa. [Internet]. Society of Petroleum Engineers (SPE).

[33] Austad T. Water-based EOR in carbonates and sandstones: new chemical understanding of the EOR potential using "smart water". Enhanc Oil Recover F Case Stud 2013;301–35.

[34] Zhang P, Tweheyo MT, Austad T. Wettability alteration and improved oil recovery by spontaneous imbibition of seawater into chalk: impact of the potential determining ions Ca2+, Mg2+, and SO42−. Colloids Surfaces A Physicochem Eng Asp 2007;301:199–208 [Elsevier].
[35] Hamouda AA, Rezaei Gomari KA. Influence of temperature on wettability alteration of carbonate reservoirs [Internet]. In: SPE/DOE symp. improv. Oil Recover. Society of Petroleum Engineers; 2006. https://doi.org/10.2118/99848-ms.
[36] Schembre JM, Tang G-Q, Kovscek AR. Wettability alteration and oil recovery by water imbibition at elevated temperatures. J Pet Sci Eng 2006;52:131–48. https://doi.org/10.1016/j.petrol.2006.03.017 [Internet. Elsevier BV].
[37] Hao J, Mohammadkhani S, Shahverdi H, Esfahany MN, Shapiro A. Mechanisms of smart waterflooding in carbonate oil reservoirs—a review. J Pet Sci Eng 2019;179:276–91. https://doi.org/10.1016/j.petrol.2019.04.049 [Internet. Elsevier BV].
[38] Bernard GG. Effect of floodwater salinity on recovery of oil from cores containing clays. In: SPE Calif reg meet. Society of Petroleum Engineers (SPE); 1967.
[39] Yildiz HO, Morrow NR. Effect of brine composition on recovery of Moutray crude oil by waterflooding. J Pet Sci Eng 1996;14:159–68 [Elsevier BV].
[40] Jadhunandan PP. Effects of brine composition, crude oil, and aging conditions on wettability and oil recovery [Ph.D. dissertation]. Socorro, NM: New Mexico Institude of Mining and Technology; 1990.
[41] Akhmetgareev V, Khisamov R. 40 years of low-salinity waterflooding in Pervomaiskoye field, Russia: Incremental oil. In: SPE Eur form damage conf exhib [Internet]. Society of Petroleum Engineers; 2015. https://doi.org/10.2118/174182-MS.
[42] Alhuraishawy AK, Bai B, Wei M, Geng J, Pu J. Mineral dissolution and fine migration effect on oil recovery factor by low-salinity water flooding in low-permeability sandstone reservoir. Fuel 2018;220:898–907. [Internet]. Available from: http://www.sciencedirect.com/science/article/pii/S0016236118301753.
[43] McGuire PL, Chatham JR, Paskvan FK, Sommer DM, Carini FH. Low salinity oil recovery: an exciting new EOR opportunity for Alaska's North Slope [Internet]; 2005. https://doi.org/10.2118/93903-MS.
[44] Soraya B, Malick C, Philippe C, Bertin HJ, Hamon G. Oil recovery by low-salinity brine injection: Laboratory results on outcrop and reservoir cores. In: SPE annu tech conf exhib [Internet]. OnePetro; 2009. https://doi.org/10.2118/124277-MS.
[45] Zhang Y, Xie X, Morrow NR. Waterflood performance by injection of brine with different salinity for reservoir cores. In: SPE annu tech conf exhib [Internet]. Society of Petroleum Engineers; 2007. https://doi.org/10.2118/109849-MS.
[46] Lebedeva E, Senden TJ, Knackstedt M, Morrow N. Improved oil recovery from tensleep sandstone—Studies of brine-rock interactions by micro-CT and AFM. In: 15th Eur symp improv oil recover 2009 [Internet]. European Association of Geoscientists & Engineers; 2009. p. 528–42. [cited 2021 Jan 27]; Available from: https://www.earthdoc.org/content/papers/10.3997/2214-4609.201404879.
[47] Xie Q, Brady PV, Pooryousefy E, Zhou D, Liu Y, Saeedi A. The low salinity effect at high temperatures. Fuel 2017;200:419–26. [Internet]. Available from: http://www.sciencedirect.com/science/article/pii/S0016236117303629.
[48] Hassenkam T, Pedersen CS, Dalby K, Austad T, Stipp SLS. Pore scale observation of low salinity effects on outcrop and oil reservoir sandstone. Colloids Surfaces A Physicochem Eng Asp 2011;390:179–88. [Internet]. Available from: http://www.sciencedirect.com/science/article/pii/S0927775711005887.
[49] Aldousary S, Kovscek AR. The diffusion of water through oil contributes to spontaneous emulsification during low salinity waterflooding. J Pet Sci Eng 2019;179:606–14. [Internet]. Available from: http://www.sciencedirect.com/science/article/pii/S0920410519303778.
[50] AlHammadi M, Mahzari P, Sohrabi M. Fundamental investigation of underlying mechanisms behind improved oil recovery by low salinity water injection in carbonate rocks. Fuel 2018;220:345–57. [Internet]. Available from: http://www.sciencedirect.com/science/article/pii/S0016236118301522.
[51] Standnes D, Austad T. Wettability alteration in chalk: 2. Mechanism for wettability alteration from oil-wet to water-wet using surfactants. J Pet Sci Eng 2000;28:123–43.

[52] Hognesen EJ, Strand S, Austad T. Waterflooding of preferential oil-wet carbonates: oil recovery related to reservoir temperature and brine composition [Internet]; 2005. https://doi.org/10.2118/94166-MS.

[53] Puntervold T, Strand S, Austad T. Water flooding of carbonate reservoirs: effects of a model base and natural crude oil bases on chalk wettability. Energy Fuels 2007;21:1606–16. https://doi.org/10.1021/ef060624b [Internet. American Chemical Society].

[54] Al SEW, Sepehrnoori K, Delshad M. Mechanisms behind low salinity water injection in carbonate reservoirs. Fuel 2014;121:11–9. [Internet]. Available from: http://www.sciencedirect.com/science/article/pii/S0016236113011897.

[55] Chandrasekhar S, Mohanty KK. Wettability alteration with brine composition in high temperature carbonate reservoirs. In: SPE annu tech conf exhib [Internet]. OnePetro; 2013. https://doi.org/10.2118/166280-MS.

[56] Seccombe JC, Lager A, Webb KJ, Jerauld G, Fueg E. Improving wateflood recovery: LoSalTM EOR field evaluation. In: SPE symp improv oil recover [Internet]. Society of Petroleum Engineers; 2008. https://doi.org/10.2118/113480-MS.

[57] Mahani H, Keya AL, Berg S, Bartels W-B, Nasralla R, Rossen W. Driving mechanism of low salinity flooding in carbonate rocks. In: Eur 2015 [Internet]. Society of Petroleum Engineers; 2015. https://doi.org/10.2118/174300-MS.

[58] den Ouden L, Nasralla RA, Guo H, Bruining H, van Kruijsdijk CPJW. Calcite dissolution behaviour during low salinity water flooding in carbonate rock. In: European association of geoscientists & engineers; 2015. [cited 2021 Jan 26]. p. cp-445-00013. Available from: http://www.earthdoc.org/publication/publicationdetails/?publication=79801.

[59] Yousef AA, Al-Saleh S, Al-Kaabi A, Al-Jawfi M. Laboratory investigation of the impact of injection-water salinity and ionic content on oil recovery from carbonate reservoirs. SPE Reserv Eval Eng 2011;14:578–93. https://doi.org/10.2118/137634-PA [Internet].

[60] Chen Y, Xie Q, Sari A, Brady PV, Saeedi A. Oil/water/rock wettability: influencing factors and implications for low salinity water flooding in carbonate reservoirs. Fuel 2018;215:171–7. [Internet]. Available from: http://www.sciencedirect.com/science/article/pii/S0016236117312747.

[61] Xie Q, Sari A, Pu W, Chen Y, Brady PV, Al Maskari N, et al. pH effect on wettability of oil/brine/carbonate system: implications for low salinity water flooding. J Pet Sci Eng 2018;168:419–25. [Internet]. Available from: http://www.sciencedirect.com/science/article/pii/S0920410518304078.

[62] Sheng JJ. Critical review of low-salinity waterflooding. J Pet Sci Eng 2014;120:216–24. https://doi.org/10.1016/j.petrol.2014.05.026 [Internet. Elsevier BV].

[63] Jackson MD, Vinogradov J, Hamon G, Chamerois M. Evidence, mechanisms and improved understanding of controlled salinity waterflooding part 1: Sandstones. Fuel 2016;185:772–93. https://doi.org/10.1016/j.fuel.2016.07.075 [Internet. Elsevier BV].

[64] Khishvand M, Oraki Kohshour I, Alizadeh AH, Piri M, Prasad S. A multi-scale experimental study of crude oil-brine-rock interactions and wettability alteration during low-salinity waterflooding. Fuel 2019;250:117–31. https://doi.org/10.1016/j.fuel.2019.02.019 [Internet. Elsevier BV].

[65] Tang G-Q, Morrow NR. Influence of brine composition and fines migration on crude oil/brine/rock interactions and oil recovery. J Pet Sci Eng 1999;24:99–111. [Internet]. Available from: http://www.sciencedirect.com/science/article/pii/S0920410599000340.

[66] Chequer L, Al-Shuaili K, Genolet L, Behr A, Kowollik P, Zeinijahromi A, et al. Optimal slug size for enhanced recovery by low-salinity waterflooding due to fines migration. J Pet Sci Eng 2019;177:766–85. https://doi.org/10.1016/j.petrol.2019.02.079 [Internet. Elsevier BV].

[67] Attar AZ. Evaluation of the impact of reservoir heterogeneity and mixing on low salinity waterflooding. Imperial College London; 2017.

[68] Bartels W-B, Mahani H, Berg S, Hassanizadeh SM. Literature review of low salinity waterflooding from a length and time scale perspective. Fuel 2019;236:338–53. https://doi.org/10.1016/j.fuel.2018.09.018 [Internet. Elsevier BV].

[69] Bennion DB, Bennion DW, Thomas FB, Bietz RF. Injection water quality-A key factor to successful waterflooding [Internet]. In: Annu. tech. meet. Petroleum Society of Canada; 1994. https://doi.org/10.2118/94-60.

[70] Blume T, Weisbrod N, Selker JS. Permeability changes in layered sediments: Impact of particle release. Ground Water 2002;40:466–74. https://doi.org/10.1111/j.1745-6584.2002.tb02530.x [Internet. Wiley].

[71] Song W, Kovscek AR. Direct visualization of pore-scale fines migration and formation damage during low-salinity waterflooding. J Nat Gas Sci Eng 2016;34:1276–83. https://doi.org/10.1016/j.jngse.2016.07.055 [Internet. Elsevier BV].

[72] Sharma MM, Filoco PR. Effect of brine salinity and crude-oil properties on oil recovery and residual saturations. SPE J 2000;5:293–300. https://doi.org/10.2118/65402-pa [Internet. Society of Petroleum Engineers (SPE)].

[73] Al-Sarihi A, Zeinijahromi A, Genolet L, Behr A, Kowollik P, Bedrikovetsky P. Effects of fines migration on residual oil during low-salinity waterflooding. Energy Fuels 2018;32:8296–309. https://doi.org/10.1021/acs.energyfuels.8b01732 [Internet. American Chemical Society (ACS)].

[74] Sarkar AK, Sharma MM. Fines migration in two-phase flow. J Pet Technol 1990;42:646–52. https://doi.org/10.2118/17437-pa [Internet. Society of Petroleum Engineers (SPE)].

[75] Valdya RN, Fogler HS. Fines migration and formation damage: influence of pH and ion exchange. SPE Prod Eng 1992;7:325–30. https://doi.org/10.2118/19413-pa [Internet. Society of Petroleum Engineers (SPE)].

[76] Grahame DC. The electrical double layer and the theory of electrocapillarity. Chem Rev 1947;41:441–501. https://doi.org/10.1021/cr60130a002 [Internet. American Chemical Society (ACS)].

[77] Tale F, Kalantariasl A, Shabani A, Abbasi S, Zohoorian AH, Khamehchi E. Experimental and simulation study of low salinity brine interactions with carbonate rocks. J Pet Sci Eng 2020;184:106497. [Internet]. Available from: http://www.sciencedirect.com/science/article/pii/S0920410519309180.

[78] Mamonov A, Puntervold T, Strand S. EOR by smart water flooding in sandstone reservoirs—Effect of sandstone mineralogy on initial wetting and oil recovery. In: SPE Russ pet technol conf [Internet]. Society of Petroleum Engineers; 2017. https://doi.org/10.2118/187839-MS.

[79] Sharma H, Mohanty KK. An experimental and modeling study to investigate brine-rock interactions during low salinity water flooding in carbonates. J Pet Sci Eng 2018;165:1021–39 [Elsevier B.V].

[80] den Ouden L, Nasralla RA, Guo H, Bruining H, van Kruijsdijk CPJW. Calcite dissolution behaviour during low salinity water flooding in carbonate rock. In: IOR 2015—18th Eur symp improv oil recover. EAGE Publications BV; 2015.

[81] Kadeethum T, Sarma HK, Maini BB. Overcome viscous fingering effect in heavy oil reservoirs by an optimized smart water injection scheme. In: SPE Canada heavy oil tech conf [Internet]. Society of Petroleum Engineers; 2017. https://doi.org/10.2118/184966-MS.

[82] Pu H, Xie X, Yin P, Morrow NR. Low salinity waterflooding and mineral dissolution. In: Proc—SPE annu tech conf exhib. Society of Petroleum Engineers (SPE); 2010. p. 1574–90.

[83] Gachuz-Muro H, Sohrabi M, Benavente D. Dissolution of rock during smart water injection in heavy oil carbonate reservoirs by natural generation of acidic water. Energy Fuels 2017;31:11852–65. https://doi.org/10.1021/acs.energyfuels.7b02163 [Internet. American Chemical Society].

[84] Mandal A, Bera A. Modeling of flow of oil-in-water emulsions through porous media. Pet Sci 2015;12:273–81. https://doi.org/10.1007/s12182-015-0025-x [Internet].

[85] Chen G, Tao D. An experimental study of stability of oil–water emulsion. Fuel Process Technol 2005;86:499–508. [Internet]. Available from: http://www.sciencedirect.com/science/article/pii/S0378382004001213.

[86] Chen K, Duan M, Fang S, Li L, Wang C, Zhang J. Nanoemulsion of AM/AA-Na/AMPS-Na aqueous solution and kerosene prepared by phase inversion emulsification and polymerization. J Dispers Sci Technol 2015;36:177–81. https://doi.org/10.1080/01932691.2014.888519 [Internet. Taylor & Francis].

[87] Kazemzadeh Y, Sharifi M, Riazi M. Mutual effects of Fe_3O_4/chitosan nanocomposite and different ions in water for stability of water-in-oil (w/o) emulsions at low–high salinities. Energy & Fuels 2018;32:12101–17. https://doi.org/10.1021/acs.energyfuels.8b02449 [Internet. American Chemical Society].

[88] Emadi A, Sohrabi M. Visual investigation of low salinity waterflooding. In: Int symp soc core anal. Aberdeen, Scotland, UK; 2012. p. 27–30.
[89] Chakravarty KH, Fosbøl PL, Thomsen K. Interactions of fines with oil and its implication in smart water flooding. In: Proceedings SPE Bergen one day semin [Internet]. Bergen, Norway; 2015. https://doi.org/10.2118/173855-MS.
[90] Chakravarty KH, Fosbøl PL, Thomsen K. Interactions of fines with base fractions of oil and its implication in smart water flooding. In: Eur 2015 [Internet]. Society of Petroleum Engineers; 2015. https://doi.org/10.2118/174335-MS.
[91] Zahid A, Sandersen SB, Stenby EH, Von Solms N, Shapiro A. Advanced waterflooding in chalk reservoirs: understanding of underlying mechanisms. Colloids Surfaces A Physicochem Eng Asp 2011;389:281–90 [Elsevier BV].
[92] Arshad MW, Loldrup Fosbøl P, Shapiro A, Thomsen K. Water-oil emulsions with fines in smart water enhanced oil recovery. In: SPE Kuwait oil gas show conf [Internet]. Society of Petroleum Engineers; 2017. https://doi.org/10.2118/187620-MS.
[93] Shindy AM, Darwich TD, Sayyouh MH, Osman A-A. Development of an expert system for EOR method selection. In: Middle East oil show conf [Internet]. Society of Petroleum Engineers; 1997. https://doi.org/10.2118/37708-MS.
[94] Babadagli T. Mature field development—a review. In: 67th EAGE conf exhib [Internet]. European Association of Geoscientists & Engineers; 2005. https://doi.org/10.2118/93884-MS.
[95] Arhuoma M, Yang D, Dong M, Li H, Idem R. Numerical simulation of displacement mechanisms for enhancing heavy oil recovery during alkaline flooding. In: Energy & fuels [Internet]. American Chemical Society, 23; 2009. p. 5995–6002. https://doi.org/10.1021/ef900690y.
[96] Bryan JL, Mai AT, Kantzas A. Investigation into the processes responsible for heavy oil recovery by alkali-surfactant flooding. In: SPE symp improv oil recover [Internet]. Society of Petroleum Engineers; 2008. https://doi.org/10.2118/113993-MS.
[97] Bryan J, Kantzas A. Enhanced heavy-oil recovery by alkali-surfactant flooding. In: Annu tech conf exhib. Anaheim, CA,: Society of Petroleum Engineers (SPE); 2007. p. 11–4.
[98] Moradi M, Alvarado V, Huzurbazar S. Effect of salinity on water-in-crude oil emulsion: evaluation through drop-size distribution proxy. Energy Fuels 2011;25:260–8. https://doi.org/10.1021/ef101236h [Internet. American Chemical Society].
[99] Jiang S, Liang P, Han Y. Effect of clay mineral composition on low-salinity water flooding. Energies 2018;11(12).
[100] Adegbite JO. Modeling, application, and optimization of engineered water injection technology in carbonate reservoirs. In: SPE annu tech conf exhib [Internet]. Society of Petroleum Engineers; 2018. https://doi.org/10.2118/194032-STU.
[101] Nasralla RA, Nasr-El-Din HA. Double-layer expansion: is it a primary mechanism of improved oil recovery by low-salinity waterflooding? SPE Reserv Eval Eng 2014;17:49–59 [OnePetro].
[102] Fang C, Yang Y, Sun S, Qiao R. Low salinity effect on the recovery of oil trapped by nanopores: a molecular dynamics study. Fuel 2020;261:116443. https://doi.org/10.1016/j.fuel.2019.116443 [Internet. Elsevier BV].
[103] Zhang T, Li Y, Li C, Sun S. Effect of salinity on oil production: review on low salinity waterflooding mechanisms and exploratory study on pipeline scaling. Oil Gas Sci Technol – Rev d'IFP Energ Nouv 2020;75:50. https://doi.org/10.2516/ogst/2020045 [Internet. EDP Sciences].
[104] Liu F, Wang M. Review of low salinity waterflooding mechanisms: wettability alteration and its impact on oil recovery. Fuel 2020;267:117112. https://doi.org/10.1016/j.fuel.2020.117112 [Internet. Elsevier BV].
[105] Rahimi A, Honarvar B, Safari M. The role of salinity and aging time on carbonate reservoir in low salinity seawater and smart seawater flooding. J Pet Sci Eng 2020;187:106739. [[Internet]. Available from: http://www.sciencedirect.com/science/article/pii/S0920410519311581.
[106] Drummond C, Israelachvili J. Surface forces and wettability. J Pet Sci Eng 2002;33:123–33. https://doi.org/10.1016/s0920-4105(01)00180-2 [Internet. Elsevier BV].
[107] Berg S, Cense AW, Jansen E, Bakker K. Direct experimental evidence of wettability modification by low salinity. Petrophysics 2010;51 [Society of Petrophysicists and Well-Log Analysts].

[108] Zhang P, Austad T. The relative effects of acid number and temperature on chalk wettability [Internet]. In: SPE int. symp. oilf. chem. Society of Petroleum Engineers; 2005. https://doi.org/10.2118/92999-ms.
[109] Standnes DC, Austad T. Wettability alteration in chalk: 1. Preparation of core material and oil properties. J Pet Sci Eng 2000;28:111–21 [Elsevier].
[110] Ding H, Rahman S. Experimental and theoretical study of wettability alteration during low salinity water flooding-an state of the art review. Colloids Surfaces A Physicochem Eng Asp 2017;520:622–39. https://doi.org/10.1016/j.colsurfa.2017.02.006 [Internet. Elsevier BV].
[111] Morrow NR, Tang G, Valat M, Xie X. Prospects of improved oil recovery related to wettability and brine composition. J Pet Sci Eng 1998;20:267–76. https://doi.org/10.1016/s0920-4105(98)00030-8 [Internet. Elsevier BV].
[112] Alameri W, Teklu TW, Graves RM, Kazemi H, AlSumaiti AM. Experimental and numerical modeling of low-salinity waterflood in a low permeability carbonate reservoir [Internet]. In: SPE west. reg. meet. Society of Petroleum Engineers; 2015. https://doi.org/10.2118/174001-ms.
[113] Kafili Kasmaei A, Rao DNN. Is wettability alteration the main cause for enhanced recovery in low-salinity waterflooding? SPE Reserv Eval Eng 2015;18:228–35. https://doi.org/10.2118/169120-pa [Internet Society of Petroleum Engineers (SPE)].
[114] Ravari RR, Strand S, Austad T. Care must be taken to use outcrop limestone cores to mimic reservoir core material in SCAL linked to wettability alteration. In: Proc 11th intenational symp reserv wettability, Calgary, CA, USA; 2010.
[115] Ayirala SC, Al-Saleh SH, Al-Yousef AA. Microscopic scale interactions of water ions at crude oil/water interface and their impact on oil mobilization in advanced water flooding. J Pet Sci Eng 2018;163:640–9. https://doi.org/10.1016/j.petrol.2017.09.054 [Internet. Elsevier BV].
[116] Shehata AMM, Alotaibi MBB, Nasr-El-Din HAA. Waterflooding in carbonate reservoirs: does the salinity matter? SPE Reserv Eval Eng 2014;17:304–13. https://doi.org/10.2118/170254-pa [Internet. Society of Petroleum Engineers (SPE)].
[117] Kiani S, Mansouri Zadeh M, Khodabakhshi S, Rashidi A, Moghadasi J. Newly prepared nano gamma alumina and its application in enhanced oil recovery: an approach to low-salinity waterflooding. Energy Fuels 2016;30:3791–7. https://doi.org/10.1021/acs.energyfuels.5b03008 [Internet. American Chemical Society].
[118] Moghadasi J, Jamialahmadi M, Müller-Steinhagen H, Sharif A, Ghalambor A, Izadpanah MR, et al. Scale formation in Iranian oil reservoir and production equipment during water injection. In: Int symp oilf scale. Aberdeen: Society of Petroleum Engineers; 2003. p. 1–14.
[119] Romanuka J, Hofman JP, Ligthelm DJ, Suijkerbuijk BM, Marcelis AH, Oedai S, et al. Low salinity EOR in carbonates. In: SPE improv oil recover symp [Internet]. Society of Petroleum Engineers; 2012. https://doi.org/10.2118/153869-MS.
[120] Abbasi P, Abbasi S, Moghadasi J. Experimental investigation of mixed-salt precipitation during smart water injection in the carbonate formation. J Mol Liq 2020;299:112131. [Internet]. Available from: http://www.sciencedirect.com/science/article/pii/S0167732219352961.
[121] Crabtree M, Eslinger D, Fletcher P, Miller M, Johnson A, King G. Fighting scale: removal and prevention. Oilf Rev 1999;11:30–45 [Elsevier, Amsterdam].
[122] Vetter OJ, Kandarpa V, Harouaka A. Prediction of scale problems due to injection of incompatible waters. J Pet Technol 1982;34:273–84. https://doi.org/10.2118/7794-PA [Internet].
[123] Liu S-T, Nancollas GH. The kinetics of dissolution of calcium sulfate dihydrate. J Inorg Nucl Chem 1971;33:2311–6. [Internet]. Available from: http://www.sciencedirect.com/science/article/pii/0022190271802051.
[124] Leone JA, Scott ME. Characterization and control of formation damage during waterflooding of a high-clay-content reservoir. SPE Reserv Eng 1988;3:1279–86. https://doi.org/10.2118/16234-PA [Internet].
[125] Lindlof JC, Stoffer KG. A case study of seawater injection incompatibility. J Pet Technol 1983;35:1256–62. https://doi.org/10.2118/9626-PA [Internet].
[126] Vitthal S, Sharma MM. A Stokesian dynamics model for particle deposition and bridging in granular media. J Colloid Interface Sci 1992;153:314–36. [Internet]. Available from: http://www.sciencedirect.com/science/article/pii/002197979290325G.

[127] Nowrouzi I, Manshad AK, Mohammadi AH. Effects of ions and dissolved carbon dioxide in brine on wettability alteration, contact angle and oil production in smart water and carbonated smart water injection processes in carbonate oil reservoirs. Fuel 2019;235:1039–51. [Internet]. Available from: http://www.sciencedirect.com/science/article/pii/S0016236118314376.

[128] Doubra P, Kamran-Pirzaman A, Mohammadi AH, Hassanalizadeh R. Thermodynamic modelling of scale (Calcite, Barite, Anhydrite and Gypsum) deposition from brine. J Mol Liq 2017;230:96–103. [Internet]. Available from: http://www.sciencedirect.com/science/article/pii/S0167732216322486.

[129] Rafiei A, Khamehchi E. Design of smart water composition based on scale minimization and its effect on wettability alteration in the presence of nanoparticles and mineral scales. J Pet Sci Eng 2021;196, 107832. [Internet]. Available from: http://www.sciencedirect.com/science/article/pii/S0920410520308925.

[130] Mirzaalian Dastjerdi A, Kargozarfard Z, Najafi B, Taghikhani V, Ayatollahi S. Microscopic insight into kinetics of inorganic scale deposition during smart water injection using dynamic quartz crystal microbalance and molecular dynamics simulation. Ind Eng Chem Res 2020;59:609–19. https://doi.org/10.1021/acs.iecr.9b05236 [Internet. American Chemical Society].

[131] Ghasemian J, Riahi S, Ayatollahi S, Mokhtari R. Effect of salinity and ion type on formation damage due to inorganic scale deposition and introducing optimum salinity. J Pet Sci Eng 2019;177:270–81. [Internet]. Available from: http://www.sciencedirect.com/science/article/pii/S0920410519301561.

[132] Ghasemian J, Mokhtari R, Ayatollahi S, Riahi S, Malekzade E. Experimental investigation of inorganic scale deposition during smart water injection—a formation damage point of view. In: IOR 2017—19th Eur symp improv oil recover [Internet]. European Association of Geoscientists & Engineers; 2017. p. 1–18. Available from: https://www.earthdoc.org/content/papers/10.3997/2214-4609.201700289.

[133] Baradari A, Hosseini E, Soltani B, Jadidi N. Experimental investigation of ionic content effect on wettability alteration in smart water for enhanced oil recovery. Pet Sci Technol 2020;1–15. https://doi.org/10.1080/10916466.2018.1550509 [Internet. Taylor & Francis].

[134] Mehraban MF, Rostami P, Afzali S, Ahmadi Z, Sharifi M, Ayatollahi S. Brine composition effect on the oil recovery in carbonate oil reservoirs: a comprehensive experimental and CFD simulation study. J Pet Sci Eng 2020;191, 107149. [Internet]. Available from: http://www.sciencedirect.com/science/article/pii/S0920410520302369.

[135] Al-Shalabi EW, Sepehrnoori K, Delshad M. Optimization of the low salinity water injection process in carbonate reservoirs. In: Int pet technol conf [Internet]. International Petroleum Technology Conference; 2014. https://doi.org/10.2523/IPTC-17821-MS.

[136] Alotaibi MB, Azmy RM, Nasr-El-Din HA. Wettability studies using low-salinity water in sandstone reservoirs. In: Offshore technol conf [Internet]. Offshore Technology Conference; 2010. https://doi.org/10.4043/20718-ms.

[137] Hosseinzade Khanamiri H, Baltzersen Enge I, Nourani M, Stensen JÅ, Torsæter O, Hadia N. EOR by low salinity water and surfactant at low concentration: impact of injection and in Situ brine composition. Energy Fuels 2016;30:2705–13. https://doi.org/10.1021/acs.energyfuels.5b02899 [Internet. American Chemical Society (ACS)].

[138] Sorbie KS, Collins I. A proposed pore-scale mechanism for how low salinity waterflooding Works. In: SPE improv oil recover symp [Internet]. Society of Petroleum Engineers; 2010. https://doi.org/10.2118/129833-ms.

[139] Alroudhan A, Vinogradov J, Jackson MD. Zeta potential of intact natural limestone: impact of potential-determining ions Ca, Mg and SO4. Colloids Surfaces A Physicochem Eng Asp 2016;493:83–98. https://doi.org/10.1016/j.colsurfa.2015.11.068 [Internet. Elsevier BV].

[140] Jackson MD, Al-Mahrouqi D, Vinogradov J. Zeta potential in oil-water-carbonate systems and its impact on oil recovery during controlled salinity water-flooding. Sci Rep 2016;6. https://doi.org/10.1038/srep37363 [Internet. Springer Science and Business Media LLC].

[141] Jackson MD, Vinogradov J. Impact of wettability on laboratory measurements of streaming potential in carbonates. Colloids Surfaces A Physicochem Eng Asp 2012;393:86–95 [Elsevier].

[142] Khalilnejad A, Moradi B. Investigation of effects of smart water injection on zeta potential and wettability of carbonate reservoirs. In: Saint Petersbg 2018. European Association of Geoscientists & Engineers; 2018. p. 1–5.

[143] Mahani H, Keya AL, Berg S, Bartels W-B, Nasralla R, Rossen WR. Insights into the mechanism of wettability alteration by low-salinity flooding (LSF) in carbonates. Energy Fuels 2015;29:1352–67. https://doi.org/10.1021/ef5023847 [Internet. American Chemical Society (ACS)].

[144] Strand S, Høgnesen EJ, Austad T. Wettability alteration of carbonates—effects of potential determining ions (Ca2+ and SO42−) and temperature. Colloids Surfaces A Physicochem Eng Asp 2006;275:1–10. https://doi.org/10.1016/j.colsurfa.2005.10.061 [Internet, Elsevier BV].

[145] Safari M, Rahimi A, Gholami R, Permana A, Siaw Khur W. Underlying mechanisms of shale wettability alteration by low salinity water injection (LSWI). J Dispers Sci Technol 2020;1–9. [Internet]. Taylor and Francis Inc.; 2020 [cited 2021 Jan 27]. Available from: https://www.tandfonline.com/doi/full/10.1080/01932691.2020.1813156.

[146] Al-Ibadi H, Stephen K, Mackay E. Heterogeneity effects on low salinity water flooding: case study. In: 82nd EAGE annu conf exhib. European Association of Geoscientists & Engineers; 2020. p. 1–30.

[147] Attar A, Muggeridge A. Low salinity water injection: impact of physical diffusion, an aquifer and geological heterogeneity on slug size. J Pet Sci Eng 2018;166:1055–70. https://doi.org/10.1016/j.petrol.2018.03.023 [Internet. Elsevier BV].

[148] Henson R, Todd A, Corbett P. Geologically based screening criteria for improved oil recovery projects [Internet]. In: SPE/DOE improv. oil recover. symp. Society of Petroleum Engineers; 2002. https://doi.org/10.2118/75148-ms.

[149] Jerauld G, Mohammadi H, Webb KJ. Interpreting single well chemical tracer tests [Internet]. In: SPE improv. oil recover. symp. Society of Petroleum Engineers; 2010. https://doi.org/10.2118/129724-ms.

[150] Jones A, Doyle J, Jacobsen T, Kjønsvik D. Which sub-seismic heterogeneities influence waterflood performance? A case study of a low net-to-gross fluvial reservoir. Geol Soc London 1995;84:5–18. https://doi.org/10.1144/gsl.sp.1995.084.01.02 [Spec Publ [Internet]. Geological Society of London].

[151] Kjonsvik D, Doyle J, Jacobsen T, Jones A. The effects of sedimentary heterogeneities on production from a shallow marine reservoir-what really matters? [Internet]. In: SPE Annu. Tech. Conf. Exhib. Society of Petroleum Engineers; 1994. https://doi.org/10.2118/28445-ms.

[152] Skrettingland K, Holt T, Tweheyo MTT, Skjevrak I. Snorre low-salinity-water injection—coreflooding experiments and single-well field pilot. SPE Reserv Eval Eng 2011;14:182–92. https://doi.org/10.2118/129877-PA [Internet].

[153] Thyne GD, Siyambalagoda Gamage PH. Evaluation of the effect of low salinity waterflooding for 26 fields in Wyoming [Internet]. In: SPE annu. tech. conf. exhib. Society of Petroleum Engineers; 2011. https://doi.org/10.2118/147410-ms.

[154] Hadia NJ, Hansen T, Tweheyo MT, Torsæter O. Influence of crude oil components on recovery by high and low salinity waterflooding. Energy Fuels 2012;26:4328–35. https://doi.org/10.1021/ef3003119 [Internet. American Chemical Society (ACS)].

[155] Hao J, Shapiro A. Effect of compaction on oil recovery under low salinity flooding in homogeneous and heterogeneous chalk. In: SPE annu tech conf exhib [Internet]. Society of Petroleum Engineers; 2019. https://doi.org/10.2118/195831-ms.

[156] Lake LW, Johns R, Rossen WR, Pope GA. Fundamentals of enhanced oil recovery. In: Society of Petroleum Engineers Richardson, TX; 2014.

[157] Rivet SM, Lake LW, Pope GA. A coreflood investigation of low-salinity enhanced oil recovery. In: SPE annu tech conf exhib [Internet]. Society of Petroleum Engineers; 2010. https://doi.org/10.2118/134297-MS.

[158] Jerauld GR, Webb KJ, Lin C-Y, Seccombe J. Modeling low-salinity waterflooding. In: SPE annu tech conf exhib [Internet]. San Antonio, Texas, USA; 2006. https://doi.org/10.2118/102239-MS.

[159] AlSofi AM, Yousef AA. Insight into smart-water recovery mechanism through detailed history matching of coreflood experiments. In: SPE Reserv Charact Simul Conf Exhib [Internet]. Society of Petroleum Engineers; 2013. https://doi.org/10.2118/166035-MS.

[160] Omekeh A, Friis HA, Fjelde I, Evje S. Modeling of ion-exchange and solubility in low salinity water flooding. In: SPE improv oil recover symp [Internet]. Society of Petroleum Engineers; 2012. https://doi.org/10.2118/154144-MS.

[161] Dang CT, Nghiem LX, Chen Z, Nguyen QP. Modeling low salinity waterflooding: Ion exchange, geochemistry and wettability alteration. In: SPE annu tech conf exhib [Internet]. Society of Petroleum Engineers; 2013. https://doi.org/10.2118/166447-MS.

[162] Koleini MM, Mehraban MF, Ayatollahi S. Effects of low salinity water on calcite/brine interface: a molecular dynamics simulation study. Colloids Surfaces A Physicochem Eng Asp 2018;537:61–8. [Internet]. Available from: http://www.sciencedirect.com/science/article/pii/S0927775717309160.

[163] Pouryousefy E, Xie Q, Saeedi A. Effect of multi-component ions exchange on low salinity EOR: coupled geochemical simulation study. Petroleum 2016;2:215–24. [Internet]. Available from: http://www.sciencedirect.com/science/article/pii/S2405656116300335.

[164] Rotondi M, Callegaro C, Masserano F, Bartosek M. Low salinity water injection: Eni's experience. In: Abu Dhabi int pet exhib conf [Internet]. Society of Petroleum Engineers; 2014. https://doi.org/10.2118/171794-MS.

[165] Kirch A, Celaschi YM, de Almeida JM, Miranda CR. Brine–oil interfacial tension modeling: assessment of machine learning techniques combined with molecular dynamics. ACS Appl Mater Interfaces 2020;12:15837–43. https://doi.org/10.1021/acsami.9b22189 [Internet. American Chemical Society].

[166] Wang L, Tian Y, Yao B, Yu X. Machine learning analyses of low salinity effect in sandstone porous media. J Porous Media 2020;23 [Begel House Inc].

[167] Ahmadi MA, Mohammadzadeh O, Zendehboudi S. A cutting edge solution to monitor formation damage due to scale deposition: application to oil recovery. Can J Chem Eng 2017;95:991–1003. https://doi.org/10.1002/cjce.22776 [Internet. John Wiley & Sons, Ltd].

[168] Hearn CL. Simulation of stratified waterflooding by pseudo relative permeability curves. J Pet Technol 1971;23(07):805–13.

[169] Chen Q, Otaibi M, Ayirala S, Yousef A. The prospects and potential opportunities of low salinity water flooding for offshore applications in sandstones. J Pet Sci Eng 2021;199:108260. https://doi.org/10.1016/j.petrol.2020.108260 [Internet. Elsevier BV].

[170] Dang CTQ, Nguyen NTB, Chen Z. Practical concerns and principle guidelines for screening, implementation, design, and optimization of low salinity waterflooding. In: SPE west reg meet [Internet]. Society of Petroleum Engineers; 2015. https://doi.org/10.2118/174008-ms.

[171] Webb KJ, Black CJJ, Al-Ajeel H. Low salinity oil recovery—log-inject-log. In: SPE/DOE symp improv oil recover [Internet]. Society of Petroleum Engineers; 2004. https://doi.org/10.2118/81460-MS.

[172] Lager A, Webb KJ, Collins IR, Richmond DM. LoSal enhanced oil recovery: evidence of enhanced oil recovery at the reservoir scale. In: SPE symp improv oil recover [Internet]. Society of Petroleum Engineers; 2008. https://doi.org/10.2118/113976-ms.

[173] Seccombe J, Lager A, Jerauld G, Jhaveri B, Buikema T, Bassler S, et al. Demonstration of low-salinity EOR at interwell scale, Endicott field, Alaska. In: SPE improv oil recover symp [Internet]. Society of Petroleum Engineers; 2010. https://doi.org/10.2118/129692-ms.

[174] Vledder P, Fonseca JC, Wells T, Gonzalez I, Ligthelm D. Low salinity water flooding: proof of wettability alteration on a field wide scale. In: SPE improv oil recover symp [Internet]. Society of Petroleum Engineers; 2010. https://doi.org/10.2118/129564-MS.

[175] Yousef AA, Liu JS, Blanchard GW, Al-Saleh S, Al-Zahrani T, Al-Zahrani RM, et al. Smartwater flooding: Industry's first field test in carbonate reservoirs. In: SPE annu tech conf exhib [Internet]. San Antonio, Texas, USA: Society of Petroleum Engineers; 2012. https://doi.org/10.2118/159526-ms.

[176] Abdulla F, Hashem HS, Abdulraheem B, Al-Naqi M, Al-Qattan A, John H, et al. First EOR trial using low salinity water injection in the Greater Burgan field, Kuwait. In: SPE Middle East oil gas show conf [Internet]. OnePetro; 2013. https://doi.org/10.2118/164341-MS.

[177] Callegaro C, Masserano F, Bartosek M, Buscaglia R, Visintin R, Hartvig SK, et al. Single well chemical tracer tests to assess low salinity water and surfactant EOR processes in West Africa. In: Int pet technol conf [Internet]. International Petroleum Technology Conference; 2014. https://doi.org/10.2523/iptc-17951-ms.

[178] Bedrikovetsky P, Zeinijahromi A, Badalyan A, Ahmetgareev V, Khisamov R. Fines-migration-assisted low-salinity waterflooding: field case analysis (Russian). In: SPE Russ pet technol conf [Internet]. Society of Petroleum Engineers; 2015. https://doi.org/10.2118/176721-ru.
[179] Al-Qattan A, Sanaseeri A, Al-Saleh Z, Singh BBB, Al-Kaaoud H, Delshad M, et al. Low salinity waterflood and low salinity polymer injection in the Wara reservoir of the Greater Burgan Field. In: SPE EOR conf oil gas West Asia [Internet]. Society of Petroleum Engineers; 2018. https://doi.org/10.2118/190481-MS.
[180] Robbana E, Buikema TA, Mair C, Williams D, Mercer DJ, Webb KJ, et al. Low salinity enhanced oil recovery—laboratory to day one field implementation—loSal EOR into the Clair Ridge project. In: Abu Dhabi int pet conf exhib [Internet]. Society of Petroleum Engineers; 2012. https://doi.org/10.2118/161750-ms.
[181] Katende A, Sagala F. A critical review of low salinity water flooding: mechanism, laboratory and field application. J Mol Liq 2019;278:627–49 [Elsevier].
[182] Awolayo AN, Sarma HK, Nghiem LX. Brine-dependent recovery processes (smart-water/low-salinity-water) in carbonate and sandstone petroleum reservoirs: review of laboratory-field studies, interfacial mechanisms and modeling attempts. Energies 2018;11. Preprints.
[183] Sorop TG, Suijkerbuijk BM, Masalmeh SK, Looijer MT, Parker AR, Dindoruk DM, et al. Integrated approach in deploying low salinity waterflooding. In: SPE enhanc oil recover conf [Internet]. Society of Petroleum Engineers; 2013. https://doi.org/10.2118/165277-MS.
[184] Reddick C, Buikema T, Williams D. Managing risk in the deployment of new technology—getting loSal EOR into the business. In: SPE improv oil recover symp [Internet]. Society of Petroleum Engineers; 2012. https://doi.org/10.2118/153933-MS.
[185] Henthorne L, Hartman M, Hayden A. Improving chemical EOR economics by optimizing water quality. In: SPE enhanc oil recover conf [Internet]. Society of Petroleum Engineers; 2011. https://doi.org/10.2118/144397-MS.
[186] Erke SI, Volokitin YE, Edelman IY, Karpan VM, Nasralla RA, Bondar MY, et al. Low salinity flooding trial at west Salym field west Salym field. In: SPE improv oil recover conf [Internet]. Society of Petroleum Engineers; 2016. https://doi.org/10.2118/179629-MS.

CHAPTER 8

A comprehensive review on the use of eco-friendly surfactants in oil industry

Hossein Mehrjoo[a], Mohsen Riazi[a,b], and Saeid Norouzi-Apourvari[a]
[a]Department of Petroleum Engineering, Shahid Bahonar University of Kerman, Kerman, Iran
[b]Enhanced Oil Recovery (EOR) Research Center, IOR/EOR Research Institute, Shiraz University, Shiraz, Iran

8.1 Overview

Only about 20%–50% of the original reserved oil can be produced by the implementation of primary and secondary oil recovery stages [1–3]. Oil adsorption on rock surfaces results in oil trapping throughout the rock pores, and consequently 50%–80% of oil is retained in the reservoir and is not produced via the primary and secondary production stages [4–6]. The reduction of oil mobility is the main reason that causes oil trapping in porous media. Reservoir depletion causes a decrease in the essential force for oil displacement. Another reason for oil trapping is alteration of wettability. As wettability of the reservoir rocks changes from water-wet to oil-wet, the adsorption of oil on the rock surface occurs, which consequently causes a reduction in the oil relative permeability [7]. Since the amount of remaining oil is high and due to the industrial developments, the world energy demand has soared and it is crucial to increase oil production from hydrocarbon reserves [8]. The future worldwide demand for energy and high costs of crude oil have resulted in the use of different enhanced oil recovery (EOR) methods [9]. Natural surfactant flooding is an environmental-friendly EOR method which is reviewed in this chapter. Different types of natural surfactants are introduced and their effects on wettability, interfacial tension, and oil recovery are discussed.

8.2 Surfactant

In 2006, the worldwide production of surfactants was around 12.5 million tons. Three million tons of surfactant were produced in Western Europe in 2007 [10]. The consumption of nonionic surfactant (polyethoxylated nonylphenol) was about 172,000 tons in the United States during 2010 [10]. Environment and health risks such as eye irritation, dermatitis, respiratory, etc., could be resulted from nonbiodegradable surfactants [10].

An amphiphile contains both hydrophilic and lipophilic groups in their structures (Fig. 8.1). A surfactant is an amphiphile that possesses surface-active properties [11]. Surfactants diminish the free energy of the system by substituting the bulk molecules

Chemical Methods
https://doi.org/10.1016/B978-0-12-821931-7.00009-2

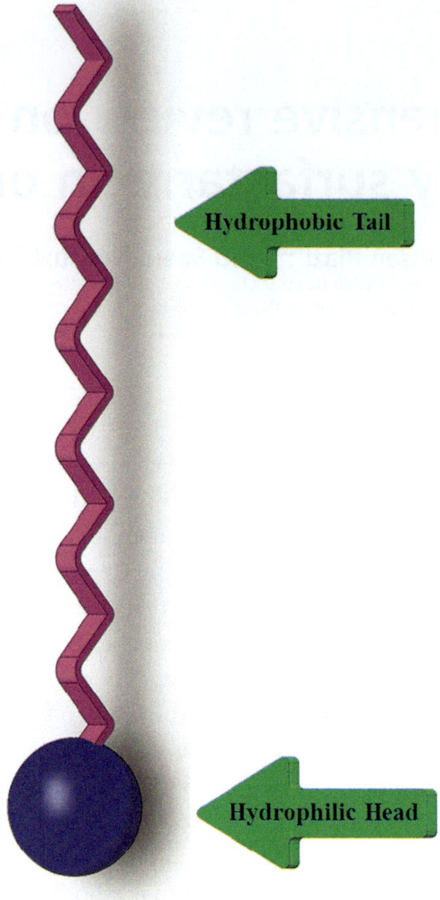

Fig. 8.1 Schematic representation of the structure of a surfactant.

of higher energy at an interface [12, 13]. Optimal curvature is the basic property of a surfactant. This property is due to opposing forces operating at both sides of the interface: repulsions among hydrophilic heads on the waterside and attractions on the hydrophobic side between hydrophobic tails [14]. Surfactants have the ability to unify immiscible phases and can yield a miscible phase, such as a miscible lipid and aqueous phase [15]. They are extensively used in food processing, chemical, agricultural, cosmetic, and pharmaceutical industries [15, 16]. A surfactant is widely utilized in industry as a foaming and flocculating agent, penetrator, wetting agent, deemulsifier, and adhesive [12, 13, 16]. A surfactant can be categorized as a nonionic, cationic, anionic, and zwitterionic agent [17]. These categories are based on their dissociation in water [18]. Cationic surfactants possess a positively charged head, while anionic ones are negatively charged. A nonionic surfactant, which has no net charge, can be used beside a cationic or anionic surfactant to provide superior performance. A zwitterionic surfactant is expensive; therefore, it is not used as much as other surfactants in the oil industry, though they are tolerant to salinity and temperature [19]. Wettability alteration of anionic surfactants [20–23], cationic surfactants [21, 24–27], and nonionic surfactants [28–32] is of great interest to the oil industry. Jafari-Sofla et al. [33] found that the anionic surfactants were not as effective as cationic surfactants in the wettability alteration of carbonates. However, for sandstone reservoirs, anionic surfactants were very effective [33].

All surfactants must have the ability to form micelles. The critical micelle concentration (CMC) is a key characteristic of a surfactant. CMC is known as the concentration of surfactants above which micelles form. At that point (i.e., CMC), surface tension remains relatively constant or varies with a lower slope [11].

Combinations of a surfactant with polymer, known as surfactant polymer (SP) flooding, and a surfactant with an alkali and polymer, known as ASP flooding, are other forms of chemical EOR. These methods have their limitations such as surfactants retention and chromatographic separation. Hence, researchers try to find polymeric surfactants (PS) that contain a polymer and surfactant in one molecule [34]. These types of surfactants have both properties of the surfactant and polymer at the same time [35]. Torrealba and Hoteit [36] showed that viscoelastic surfactants contain elastic and viscous properties that could increase conformance efficiency once applied in injector-producer pairs with low sweep efficiency resulting from the thief zone or fractures [37].

Surfactants are used not only in the chemical process for enhancing oil recovery, but also in hydraulic fracturing or completion fluids due to the wettability alteration and interfacial tension (IFT) reduction [38]. Also, the effects of combining nanoparticles and surfactants on oil recovery and surfactant flooding efficiency have been recently investigated by some researchers [39, 40]. The extensive use of surfactants in diverse areas has led to the emergence of surfactant science. Different types of surfactants are introduced for various applications. Although surfactants are useful in many areas, they have their own concerns such as serious environmental problems. Environmental concern on the use of surfactant is the starting point for the researchers to introduce and use alternative to synthetic surfactants. The main alternative to synthetic surfactants is natural surfactants [14].

8.2.1 Synthetic surfactants

Synthetic surfactants are those derived from petrochemical compounds [41]. Availability is the advantage of synthetic surfactants, which causes these surfactants to be used recently [15]. They are chemically synthesized from feedstocks such as crude oil, coal, and natural gas. These surfactants are used extensively and have a high consumption rate. The global production of these surfactants has been stated to be 7.2 million tons in 1993 [42]. These surfactants have applications in diverse fields such as food preparation, environmental

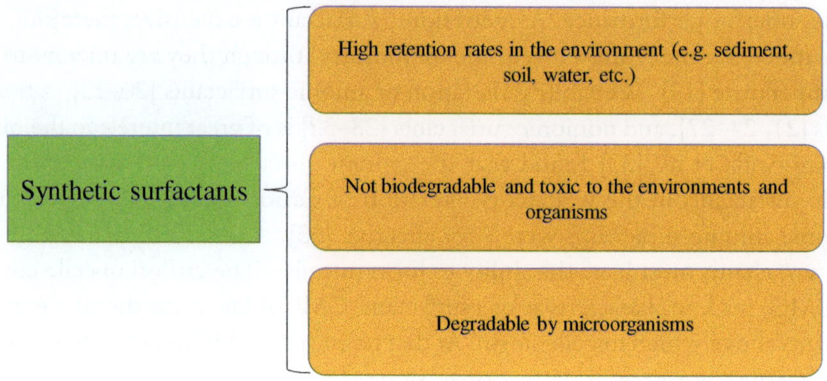

Fig. 8.2 Types of synthetic surfactants.

remediation, industrial, and human health areas [43]. Synthetic surfactants exist in three main groups (Fig. 8.2). Synthetic surfactants are dispersed in various compartments of the environment (water, soil, and sediment); therefore, the extensive use of synthetic surfactants in industries and households imposes severe environmental problems [11].

Although the effective performance of chemical surfactants was proved, their wide application could introduce challenges, especially in a field scale [44]. The main disadvantage of synthetic surfactants is their cost [44]. The high cost of synthetic surfactants converts this chemical method to an uneconomical method [33]. Environmental issues are known as another disadvantage of an unfriendly surfactant to the environment. Most of them are toxic and corrosive (for instance, when ionic liquids are employed as a surfactant), apart from being hazardous to the equipment, ecosystem, and also personnel [45, 46]. Hence, there is good motivation on the usage of green surfactants.

8.2.2 Natural surfactant

Natural surfactants or bio-based surfactants are surfactants that are produced directly from natural sources. They can be obtained from sources with plant or animal origins. No organic synthesis should be involved during the collection of surfactants from the source, not even after the treatment. Filtration, extraction, precipitation, and distillation are some sort of separation techniques. These techniques should be adopted to acquire surfactant products [47]. By-products of the food industry are other sources of natural surfactants [48, 49]. Natural raw materials and polyols [50], such as glucose, amino acid residues, and simple sugars, are used to produce eco-friendly surfactants. Many examples of natural surfactants can be found in the literature.

Another alternative for synthetic surfactants includes fatty acid esters (FAE) of sugars and FAE or amides of amino acids [41, 47, 51]. These are some examples of natural surfactants.

A natural surfactant is classified into three groups, as stated by Holmberg [47]: (1) surfactant with a natural hydrophilic head group, (2) surfactant with a natural hydrophobic tail, and (3) surfactant produced by the fermentation of natural raw material [47, 52]. The first group of surfactants is obtained directly from a natural source. These contain products such as humic and saponin compounds. Humic acid is obtained from biomass, and as a green surfactant, it has many industrial uses [41]. Enhanced oil recovery [53], soil remediation [52, 54–56], washing of soil [57], and removal of heavy metals from soil [58] are some applications of saponins.

Low environmental risk due to their natural origins, low price, biocompatibility, specificity, low toxicity, availability, and biodegradability are the advantages of natural surfactants over synthetic surfactants. As it is natural, it does not have any environmental risk to both human health and human welfare compared to synthetic surfactants [59]. The two main advantages of natural surfactants are availability and reasonable cost. These two advantages make them economically viable for surfactant flooding. Another main advantage of the newly introduced surfactants is their lower risk to the environment [7]. Green surfactants have been considered to be biodegradable [57, 60]. This can also be seen from their low CMC [61, 62]. In extreme conditions such as temperature, salinity, and pH, a green surfactant is a good choice and is available in large quantities. Green surfactants can be applied in environmental control [53, 55, 58, 63]. As bio-surfactants are amphiphilic compounds with both hydrophilic and hydrophobic portions, they can increase solubilities of hydrophobic organic compounds (HOCs) in the aqueous phase [58].

Although the mechanisms of a green surfactant are not completely known, it seems they are a suitable alternative to synthetic surfactants for EOR operations [33].

8.2.2.1 Sources of green surfactant

Different sources of natural surfactants have been introduced in the literature. Bacteria, different parts of plants, animals, etc., are some examples. Triglyceride fatty acid residue can be found in most vegetable oils [64]. The reaction of ethanol with fatty acid generates ethyl ester, whereas the reaction of methanol with fatty acid creates methyl ester [65]. Amino acids are known as another source of a natural surfactant that can be obtained from both animals and plants [64]. Amino acids can produce different types of surfactants and they have a switchable nature [66–69]. Ionic liquids are other materials that are explored by several researchers as EOR agents. They have an aromatic and ionic nature [70, 71].

8.2.2.2 Classification of natural surfactant

The same classification as synthetic surfactants can be used for natural ones. Beside cationic, zwitterionic, anionic, and nonionic surfactants, natural Gemini and polymeric surfactants are regarded as another type of ecó-friendly surfactants.

Cationic green surfactants

A schematic representation of a cationic surfactant is shown in Fig. 8.3. As the adsorption of cationic surfactants in a sandstone reservoir is high, they are less employed in this reservoir; however, they are useful in a carbonate reservoir. Olive leaves, mulberry leaves, *Seidlitz rosmarinus*, and henna leaf surfactants are some examples of natural cationic surfactants and all of them are plant extracts. The powder of leaves of the aforementioned plants can be used for washing dishes and clothes. The presence of Na^+ and K^+ [72] in the structure of *Seidlitz rosmarinus*, $RNH_3^+Cl^-$ radical in henna leaf, and diammonium salt in mulberry makes these surfactants cationic ones. The high pH of the mulberry leaf could help to reduce the IFT [64]. The CMC of the aforementioned natural cationic surfactant is presented in Table 8.1. Different CMC values have been reported for mulberry. The procedure of extraction and experimental error are the two main reasons for this.

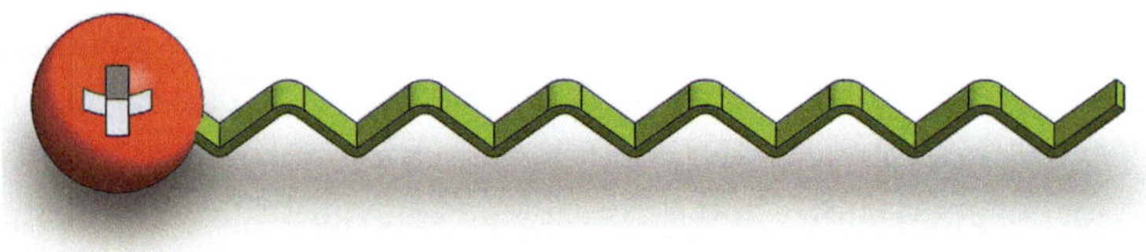

Fig. 8.3 Schematic representation of a cationic surfactant.

Table 8.1 Summary of information of cationic surfactant.

Ref.	Name	CMC (wt.%)	Extraction's method	Properties
[73]	Olive extract	2.1	Extraction from Soxhlet	Suitable surface-activity
[73]	Spistan extract	1.95	Extraction from Soxhlet	Suitable surface-activity
[74–76]	Mulberry leaves extract	1.2–1.6, 2.6	Ultrasonic extraction Soxhlet extraction	Ultimate recovery of 2% and 7% in sandstone and carbonate, respectively Alkaline in nature Wettability alteration ability Suitable surface-activity
[77]	*Seidlitzia rosmarinus* extract	8	Spray dryer method	Can reduce IFT, however it does not have EOR application
[74]	Henna leaf extract	2.8	Extraction from Soxhlet	Suitable good wetting ability and surface activity
[73]	Prosopis leaf extract	2.3	Extraction from Soxhlet	Suitable surface-activity

Zwitterionic green surfactants

Although a zwitterionic surfactant (Fig. 8.4) has suitable properties for EOR, it is not used much in the oil industry [19, 78]. The zwitterionic surfactant from castor oil, *N*-phenyl-fatty-amido-propyl-*N*, *N*-dimethyl-carboxyl betaine (CPDB) has strong electrolyte tolerance, well thermal stability, suitable foaming and wetting performance, and good dispersion efficiency. The mass weight of CPDB is 497.01 g/mol and it has a CMC of 0.00035 wt.% [64, 79].

EOR applications of the zwitterionic surfactant that was derived from waste oils used in cooking, known as *N*-phenyl-fatty-amido-propyl-*N*, *N*-dimethyl-carboxyl-betaine (PFAPMB), were investigated. It has some excellent properties such as great emulsifying property, hardness tolerant, high foaming property, biodegradability, strong electrolyte resistance, and superior wetting property [64]. The CMC of PFAPMB was 7.34×10^{-5} [80]. Another zwitterionic surfactant that synthesized from waste oils used in cooking is *N*-aliphatic-amide-*N*, *N*-diethoxy-propyl-sulfonate (NDPSS). NDPSS is an anionic-nonionic amphoteric sulfonate surfactant, which is created by the reaction of diethanolamine with fatty acid methyl ester (FAME). Reaction between 1,3-propanesultone and fatty acid alkanolamide generates this surfactant. The CMC of NDPSS is 0.4 wt.% and an ultralow IFT can be obtained at this concentration [81].

Bio-sulfonic ZS (BSZS) is a ZS that is produced from soybean oil. The CMC of BSZS is 0.003334 wt.% and it has great wetting, foaming, and emulsifying properties. BSZS has a lower CMC than synthetic surfactants such as sodium dodecyl sulfate (SDS) and cetyl trimethyl ammonium bromide (CTAB). As its Krafft is lower than 0°C, it has better solubility than normal alkyl benzene sulfonates [82].

Lignin is a biodegradable, very cheap, and renewable material and lignin sulfobetaine zwitterionic (LSZ) surfactant is derived from it [83]. Alkylation reaction, quaternization, and sulfonation reaction can create LSZ with varying alkyl chain length (LSBA-3, LSBA-2, and LSBA-1). 3, 2, and 1 alkyl chains have hydrophilic-lipophilic balance (HLB) values 10, 11, and 14, respectively, and their optimum concentrations are 1.04×10^{-7}, 4.45×10^{-8}, and 4.36×10^{-8}. Luan et al. [84] investigated the potential of the cardanol-betaine surfactant for IFT reduction. Ultralow interfacial tension without the presence of cosurfactants and alkali in not suitable condition of hardness, salinity, and temperature can be achieved by a cardanol-betaine surfactant. *N*,*N*-Dimethyl-*N*-[2-hydroxy-3sulfopropyl]-*N*′-phenyloctadecanoyl-1,3-diamino propane (SPODP) is a bio-based zwitterionic surfactant derived from waste oils used in cooking via the

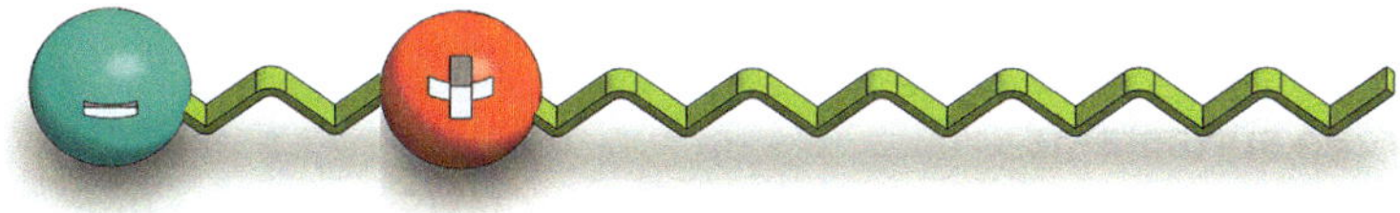

Fig. 8.4 Schematic representation of a zwitterionic surfactant.

Table 8.2 Summary of information of zwitterionic surfactant.

Ref.	Name	CMC	Source	Properties
[82]	Soybean oil-based sulfonic ZS	0.003334	Soybean oil	Good wettability, emulsion ability, and foamability
[86]	CPDB	0.00035	Castor oil	Strong electrolyte tolerance, thermostability, and temperature resistance Better foaming and wetting performance
[80]	PFAPMB	0.0000734	Waste oils used in cooking	Excellent foaming, wetting, emulsifying property Biodegradable Strong electrolyte resistance Hardness tolerant Strong interfacial activity
[85]	SPODP	0.0043	Waste oils used in cooking	Salinity tolerant Suitable interfacial activity behavior at adverse condition (high salinity and temperature) Stable in hard water Strong wetting ability
[83]	LSBA-3, LSBA-2, LSBA-1	1.04E-7 4.36E-8 4.45E-8	Lignin	Good emulsifications High surface-activity Good wetting performance Low CMC
[81]	NDPSS	0.4	Waste cooking oil	Reached ultralow IFT
[84]	Sugar-based anionic-nonionic surfactant (DAGA-ES)	5.0[a]	Sugar	Good surface-activity HLB value of 11.72 Great recovery factor anionnonionic surfactant Suitable wetting ability
[84]	Cardanol-betaine	0.03	Cardanol	Low IFT in adverse condition

[a]mmol/L.

quaternization, amidation, and alkylation process [85]. The CMC and molecular mass of this surfactant are 0.0043 wt.% and 605.4019 g/mol, respectively. SPODP is degradable, could withstand the hardness, and is tolerant to high brine concentration. Various zwitterionic surfactants are presented in Table 8.2.

Anionic green surfactants

A schematic representation of an anionic surfactant is shown in Fig. 8.5. They are the widely used and produced surfactants among other ones. Amino acids and fatty acid

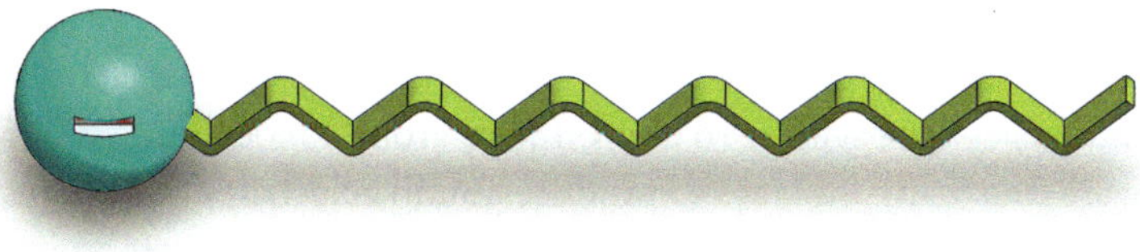

Fig. 8.5 Schematic representation of an anionic surfactant.

of natural oil (e.g., ethyl and methyl ester sulfonates) are the two main sources of this type of surfactant. Sulfonates, as the most used surfactant, are produced through sulfonation and transesterification or esterification reaction [87]. Their structure contains ethyl and methyl, R group, and sulfonate group. The thermal stability of these types of surfactants are due to the sulfonate group [88]. The phi or alpha type of compound is achieved with the position of carbon in the ester-linked sulfonates [89].

One of the main properties of these types of surfactants is stability in high brine [90, 91]. Attachment of calcium to alkylbenzene sulfonate needs more energy than that of methyl ester sulfonate [92]. Softening, wetting, and emulsifying properties, better HLB, and notable detergency are some of the suitable properties of sulfonates that are achieved from natural oil [93, 94]. Some examples of anionic surfactants are waste chicken [95], EES and sodium methyl ester sulfonate (SMES) from castor and jatropha oil [96–101], Mahua oil [102], soap nut oil [103], and palm oil [104]. Sodium epoxidized MES (SEMES), SEES, SMES, and alpha sulfonated ethyl ester (ASEE) are some examples of anionic surfactants extracted from seeds, roots, fruits, and leaves.

Sulfate-type surfactants, as high salinity tolerant surfactants, are another type of anionic surfactants. Alkoxylate sulfate is a product of the cashew nutshell, which is an example of a sulfate surfactant. Alkoxylate sulfate is known as a mobilizing and solubilizing agent of oil that consists of naphthenate, wax, and asphaltene. Its effect on wettability and IFT has not been investigated. Alkyl ethoxy carboxylate (AEC), another anionic surfactant, is a carboxylate which has high foaming stability and good solubilization, emulsifying and wetting capacity [105]. A combination of polymer, nonionic ethoxylate, and ethoxy carboxylate anionic surfactant that is produced from palm could result in a reduction of the IFT of brine-oil to an ultralow value [106].

Sulfonation reaction can modify alkyl polyglucoside and it is results in alkyl polyglucoside sodium hydroxypropyl sulfonate (APGSHS) which is an anionic surfactant. Although APGSHS's surface activity is superior to that of CTAB, once it is mixed with dodecyl trimethyl ammonium bromide (DTAB) and CTAB, it has better solubility. A reservoir with high salinity and low permeability is a good candidate for a mixture of CTAB and APGSHS as they can achieve ultralow IFT and therefore increase oil recovery [107]. Lipopeptides-based surfactants (C16 and C12) [108], hydrophobins [109], sodium cocoyl alaninate (SCA), *N*-lauroyl-L-cysteine, *N*-lauroyl-L-arginine, and *N*-lauroyl-L-lysine are other types of anionic surfactants from proteins [86, 110–112].

There are two classifications of hydrophobins, hydrophobins I and II. Hydrophobins II are incompatible with salt and they precipitate when they come in contact with brines. The CMC of peptides acylated with C16 and C12 are around 0.004 and 0.0075, respectively [108]. There are four synthesis ways to produce amino acid-based surfactants (AAS) [113]. Reaction of carboxylic with fatty alcohol converts this group to ester amines and its reaction with amine with a higher chain length converts it to amidoamine [114]. The reaction of an amino group and a fatty acid creates a surface-active amidoamine. In order to achieve a longer *N*-alkyl amino acid chain, a reaction of the amino group with alkyl halides is needed. *N*-lauroyl-lysine, $C_{18}H_{36}N_2O_3$, with a molecular weight of 328.49 g/mol, has a CMC of 0.4 wt% [111, 115]. The reaction of dodecanoyl chloride and L-arginine in methanol generates (*S*)-dodecanoyl-L-arginine with 0.9 wt% as an optimal concentration [116]. Asl et al. tested *N*-lauroyl-cystine (LCyst) with a CMC of 0.45 wt.% and *N*-lauroyl-arginine (L-arg) with a CMC of 0.2 wt.% [112, 117]. The presence of SiO_2 resulted in better surface activity of these two surfactants. One acylated AAS, which is produced from coconut fatty acid and L-alanine, is SCA. The presence of silk protein causes this surfactant to be functional and biocompatible. This yellowish liquid, which had better performance at neutral pH, has a CMC of 0.233 wt.% [86]. An example of anionic surfactants is presented in Table 8.3.

Nonionic green surfactants

A nonionic surfactant (Fig. 8.6) does not ionize in water as they have no charge on their head. The solubility of nonionic surfactants is affected by two main interactions: van der Waals and hydrogen bonding. Nonionic green surfactants can be extracted from bio-sources and plants (roots, leaf, etc.). Therefore, they are both cheap and biodegradable. Some examples of nonionic surfactants that are used in EOR are presented in Table 8.4. One common nonionic natural surfactant which is produced from a renewable and natural source is alkyl polyglucosides (APG) [132]. This is a carbohydrate-based surfactant [132]. The hydrophilic part of APG is a saccharide structure, obtained from D-glucose connected via a bond of glycosidic, and its hydrophobic component is from fatty alcohol. The existence of alcohol beside a bio-based surfactant resulted in an ultralow IFT.

Saponin is a natural surfactant and could be an alternative for synthetic surfactants. In other words, saponin is the main source of plant-based natural surfactants [33]. Marine organisms and plant species are a good source of saponins. It is derived from plant belonging to the genus Sapindaceae [52, 57]. Saponin's well solubilization, foaming, and emulsification are the main reasons for it to be widely used in industry [52, 53, 57, 58]. The surface activity of saponin is due to the simultaneous presence of sugar chains and aglycone. The nature of saponin is considered to be weakly acidic and the main cause of this nature is the hydrolysis of glycosides [56]. Saponin is a high-molecular-weight glycoside with detergent properties and is considered as a natural surface-active material. It can also produce stable foam in water [133]. The chemical formula for saponins obtained from the

Table 8.3 Summary of information of anionic surfactants.

Ref.	Name	CMC	Source	Properties
[104]	Methyl ester sulfonate (MES)	0.4[a]	Palm oil	Can resist brine containing Mg^{2+} and Ca^{2+} Thermal stable
[101]	Sodium methyl ester sulfonate	0.7	Castor oil	Optimum salinity of 3 wt.% Thermal stable Hardness tolerant
[96–99]	Sodium methyl ester sulfonate	0.8	Jatropha oil	Hardness tolerant Optimum salinity of 3.2 wt.% Thermal stable Up to 8 wt.% NaCl tolerant
[118]	SEMES	–	Palm oil	Suitable thermal stability
[103]	MES	9.3[a]	Soapnut oil	Suitable thermal stability
[118]	SMES	–	Mahua oil	2% is its optimum salinity Thermally stable
[119]	Sodium ethyl ester sulfonate	0.5	Coconut oil	≤18% NaCl tolerant Stable
[120]	Ethyl ester sulfonate	0.55	Waste chicken fat	Combination of ethyl ester sulfonate and alkali resulted in more reduction in IFT 1.5% is its optimal salinity Resistant up to 13 wt.% of salt concentration Thermally stable Good foaming, emulsifying property and wetting
[121]	ASEE	0.8	Palm oil	≤7% NaCl tolerance Thermally stable 2% is its optimal salinity
[105, 106]	AEC	0.008	–	Low critical micelle concentration
[122]	Cashew nutshell liquid alkoxylate sulfate	<0.1	Cashew nutshell	5% is its optimal salinity
[108]	C16 lipopeptides	0.004	–	Low CMC
[108]	C12 lipopeptides	0.0075	Powder of pea proteins	Low critical micelle concentration
[75]	(*S*)-Dodecanoyl-L-arginine	0.9	Arginine	Good wetting property Recovery factor is improved around 2.49%
[111, 115]	*N*-lauroyl-L-lysine	0.4	Lysine	About 29.61% of recovery factor

Continued

Table 8.3 Summary of information of anionic surfactants—cont'd

Ref.	Name	CMC	Source	Properties
				Wettability alteration occurred at high salt concentration >40% reduction in IFT
[112]	*N*-lauroyl-*L*-arginine	0.2	Arginine	Good as a wettability modifier >40% of IFT reduction In carbonate reservoir resulted in 11.9% recovery factor
[86]	SCA	0.233	Coconut fatty acid, alanine (alpha-amino acid)	Sodium cocoyl alaninate is better than SDS in wettability alteration and decreased more IFT Resistant up to 29 wt.% of NaCl concentration Temperature tolerant Amount of its HLB is 29.6
[112]	*N*-lauroyl-L-cysteine	0.45	Cysteine	Suitable as a wettability alteration In carbonate reservoir resulted in 8.9% recovery factor Suitable IFT reducer
[107]	APGSHS	2.06×10^{-4}[b]	Glucosides	Low CMC Reduce IFT to low value Good candidate in low permeability reservoirs

[a]mol/L.
[b]mM.

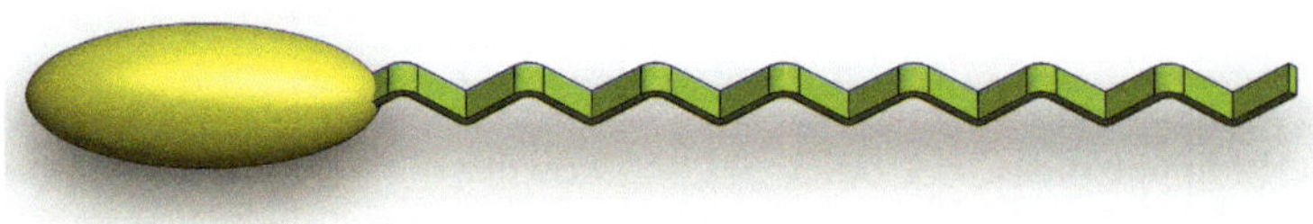

Fig. 8.6 Schematic representation of a nonionic surfactant.

soapnut fruit, as reported in the literature, is presented in Table 8.5 [53, 134–136]. 6%H, 51%O, and 44%C are present in saponin. High endurance to pressure and temperature is another aspect of soapnut saponin [137].

A tannic acid-based surfactant is a nonionic surfactant that has good surface properties and can reduce the IFT better than can a conventional anionic surfactant (i.e., SDS)

Table 8.4 Summary of information of nonionic surfactant.

Ref.	Name	CMC (wt.%)	Extraction method	Properties
[123, 124]	Zizyphus Spina-Christy saponin	3.7 and 3.4, 3	Spray drying method	It can reduce IFT slightly Combination of Zizyphus Spina Christy saponin and nanoparticle can improve the surface-activity behavior Suitable wetting ability Suggested for use in fractured reservoirs
[125]	Ginseng medical plant saponin	0.083	Soxhlet extraction	Low CMC Suitable surface-activity
[59]	*Matricaria chamomilla* extract	5.5	–	Suitable IFT reduction ability
[124]	*Glycyrrhiza Glabra* saponin	3.5	Spray drying method	IFT reduction ability Wettability alteration ability
[126]	Soapnut extracts	1.5 0.2	Cold and hot extraction	Good foaming and wetting at high reservoir pressure and temperature conditions
[126]	Soapwort plant extracts	0.225	Ultrasonic extraction	Reduce IFT sharply at optimum salt concentration of formation water, $MgCl_2$ and NaCl High oil recovery in the presence of alkali-polymer Thermal stable Suitable wetting ability even in the existence of Ca^{2+}, Mg^{2+}, and SO_4^{2-} IFT of Soapwort extracts synergic with alkali for $NahCO_3$ and NaOH is 10^{-2} IFT of Soapwort extracts synergic with alkali for Na_2CO_3 is 10^{-1} at high temperature
[127]	Alfalfa plant extract	4	Spray drying method	Mix of alfalfa plant extract resulted in recovery factor of 62% 63.39% IFT reduction Good wetting ability
[128]	Jatropha oil-based surfactants (E400S, E400V, E600S, E600V)	0.09–0.42[a]	Hydrolysis, Esterification	HLB ranged between 7 and 11 Good surface activity Good emulsifiers

Continued

Table 8.4 Summary of information of nonionic surfactant—cont'd

Ref.	Name	CMC (wt.%)	Extraction method	Properties
[129]	Gallic acid-based surfactant (GA 1500, (GA-600, GA-400)	0.75[a] 4.47[a] 5.62[a]	–	Suitable emulsifiers Better surface-activity
[130]	Alkanolamide (epoxidized di-ethanolamide)	–	Transamidation Epoxidation	Suitable foaming and wetting ability Able to form stable emulsions Better surface-activity
[131]	Tannic acid-based surfactant	2.5[a]	Microfluidic technology	Better sweep efficiency Resulted to decreased viscosity fingering Better IFT reduction than SDS
[132]	Alkyl poly-glucosides	–	–	More IFT reduction because of longer alkyl chain Increase oil recovery Temperature and salinity and tolerant Reduce IFT by the presence of alkyl poly-glucosides and alcohol

[a]mM.

Table 8.5 Chemical formula, molecular ion, mass to charge ratio, and molecular weight of triterpenoid saponins from the pericarp of *Sapindus mukorossi* Gaerten.

Formula	Molecular ion	Mass to charge ratio (*m/z*)	Molecular weight
$C_{59}H_{92}O_{25}$	$[C_{59}H_{92}O_{25}+Na]^+$	1229	1200
$C_{58}H_{94}O_{26}$	$[C_{58}H_{94}O_{26}+Na]^+$	1223	1206
$C_{53}H_{86}O_{22}$	$[C_{53}H_{86}O_{22}+Na]^+$	1097	1074
$C_{50}H_{78}O_{18}$	$[C_{50}H_{78}O_{18}+Na]^+$	989.509	966
$C_{48}H_{76}O_{17}$	$[C_{48}H_{76}O_{17}+Na]^+$	947.498	924
$C_{46}H_{74}O_{15}$	$[C_{46}H_{74}O_{15}+Na]^+$	889	866
$C_{41}H_{66}O_{12}$	$[C_{41}H_{66}O_{12}+Na]^+$	773	750

[131]. A gallic acid-based surfactant is another nonionic surfactant that has good emulsifying and corrosion inhibition properties.

Soybean oil [130], *Gliricidia sepium* oil [120, 138], laurel oil [139], and castor oil are some examples of nonionic surfactants that are produced from natural oils.

Sodium lauryl sulfate has a worse emulsion stability than epoxidized di-ethanolamide, while its foaming and wetting ability is much better.

Natural Gemini surfactant

As natural Gemini surfactant (NGS) has a lower Krafft point, excellent foaming, lower CMC, excellent solubilizing, foaming, and wetting properties than other monomeric surfactants, it become interesting surfactant for researchers to investigate their application in EOR [140]. Two or more hydrophilic (head) groups that are linked to a hydrophobic (tail) with a spacer exist in their structure. The shape and length of the micelle is determined with the spacer. The head part could be nonionic [141], anionic [142], zwitterionic, and cationic [143]. Some Gemini surfactants from sugar, resin, oil, and amino acids have been applied in other fields, while their application in EOR is few [144]. Some examples of these types of surfactants can be found in Table 8.6.

Compared to its single-chained counterpart, diacyl amino acid-based surfactant (DAAS), produced from lysine, has better surface properties [150]. DAAS can reduce surface energy to 0.01 dyn/cm at 0.3 wt.% of its concentration [146]. Bis (monoglyceride-1-hydroxymethyl-2-FAE)-a,ω-alkane-diether), a Gemini surfactant derived from sunflower oil, is a nonionic surfactant and its CMC, when it has 4 spacers, is 0.1496 mM, and with 6 spacer its CMC is 0.2409 mM. An increase in the spacer chain results in a higher CMC and could increase the surfactant's hydrophobic character. In addition, a longer chain results in better miscibility of the surfactant [145].

The CMC of 1,2 ethane bis (dimethyl 2-*N*-2-oxoethyl ammonium chloride), guava-based cationic Gemini surfactants, varies from 0.16 to 6.3×10^{-5}. The foaming and emulsifying ability and surface properties of these surfactants are acceptable and their Krafft point is below 0°C. The CMCs of the Gemini imidazolium surfactant [C4im-Cn-imC4]Br2, $n = 6$, 4, and 2 are 0.905, 0.999, and 0.905 mol/dm^3, respectively. Unlike other Gemini surfactants in which more spacer lengths resulted in better surface activity, the Gemini imidazolium surfactant [C4im-C6-imC4]Br2 and Gemini imidazolium surfactant [C4im-C2-imC4]Br2 have worse surface activity than the Gemini imidazolium surfactant [C4im-C4-imC4]Br2 [149]. The Gemini imidazolium ionic liquid surfactant is an ionic liquid Gemini surfactant. In this type of surfactant, an increase in chain length results in better surface activity [148].

Polymeric green surfactants

Hydrophobic and hydrophilic parts are two main parts of the polymeric surfactants' structure. Copolymerization of hydrophilic and hydrophobic monomers and polymerization of monomer are two ways for the synthesis of polymeric surfactants. These types of surfactants combine the property of polymer, high viscosity, and surfactant interfacial property and can be a good choice for EOR [151]. Ultralow interfacial tension could not be achieved, because they do not have suitable surface activities. The IFT reduction for

Table 8.6 Example of natural Gemini surfactant.

Ref.	Gemini surfactant	CMC (wt.%)	Sources	Properties
[145]	Bis (monoglyceride-1-hydroxymethyl-2-FAE)-α,ω-alkane-diether (SF-*n*-SF, $n=4$, 6)	0.1496 and 0.2409 (M)	Sunflower oil	Hardness and salinity and tolerant The presence of salt resulted in ultralow IFT Emulsion ability Good dispersing ability
[146]	Diacyl amino acid surfactant	–	Lysine	17.8% oil recovery for AS flooding Mixture of surfactant and sulfonate resulted in ultralow IFT
[147]	1,2-Ethane bis (dimethyl 2-*N*-2-oxoethyl ammonium chloride	$0.16–6.3\times10^{-5}$	Guava	Krafft point is below 0°C Suitable emulsifying ability Show suitable tolerant in low pH Suitable oil-dispersing ability Thermal stable High foaming ability
[148]	Gemini imidazolium surfactants (C12D, C10D, C8D)	0.04 ($n=12$) 0.2 ($n=10$)	Ionic liquids	Better HLB value Unusual salt effect on interfacial tension Contain of 4 space group of methylene
[149]	Gemini Imidazolium surfactant [C4im-Cn-imC4] Br2, $n=6$, 4, and 2	$0.820–0.990\,\text{mmol/dm}^3$	Ionic liquids	Thermal stable Good surface activity Reduce IFT

functionalized polymeric surfactants is in the order of magnitude of 1–2. Hydrolyzed polyacrylamide solution revealed water-oil interfacial tension of 20–30 dyn/cm [152]. Polymeric surfactants act as an emulsifying and thickening agent and reduce IFT [153].

The trapping and shearing action of hydrophobic micro-domains on the droplet of residual oil supported by the colloidal asphaltene effect is the main mechanism of emulsification of polymeric surfactants [154]. Several examples of polymeric surfactants can be found in Table 8.7. The presence of a hydrophobic group improves their surface activities. The surface tension of water can be reduced to 26.33 dyn/cm by an epoxidized

Table 8.7 Example of natural polymeric surfactants.

Ref.	Name	CMC	Sources	Properties
[101]	PMES	0.5	Castor oil	>20% additional oil recovery Suitable wetting ability Thermal stable Good viscosifier agent Salinity and hardness tolerant At the salinity of 4 wt.% (optimal salinity), ultralow IFT achieved
[99]	PMES	0.6	Jatropha oil	Salinity tolerant Good rheological behavior Improved EOR more than 20% hardness tolerant Thermal stable Good wetting ability Achieved IFT of 10^{-1} mN/m 2.5 wt% concentration of salt is its optimum salinity
[155, 156]	Hydrophobically modified polysaccharides (CMP35C8 and CMP10C8)	–	Carboxymethyl pullulan	Show suitable interfacial and associative properties
[157]	H-ESOHEC-Na	From 0.1053 to 0.0157 wt.%	Soyabean oil	HLB value between 9.6 and 10.6 Suitable surface-activity Greater shear rate resulted in alteration of H-ESOHEC-Na solution from pseudoplastic

Continued

Table 8.7 Example of natural polymeric surfactants—cont'd

Ref.	Name	CMC	Sources	Properties
				state to Newtonian
[151]	Carboxymethyl cellulose-derived polymeric surfactants	–	Carboxymethyl cellulose	The presence of alkali resulted in ultralow IFT Carboxymethyl cellulose derived polymeric surfactants can resulted in IFT magnitude of 1–2 dyn/cm Suitable surface activity
[158]	Poly [1-hexadecyl-3-vinyl-imidazolium bromide] (poly [C16VIm +] [Br^-])	0.02	Ionic liquids	Good wetting property Reached low IFT Thermal stable 31.25% additional oil recovery Hardness tolerance
[123, 159]	Polyacrylic acid rosin poly ethylene glycol esters P-AAR-PEG (4000, 2000, 1000, 600, and 400)	0.033–0.087	Diterpenic rosin	HLB values varied largely (from 13.3 to 19.1) Molecular weight between 1770 and 5071 Suitable surface-activity Greater shear rate resulted in alteration of introduced surfactant solution from pseudoplastic state to Newtonian Suitable foaming generation Thermal stable Good emulsion stability

soybean oil-based polymeric surfactant (H-ESO-HEC-Na). Based on its molecular weight, its CMC can vary from 0.1053 wt.% to 0.0157 wt.% [157].

Polymeric sodium methyl ester sulfonate (PMES), polymerization of the SMES surfactant, and a monomer of acrylamide synergy with potassium metabisulfite created PMES have been tested in the oil industry [98]. Through polymerization reaction, monomers of acrylamide and sulfonate groups in sodium methyl ester sulfonate are attached to generate PMES. PMES derived from castor oil has a CMC of 0.5 wt.% and one that is derived from jatropha has a CMC of 0.6 wt.% [101]. In contrast to most polymeric surfactants, PMES can result in an ultralow interaction tension, as it has a suitable surface activity [101]. Hence, this type of surfactant is comparable to any synthetic surfactants [160].

8.3 Surfactant flooding

One of the most popular and effective chemical methods is surfactant flooding in oil-wet rock [20, 161–163]. Oil price, environmental impacts, and surfactant cost are three main parameters which affect the result of surfactant flooding [124]. More details on surfactant flooding are given below.

In surfactant flooding, the remaining trapped oil will be recovered by decreasing the interfacial tension between the reservoir's fluid and injection fluid. The wettability alteration of the reservoir rock surfaces by means of the adsorption process is another mechanism used to recover the trapped oil [164]. Surfactant flooding has emerged as an interesting EOR method by modifying wettability and reducing the oil-water IFT [33]. By alteration of wettability or creation of an ultralow IFT, a high capillary number is achieved that results in more oil production [64]. Emulsification is other important mechanism of surfactant in EOR processes [64].

8.4 Surfactant EOR mechanisms

8.4.1 Reduction in the IFT

The relationship between N_c and S_{or}, known as the capillary desaturation curve, can explain the role of the IFT [165]. Demonstrated by a lot of experimental data, increasing the capillary number leads to decreased residual saturations [166]. The capillary number must be increased by 1000 times from a typical number of 10^{-7} for water flooding to reduce the water-flooding residual oil saturation by half [167]. There are three methods to increase the capillary number, including (1) increasing the injection fluid viscosity, μ, (2) increasing the injection fluid velocity, u, and (3) reducing the IFT, σ. It is obvious that N_c cannot be practically increased by 1000 times by the first two methods. It is known that the IFT between an injection solution and reservoir fluid (surfactant solution oil) can be reduced from 30 mN/m to about 10^{-3} mN/m. It means that by adding surfactants, N_c

can be practically increased by more than 1000 times [165]. An appropriate surfactant system can produce a microemulsion at the interface between water and crude oil to reduce the interfacial tensions and consequently mobilize the residual oil and enhance the oil recovery [33]. At a critical micelle concentration (CMC), the most efficient IFT reduction will occur. The surfactant monomers present in the emulsion have the highest possible concentration at the CMC point. Therefore, the most reduction of IFT will occur [168]. The spinning drop method and the pendant drop method are two methods to measure the IFT.

Surfactant injection reduces S_{or} and increases k_{ro}. Examining the permeability ratio of the aqueous phase to the oleic phase showed that the relative permeability ratio declined as the aqueous phase saturation was high and it happened once the IFT decreased [169]. Hence, the oil sweep efficiency is enhanced due to the surfactant injection [165].

Gibbs' adsorption theory can explain the IFT reduction that results from surfactant:

$$\Gamma = -\left(\frac{C}{RT}\right)\left(\frac{\partial \sigma}{\partial C}\right) \tag{8.1}$$

where Γ and C are the surface excess concentration and chemical activity of solution, respectively. In addition, σ, T, and R show the IFT, absolute temperature, and universal constant. Based on Eq. (8.1), the reason for the IFT reduction between water and oil is the adsorption of surfactant into the interface. As the adsorption of surfactant molecules increases, the reduction in the IFT will be increased. At a specific concentration of surfactant, i.e., CMC, reduction of the IFT will cease.

8.4.2 Wettability alteration

Surface wetting condition is an important parameter that affects the costs of the EOR process and controls the final production [170]. A reservoir may have a mix-wet condition. In this situation, water exists in smaller pores, whereas oil occupies the larger pores and makes them oil-wet [171]. The production of petroleum is affected significantly by wettability. The residual oil saturation can be decreased by changing the wettability condition from oil to water wet. The wettability alteration mechanism is more favorable in carbonate reservoirs, which are more oil-wet [172]. When the rock becomes more water-wet, water imbibition is increased, and thus S_{or} is decreased. It is expected that the water-wet rock yields more oil recovery [165]. Therefore, it is essential to alter the wettability of a formation to an ideal wetting state. This will result in more oil recoveries. In fact, the surfactant can change wettability to the desired wettability for more production [21, 23, 26, 173–175]. In carbonate reservoirs that contain natural fracture, the surfactant injection alters the matrix to a more water-wet state. Afterward, water can be imbibed from fracture into matrix blocks and push oil out of the matrices [165]. Sheng's numerical simulation results showed that as the IFT is high, wettability alteration plays a key role. While the IFT is effective during the whole EOR method, wettability alteration is just effective at an early time [176]. Oil composition, type and concentration of surfactant, type of the rock, and brine salinity are the key variables in the wettability alteration with surfactants [177].

8.4.3 Natural surfactant adsorption

Adsorption of surfactant on the rock surface during surfactant flooding is an important issue. As surfactant is lost due to adsorption on the surface of the reservoir rock, it damages the performance of chemical flooding and therefore renders the process economically [164]. During surfactant adsorption, transformation of surfactant molecules from bulk solution to the surface occurs [31]. The more the surfactant adsorption, the less the oil that is produced [31].

The arrangement of surfactant particles in the bulk fluid phase differs from that in the adsorbed phase. Two-dimensional hemimicelles are produced as a result of aggregation of surfactant molecules on the adsorbed phase. Interaction between adsorbed surfactants and the hydrophobic tail of other surfactants in the bulk fluid is the main reason for production of two-dimensional (2D) hemimicelles. Surfactant monomers in aqueous solution are in equilibrium with both adsorbate and micelles [5, 178–180].

Different factors such as interaction of the surfactants with formation fluids, surfactant adsorption onto the solid surfaces, coalescence of the oil droplets, and porous medium heterogeneity make chemical flooding more complex. Sorption and retention of expensive surface agent in porous media is an important issue in surfactant flooding. Surfactant loss during surfactant flooding affects both the efficiency and economy of the process. Type of surfactant, type of electrolytes existing in aqueous solution, properties of alcohols and cosurfactants, and mineralogical and morphological properties of the formation strongly affect the adsorption isotherm [5, 178–181]. The amount of surfactants adsorbed is affected by the charge on the fluid interfaces and the rock surface. Van der Waals interaction between hydrocarbon chains of the adsorbed surfactant, electrostatic interaction, nonpolar chain-solid interactions, salvation and desalvation of adsorbate and adsorbent materials, covalent bonding, and hydrogen bonding are some of the mechanisms that might be involved in surfactant adsorption. The two main factors for understanding the interactions between formation rocks and surfactant are kinetics and adsorption equilibrium. These two factors are also important to optimize the chemical flooding process. The information of adsorption on rock surfaces is obtained by equilibrium results and adsorption rate that correspond to kinetic data. After a few days or even a few weeks of adsorptions, equilibrium on the surface of rock will occur. Understanding the mechanisms of surfactant transport in the reservoir will be facilitated by studying the equilibrium and kinetic rate and this helps to have cost-effective injection schemes [182–184]. The properties or nature of surfactants and solid-liquid interface have an effect on the equilibrium and adsorption kinetics of surfactants [185].

Several researchers have investigated the adsorption behavior of surfactant [182–193]. The adsorption of a dilute solution of sodium alkyl benzene sulfonate (SDDBS) as a surfactant on several clay minerals was investigated by Wayman [192] and Scamehorn [190]. Their results show that the Langmuir equation is a suitable equation to explain the behavior of the aforementioned surfactant. Trogus et al. [191] and Meader [189] studied the adsorption of sodium alkyl benzene sulfonate on Berea sand core, kaolin, and bioglass. Based on their investigation, the Langmuir isotherm is not suitable to explain the equilibrium adsorption of these specific materials [189, 191]. The adsorption equilibrium of Na-lignosulphonate onto the porous media of the Oil India Limited (OIL) petroleum reservoir rocks was reported by Gogoi [187]. Once the concentration of NaCl increased, the amount of adsorption increased too. However, the amount of adsorption decreased when the pH increased [187]. Liu [194] showed that if the adsorption was calculated by using the surfactant adsorption amount per porous media surface area, the adsorption on the three samples is the same. The adsorption of nonionic surfactants on the sandstone rock type is more than that of anionic ones. However, based on Liu's result, the adsorption of anionic surfactant on calcite rock type is more than that of the anionic one, as Na_2CO_3 is not present. If Na_2CO_3 is present, the adsorption of both surfactants on the calcite rock type is the same. The effect of temperature is more important for the nonionic surfactant than for the anionic one, while the effect of salinity for the nonionic surfactant is less important.

The attraction force between the positively charged surface of carbonate rock and negatively charged ZSC (Zizyphus Spina-Christi) solution resulted in a higher required dose to achieve micellization compared to an aqueous phase without carbonate [164].

The collection of dynamic and static adsorption tests for eco-friendly surfactants is summarized in Table 8.8.

8.4.4 Natural surfactant emulsification

Emulsions are generated once two immiscible fluids are mixed and one of them dispersed in the other one. Between the two common types of emulsions, oil in water and water in oil, water in oil is more suitable for the EOR application. Emulsion can be injected into the reservoir or be generated inside the reservoir. Mechanical agitations are one common way in a laboratory to create emulsion, while spontaneous emulsion can be created with less energy. One of the methods to investigate emulsification is the method that was explained by Kothekar et al. [204]. In this method, 20 mL of refined oil and 20 mL of surfactant solution was shaken in a measuring cylinder. The time that was needed for separation of 10 mL of surfactant solution from the emulsion was used to determine the emulsion stability [204]. The shearing and snapping action of emulsifier and residual oil can result in the generation of emulsion in porous media.

Table 8.8 Adsorption tests for eco-friendly surfactant.

Ref.	Natural surfactant	Type of experiment	Absorbent	Parameters considered	Method for equilibrium concentration determination	Findings
[195]	ZSC extracts	Static	Carbonate rock	–	Conductivity method	More reduction in adsorption of ZSC extract by hydrophobic nano-silica
[196]	ZSC extracts	Static	Carbonate rock	Temperature = 28°C Concentration = 0.1–8 wt.%	Conductivity method	Suitable range for ZSC adsorption
[164]	ZSC extracts	Static	Carbonate rock	Temperature = 28–75°C Concentration = 0.1–8 wt.%	Conductivity method	Freundlich isotherm has the lowest error (R2 > 0.99) Exothermic adsorption process
[197]	ZSC extracts	Static adsorption	Carbonate rock	–	Conductivity method	$MgCl_2$ affect the adsorption of ZSC more than KCl
[196]	ZSC extracts	Dynamic	Shale sandstone	Temperature = 90°C Pressure = 2500 psi PV:10	Core displacement	At 8 wt.%, 6.881 mg/g At 4 wt.%, 3.316 mg/g
[196]	ZSC extracts	Static	Shale sandstone	Concentration = 0.1–8 wt.% ZSc Temperature = 25°C Surfactant mass ratio = 1:5	Conductivity method	At 8 wt.%, 35.35 mg/g At 4 wt.%, 24.70 mg/g
[7]	Glycyrrhiza Glabra extracts	Static adsorption	Sandstone rock	Temperature = 27–150°C Concentration = 0.1–8 wt.% Surfactant mass ratio = 1:5	Conductivity method	Best-fitting isotherm is Langmuir isotherm At CMC adsorption is 4 mg/g Pseudo-second-order kinetic model is the best kinetic model The adsorption process is exothermic

Continued

Table 8.8 Adsorption tests for eco-friendly surfactant—cont'd

Ref.	Natural surfactant	Type of experiment	Absorbent	Parameters considered	Method for equilibrium concentration determination	Findings
[30]	Glycyrrhiza Glabra extracts	Static	Carbonate rock	Surfactant mass ratio = 1:10 Concentration = 1.0 to 8.0 wt.%	Conductivity method	The best kinetic model is Brouers-Sotolongo model Best-fitting isotherm are Jovanovic and Two-parameter isotherms models
[32]	Trigoonella foenum-graceum (TFG) extract	Static adsorption	Sandstone rock	0.5–8 wt.% TFG	Conductivity method	Pseudo-second order is the best kinetic model Best-fitting isotherm are Jovanovic and Brouers-Sotolongo
[198]	Imidazolium-based ILs	Static adsorption	Berea Sandstone	Concentration = from 0.05 to 1 wt.% Ambient pressure Salinity = 2 wt.% Surfactant mass ratio = 1:10 Temperature = 25°C	UV-vis Spectrophotometry	Higher density of imidazolium-based ILs than MFOMAX and IOS Adsorption of MFOMAX and IOS reduced by imidazolium-based ILs
[199]	Trigoonella foenum-graceum (TFG) extract	Dynamic	Carbonate rock	Temperature = 50°C Pressure = 3000	Core displacement	4.3 mg/g is saturated adsorption
[200]	Henna extract	Static adsorption	Kaolinite	Concentration = 0.3–0.8 wt.% Salinity = 0–5 wt.%	UV-vis Spectrophotometry	Adsoprtion of SDS reduced by introduced surfactant Increase in salinity resulted in increase in adsoprion of natural surfactant

[201]	Palm kernel alkanolamide	Static adsorption	Sandstone (silica)	Concentration = 0.162–15.6 mmol/L Temperature = 30°C	UV-vis Spectrophotometry	15.20 mmol/g is maximum adsorption density
[202]	Surface-active ionic liquids	Static adsorption	Sandstone and carbonate rock	15 wt.% TDS	Spectrophotometry, two-phase titration	Less adsorption of SAILs than CTAB
[203]	Henna extract	Static adsorption	Quartz sand	Concentration = 0.2 wt.% Surfactant ratio = 1:5	UV-vis Spectrophotometry	Increasing in salinity resulted in more Henna extract adsorption SDS adsorption decreased in the presence of the henna extract
[85]	SPODP	Static	Quartz sand	Concentration = 0–0.5 wt.% Temperature = 45°C surfactant ratio = 1:9	Two-phase extraction method	Saturated adsorption of CAO is more than SPODP one

Although ultralow IFT is not necessary to generate spontaneous emulsion, low IFT is needed to emulsify the water and oil system spontaneously. Stable emulsion can help to recover more oil during surfactant, alkali, and polymer flooding via both wettability alteration and IFT reduction. More displacement efficiencies are recorded for surfactant flooding with spontaneous emulsification than for single surfactant flooding. During surfactant flooding, once the surfactant partitions from the oil phase, spontaneous emulsion is generated. Solubilization and spontaneous emulsion can result in more oil recovery. IFT, hydrophilic-lipophilic balance (HLB), bulk viscosity, phase transition region, interfacial viscosity, and hydrophilic-lipophilic deviation are the main factors that affect the emulsification process.

By reduction of the IFT of oil and water, increasing surface viscosity, electric double layer repulsion, and surface elasticity, natural surfactants make emulsion more stable. Via HLB values, natural surfactants specify which phase disperses in another one. If HLB values vary between 3 and 8, water in oil emulsion and between 8 and 18, oil in water emulsion will be generated.

Emulsification is known as a significant property of a surfactant [10]. Water insoluble substances, e.g., hydrocarbons, are solubilized because of the amphiphilic nature of the surfactant [10]. Adsorption of surfactant at the interface of oil-water results in the reduction of interfacial energy [10]. Therefore, less energy is spent to make the oil-water interfaces needed for an emulsion and stable emulsion is generated [10]. Pradhan and Bhattacharyya [10] showed that the emulsion stability of Ritha is much better than that of Henko and Pyagi Phool. Stable emulsions are generated once the adsorbed surfactant induces repulsive interactions between the droplets and generates an energy barrier against rupture [10].

Some natural surfactants that can act as emulsifiers are polyacrylic acid rosin polyethylene glycol esters P-AAR-PEG (4000, 2000, 1000, 600, and 400), soybean oil-based sulfonic zwitterionic surfactant, gallic acid-based surfactant, *N*-phenyl fatty amido-propyl-*N*, jatropha oil-based surfactants (E600V, E400V, E600S, E400S), *N*-dimethyl carboxyl betaine (PFAPMB), ethyl ester sulfonate, alkanolamide (epoxidized di-ethanolamide), and lignin sulfo-betaine LSBA (lignin sulfobetaine zwitterionic surfactants)-1, 2 and 3.

8.5 Effect of eco-friendly surfactants on the IFT, wettability alteration, and oil recovery

Different types of eco-friendly surfactants and their effect on IFT, wettability, and oil recovery are shown in Tables 8.9 and 8.10. Similar to eco-unfriendly surfactants, natural ones can affect both the IFT and wettability, the two main factors for EOR application; therefore, at first sight, they are useful and suitable for the EOR process. Although the condition was not the same for all surfactants, most of the IFT reduction belonged to

Table 8.9 Comparison of natural surfactants in terms of CMC and IFT reduction.

Ref	Name of surfactant	CMC	IFT reduction	Type of oil	Oil property
[53]	Soap nut (*Sapindus mukorossi*)	1.5 wt.%	15.75 mN/m	Crude oil	Specific gravity = 0.7–0.95
[205]	Soap nut	8000 ppm	0.22877 mN/m	Crude oil	API = 18.9
[7, 30]	*Glycyrrhiza Glabra*	3.5 wt.%	25.1 mN/m	Crude oil	
[164, 206]	Zizyphus Spina Christi	3.7 wt.%	22.5 mN/m	Crude oil	API = 40
[207]	Zizyphus Spina Christi	4%	23 mN/m	Crude oil	API = 43
[208]	Zizyphus Spina Christi	–	39 dyn/m	Kerosene	density = 801.7 kg/m^3
[75]	Mulberry leaf		26.1 mN/m	Kerosene	
[59]	*Matricaria chamomilla*	5.50%	18.06 mN/m	Kerosene	density = 801.7 kg/m^3
[8]	Dodecanoyl glucosamine	800 ppm	11.36 mN/m	Crude oil	API = 32.6
[209]	*Tribulus terrestris*	3 wt.%	31.8 mN/m	Crude oil	
[102]	*Madhunca longifolia* (Mahua)	9000 ppm	0.48 mN/m	Crude oil	API = 18.9
[210]	Reetha (*Sapindus mukorossi*)	2 wt.%	11.58 mN/m	Crude oil	API = 21.2
[44]	Anabasis setifera	3000 ppm	24.542 mN/m	Crude oil	Mw = 247
[32]	Trigoonella foenum-graceum	3.23%	~55 mN/m	Kerosene	
[77]	Seidlitzia rosmarinus	8%	23 mN/m	Kerosene	Density = 801.7 kg/m^3
[211]	Spistan (*Cordia myxa* leaf)	3.20%	47.04 mN/m	Kerosene	Density = 821 Kg/m^3
[73]	Olive	1.95%	22.5 mN/m	Kerosene	Density = 776.9 kg/m^3
	Spistan (*C. myxa* leaf)	2.10%	16.35 mN/m	Kerosene	Density = 776.9 Kg/m^3
	Prosopis	2.30%	21.39 mN/m	Kerosene	Density = 776.9 kg/m^3

Continued

Table 8.9 Comparison of natural surfactants in terms of CMC and IFT reduction—cont'd

Ref	Name of surfactant	CMC	IFT reduction	Type of oil	Oil property
[76]	Mulberry leaf	1.2–1.6 wt.%	22 dyn/cm	Kerosene	Density = 786.3 kg/m^3
[95]	Used chicken skin fat as primary substance for the synthesis of surfactant	5500 ppm	9.957 mN/m	Crude oil	Mw = 247
[101]	Sodium methyl ester synthesized from castor oil was used as a surfactant	6000 ppm	80×10^{-3} mN/m	Crude oil	API = 38.86
[40]	Ceder extraction	5%	24.1 mN/m at 10 wt.% of surfactant 2.58 mN/m in the presence of nano	Kerosene	Density = 800 kg/m^3
[128]	E600v E600S 6 nonionic surfactants derived from Jatropha oil (E400v) 7 nonionic surfactants derived from Jatropha oil (E400S)	0.00042 M 0.00032 M 0.00021 M 0.00009 M			
[212]	Tragacanth gum	3000 ppm	Depend on type of salts and their concentration. In addition in the presence of sodium hydroxide and monoethylamine at different concentration, IFT will be affect	Crude oil	API = 31.14
[119]	Synthesized sodium ethyl ester sulfonate (SEES) surfactant	0.5 wt.% at 303 K 0.6 wt.% at 323 K 0.7 wt.% at 343 K	0.735 mN/m 0.23571 mN/m	Crude oil	API = 23.55

Table 8.10 Wettability reduction at CMC and oil recovery resulted by natural surfactants.

Ref.	Name of surfactant	Wettability alteration	Type of core	Increasing oil recovery
[53]	Soap nut (*Sapindus mukorossi*)	–		Depend on the concentration of natural surfactant, polymer, and silica nanoparticles
[205]	Soap nut	62.7 degree	Sandstone for contact angle and sandpack for flooding test	30%
[124]	*Glycyrrhiza glabra*	40 degree at CMC	Limestone	Depend on the concentration of natural surfactant (16.5%, 22.7%, 29.5%, 36%)
[33]	Zizyphus Spina Christi	Depend on the type of rock	Carbonate	15%
[207]	Zizyphus spina Christi		Carbonate	13.68%
[75]	Mulberry leaf		Sandstone/ carbonate	19.60%
[8]	Dodecanoyl glucosamine	20.06 degree at CMC	16%	Carbonate
[209]	*Tribulus terrestris*	79.7 degree at CMC	Carbonate	7.80%
[102]	*Madhunca longifolia* (Mahua)	86 degree at CMC	Sandstone	17.74%
[210]	Reetha (*Sapindus mukorossi*)	Depend on the concentration of natural surfactant, polymer, and silica nanoparticles	Berea core	
[44]	Anabasis setifera	65.76 degree at CMC	Carbonate	15.40%
[211]	Spistan (*C. myxa* leaf)	6.4 degree at 5.5 wt. %	Carbonate	27%
[76]	Mulberry leaf	120 degree	Carbonate and sandstone	7% for carbonate and 1.5% for sandstone
[95]	Used chicken skin fat as primary substance for the	96.1 degree	Carbonate	17.8% by ASP injection (surfactant solution at CMC + optimum salinity

Continued

Table 8.10 Wettability reduction at CMC and oil recovery resulted by natural surfactants—cont'd

Ref.	Name of surfactant	Wettability alteration	Type of core	Increasing oil recovery
	synthesis of surfactant			and alkali and 1000 ppm PHPA polymers)
[101]	Sodium methyl ester synthesized from castor oil was used as a surfactant	54 degree for SMES	Quartz	
[40]	Ceder Extraction		Sandstone	15% only surfactant exist in injection solution 21% after adding nano to surfactant solution
[212]	Tragacanth Gum		Carbonate	21.40%
[119]	Synthesized sodium ethyl ester sulfonate (SEES) surfactant	90 degree	Sandpack	17.49% 18.97%

Trigoonella foenum-graceum when kerosene was used as the oil phase and *Tribulus terrestris* once the oil phase was crude oil. The most wettability alteration was achieved by the mulberry leaf and higher recovery was recorded as soap nut was used in the injection solution. Although different factors must be considered to choose a suitable surfactant, if we want to choose the best one based on the oil recovery, soap nut could be a good choice for the EOR application. Some of the researchers (e.g., [75, 123, 124, 209]) believed that the main factor in enhanced oil recovery is the reduction in the IFT.

In addition to the EOR application of natural surfactants, as previously mentioned, they can be employed in different industries. Surfactants can also be used for soil remediation. Depending on the nature of contaminants, different types of surfactant must be used [213–217]. The toxicity of some synthetic surfactants convinced some researchers to try to find a natural surfactant that has the same efficiency as the chemical one in soil remediation [54]. They used two eco-unfriendly surfactants (Triton X-100 and sodium dodecyl sulfate) and a solution of humic acid. Their results showed that both the synthetic surfactants were efficient in soil washing. However, because of their biological toxicity, they might become a further environmental problem. They claimed that natural surfactants such as humic acid solution could be used in the washing of contaminated soil with the same efficiency as a synthetic surfactant [54]. In addition, an eco-friendly surfactant can be used in drilling. Water-based drilling muds are widely used for drilling oil and gas wells and are more environmentally friendly than oil-based mud. However, if the formation contains an active clay, preserving the stability of the borehole would be challenging. The feasibility of employing the Korean red ginseng root extract as a clay swelling inhibitor was investigated by Ghasemi [218]. Nonionic surfactants such as ginsenosides, Rg3, RK1, and Rg5 exist in the root extract. Their results showed that this bio-surfactant provides better clay swelling inhibition properties compared to cetrimonium bromide (a cationic surfactant) and potassium chloride (two high-performance clay swelling inhibitors) [218].

8.6 Other applications of surfactant

8.6.1 Foaming ability

Bikerman and Bartsch tests are used for checking the foaming ability of surfactants. A cylindrical column with a height of 100 cm, a fritted glass filter is used in the Bikerman test. In all, 40 mL of surfactant solution is poured in the cylindrical column. Nitrogen gas at a constant rate of 1 L/min is passed through the surfactant solution. Surfactant solutions are checked from 2.5×10^{-5} to 1.53×10^{-2} g/cc and the maximum foam height achieved gives the foaming ability [219, 220]. Foam decay is determined using the same column with 100 mL of surfactant solution [221, 222].

Although the cleansing ability and foaming are not much correlated, foaming is a critical criterion in detergent assessment by customers. Foam formation consists of the continuous establishment of new interfaces. A high foaming power necessitates high surface elasticity, high surface viscosity, and quick adsorption [223]. A large amount of saponin in Ritha solution, known as a natural surfactant, may result in the production of a high-quality thick foam. When saponin is present in solution, dynamic surface tension is reduced and large surface areas needed for foaming are generated [224]. An increase in concentration of the natural surfactant increases the foaming ability and results in more stable foams [225].

The trend of both tests is the same; however, the Bikerman test produces more foam because more gas is involved [10]. Foam that is produced by mechanical agitation is not stable. The foam stability is computed as follows [222]:

$$R5 = \frac{h_5}{h_0} \times 100 \tag{8.2}$$

In the above equation, h_5 and h_0 show foam height at 5 min and maximum foam height, respectively. An $R5$ value of 50% makes a cutoff between low stability foam and metastable foam [226].

8.6.2 Wetting

Wetting plays an essential role in the removal of oil, dirt, and soils. Wetting is known as a complex phenomenon. Concentration, nature of the surface, and surface tension can affect wetting. At low surface tension, the penetration of a solution through a surface and wetting completely happens [227]. The investigation of Pradhan and Bhattacharyya indicated that Henko had the least wetting time, followed by Ritha and Pyagi Phool. The better wetting of Ritha than Pyagi Phool was attributed to the low surface tension and hence good penetration of solution to the surface. Wetting behavior presents good agreement with surface tension measurements. [10]

8.6.3 Cleaning

The removal of grease, soil, and dirt is the primary goal of a surfactant. The equation below was used to calculate the percentage of cleaning:

$$C=\left[\frac{W_2-W_3}{W_2-W_1}\right]\times 100\% \tag{8.3}$$

where W_1, W_2, and W_3 are the initial weight of the cloth, weight of the cloth with simulated dirt, and weight after being cleaned with surfactant solution and water, respectively. Although the amount of cleaning are different for three surfactants, all the surfactants show a similar trend in cleaning. At high concentration of the surfactant, the cleaning ability of Ritha and Henko increased. Suitable surface reduction may be the main reason for this event. [10]

8.6.4 Dirty dispersion

Dirt dispersion in foam is known as a test for surfactant effect when the dirt remaining in the foam is hard to rinse and may redeposit at a cleaned surface.

At low concentration, there is not enough surfactant to form foam; hence, less dirt will be carried, and consequently the amount of dirt in the foam is very low. By increasing concentration, foam is created with more surfactants, so that more dirt is attracted to the foam and it is continued until the micelle begins to form. After a CMC, there is surfactant in solution and it can keep dirt in suspension. Therefore, by increasing the concentration of the surfactant, the amount of dirt in solution decreases and it is a cheap method to estimate the CMC of the surfactant.

8.7 Future perspectives, recommendations, and challenges

The successful experience on using of natural surfactant in different industries such as medicine, cosmetic and food, encouraged many researchers to investigate their application in the oil industry. Several natural surfactants were introduced in the literature. As shown in the previous sections, most of them are effective for reducing the IFT, changing the wettability to a desired situation, and enhancing oil recovery. However, widespread investigations are required to achieve more information about natural surfactants' efficiency.

Adsorption of surfactant on the surface of rock is one of the important issues that must be investigated before conducting surfactant flooding [124]. The efficiency of surfactants on the IFT would be decreased once they get adsorbed onto the solid surface [32]. Although the adsorption behavior of the surfactant onto the rock surfaces plays a fundamental role in the performance of oil recovery [5, 179, 228–242], research studies related to the adsorption of new natural surfactants are lacking.

Another main factor that can affect the amount of IFT reduction and consequently the oil recovery is temperature. Therefore, investigation of the effect of temperature on the performance of eco-friendly surfactants and consequently IFT reduction is necessary. The effect of heat on IFT was investigated at 50°C and great IFT reduction resulted [53]. In another study by Saxena et al. [102], TGA analysis disclosed that the natural surfactant that was used in their study, *Madhuca longifolia*, was stable at high temperatures so that it can be used in the high-temperature reservoirs [102]. Several eco-friendly surfactants, especially those extracted from the plant extracts, may degrade at high-temperature reservoir [64]. In addition to the adsorption of a new natural surfactants on the rock surface, the effect of temperature on their performance needs to be investigated widely.

Increasing the salt concentration postponed the formation of the surfactant micelles [111]. A high salinity reservoir resulted in the precipitation of some natural surfactants [64]. Hence, investigation of the effect of salinity on the introduced natural surfactant is crucial.

Although the effect of eco-friendly surfactants on oil recovery has been widely investigated, studying the synergic effect of these surfactants with new and emerging technologies is worthwhile.

Most of the oil reserves are in carbonate formations for which cathionic surfactants are suitable for surfactant flooding. Since conventional cationic surfactants are so expensive, finding a less expensive way to synthesize these types of surfactants from less expensive renewable materials would be necessary [64].

In most of the literature, the performance of natural surfactants on the oil recovery was used as a criterion to show that natural surfactants are useful in the EOR application. However, a comparison of the newly introduced natural surfactants with eco-unfriendly ones can get a better view of the performance of natural surfactants.

As shown in this study, most of the studies about natural surfactants are at the laboratory scale. Therefore, examining an eco-friendly surfactant at the field scale can give us more information about the ability of these types of surfactants.

In spite of the fact that eco-friendly surfactants are low in cost and environmentally friendly, their applications face some limitations, e.g., large expense of workup, ineffective and time-consuming separation process, and existence in minor amounts in natural products [47, 243, 244]. Despite the aforementioned drawbacks, the merits of using natural surfactants might seem to be more significant.

References

[1] Taber JJ. Research on enhanced oil recovery: past, present and future. In: Surface phenomena in enhanced oil recovery. Springer; 1981. p. 13–52.
[2] Donaldson EC, Chilingarian GV, Yen TF. Enhanced oil recovery, I: fundamentals and analyses. Elsevier; 1985.
[3] Lake LW, Venuto PB. A niche for enhanced oil recovery in the 1990s. Oil Gas J 1990;88(17):62–7.
[4] Ehrlich R, Wygal Jr RJ. Interrelation of crude oil and rock properties with the recovery of oil by caustic waterflooding. Soc Pet Eng J 1977;17(04):263–70.
[5] Morrow NR. Wettability and its effect on oil recovery. J Petrol Tech 1990;42(12):1476–84.
[6] Dandekar AY. Petroleum reservoir rock and fluid properties. CRC Press; 2013.
[7] Arabloo M, Ghazanfari MH, Rashtchian D. Spotlight on kinetic and equilibrium adsorption of a new surfactant onto sandstone minerals: a comparative study. J Taiwan Inst Chem Eng 2015;50:12–23.
[8] Haghighi OM, Firozjaii AM. An experimental investigation into enhancing oil recovery using combination of new green surfactant with smart water in oil-wet carbonate reservoir. J Pet Explor Prod Technol 2019;1–9.
[9] Satter A, Iqbal GM, Buchwalter JL. Practical enhanced reservoir engineering: assisted with simulation software. Pennwell Books; 2008.
[10] Pradhan A, Bhattacharyya A. Quest for an eco-friendly alternative surfactant: surface and foam characteristics of natural surfactants. J Clean Prod 2017;150:127–34.
[11] Muntaha S-T, Khan MN. Natural surfactant extracted from Sapindus mukurossi as an eco-friendly alternate to synthetic surfactant–a dye surfactant interaction study. J Clean Prod 2015;93:145–50.
[12] Mulligan CN. Environmental applications for biosurfactants. Environ Pollut 2005;133(2):183–98.
[13] Mulligan CN. Recent advances in the environmental applications of biosurfactants. Curr Opin Colloid Interface Sci 2009;14(5):372–8.
[14] Chevalier Y. New surfactants: new chemical functions and molecular architectures. Curr Opin Colloid Interface Sci 2002;7(1–2):3–11.
[15] Deotale SM, et al. Coffee oil as a natural surfactant. Food Chem 2019;295:180–8.
[16] Gharaei-Fathabad E. Biosurfactants in pharmaceutical industry: a mini-review. Am J Drug Discovery Dev 2011;1(1):58–69.
[17] Deleu M, Paquot M. From renewable vegetables resources to microorganisms: new trends in surfactants. C R Chim 2004;7(6–7):641–6.
[18] Salager J. Surfactant's types and uses. Firp Booket-E300-Attaching Aid in surfactant science and engineering in english. Meridia, Venezuela: University of Andes; 2002.
[19] Negin C, Ali S, Xie Q. Most common surfactants employed in chemical enhanced oil recovery. Petroleum 2017;3(2):197–211.
[20] Seethepalli A, Adibhatla B, Mohanty K. Wettability alteration during surfactant flooding of carbonate reservoirs. In: SPE/DOE symposium on improved oil recovery. Society of Petroleum Engineers; 2004.
[21] Babadagli T. Analysis of oil recovery by spontaneous imbibition of surfactant solution. Oil Gas Sci Technol 2005;60(4):697–710.
[22] Zhang DL, et al. Wettability alteration and spontaneous imbibition in oil-wet carbonate formations. J Petrol Sci Eng 2006;52(1–4):213–26.
[23] Mohammadi MS, Moghadasi J, Kordestany A. A laboratory investigation into wettability alteration of carbonate rock by surfactants: the effect of salinity, pH, and surfactant concentration. Iran J Oil Gas Sci Technol 2014;3(3):1–10.
[24] Austad T, Standnes DC. Spontaneous imbibition of water into oil-wet carbonates. J Petrol Sci Eng 2003;39(3–4):363–76.

[25] Xie X, et al. Improved oil recovery from carbonate reservoirs by chemical stimulation. SPE J 2005;10(03):276–85.
[26] Jarrahian K, et al. Wettability alteration of carbonate rocks by surfactants: a mechanistic study. Colloids Surf A Physicochem Eng Asp 2012;410:1–10.
[27] Hou B-F, Wang Y-F, Huang Y. Study of spontaneous imbibition of water by oil-wet sandstone cores using different surfactants. J Dispers Sci Technol 2015;36(9):1264–73.
[28] Rao DN, et al. Impact of low-cost dilute surfactants on wettability and relative permeability. In: SPE/DOE symposium on improved oil recovery. Society of Petroleum Engineers; 2006.
[29] Gupta R, Mohanty K. Temperature effects on surfactant-aided imbibition into fractured carbonates. SPE J 2010;15(03):588–97.
[30] Ahmadi MA, et al. Nonionic surfactant for enhanced oil recovery from carbonates: adsorption kinetics and equilibrium. Ind Eng Chem Res 2012;51(29):9894–905.
[31] Safian-Boldaji M, et al. New surfactant extracted from Zizyphus spina-christi for enhanced oil recovery: experimental determination of static adsorption isotherm. J Jpn Pet Inst 2013;56(3):142–9.
[32] Barati-Harooni A, et al. Experimental and modeling studies on adsorption of a nonionic surfactant on sandstone minerals in enhanced oil recovery process with surfactant flooding. J Mol Liq 2016;220:1022–32.
[33] Sofla SJD, Sharifi M, Sarapardeh AH. Toward mechanistic understanding of natural surfactant flooding in enhanced oil recovery processes: the role of salinity, surfactant concentration and rock type. J Mol Liq 2016;222:632–9.
[34] Yu F, et al. Study on synthesis and applied performance of petroleum sulfonate for enhanced oil recovery. J Dispers Sci Technol 2008;29(7):975–84.
[35] Raffa P, Broekhuis AA, Picchioni F. Polymeric surfactants for enhanced oil recovery: a review. J Petrol Sci Eng 2016;145:723–33.
[36] Torrealba V, Hoteit H. Conformance improvement in oil reservoirs by use of microemulsions. SPE Reser Eval Eng 2019;22(03):952–70.
[37] van Santvoort J, Golombok M. Viscoelastic surfactants for diversion control in oil recovery. J Petrol Sci Eng 2015;135:671–7.
[38] Zhang F, et al. Scaling for wettability alteration induced by addition of surfactants in completion fluids: surfactant selection for optimum performance. In: Unconventional resources technology conference (URTEC); 2018.
[39] Bagrezaie MA, Pourafshary P. Improvement of surfactant flooding performance by application of nanoparticles in sandstone reservoirs. J Jpn Pet Inst 2015;58(2):97–102.
[40] Emadi S, et al. Effect of nano silica particles on interfacial tension (IFT) and mobility control of natural surfactant (Cedr extraction) solution in enhanced oil recovery process by nano-surfactant flooding. J Mol Liq 2017;248:163–7.
[41] Salati S, Papa G, Adani F. Perspective on the use of humic acids from biomass as natural surfactants for industrial applications. Biotechnol Adv 2011;29(6):913–22.
[42] Di Corcia A. Characterization of surfactants and their biointermediates by liquid chromatography–mass spectrometry. J Chromatogr A 1998;794(1–2):165–85.
[43] Ying G-G. Fate, behavior and effects of surfactants and their degradation products in the environment. Environ Int 2006;32(3):417–31.
[44] Nowrouzi I, Mohammadi AH, Manshad AK. Water-oil interfacial tension (IFT) reduction and wettability alteration in surfactant flooding process using extracted saponin from anabasis Setifera plant. J Petrol Sci Eng 2020;189:106901.
[45] Manshad AK, et al. Wettability alteration and interfacial tension (IFT) reduction in enhanced oil recovery (EOR) process by ionic liquid flooding. J Mol Liq 2017;248:153–62.
[46] Emadi S, et al. Effect of using Zyziphus spina christi or Cedr extract (CE) as a natural surfactant on oil mobility control by foam flooding. J Mol Liq 2019;293:111573.
[47] Holmberg K. Natural surfactants. Curr Opin Colloid Interface Sci 2001;6(2):148–59.
[48] Adlercreutz P, Hatti-Kaul R. Synthesis of surfactants using enzymes. In: Kjellin M, Johansson I, editors. Surfactants from renewable resources. Sweden: John Wiley & Sons; 2010. p. 143–65.
[49] Moran C, et al. Amino acids, lactic acid and ascorbic acid as raw materials for biocompatible surfactants. In: Kjellin M, Johansson I, editors. Surfactants from renewable resources. Sweden: John Wiley & Sons; 2010. p. 85–111.

[50] Foley P, Beach ES, Zimmerman JB. Derivation and synthesis of renewable surfactants. Chem Soc Rev 2012;41(4):1499–518.
[51] Johansson I, Svensson M. Surfactants based on fatty acids and other natural hydrophobes. Curr Opin Colloid Interface Sci 2001;6(2):178–88.
[52] Kommalapati RR, et al. Aqueous solubility enhancement and desorption of hexachlorobenzene from soil using a plant-based surfactant. Water Res 1997;31(9):2161–70.
[53] Chhetri A, et al. Soapnut extract as a natural surfactant for enhanced oil recovery. Energy Sources, Part A 2009;31(20):1893–903.
[54] Conte P, et al. Soil remediation: humic acids as natural surfactants in the washings of highly contaminated soils. Environ Pollut 2005;135(3):515–22.
[55] Kosaric N. Biosurfactants and their application for soil bioremediation. Food Technol Biotechnol 2001;39(4):295–304.
[56] Hong K-J, Tokunaga S, Kajiuchi T. Evaluation of remediation process with plant-derived biosurfactant for recovery of heavy metals from contaminated soils. Chemosphere 2002;49(4):379–87.
[57] Roy D, et al. Soil washing potential of a natural surfactant. Environ Sci Technol 1997;31(3):670–5.
[58] Song S, Zhu L, Zhou W. Simultaneous removal of phenanthrene and cadmium from contaminated soils by saponin, a plant-derived biosurfactant. Environ Pollut 2008;156(3):1368–70.
[59] Shadizadeh S, Kharrat R. Experimental investigation of matricaria chamomilla extract effect on oil-water interfacial tension: usable for chemical enhanced oil recovery. Pet Sci Technol 2015;33 (8):901–7.
[60] Sarma MB, et al. Degumming of muga silk fabric by biosurfactant. India: NISCAIR-CSIR; 2012.
[61] Balakrishnan S, Varughese S, Deshpande A. Micellar characterisation of saponin from Sapindus mukorossi. Tenside Surfactant Deterg 2006;43(5):262–8.
[62] Schmitt C, et al. Saponins: a renewable and biodegradable surfactant from its microwave-assisted extraction to the synthesis of monodisperse lattices. Biomacromolecules 2014;15(3):856–62.
[63] Rahman P, Gakpe E. Production, characterisation and applications of biosurfactants-review. Biotechnology 2008;7(2):360–70.
[64] Atta DY, et al. A state-of-the-art review on the application of natural surfactants in enhanced oil recovery. J Mol Liq 2020;114888.
[65] Yusoff MFM, Xu X, Guo Z. Comparison of fatty acid methyl and ethyl esters as biodiesel base stock: a review on processing and production requirements. J Am Oil Chem Soc 2014;91(4):525–31.
[66] Mujumdar S, Joshi P, Karve N. Production, characterization, and applications of bioemulsifiers (BE) and biosurfactants (BS) produced by Acinetobacter spp.: a review. J Basic Microbiol 2019;59 (3):277–87.
[67] Perinelli DR, et al. Chemical–physical properties and cytotoxicity of N-decanoyl amino acid-based surfactants: effect of polar heads. Colloids Surf A Physicochem Eng Asp 2016;492:38–46.
[68] Bustelo M, et al. Monocatenary histidine-based surfactants: role of the alkyl chain length in antimicrobial activity and their selectivity over red blood cells. Colloids Surf A Physicochem Eng Asp 2017;532:501–9.
[69] Fawzy A, et al. New synthesized amino acids-based surfactants as efficient inhibitors for corrosion of mild steel in hydrochloric acid medium: kinetics and thermodynamic approach. Int J Electrochem Sci 2018;13(5):4575–600.
[70] Pereira JF, et al. Ionic liquid enhanced oil recovery in sand-pack columns. Fuel 2014;134:196–200.
[71] Hezave AZ, et al. Dynamic interfacial tension behavior between heavy crude oil and ionic liquid solution (1-dodecyl-3-methylimidazolium chloride ([C12mim][Cl]+ distilled or saline water/heavy crude oil)) as a new surfactant. J Mol Liq 2013;187:83–9.
[72] Hadi M. Biotechnological potentials of Seidlitzia rosmarinus: a mini review. Afr J Biotechnol 2009;8 (11).
[73] Ghahfarokhi AK, et al. Feasibility study of new natural leaf-derived surfactants on the IFT in an oil–aqueous system: experimental investigation. J Pet Explor Prod Technol 2015;5(4):375–82.
[74] Rahmati M, et al. Effect of natural leaf-derived surfactants on wettability alteration and interfacial tension reduction in water-oil system: EOR application. J Jpn Pet Inst 2015;58(4):245–51.
[75] Ahmadi MA, et al. Preliminary evaluation of mulberry leaf-derived surfactant on interfacial tension in an oil-aqueous system: EOR application. Fuel 2014;117:749–55.

[76] Ravi S, Shadizadeh S, Moghaddasi J. Core flooding tests to investigate the effects of IFT reduction and wettability alteration on oil recovery: using mulberry leaf extract. Pet Sci Technol 2015;33(3):257–64.
[77] Deymeh H, Shadizadeh S, Motafakkerfard R. Experimental investigation of Seidlitzia rosmarinus effect on oil–water interfacial tension: usable for chemical enhanced oil recovery. Sci Iran 2012;19 (6):1661–4.
[78] Kamal MS, Sultan AS, Hussein IA. Screening of amphoteric and anionic surfactants for cEOR applications using a novel approach. Colloids Surf A Physicochem Eng Asp 2015;476:17–23.
[79] Zhang Q-Q, et al. Novel zwitterionic surfactant derived from castor oil and its performance evaluation for oil recovery. Colloids Surf A Physicochem Eng Asp 2015;483:87–95.
[80] Zhang Q-Q, et al. The rebirth of waste cooking oil to novel bio-based surfactants. Sci Rep 2015; 5(1):1–7.
[81] Zhang Y, et al. Investigation on interfacial/surface properties of bio-based surfactant N-aliphatic amide-N, N-diethoxypropylsulfonate sodium as an oil displacement agent regenerated from waste cooking oil. J Mol Liq 2016;223:68–74.
[82] Xu W, et al. A novel bio-based sulfonic zwitterionic surfactant derived from transgenic soybean oil and its performance in surface and interfacial activities. J Pet Sci Technol 2018;8(1):32–44.
[83] Chen S, et al. Synthesis and physiochemical performance evaluation of novel sulphobetaine zwitterionic surfactants from lignin for enhanced oil recovery. J Mol Liq 2018;249:73–82.
[84] Luan H, et al. Study on cardanolbetaine surfactants for ultralow interfacial tension in a low range of surfactant concentration and wide range of temperature applied in compound flooding. Tenside Surfactant Deterg 2015;52(1):29–34.
[85] Chen Z-Z, et al. A thermal-stable and salt-tolerant biobased zwitterionic surfactant with ultralow interfacial tension between crude oil and formation brine. J Petrol Sci Eng 2019;181:106181.
[86] Tackie-Otoo BN, Mohammed MAA. Experimental investigation of the behaviour of a novel amino acid-based surfactant relevant to EOR application. J Mol Liq 2020;316:113848.
[87] Hutchinson J, et al. Method of making methyl ester surfactants; 2006 [Google Patents].
[88] Hirasaki GJ, Miller CA, Puerto M. Recent advances in surfactant EOR. SPE J 2011;16(04):889–907.
[89] Cohen L, Roberts DW, Pratesi C. Φ-Sulfo fatty methyl Ester sulfonates (Φ-MES): a novel anionic surfactant. Chem Eng 2010;21.
[90] Shupe R, Maddox Jr J. Surfactant oil recovery process usable in high temperature, high salinity formations; 1978. US Patent 4077471.
[91] Tai X-M, et al. The performance test of fatty acid methyl ester sulfonates and application in the dishwashing liquid detergent. J Dispers Sci Technol 2018;39(10):1422–6.
[92] Ivanova VI, et al. Sulfonated methyl esters, linear alkylbenzene sulfonates and their mixed solutions: micellization and effect of Ca2+ ions. Colloids Surf A Physicochem Eng Asp 2017;519:87–97.
[93] Permadani RL. Development of nanofluid detergent based on methyl ester sulfonates surfactant from waste cooking oil and titanium dioxide nanoparticles. In: IOP conference series: materials science and engineering. IOP Publishing; 2019.
[94] Asselah A, et al. Self-aggregation and emulsifying properties of methyl ester sulfonate surfactants. J Surfactant Deterg 2017;20(6):1453–65.
[95] Nowrouzi I, Mohammadi AH, Manshad AK. Primary evaluation of a synthesized surfactant from waste chicken fat as a renewable source for chemical slug injection into carbonate oil reservoirs. J Mol Liq 2020;112843.
[96] Kumar S, Kumar A, Mandal A. Characterizations of surfactant synthesized from Jatropha oil and its application in enhanced oil recovery. AIChE J 2017;63(7):2731–41.
[97] Majidaie S, et al. Green surfactant for enhanced oil recovery. In: 2011 national postgraduate conference. IEEE; 2011.
[98] Kumar S, Saxena N, Mandal A. Synthesis and evaluation of physicochemical properties of anionic polymeric surfactant derived from Jatropha oil for application in enhanced oil recovery. J Ind Eng Chem 2016;43:106–16.
[99] Elraies KA, et al. The synthesis and performance of sodium methyl ester sulfonate for enhanced oil recovery. Pet Sci Technol 2010;28(17):1799–806.
[100] Pal N, Babu K, Mandal A. Surface tension, dynamic light scattering and rheological studies of a new polymeric surfactant for application in enhanced oil recovery. J Petrol Sci Eng 2016;146:591–600.

[101] Babu K, et al. Studies on interfacial tension and contact angle of synthesized surfactant and polymeric from castor oil for enhanced oil recovery. Appl Surf Sci 2015;353:1126–36.

[102] Saxena N, Saxena A, Mandal A. Synthesis, characterization and enhanced oil recovery potential analysis through simulation of a natural anionic surfactant. J Mol Liq 2019;282:545–56.

[103] Saxena N, et al. Synthesis, characterization, physical and thermodynamic properties of a novel anionic surfactant derived from Sapindus laurifolius. RSC Adv 2018;8(43):24485–99.

[104] Mahendran S, et al. Exploring the potential application of palm methyl ester sulfonate as an interfacial tension reducing surfactant for chemical enhanced oil recovery. In: Key engineering materials. Trans Tech Publications; 2019.

[105] Lifen HLLJH, Qiufeng CYA. Preparation and properties of a novel alkyl ethoxy carboxylate surfacant. Spec Petrochem 2008;6.

[106] Hakiki F, Maharsi DA, Marhaendrajana T. Surfactant-polymer coreflood simulation and uncertainty analysis derived from laboratory study. J Eng Technol Sci 2015;**47**(6).

[107] Li Z, et al. Ultra-low interfacial tension biobased and catanionic surfactants for low permeability reservoirs. J Mol Liq 2020;309:113099.

[108] Rondel C, et al. Green production of anionic surfactant obtained from pea protein. J Surfactant Deterg 2011;14(4):535–44.

[109] Szilvay GR, Nakari-Setälä T, Linder MB. Behavior of Trichoderma reesei hydrophobins in solution: interactions, dynamics, and multimer formation. Biochemistry 2006;45(28):8590–8.

[110] Infante M, Pinazo A, Seguer J. Non-conventional surfactants from amino acids and glycolipids: structure, preparation and properties. Colloids Surf A Physicochem Eng Asp 1997;123:49–70.

[111] Rostami A, et al. Experimental assessment of a lysine derivative surfactant for enhanced oil recovery in carbonate rocks: mechanistic and core displacement analysis. J Mol Liq 2017;232:310–8.

[112] Asl HF, et al. Experimental investigation into l-Arg and l-Cys eco-friendly surfactants in enhanced oil recovery by considering IFT reduction and wettability alteration. Pet Sci 2020;17(1):105–17.

[113] Bordes R, Holmberg K. Amino acid-based surfactants–do they deserve more attention? Adv Colloid Interface Sci 2015;222:79–91.

[114] Yan T, Feringa BL, Barta K. Direct N-alkylation of unprotected amino acids with alcohols. Sci Adv 2017;3(12):eaao6494.

[115] Takassi MA, et al. A lysine amino acid-based surfactant: application in enhanced oil recovery. Pet Sci Technol 2016;34(17–18):1521–6.

[116] Madani M, et al. Fundamental investigation of an environmentally-friendly surfactant agent for chemical enhanced oil recovery. Fuel 2019;238:186–97.

[117] Asl HF, et al. Effect of SiO2 nanoparticles on the performance of L-Arg and L-Cys surfactants for enhanced oil recovery in carbonate porous media. J Mol Liq 2020;300:112290.

[118] Rabiu AM, Elias S, Oyekola O. Evaluation of surfactant synthesized from waste vegetable oil to enhance oil recovery from petroleum reservoirs. Energy Procedia 2016;100:188–92.

[119] Pal N, et al. Interfacial behaviour, wettability alteration and emulsification characteristics of a novel surfactant: implications for enhanced oil recovery. Chem Eng Sci 2018;187:200–12.

[120] Kaniampady MM, et al. Essential oil composition of *Gliricidia sepium* (Leguminosae) leaves and flowers. India: NISCAIR; 2007.

[121] Saxena N, et al. Characterizations of surfactant synthesized from palm oil and its application in enhanced oil recovery. J Taiwan Inst Chem Eng 2017;81:343–55.

[122] Leinweber D, et al. Cashew nutshell liquid alkoxylate sulfate as a new renewable surfactant composition for enhanced oil recovery applications; 2018 [Google Patents].

[123] Shahri MP, Shadizadeh SR, Jamialahmadi M. Applicability test of new surfactant produced from Zizyphus spina-Christi leaves for enhanced oil recovery in carbonate reservoirs. J Jpn Pet Inst 2012;55(1):27–32.

[124] Ahmadi MA, Galedarzadeh M, Shadizadeh SR. Wettability alteration in carbonate rocks by implementing new derived natural surfactant: enhanced oil recovery applications. Transp Porous Media 2015;106(3):645–67.

[125] Rashidi M, et al. Extraction of nonionic natural surfactants (saponin) from ginseng medical plant. In: The 18th International Electronic Conference on Synthetic Organic Chemistry. MDPI; 2014.

[126] Nowrouzi I, Mohammadi AH, Manshad AK. Characterization and evaluation of a natural surfactant extracted from soapwort plant for alkali-surfactant-polymer (ASP) slug injection into sandstone oil reservoirs. J Mol Liq 2020;318:114369.
[127] Eslahati M, et al. Experimental investigation of alfalfa natural surfactant and synergistic effects of Ca2+, Mg2+, and SO42− ions for EOR applications: interfacial tension optimization, wettability alteration and imbibition studies. J Mol Liq 2020;310:113123.
[128] Negm NA, et al. Synthesis, characterization, biodegradation and evaluation of the surface active properties of nonionic surfactants derived from Jatropha oil. J Surfactant Deterg 2013;16(6): 857–63.
[129] Negm NA, et al. Synthesis and surface activity of nonionic surfactants derived from gallic acid. Arab J Sci Eng 2016;41(1):67–73.
[130] Adewuyi A, et al. Synthesis of Alkanolamide: a nonionic surfactant from the oil of Gliricidia sepium. J Surfactant Deterg 2012;15(1):89–96.
[131] Seo S, et al. Performance evaluation of environmentally benign nonionic biosurfactant for enhanced oil recovery. Fuel 2018;234:48–55.
[132] Iglauer S, et al. Alkyl polyglycoside surfactant–alcohol cosolvent formulations for improved oil recovery. Colloids Surf A Physicochem Eng Asp 2009;339(1–3):48–59.
[133] Hostettmann K, Marston A. Saponins. Cambridge University Press; 2005.
[134] Li R, et al. Separation of total saponins from the pericarp of *Sapindus mukorossi* Gaerten. by foam fractionation. Ind Crop Prod 2013;51:163–70.
[135] Huang H-C, et al. Molluscicidal saponins from Sapindus mukorossi, inhibitory agents of golden apple snails, Pomacea canaliculata. J Agric Food Chem 2003;51(17):4916–9.
[136] Huang H-C, et al. Triterpenoid saponins from the fruits and galls of Sapindus mukorossi. Phytochemistry 2008;69(7):1609–16.
[137] Yekeen N, et al. Foaming properties, wettability alteration and interfacial tension reduction by saponin extracted from soapnut (*Sapindus Mukorossi*) at room and reservoir conditions. J Petrol Sci Eng 2020;195:107591.
[138] Chaverri Chaverri C, Alberti JFC. Leaf and flower essential oil compositions of *Gliricidia sepium* (Fabaceae) from Costa Rica. Am J Essent Oils Nat Prod 2015;2:18–23.
[139] Kolancilar H. Preparation of laurel oil alkanolamide from laurel oil. J Am Oil Chem Soc 2004; 81(6):597–8.
[140] Kamal MS, Hussein IA, Sultan AS. Review on surfactant flooding: phase behavior, retention, IFT, and field applications. Energy Fuel 2017;31(8):7701–20.
[141] Castro MJ, Kovensky J, Fernández Cirelli A. New family of nonionic gemini surfactants. Determination and analysis of interfacial properties. Langmuir 2002;18(7):2477–82.
[142] Faustino ClM, et al. Mixed micelle formation between an amino acid-based anionic gemini surfactant and bile salts. Ind Eng Chem Res 2014;53(24):10112–8.
[143] Banno T, et al. Synthesis and properties of gemini-type cationic surfactants containing carbonate linkages in the linker moiety directed toward green and sustainable chemistry. J Surfactant Deterg 2009; 12(3):249–59.
[144] WANG J, WANG D, SHANG S. Research progress in Gemini surfactants from natural products. Chem Ind Eng Prog 2012;31(12):2761–5.
[145] Pal N, et al. Performance evaluation of novel sunflower oil-based gemini surfactant (s) with different spacer lengths: application in enhanced oil recovery. Energy Fuel 2018;32(11):11344–61.
[146] Ren H, et al. Synthesis of diacyl amino acid surfactant and evaluation of its potential for surfactant–polymer flooding. Appl Petrochem Res 2016;6(1):59–63.
[147] Abo-Riya M, Tantawy AH, El-Dougdoug W. Synthesis and evaluation of novel cationic gemini surfactants based on Guava crude fat as petroleum-collecting and dispersing agents. J Mol Liq 2016; 221:642–50.
[148] Zhou H, et al. Systematic investigation of ionic liquid-type gemini surfactants and their abnormal salt effects on the interfacial tension of a water/model oil system. J Mol Liq 2018;249:33–9.
[149] Kharazi M, et al. Different spacer homologs of gemini imidazolium ionic liquid surfactants at the interface of crude oil-water. J Mol Liq 2019;296:111748.

[150] Fan H, et al. Active control of surface properties and aggregation behavior in amino acid-based Gemini surfactant systems. J Colloid Interface Sci 2008;321(1):227–34.

[151] Cao Y, Li H. Interfacial activity of a novel family of polymeric surfactants. Eur Polym J 2002;38(7):1457–63.

[152] Co L, et al. Evaluation of functionalized polymeric surfactants for EOR applications in the Illinois Basin. J Petrol Sci Eng 2015;134:167–75.

[153] Tadros TF, et al. Stabilization of emulsions using polymeric surfactants based on inulin. Adv Colloid Interface Sci 2004;108:207–26.

[154] Lee Y, et al. Oil recovery by low-salinity polymer flooding in carbonate oil reservoirs. J Petrol Sci Eng 2019;181:106211.

[155] Henni W, et al. Aggregation of hydrophobically modified polysaccharides in solution and at the air–water interface. J Colloid Interface Sci 2005;281(2):316–24.

[156] Henni-Silhadi W, et al. Influence of alkyl chains length on the conformation and solubilization properties of amphiphilic carboxymethylpullulans. Colloid Polym Sci 2008;286(11):1299–305.

[157] Huang X, et al. Preparation and characterization of polymeric surfactants based on epoxidized soybean oil grafted hydroxyethyl cellulose. J Agric Food Chem 2015;63(41):9062–8.

[158] Pillai P, Mandal A. A comprehensive micro scale study of poly-ionic liquid for application in enhanced oil recovery: synthesis, characterization and evaluation of physicochemical properties. J Mol Liq 2020;302:112553.

[159] Yan X, et al. Synthesis and properties of polyester-based polymeric surfactants from diterpenic rosin. Ind Crop Prod 2017;108:371–8.

[160] Babu K, et al. Synthesis and characterization of a new polymeric surfactant for chemical enhanced oil recovery. Korean J Chem Eng 2016;33(2):711–9.

[161] Strand S, Standnes DC, Austad T. Spontaneous imbibition of aqueous surfactant solutions into neutral to oil-wet carbonate cores: effects of brine salinity and composition. Energy Fuel 2003;17(5):1133–44.

[162] SayedAkram NI, Mamora D. Simulation study on surfactant-polymer flood performance in fractured carbonate reservoir. In: SPE/DGS Saudi Arabia section technical symposium and exhibition. Society of Petroleum Engineers; 2011.

[163] Mohsenatabar Firozjaii A, Derakhshan A, Shadizadeh SR. An investigation into surfactant flooding and alkaline-surfactant-polymer flooding for enhancing oil recovery from carbonate reservoirs: experimental study and simulation. Energy Sources, Part A 2018;40(24):2974–85.

[164] Zendehboudi S, et al. Experimental study on adsorption of a new surfactant onto carbonate reservoir samples—application to EOR. Can J Chem Eng 2013;91(8):1439–49.

[165] Sheng J. Status of surfactant EOR technology. Petroleum 2015;1(2):97–105.

[166] Lake LW. Enhanced oil recovery. Englewood Cliffs: Prentice-Hall Inc.; 1989.

[167] Sheng J. Optimum phase type and optimum salinity profile in surfactant flooding. J Petrol Sci Eng 2010;75(1–2):143–53.

[168] Puerto MC. Surfactants: fundamentals and applications in the petroleum industry. Cambridge University Press; 2000. p. 621 [@ $85.00 (US $140.00)(hardback), ISBN 0-521-64067-9. Chemical Engineering Journal, 2001. **1**(83): p. 63].

[169] Sheng JJ. Modern chemical enhanced oil recovery: theory and practice. Gulf Professional Publishing; 2010.

[170] Hirasaki G, Zhang DL. Surface chemistry of oil recovery from fractured, oil-wet, carbonate formations. SPE J 2004;9(02):151–62.

[171] Mittal K, Etzler FM. Is the world basic? Lessons from surface science. In: Contact angle, wettability and adhesion, 6; 2009. p. 111–23.

[172] Sun X, et al. Application of nanoparticles in enhanced oil recovery: a critical review of recent progress. Energies 2017;10(3):345.

[173] Boneau D, Clampitt RL. A surfactant system for the oil-wet sandstone of the North Burbank unit. J Petrol Tech 1977;29(05):501–6.

[174] Bae J, Petrick C. Glenn pool surfactant flood pilot test: comparison of laboratory and observation-well data. SPE Reserv Eng 1986;1(06):593–603.

[175] Standnes DC, Austad T. Wettability alteration in chalk: 2. Mechanism for wettability alteration from oil-wet to water-wet using surfactants. J Petrol Sci Eng 2000;28(3):123–43.
[176] Sheng JJ. Comparison of the effects of wettability alteration and IFT reduction on oil recovery in carbonate reservoirs. Asia Pac J Chem Eng 2013;8(1):154–61.
[177] Kumar S, et al. Interfacial interaction of cationic surfactants and its effect on wettability alteration of oil-wet carbonate rock. Energy Fuel 2016;30(4):2846–57.
[178] Hirasaki GJ, Miller CA, Pop GA, Jackson RE. Surfactant based enhanced oil recovery and foam mobility control. United States: Rice Univ., Houston, TX; 2006.
[179] Buckley J, Liu Y. Some mechanisms of crude oil/brine/solid interactions. J Petrol Sci Eng 1998;20 (3–4):155–60.
[180] Moritis G. EOR continues to unlock oil resources. Oil Gas J 2004;102(14):45–8.
[181] Thomas MM, Clouse JA, Longo JM. Adsorption of organic compounds on carbonate minerals: 1. Model compounds and their influence on mineral wettability. Chem Geol 1993;109(1–4):201–13.
[182] Bianco Prevot A, et al. Microwave-assisted extraction of polycyclic aromatic hydrocarbons from marine sediments using nonionic surfactant solutions. Anal Chem 2001;73(15):3790–5.
[183] Pino V, et al. Determination of polycyclic aromatic hydrocarbons in marine sediments by high-performance liquid chromatography after microwave-assisted extraction with micellar media. J Chromatogr A 2000;869(1–2):515–22.
[184] Pino V, et al. Cloud-point preconcentration and HPLC determination of polycyclic aromatic hydrocarbons in marine sediments. Fresenius J Anal Chem 2001;371(4):526–31.
[185] Paria S, Khilar KC. A review on experimental studies of surfactant adsorption at the hydrophilic solid–water interface. Adv Colloid Interface Sci 2004;110(3):75–95.
[186] Ayranci E, Duman O. Removal of anionic surfactants from aqueous solutions by adsorption onto high area activated carbon cloth studied by in situ UV spectroscopy. J Hazard Mater 2007;148(1–2):75–82.
[187] Gogoi SB. Adsorption of non-petroleum base surfactant on reservoir rock. Curr Sci 2009;1059–63.
[188] Kjellin M, Johansson I. Surfactants from renewable resources. John Wiley & Sons; 2010.
[189] Meader Jr AL, Fries BA. Adsorption in the detergent process. Ind Eng Chem 1952;44(7):1636–48.
[190] Scamehorn JF. Equilibrium adsorption of surfactants on mineral oxide surfaces from aqueous solutions. PhD thesis, The University of Texas at Austin; 1981.
[191] Trogus F, Schechter R, Wade W. A new interpretation of adsorption maxima and minima. J Colloid Interface Sci 1979;70(2):293–305.
[192] Wayman C. Surfactant sorption on heteroionic clay minerals. In: International clay conference, Stockholm; 1963.
[193] Yeom IT, et al. Micellar solubilization of polynuclear aromatic hydrocarbons in coal tar-contaminated soils. Environ Sci Technol 1995;29(12):3015–21.
[194] Liu S. Alkaline surfactant polymer enhanced oil recovery process. PhD thesis, Rice University; 2008.
[195] Ahmadi MA, Shadizadeh SR. Adsorption of novel nonionic surfactant and particles mixture in carbonates: enhanced oil recovery implication. Energy Fuel 2012;26(8):4655–63.
[196] Ahmadi MA, Shadizadeh SR. Experimental investigation of adsorption of a new nonionic surfactant on carbonate minerals. Fuel 2013;104:462–7.
[197] Ahmadi MA, Shadizadeh SR. Spotlight on the new natural surfactant flooding in carbonate rock samples in low salinity condition. Sci Rep 2018;8(1):1–15.
[198] Hanamertani AS, et al. Ionic liquids as a potential additive for reducing surfactant adsorption onto crushed Berea sandstone. J Petrol Sci Eng 2018;162:480–90.
[199] Barati-Harooni A, et al. Experimental investigation of dynamic adsorption–desorption of new nonionic surfactant on carbonate rock: application to enhanced oil recovery. J Energy Resour Technol 2017;139(4).
[200] Musa MSM, et al. Henna extract as a potential sacrificial agent in reducing surfactant adsorption on kaolinite: the role of salinity. J King Saud Univ-Eng Sci 2020;32(8):543–7.
[201] Traiwiriyawong P, Kungsanant S. Potential of palm kernel alkanolamide surfactant for enhancing oil recovery from sandstone reservoir rocks. Environ Nat Resour J 2020;18(4):333–44.
[202] Nandwani SK, Chakraborty M, Gupta S. Adsorption of surface active ionic liquids on different rock types under high salinity conditions. Sci Rep 2019;9(1):1–16.

[203] Musa MSM, et al. Application of henna extract in minimizing surfactant adsorption on quartz sand in saline condition: a sacrificial agent approach. SN Appl Sci 2019;1(8):1–11.
[204] Kothekar SC, et al. Comparative analysis of the properties of tween-20, tween-60, tween-80, Arlacel-60, and Arlacel-80. J Dispers Sci Technol 2007;28(3):477–84.
[205] Saxena N, et al. Bio-based surfactant for enhanced oil recovery: interfacial properties, emulsification and rock-fluid interactions. J Petrol Sci Eng 2019;176:299–311.
[206] Ahmadi MA, Shadizadeh SR. Induced effect of adding nano silica on adsorption of a natural surfactant onto sandstone rock: experimental and theoretical study. J Petrol Sci Eng 2013;112:239–47.
[207] Ahmadi MA, Shadizadeh SR. Implementation of a high-performance surfactant for enhanced oil recovery from carbonate reservoirs. J Petrol Sci Eng 2013;110:66–73.
[208] Pordel Shahri M, Shadizadeh S, Jamialahmadi M. A new type of surfactant for enhanced oil recovery. Pet Sci Technol 2012;30(6):585–93.
[209] Moradi S, et al. Combination of a new natural surfactant and smart water injection for enhanced oil recovery in carbonate rock: synergic impacts of active ions and natural surfactant concentration. J Petrol Sci Eng 2019;176:1–10.
[210] Saha R, Uppaluri RV, Tiwari P. Impact of natural surfactant (Reetha), polymer (xanthan gum), and silica nanoparticles to enhance heavy crude oil recovery. Energy Fuel 2019;33(5):4225–36.
[211] Jalali A, Mohsenatabar Firozjaii A, Shadizadeh SR. Experimental investigation on new derived natural surfactant: wettability alteration, IFT reduction, and core flooding in oil wet carbonate reservoir. Energy Sources, Part A 2019;1–11.
[212] Nowrouzi I, Manshad AK, Mohammadi AH. Effects of Tragacanth gum as a natural polymeric surfactant and soluble ions on chemical smart water injection into oil reservoirs. J Mol Struct 2020;1200:127078.
[213] Noordman WH, et al. Facilitated transport of a PAH mixture by a rhamnolipid biosurfactant in porous silica matrices. J Contam Hydrol 2000;44(2):119–40.
[214] Mata-Sandoval JC, Karns J, Torrents A. Influence of rhamnolipids and Triton X-100 on the desorption of pesticides from soils. Environ Sci Technol 2002;36(21):4669–75.
[215] Garon D, et al. Influence of surfactants on solubilization and fungal degradation of fluorene. Chemosphere 2002;47(3):303–9.
[216] Cuypers C, et al. The estimation of PAH bioavailability in contaminated sediments using hydroxypropyl-β-cyclodextrin and Triton X-100 extraction techniques. Chemosphere 2002;46(8):1235–45.
[217] Chun CL, Lee J-J, Park J-W. Solubilization of PAH mixtures by three different anionic surfactants. Environ Pollut 2002;118(3):307–13.
[218] Ghasemi M, et al. Primary evaluation of a natural surfactant for inhibiting clay swelling. J Petrol Sci Eng 2019;178:878–91.
[219] Bhattacharyya A, et al. Surface rheology and foam stability of mixed surfactant – polyelectrolyte solutions. Langmuir 2000;16(23):8727–32.
[220] Waltermo Å, et al. Foam and thin-liquid-film studies of alkyl glucoside systems. Langmuir 1996;12(22):5271–8.
[221] Iglesias E, et al. A new method to estimate the stability of short-life foams. Colloids Surf A Physicochem Eng Asp 1995;98(1–2):167–74.
[222] Lunkenheimer K, Malysa K. Simple and generally applicable method of determination and evaluation of foam properties. J Surfactant Deterg 2003;6(1):69–74.
[223] Piispanen PS, et al. Surface properties of surfactants derived from natural products. Part 2: structure/property relationships—foaming, dispersion, and wetting. J Surfactant Deterg 2004;7(2):161–7.
[224] Birdi K. Surface and colloid chemistry: principles and applications. CRC Press; 2009.
[225] Pradhan A, Bhattacharyya A. Shampoos then and now: synthetic versus natural. J Surf Sci Technol 2014;30(1–2):59–76.
[226] Mata J, Varade D, Bahadur P. Aggregation behavior of quaternary salt based cationic surfactants. Thermochim Acta 2005;428(1–2):147–55.
[227] Yang C-H, et al. Foam properties, detergent abilities and long-term preservative efficacy of the saponins from Sapindus mukorossi. J Food Drug Anal 2010;18(3):155–60.

[228] Gale WW, Sandvik EI. Tertiary surfactant flooding: petroleum sulfonate composition-efficacy studies. Soc Pet Eng J 1973;13(04):191–9.

[229] Pursley S, Graham H. Borregos field surfactant pilot test. J Petrol Tech 1975;27(06):695–700.

[230] Shah DO. Improved oil recovery by surfactant and polymer flooding. Elsevier; 2012.

[231] Cole EL. An evaluation of the big muddy field low-tension flood demonstration project. Tulsa, OK: K and A Technology; 1988.

[232] Cole E. An evaluation of the Robinson M-1 commerical scale demonstration project of enhanced oil recovery by micellar-polymer flood; 1988 [Report DOE/BC/10830-10, US DOE, Bartlesville, OK, USA. December].

[233] Reppert T, et al. Second Ripley surfactant flood pilot test. In: SPE/DOE enhanced oil recovery symposium. Society of Petroleum Engineers; 1990.

[234] Maerker J, Gale W. Surfactant flood process design for Loudon. SPE Reser Eng 1992;7(01):36–44.

[235] Buckley JS. Effective wettability of minerals exposed to crude oil. Curr Opin Colloid Interface Sci 2001;6(3):191–6.

[236] Wu Y, et al. Synergistic effect of alkyl polyglycoside and sorbitan mixtures on lowering interfacial tension and enhancing oil recovery. Prepr Am Chem Soc Div Petrol Chem 2004;49(2):208–10.

[237] Wu Y, et al. A study of branched alcohol propoxylate sulfate surfactants for improved oil recovery. In: SPE annual technical conference and exhibition. Society of Petroleum Engineers; 2005.

[238] Farajzadeh R, Andrianov A, Zitha P. Investigation of immiscible and miscible foam for enhancing oil recovery. Ind Eng Chem Res 2010;49(4):1910–9.

[239] Hassenkam T, et al. Pore scale observation of low salinity effects on outcrop and oil reservoir sandstone. Colloids Surf A Physicochem Eng Asp 2011;390(1–3):179–88.

[240] Barzegar Alamdari B, Kiani M, Kazemi H. Experimental and numerical simulation of surfactant-assisted oil recovery in tight fractured carbonate reservoir cores. In: SPE improved oil recovery symposium. Society of Petroleum Engineers; 2012.

[241] Golabi E, et al. Experimental study of wettability alteration of limestone rock from oil wet to water wet by applying various surfactants. In: SPE heavy oil conference Canada. Society of Petroleum Engineers; 2012.

[242] Puerto M, et al. Surfactant systems for EOR in high-temperature, high-salinity environments. SPE J 2012;17(01):11–9.

[243] Silva SG, et al. Towards novel efficient monomeric surfactants based on serine, tyrosine and 4-hydroxyproline: synthesis and micellization properties. Tetrahedron 2009;65(21):4156–64.

[244] Van Hamme J, Singh A, Ward O. Surfactants in microbiology and biotechnology: part 1. Physiological aspects. Biotechnol Adv 2006;24:604–20.

CHAPTER 9

Combination of chemical methods

Morteza Asemani[a] and Amin Rezaei[b]
[a]Department of Geology, Faculty of Earth Sciences, Kharazmi University, Tehran, Iran
[b]Department of Petroleum Engineering, School of Chemical and Petroleum Engineering, Shiraz University, Shiraz, Iran

9.1 Introduction

The chemical EOR methods have been receiving considerable attention in recent years [1]. These methods are suitable to decrease both bypassed and residual oils in the reservoir to increase ultimate oil recovery [2]. Despite the inherent advantages of the chemical EOR methods, the conventional methods are bound by some limitations, field challenges, or low sweep efficiencies (i.e., macroscopic and microscopic). Macroscopic or volumetric sweep efficiency measures the displacing effectiveness in the reservoir on a volumetric scale. In other words, the volumetric sweep efficiency indicates how the injected phase is sweeping out the volume of the reservoir, both vertically or areally, and pushing the reservoir oil toward the producing wells [3, 4]. Fig. 9.1 shows a schematic representation of volumetric sweep efficiency, which consists of areal and vertical sweep efficiency. Microscopic or displacement efficiency measures mobilization and moving oil at the pore scale upon contact of the displacing phase and oil [3, 4]. Microscopic sweep efficiency is schematically represented in Fig. 9.2. Some examples of the shortcomings of chemical methods are as follows. Polymer flooding is a valuable technique to control the mobility ratio and increase the macroscopic sweep efficiency, but this technique failed to improve the microscopic sweep efficiency. Surfactant injection has a lower volumetric displacement due to an unfavorable mobility ratio. Such limitations are conventional in all routine methods because each chemical has limited potential and capabilities for enhancing oil recovery. Hence, researchers propose combining different chemical methods to extend their applicability, benefit from the synergy mechanisms of different injectants, increasing reservoir performance, and achieving higher ultimate oil recovery [7, 8].

9.2 Key components of combined chemical methods

The combined chemical methods include several components to advantage from the individual and synergy mechanisms of different components to achieve a higher sweep efficiency and incremental oil recovery. The chemicals are solubilized in water and then simultaneously or alternatively injected with gas into the reservoir. The gas injection may

Chemical Methods
https://doi.org/10.1016/B978-0-12-821931-7.00005-5

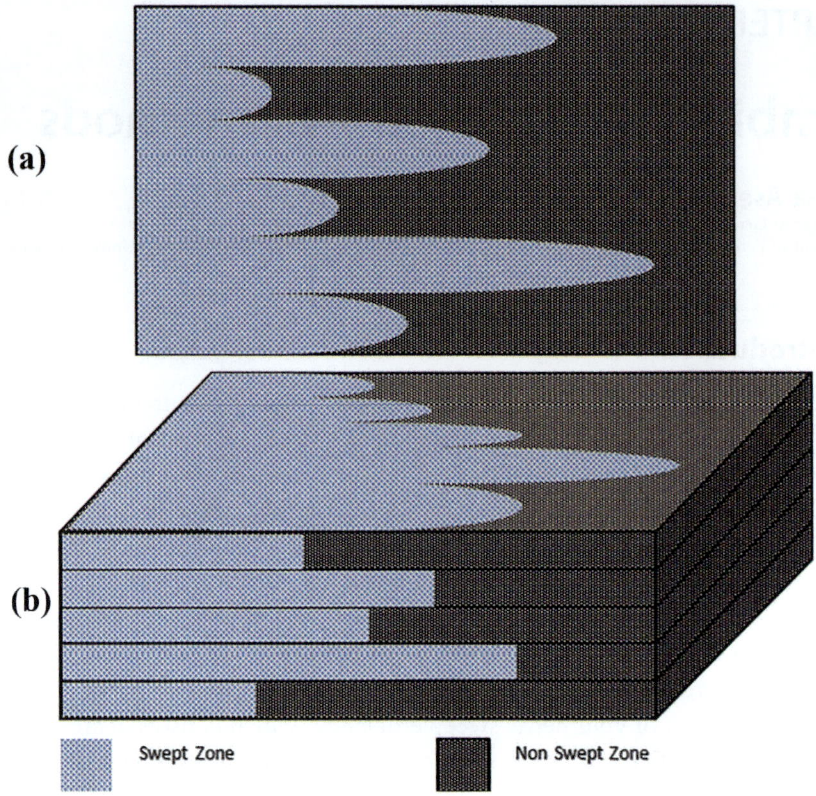

Fig. 9.1 Schematic representation of areal (A) and vertical (B) elements of volumetric sweep efficiency [5].

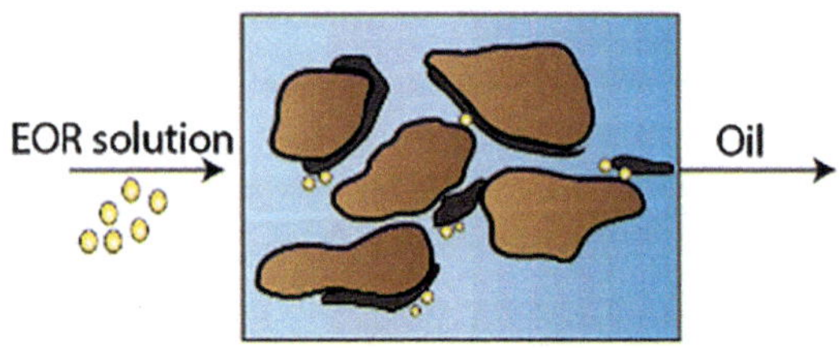

Fig. 9.2 Schematic representation of microscopic sweep efficiency [6].

be miscible or immiscible and rely on the injection pressure and minimum miscibility pressure (MMP) of the reservoir. Miscibility is achieved when the operating pressure is at or above the MMP of the reservoir. In this condition, the gas is mixed with the oil phase and improves the microscopic sweep efficiency. The key elements of the combined chemical methods are schematically illustrated in Fig. 9.3. Based on a different

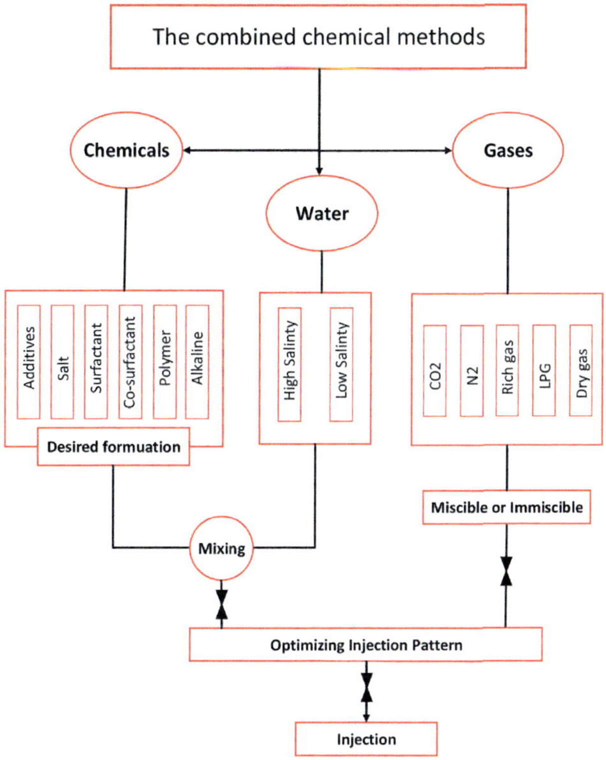

Fig. 9.3 Schematic representation of the combined chemical methods scheme. *(Modified after Kumar S, Mandal A. A comprehensive review on chemically enhanced water alternating gas/CO2 (CEWAG) injection for enhanced oil recovery. J Petrol Sci Eng 2017;157:696–715.)*

combination of these components, various combined chemical EOR methods have appeared in the literature and oil industry.

9.2.1 Water

Waterflooding is a routine process in the oil fields, and most of the fields subjected to chemical flooding have experienced waterflooding. Water is the principal part of every chemical flooding. Chemical agents are dissolved in the water and then injected into the reservoir. High salinity brine is used routinely in the chemical methods. However, over the last decades, numerous studies have indicated that brine with salinity in the range of 1000–2000 ppm provides better crude oil/brine/rock (COBR) interactions and causes a decrease in the remaining oil saturation [9, 10]. It is noteworthy to mention that freshwater is not used in low salinity water (LSW) flooding. Freshwater causes pores block due to clay swelling and leads to adverse effects on oil recovery. Brine modification should be modeled to achieve optimal composition, which has higher oil recovery [11, 12].

The low salinity water can be used with other chemicals/gases to employ the synergy mechanisms of low salinity water and chemicals or gases. LSW has successfully combined with the surfactant, polymer, and CO_2 gas. The aqueous phase is mainly regarded as a high-density phase in EOR methods. The aim of the injection in this phase is to improve the macroscopic sweep efficiency.

9.2.2 Chemicals

Chemicals include special synthetic or natural compounds designed to dissolve in water and improve its ability in the EOR process. The water that contains dissolved chemicals is named chemically enhanced water (CEW). CEW has a higher sweep efficiency in EOR processes and depends on the dissolved chemicals, providing certain mechanisms for improving oil recovery. Surfactants, cosurfactants, alkalis, salts, polymers, and nanoparticles are common chemical compounds. These compounds are used separately or combine with other compounds to achieve target formulation for oil recovery and sweep efficiency [13, 14]. Three frequently used chemical compounds in EOR methods are surfactant, polymer, and alkali. Surfactants are a known and efficient chemical agent in EOR methods. The main mechanism of surfactants is reducing the interfacial tension (IFT) between the aqueous and oil phases, but other mechanisms have also been reported for surfactants. The wettability alteration, lowering capillary forces, foam generation, and remobilization of residual oils are the other important mechanisms of surfactants for improving sweep efficiency and enhancing oil recovery. Polymers are used to control the mobility of the aqueous phase. Usually, the aqueous phase has a higher mobility than the oil phase, especially in semiheavy and heavy oil reservoirs. Under this condition, the mobility ratio is large in which the viscous fingering and early breakthrough will occur. Polymers are used to reduce the mobility ratio and provide a stable displacement front in the reservoir to increase sweep efficiency [15]. Alkalis produce an in situ surfactant upon interaction with fatty acids in the oil to reduce the IFT in the reservoir. Also, alkalis are an efficient agent to reduce surfactant and polymer adsorption on grain surfaces [14].

9.2.3 Gases

The gas phase is used to improve the microscopic sweep efficiency and overcome the poor displacement efficiency of water and chemicals in pore scales [5]. Hence, the gas phase is injected alternatively in front of the aqueous phase or simultaneously with the water phase. The IFT reduction at the oil/gas interface, oil swelling, and viscosity reduction are the important mechanisms of gas injection to increase sweep efficiency and oil recovery [16, 17]. CO_2 is a commonly used gas in EOR methods and has a high solubility in both water and oil. The solubility of CO_2 in oil causes the remobilization of residual oil and increases oil recovery [18]. Unlike CO_2 gas, other gases, including N_2, liquefied petroleum gases (e.g., propane), acid gas, flue gas, and air, have a lower contribution to total oil recovery in EOR. N_2 gas has a lower density and viscosity than other gases. These characteristics cause gravity override and viscous fingering during the injection [18]. Hydrocarbon gases are an economical and accessible option in most oil fields for injection as a water alternating gas (WAG) or miscible process. These gases have a higher compatibility with reservoir fluid and rock [18].

9.3 Combined chemical EOR methods

9.3.1 Water alternating gas (WAG) injection

Water alternating gas (WAG) is routinely regarded as a base process in EOR methods. The feasibility and efficiency of other methods are compared to this method. Also, since this method's aqueous phase may contain chemicals, this technique is explained as a combined chemical EOR method. In this method, water and gas are alternatively injected into the reservoir (Fig. 9.4). Generally, the WAG projects' primary purpose is to improve oil recovery by eliminating water or gas injection shortcomings. Due to the significant density difference between the injected gas phase and the in-place oil phase and low gas viscosity, the gas injection process leads to an inadequate microscopic sweep efficiency, early gas breakthrough, unstable displacement front, and bypassing oil regions. The WAG injection was first applied in 1957 in one of the Canadian oil reservoirs located

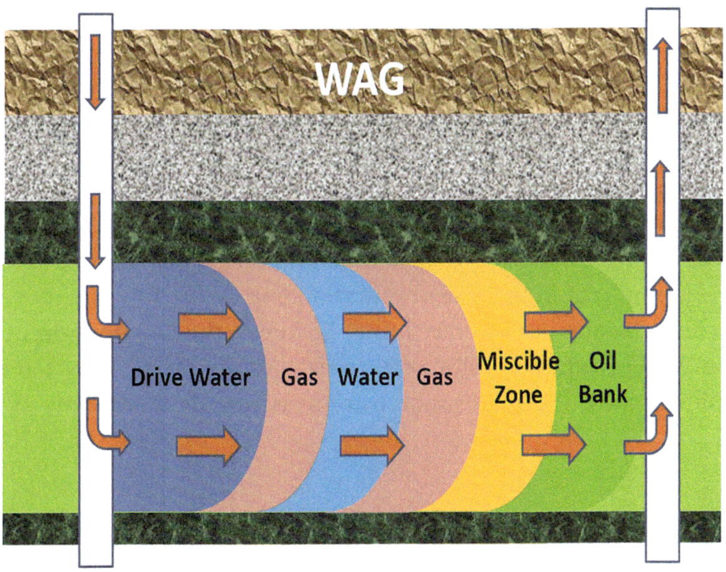

Fig. 9.4 Schematic representation of the WAG method. *(Reprinted with permission from Dai Z, Middleton R, Viswanathan H, Fessenden-Rahn J, Bauman J, Pawar R, et al. An integrated framework for optimizing CO2 sequestration and enhanced oil recovery. Environ Sci Technol Lett 2014;1:49–54. Copyright © 2014, American Chemical Society.)*

in Alberta [19]. The WAG injection is currently a well-established EOR method, in which the produced gas is reinjected in a cyclic process with water injection [20]. On the other side, the WAG injection in the field scale is usually associated with difficulties, such as high operational and completion costs, that limit this technique's applicability [21]. However, the injection of water cycles after gas injection will increase the sweep efficiency and front stability, which reduces the costs by lowering the required gas volume for the process [22]. According to reports, almost 80% of the USA's field-scale WAG projects have led to enhancing oil recovery [23]. Probably, an average increase of 10% OOIP (original oil in place) in oil recovery has been reported for 59 WAG injection projects performed in the Snorre, South Brae, Statfjord, Oseberg Ost, and Gullfaks oil fields in the North Sea [24]. The main factors that affect the WAG performance are the WAG ratio, injection rate and pressure, volume and number of each cycle, and injection wells pattern [25]. Hence, knowing the critical parameters and mechanisms during the WAG injection is of the utmost importance in designing a successful project. In this regard, there are several reports about the failed WAG projects in which the main reason for being unsuccessful was improper process designs (volume/number of cycles, water/gas injection rates, etc.) [26, 27].

On the other hand, due to the sequential imbibition and drainage processes, which are associated with the cycles of water and gas injection, respectively, simulating the WAG injection scenarios requires a comprehensive understanding of the capillary pressure and three-phase relative permeability [28]. Also, due to mass transfer between the phases, the experimental correlations to obtain the relative permeability values would be associated with many errors. Also, these complexities will increase in the WAG simulation process when we consider the dynamic effects of residual oil evaporation and the dissolution of the injected gas in the residual oil that remains in the reservoir [29, 30]. Several theoretical, experimental, and simulation studies have been performed from pore to field scale in the literature. Different gas types, such as air, N_2, O_2, and CO_2, have been used as the gaseous phase, among which CO_2 is the most common gas recommended by several scientists [31, 32]. Similarly, in the case of waterflooding cycles, low salinity waterflooding has been preferred to the high salinity brine; however, the composition of low salinity water in the WAG method is still controversial. Additionally, many researchers have experimentally proposed the five-spot pattern as the most efficient WAG process pattern [32].

9.3.2 Surfactant alternating gas (SAG) injection

The surfactant alternating gas (SAG) technique has been gaining more attention due to reduction in the IFT and changes in wettability; however, the former mechanism has a higher impact in increasing oil recovery. The foam generation mechanism in this technique causes increasing oil recovery. SAG injection benefits from the numerous

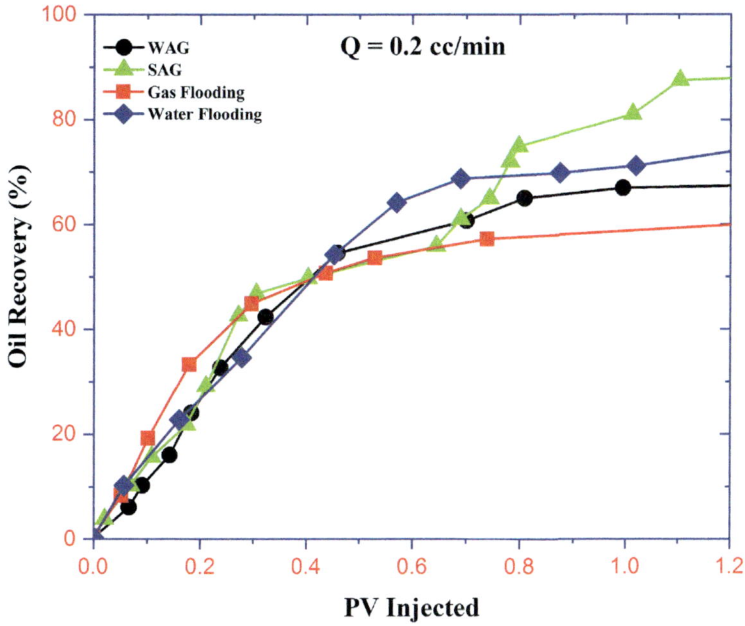

Fig. 9.5 Comparison of the obtained oil recovery by various injection scenarios [35].

potentials of the foam injection process, such as higher sweep efficiency, less gravity override, sweeping low permeable layers, less viscous fingering, etc. The injected foam can be a pregenerated foam or be formed by alternate injection of surfactant and gas; the latter is known as the SAG injection [33]. Renkema and Rosson performed a comparison of the WAG and SAG methods in both homogeneous and heterogeneous formations [34]. Their simulation studies showed that the SAG injection moves more pistonlike through the core samples and renders a higher oil recovery factor in porous media. Additionally, due to the less water volume required for the SAG injection (less than 15%–20% is water), this method is superior to the WAG injection technique in EOR projects [35]. According to Salehi et al.'s studies, at a constant injection rate of 0.2 cc/min, the amount of oil recovered by the SAG injection was about 17% more than that for the WAG injection process (70% oil recovery). In contrast, the oil recovery for the water and gas injection approaches was measured at 66% and 59% of OOIP, respectively [35]. Fig. 9.5 shows the oil recovery versus pore volume of the injected phase obtained by Salehi et al. during various flooding scenarios.

9.3.3 Alkaline-surfactant-gas (ASG) injection

Numerous studies have recommended implementing the alkaline-surfactant-gas (ASG) injection as an efficient EOR approach, which takes advantage of the synergistic combination of both chemical and gas-based methods. The injection of chemical slugs (alkali/surfactant) followed by gas injection prevents gas fingering/gas override by controlling the gas mobility due to the in situ generation of foam. Consequently, the apparent gas viscosity will increase and gas relative permeability will decrease, which will lead to a better volumetric sweep efficiency. Furthermore, the simultaneous presence of the alkali and surfactant as chemical additives can reduce the oil–water IFT to ultralow values, which improves the efficiency of the applied EOR technique to achieve higher oil recovery. Many scientists refer to this approach as a modified WAG method to overcome the tremendous limitations associated with the WAG injection, such as injectivity loss [36], early gas breakthrough [37], asphaltene precipitation [24], hydrate formation [24], and corrosion [21]. For instance, corrosion occurrence is reduced in the ASG injection compared to the foam injection or WAG method due to fewer contacts between water and gas in the ASG injection [38].

The presence of surfactant in the injection process will improve oil recovery through various mechanisms, such as foam generation, IFT reduction, rock surface wettability alteration, etc. However, surfactants are prone to be adsorbed by clay minerals and reduce the EOR efficiency. Thus, as a remedy to this problem, alkalis are applied to limit surfactant adsorption on rock surfaces and improve the efficiency of the process [14]. Also, in the case of naphthenic acid crude oils, alkalis will react with the present organic acids and generate in situ surfactants (soap) to decrease the IFT of the oil–water system to ultralow values [39].

So far, the ASG approach has been of interest to many researchers. Several aspects, including screening laboratory investigations to find appropriate surfactants, pHase behavior (in the presence/absence of alkali), foam generation studies and properties of the generated foam, amount of residual oil after core flooding in various types of rocks, etc., were considered in the previous publications. Cottin et al. used the ASG method, with nitrogen and methane as the gaseous phase, to improve oil recovery from carbonate core samples [40]. The system could achieve ultralow IFT values (less than 0.003 mN/m) and recovered about 76% of ROIP (residual oil in place) after the synthetic formation water injection. Phukan et al. utilized a combination of 0.3 wt% of sodium dodecyl sulfate (SDS) and 1.0 wt% sodium carbonate as the aqueous phase, and CO_2 as the gaseous phase, to improve oil recovery from sandstone core samples during ASG flooding [39]. They also compared the results of oil recovery obtained from implementing ASG flooding with the SAG and WAG injections. They showed that the higher oil recovery (approximately 23% OOIP compared to 18.29 and 8.84% OOIP for SAG and WAG, respectively) was achieved from the sandstone core plug by the ASG method.

9.3.4 Heated chemically enhanced water alternating gas (CEWAG)/water alternating gas (WAG) injection

The combination of two well-known thermal and chemically enhanced water alternating gas (CEWAG)/ water alternating gas (WAG) methods is named heated CEWAG/WAG or Hot CEWAG/WAG. This process is specially designed for semiheavy to heavy crude oils to reduce the viscosity of crude oil in addition to the specific benefits of the CEWAG/WAG process [41]. The WAG process is disappointing in these reservoirs due to the gravity overriding and poor mobility ratio. The CEWAG/WAG injection routinely causes a reduction in the temperature of the reservoir, especially near the injection wells. This phenomenon is expected because the temperature of injected fluids (e.g., water, gases, or chemicals) is lower than that of the reservoir. Also, these fluids have a high specific heat capacity to reduce the temperature. The reduced temperature leads to an increase in oil viscosity and a decrease in the volumetric sweeping efficiency of the process. The increasing viscosity is more intense in heavy crude oil reservoirs. Hence, the use of thermal processes in the conjugation of other methods is highly recommended for semiheavy to heavy crude oil reservoirs.

On the other side, the injected water usually contains biologic cultures, especially sulfate-reducing bacteria (SRB), or nutrients suitable for in situ bacteria communities. Moreover, the waterflooding process that lowers the reservoir temperature provides favorable conditions for bacterial growth and activities. The colonization of biogenetic cultures, especially SRB bacteria, in the cooler zones is a major concern in the petroleum industry due to the generation of hydrogen sulfide (H_2S) gas in the waterflooded reservoirs. The increasing operating costs and corrosion, iron sulfide formation, increasing environmental and health hazards, and reducing profits are the serious difficulties of the presence of H_2S. The generated insoluble iron sulfides are the main culprits responsible for the reducing porosity and permeability in the reservoir. The production of corrosive chemical agents (e.g., acids) and direct attack of iron via withdrawal of electrons are the two principal procedures of microorganisms for corrosion. They impose a significant economic loss in the oil industry, especially the offshore industry and gas pipelines. The growth of bacterial biofilms related to SRB bacteria produces acids and results in severe corrosion in metallic production facilities. About 15%–30% of corrosion failures in the oil industry are related to microbiologically influenced corrosion [42, 43]. There are several strategies for controlling SRB activities during waterflooding, such as using biocides or nonbiocidal techniques (i.e., biocompetitive exclusion technology, periodic change in ionic strengths, and bacterial film inhibitors). The efficiency of these procedures is usually unsatisfactory due to microbial resistance, human health risk, and environmental hazards [42–44]. Temperature is a vital controlling factor in bacterial growth, and its effect has been proven in several studies. The optimal temperature growth has been positively related to the strains of bacteria present in the system. The highest temperature for a strain isolated from an Alaska oil was estimated at 37°C at 100 atm. The coliform occurrences are highly reduced with increasing temperature, and after a specific temperature (mainly 80°C), the microorganisms become inactive. Unlike temperature, the pressure of the system has an insignificant influence on bacterial growth [44].

The use of heated CEWAG/WAG is an efficient way for increasing reservoir temperature to reduce the viscosity of crude oil and microorganisms' activity. These factors cause an increasing sweep efficiency of the EOR process, higher oil production recovery, lower corrosion effects, and high microscopic and macroscopic sweep efficiency. In this way, both the aqueous and gas phases are heated up to 90°C before injection into the reservoir. Instead of heated water, steam can also be used, whose temperature has highly increased even to 300°C. Although steam is a routine process in the recovery of heavy crude oil reservoirs such as the steam-assisted gravity drainage (SAGD) and vapor-extraction (VAPEX) methods, the heated CEWAG/WAG has been tested only in lab models or numerical simulations, and there is no field experience related to this process [45–47].

9.3.5 Alkaline surfactant polymer (ASP) and ASP-gas injection

ASP flooding serves the benefits of alkali, surfactant, and polymer injections simultaneously [3]. This technique is designed to reduce the IFT, decrease the mobility of the displacing phase and improve the mobility ratio of flooding, improving the displacement (i.e., microscopic) and volumetric (i.e., macroscopic) sweep efficiencies, enhancing the capillary number, and wettability reversal to produce the remaining residual and bypassed crude oils [3, 48, 49]. The reaction of the injected alkali and organic acids in crude oil produces an in situ surfactant (i.e., soap). It causes further reduction in IFT in conjugation with an injected surfactant to supply an ultralow IFT condition [3, 50, 51]. The generated soap plays a crucial role in lowering surfactant adsorption on the grain surfaces in the reservoir, decreasing the deleterious effect of divalent ions on the injected surfactant, controlling salinity, and improving surfactant efficiency [3, 52, 53]. The addition of polymer increases the sweep efficiency by increasing the aqueous phase's viscosity, decreasing its mobility, and controlling the mobility ratio [3, 54]. In other words, ASP processes improve the microscopic sweep efficiency by increasing the capillary number (via the addition of surfactant and alkali) and macroscopic sweep efficiency by reducing the mobility ratio (via the addition of polymer) [3]. Farad et al. carried out the ASP-Gas technique to benefit from both ASP and CO_2 injections. They investigated the effect of the pH and slug ratio of ASP/gas. This technique is similar to the WAG process in which water is replaced with the ASP [41]. The use of CO_2 before ASP injection increases the microscopic sweep efficiency, and higher incremental oil recovery will be achieved.

The mechanisms of each component of the ASP process and their synergy are further reviewed in the following four subsections.

9.3.5.1 Alkali flooding mechanisms

The ionic salt of the groups I (i.e., alkali metals) and II (i.e., alkaline metals) elements of the periodic table is named alkali or alkaline, dependent on its cation type. In this chapter, alkali and alkaline are used equivalently, meaning "the ionic salt compounds that their solution has pH values higher than 7." The injected alkali reacts with the acid compounds (naphthenic acids) present in the reservoir crude oil, and in situ surfactant (also called soap) is generated based on Eq. (9.1) [52].

$$HA^- + OH^- \rightarrow A^- + H_2O \tag{9.1}$$

where HA is acid compound in crude oil, OH is produced hydroxide ion upon dissolution of alkali in the water, and A is the generated soap.

The generated soap provides a similar function to the injected surfactant in reducing the interfacial tension (IFT) between oil/water and mobilizing the remaining oils in the pores. The generated soap acts multifunctional and lowering the IFT is known as a primary function. Alkali solution has a key role in the formation and sequestering of divalent ions that cause a reduction in the adsorption of injected surfactant in the reservoir [3]. Alkali has a lower price than the surfactant. When an alkali is used, a less amount of expensive surfactant is required, making the EOR process cost-effective. Wettability reversal has been reported due to the presence of alkali in the reservoir. The formation of emulsion is another function of alkali during the ASP process [55]. Emulsification has a close relationship to increasing incremental oil recovery. The higher incremental ASP projects have been reported from reservoirs with strong emulsification [56–58]. The adsorption of surfactant and polymer on the grain surface is an undesirable phenomenon in the ASP process. Alkali highly reduces polymer and surfactant adsorption by increasing the reservoir pH, increasing the negative charge density on the grain surfaces, and wettability changes [3, 59, 60]. The optimum condition for the injected surfactant to achieve ultralow IFT is a high salinity medium. The presence of alkali lowers the optimal salinity for the surfactants [61].

9.3.5.2 Surfactant flooding mechanisms

The primary goal in surfactant injection is reducing the oil–water IFT and in turn increasing the capillary number. The capillary number (N_c) is defined as a ratio of viscous to capillary forces (Eq. 9.2).

$$N_c = \frac{\nu\mu}{\sigma \cos\theta} \tag{9.2}$$

where ν is the interstitial velocity, μ is the viscosity of the displacing phase, σ is the interfacial tension between the displacing/displaced phases, θ is the contact angle of the wetting phase and rock surface [4].

As the capillary number increases, residual oil saturation decreases because the capillary force, the main factor in the trapping of oil droplets, tends to lower values in the denominator of Eq. (9.2). The low values of capillary number in the waterflooding process are the main reason for trapping significant crude oils in reservoir pores. The common order of capillary number in waterflooding operations is 10^{-6}–10^{-7}. When N_c's value increases to 10^{-5} in flood, the remaining oil will decrease. At least three possible strategies are available to increase N_c in EOR processes. (1) Increasing the flow rate of flooding (i.e., displacing phase), (2) increasing the viscosity of the displacing phase, and (3) lowering the IFT between the displaced and displacing phases [4]. The first strategy has some limitations related to the capacity of pumps and other facilities. The second and third strategies are most straightforward and practical. Surfactant flooding increases N_c through the lowering IFT.

Surfactants are amphiphilic organic compounds with a hydrophobic (water-insoluble) and hydrophilic (water-soluble) group in their molecules. However, the hydrophobic tail of the surfactant comprises long-chain hydrocarbons (C8–C18), whereas the hydrophilic head consists of different compounds depending on the type of surfactant. The four different surfactants are anionic, cationic, zwitterionic (or amphoteric), and nonionic [3]. The schematic representation of different types of surfactants is shown in Fig. 9.6. When the surfactant is injected into the reservoir at the oil and water interface, the IFT decreases. Decreasing the IFT facilitates a microemulsion of oil in water (Fig. 9.7) [63].

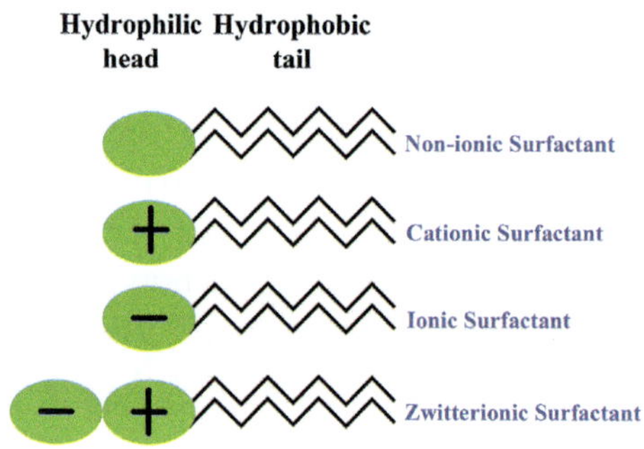

Fig. 9.6 Schematic representation of four different types of surfactants depending on the nature of the hydrophilic group [62].

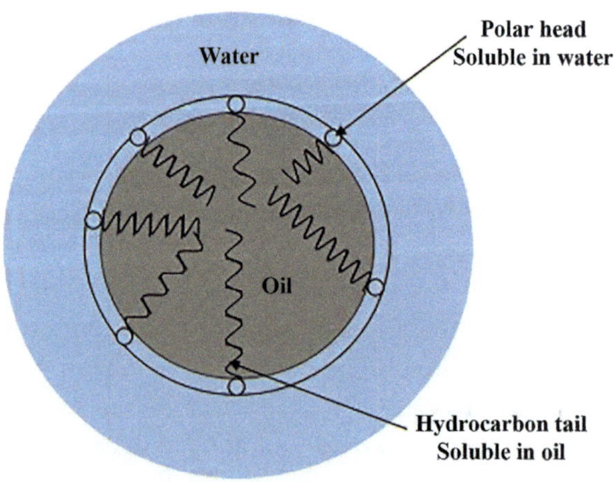

Fig. 9.7 Role of surfactant for emulsification of oil in water [3].

Hence, oil droplets are surrounded by water, and the coalescence of emulsion droplets mobilizes the residual oil.

9.3.5.3 Polymer flooding mechanisms

Polymers incorporate increasing N_c through the enhancing viscosity of the displacing phase (Eq. 9.2). The increase of viscous forces by the polymer is not sufficient to remobilize the residual oils. Capillary forces trap the residual oil in pores. Polymer flooding is mainly designed to recover bypassed oils in waterflooding, not including residual oils. These oils were bypassed due to the unfavorable mobility ratio in waterflooding or excessive heterogeneities in the reservoir [64]. However, Wang et al. showed that polymer's viscoelastic properties cause a viscous dragging force parallel and perpendicular to the oil–water interface, which mobilizes residual oils and decreases residual oils [65, 66]. Polymer alleviates the mobility of the displacing phase and improves the mobility ratio. The mobility ratio is defined as the ratio of the mobility of the displacing phase to the mobility of the displaced phase (Eq. 9.3).

$$M = \frac{\text{Mobility of displacing phase}}{\text{Mobility of displaced phase}} = \frac{\lambda_{\text{water}}}{\lambda_{\text{Oil}}} = \frac{K_{rw}/\mu_w}{K_{ro}/\mu_o} = \frac{K_{rw} \times \mu_o}{K_{ro} \times \mu_w} \tag{9.3}$$

where water and oil are the displacing and displaced phases, respectively, λ is mobility, Kr is the relative permeability, and μ is the viscosity.

As the displaced phase's mobility becomes larger than the mobility of the displacing phase, the volumetric or macroscopic sweep efficiency will increase. The stable displacement process occurs when M is lower than 1. When M is higher than 1, the displacement is unstable and viscous fingering or a nonuniform displacement front will happen (Fig. 9.8). Hence, when the injected water has larger mobility regarding reservoir oil, the mobility ratio is unfavorable. The ASP process is a suitable way to achieve the $M < 1$ process. In this way, the viscosity of oil decreases, the viscosity of water increases, the relative permeability of water decreases, and the relative permeability of oil increases [3]. The latter mechanism is also named disproportionate permeability reduction, in which the relative permeability of water (K_{rw}) is largely decreased regarding the relative permeability of oil (K_{ro}) during the addition of polymer [52]. Increasing the viscosity of oil causes decrease in water cut (f_w) and improves sweep efficiency (Eq. 9.4).

$$f_{w} = \frac{1}{1 + \left(\frac{K_{ro}}{K_{rw}}\right)\frac{\mu_w}{\mu_o}} = \frac{1}{1 + \frac{\lambda_o}{\lambda_w}} = \frac{1}{1 + \frac{1}{M}} \tag{9.4}$$

Viscous fingering is severe in heterogeneous reservoirs with higher permeability streaks when the mobility ratio is unfavorable. In this situation, the early breakthrough of water from the thief zone will occur, and displacement efficiency largely decrease, even when the mobility ratio is close to one. In polymer flooding, the fluid cross-flow

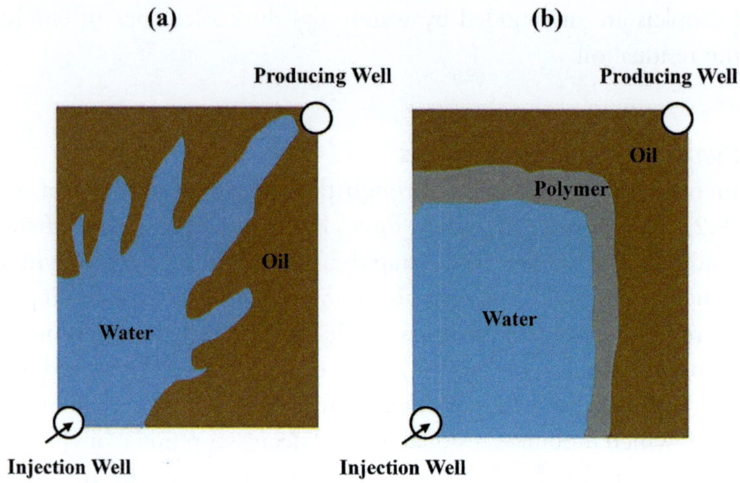

Fig. 9.8 Schematic representation for showing the effect of mobility ratio in flooding. (A) Viscous fingering of water through the oil in the condition of unfavorable mobility ratio ($M > 1$), and (B) uniform displacement due to the addition of polymer to injected water and providing a favorable mobility ratio ($M < 1$). *(Modified after Xia S, Zhang L, Davletshin A, Li Z, You J, Tan S. Application of polysaccharide biopolymer in petroleum recovery. Polymers (Basel) 2020;12:1860.)*

between reservoir layers has a primary role in improving uniform displacement in heterogeneous reservoirs [2].

9.3.5.4 The synergy of mechanisms in ASP flooding

The ASP process benefits from the synergy of the individual mechanisms of the alkali, surfactant, and polymer. This process improves both the microscopic and macroscopic sweep efficiencies by remobilization of the residual and bypassed oils. The synergy between the generated in situ soap and synthetic surfactant allows to achieving ultra-low IFT conditions. The optimum salinity for soap and the synthetic surfactant is mutually opposed in which the soap has the higher efficiency in low salinity. They function together to provide more accessible optimum salinity conditions for better performance in the reservoir. Hence, the soap and surfactant synergistically decrease the IFT between water and oil, and the polymer improves the displacement efficiency [3]. Different injection schemes have been employed in the literature. The typical scheme of the ASP process has several injection cycles (Fig. 9.9). These cycles comprise a preflush of brine, injection of alkali and surfactant combinations, polymer solution injection, freshwater injection to optimize chemicals recovery, and finally a flood with drive water. The main part of the ASP process has a standard formulation of the injection of 0.2 pore volume (PV) from the mixture of 1% surfactant and 0.5% alkali. This slug is followed by the 0.3 PV or more injection from the 1000 ppm polymer solution [3].

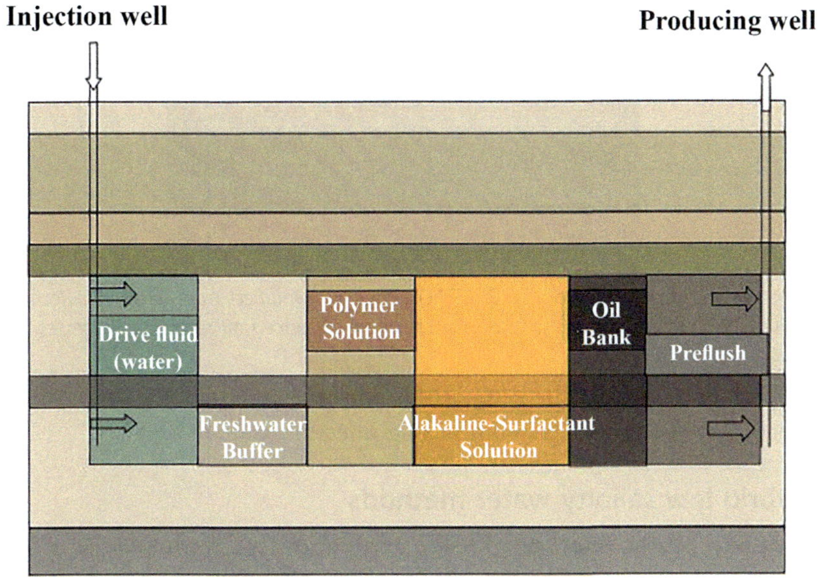

Fig. 9.9 Schematic representation of different injection cycles in the ASP process [3].

9.3.6 Polymer alternating gas (PAG) injection

The PAG technique is a technique of applying a polymeric solution instead of water in the conventional WAG process [67]. In this process, polymer flooding is alternatively injected with the gas to improve the volumetric sweep efficiency and final recovery in the WAG process, especially in the high heterogeneity reservoirs. This technique is designed to surpass WAG or gas flooding drawbacks in naturally fractured reservoirs or reservoirs with high permeability zones, such as early gas breakthrough and gravity segregation [68]. CO_2 is a preferred gas for heavy reservoir oils or in the PAG process, which can be injected as immiscible or miscible [68–71]. This gas has lower injectivity problems, a higher oil recovery efficiency, a lower formation volume factor, acceptable microscopic sweep efficiency, and large reserves of this gas available in nature [71, 72]. A miscible method has higher advantages than the immiscible method because the dissolution of gas into oil reduces viscosity, swells crude oil, increases mobility, and improves the internal solution gas drive [73–75]. Viscosity fingering is an unfavorable prevailing phenomenon in waterflooding or WAG processes due to the higher mobility of the displacing phase (i.e., water) than the displaced phase (i.e., crude oil). A combination of polymer with water increases the displacing phase viscosity and provides a favorable mobility ratio [69, 75]. Furthermore, adsorption of polymer in high permeability thief zones causes a lowering permeability and increasing viscosity of water. Hence, the areal sweep efficiency is improved and the incremental oil recovery will be

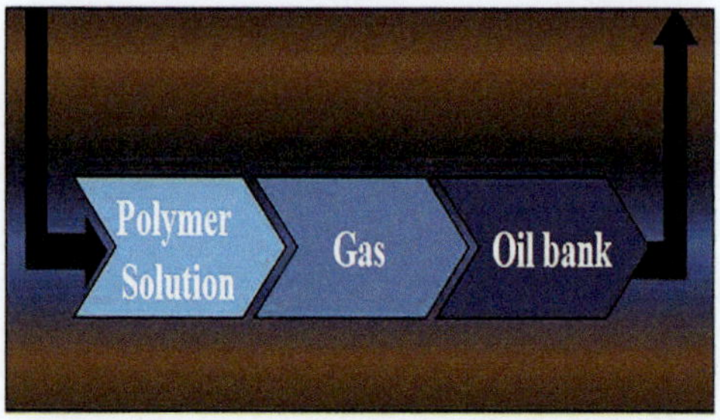

Fig. 9.10 Schematic representation of the PAG process. *(Modified from Scott AJ, Romero-Zerón L, Penlidis A. Evaluation of polymeric materials for chemical enhanced oil recovery. Processes 2020;8:361.)*

enhanced [70]. Each cycle of the PAG process includes an initial step of gas slug injection and the second step of polymer solution slug injection (Fig. 9.10).

9.3.7 Hybrid low salinity water methods

The low salinity water injection (LSWI) application has been successfully tested in sandstone and carbonate reservoirs. However, most of the studies have been conducted in sandstone reservoirs [76]. Numerous laboratory investigations and pilot studies have shown that using low salinity water instead of high salinity brine improves oil recovery in the secondary or tertiary EOR processes [77–83]. This technique has received much attention because of its low cost, environmental friendliness, and easy implementation [77]. LSW is also used in conjunction with other agents, named hybrid LSW, especially chemicals (e.g., surfactants or polymers) or gases. The CO_2 low salinity water alternating gas (CO_2 LSWAG) method includes using low salinity water instead of brine and CO_2 as a gas in the routine WAG process. LSW has been successfully implemented with surfactant injection. The optimum salinity for surfactants is low, and using LSW improves the solubility of the surfactant, reduces adsorption and retention during the flooding process [84]. Results showed that low salinity surfactant (LS-S) flood has a higher incremental oil recovery than conventional surfactant injection with high salinity brine [85]. LSW also showed several economic advantages when used in polymer flooding. In the LSW method, a lower amount of polymer is needed for achieving the target viscosity, and the oil recovery has higher than using high salinity water [86, 87].

The mechanisms of LSW have been controversial for many years. The known mechanisms include fine migration, multicomponent ionic exchange (MIE), wettability alteration, pH increase, and salting-in effect [76, 77]. These mechanisms are combined and synergized by the miscible CO_2 effects in the CO_2 LSWAG process to improve the sweep efficiency and increase oil recovery. The wettability alteration is a critical and unanimously acceptable mechanism in this process's effectiveness [77]. This phenomenon directly impacts the relative permeability in which the relative permeability of water decreases and the relative permeability of oil increases [88–90]. The wettability alteration occurs due to the ion exchange and geochemical reactions [77]. The combination of LSW and polymer is a successful method for improving both the macroscopic and microscopic sweep efficiency. Although polymer improves the displacement mobility ratio, LSW changes the CBR interactions and wettability. It is believed that the injection of LSW as preflush before polymer flooding has a higher efficiency than injection after polymer flooding. In this way, the porous medium wettability is changed to more water-wet, in which oil droplets are detached from small pores and pore walls, and subsequent polymer flooding can displace these oils to improve oil recovery [91].

9.4 Comparison of different methods

Different combined chemical methods were reviewed in the previous sections. These methods help to increase oil recovery through specific mechanisms, depending on the different components available in each method. Table 9.1 summarized the important mechanisms active in each method and compared the advantages of these methods. Although a number of these methods were implemented on the field scale, a few were merely carried out on the lab scale or simulations.

9.5 Screening criteria for selecting the combined EOR methods

The economic and technical success in implementing EOR methods in every field is achieved when a suitable and efficient recovery process is designed and implemented. Recovery methods are diverse and expensive and should be selected cautiously based on the nature of the reservoir fluid and rock properties. The descriptive screening criteria for EOR methods have been developed in the literature for field appraisal purposes. These standards are established from real fields and provide a general rule of thumb to consider suitable EOR techniques for a candidate reservoir. The assigned screening criteria are useful as a general reference but cannot provide any information about reservoir performance quantification. The expected oil recovery can be determined from laboratory experiments, numerical simulations, or machine learning models in the study's appraisal phase [103]. Some of the abovementioned combined chemical methods have not been applied in real fields but they have only been conducted in the lab or modeled in simulations. In the literature, the screening criteria are available for methods performed in several real fields around the world. Here the criteria for the combined methods that were not available in the literature were determined based on individual methods. For example, the heated CEWAG/WAG's proposed screening criteria were a combination of hot water and WAG methods. Table 9.2 reported the screening criteria for the combined chemical methods.

Table 9.1 Comparison of different combined chemical EOR methods.

Methods	Mechanisms	Merits	Field experiences	Reference(s)
Water alternating gas injection (WAG)	Mobility control, reducing viscosity, IFT reduction	Improving both microscopic and macroscopic sweep efficiency	Canada, India, United States, China, North Sea, Middle East, Indonesia, Russia, Malaysia	[18, 22, 23, 92–94]
Surfactant alternating gas injection (SAG)	Reducing IFT, foam formation, mobility control	Less gravity overrides, less viscous fingering, reducing corrosion	Colombia, North Sea	[95, 96]
Alkaline-surfactant gas (ASG)	Reducing viscosity, IFT reduction, foam formation, mobility control, wettability alteration	Suitable for low permeability-porosity reservoirs		
Heated CEWAG/WAG injection	Lowering viscosity, improving microscopic and macroscopic sweep efficiency	Suitable for semiheavy and heavy crude oils, inhibiting bacterial activities		
Alkaline-surfactant-polymer (ASP) and ASP alternating gas (ASP-gas)	Lowering IFT, decreasing mobility ratio, increasing viscosity of displacing phase, increasing the flow rate of flooding, wettability alteration, emulsification	Increasing both microscopic and macroscopic sweep efficiency, suitable for heterogeneous reservoirs	China, United States, India, Venezuela, Canada	[52, 56, 97–100]
Polymer alternating gas (PAG) injection	Reducing oil viscosity, swelling crude oil, increasing mobility, improving internal solution gas drive, decreasing mobility ratio	Improving both microscopic and macroscopic sweep efficiency	United States, China	[68, 101, 102]

Hybrid low salinity water (hybrid LSW) methods	CO_2 LSWAG	Fine migration, multicomponent ionic exchange (MIE), wettability alteration, pH increase, salting-in effect, reducing IFT, reducing viscosity, wettability alteration, decreasing relative permeability of water phase	Environmentally friendly, cost effective		
	Low salinity surfactant (LS-S) flood	Fine migration, multicomponent ionic exchange (MIE), wettability alteration, pH increase, salting-in effect, reducing IFT, wettability alteration	Environmentally friendly, cost effective		
	LSW polymer	Fine migration, multicomponent ionic exchange (MIE), wettability alteration, pH increase, salting-in effect, wettability alteration, decreasing mobility ratio	Environmentally friendly, cost effective	North Africa, United States	[100]

9.6 Modeling and simulations

Numerical reservoir simulation uses the understanding of the past performance to predict the future performance of fluid flow in the reservoir. It is a helpful and efficient way to evaluate the sensitivity of different parameters and various scenarios about the reservoir's future life cycle [104]. Evaluation of incremental oil recovery is a primary goal of reservoir simulation in the candidate field. The simulation of chemical methods is highly crucial to assess the technical and economic feasibility of different scenarios. The combined chemical methods are complex processes in which the impacts of different chemicals and phases should be considered simultaneously. Hence, multicomponent multiphase models should be considered for simulation of these methods. In the combined chemical process simulation, all possible effects should be considered for the comprehensive modeling of fluid flow in porous media.

Christensen et al. investigated the effect of relative permeability hysteresis due to the gas phase's trapping in the WAG process. They used the relative permeability hysteresis model in a compositional reservoir simulator for a sandstone reservoir in the North Sea. The formation of a trapped gas zone in the vicinity of the injector well was observed in the simulated model. The real ultimate recovery due to hysteresis is about 5% lower than the standard compositional model. In this condition, free gas is not produced due to trapping gas around the injector and causes lowering gas injectivity [29]. The modeling of polymer-based flooding methods (i.e., ASP and PAG) should include different items. Polymer adsorption and retention, elastic behavior, elongation viscosity, reduction in the relative permeability of the water phase, viscosity, salt effects, dispersion-diffusion phenomena, thermal effects, inaccessible pore volume, an effective concentration of polymer, equivalent shear rate in porous media, polymer rheology, and pH effects are important parameters for the detailed and comprehensive model [68, 105–107].

Different commercial simulators are available for simulating combined chemical methods. STARS by CMG, UTCHEM developed by the University of Texas at Austin, and ECLIPSE by Schlumberger are three known commercial simulators for modeling complex chemical reactions in both the lab scale and field scale [107, 108]. Also, scientific software based on the computational fluid dynamics (CFD) method (e.g., COMSOL, ANSYS, and Fluent) has been successfully used for simulating chemical methods [109–111].

9.7 Application of machine learning

Many parameters should be considered in designing, implementing, and predicting every EOR process, especially the combined chemical methods. On the one hand, each reservoir has its specific characteristics that impact certain EOR methods' performance. On the other hand, there are different combined chemical EOR methods. Besides the high degree of uncertainties in the reservoir, the items mentioned above provide difficulties in selecting a proper method for the studied reservoir. Numerical simulations in reservoir engineering studies are accurate and efficient for predicting each scenario's future performance. But these simulations are time-consuming to evaluate all possible scenarios to select the optimum one [112]. This is especially important when several reservoirs should be evaluated for performing the EOR process [113]. Also, chemical methods require costly and time-consuming laboratory experiments to achieve optimum criteria for effective implementation of chemical injection [114]. In recent years, machine learning has been employed successfully in EOR studies and provides beneficial results in a short time. Machine learning uses automated learning from the training or sample data to find the relevant patterns. The obtained patterns are employed to make a decision or accurate predictions [115, 116]. Machine learning has been efficiently applied to screen or predict EOR methods' performance [112–114, 117, 118]. Also, this technique can be used to optimize chemicals before injections. Alqahtani developed new models to predict HPAM polymers' viscosity by machine learning before numerical reservoir simulation of chemical flooding [119]. Ramos and Akanji used a Neuro-fuzzy model with a feed-forward backpropagation algorithm for the technical screen of EOR methods to select an appropriate method for implementation in Block T of the Angolan oil field. The input data for screening were porosity, permeability, depth, viscosity, oil saturation, and reservoir oil API. The obtained results of the sensitivity analysis confirmed the robustness of the proposed model [120]. Giro et al. used the artificial intelligence (AI) methods to select EOR materials in the specific reservoir. Their model was successfully correlated physical and chemical characteristics of injecting fluids (i.e., EOR materials) with rock and fluid properties of the reservoir [121]. Ahmadi and Pournik successfully applied a model based on the least square support vector machine to monitor both technical [i.e., recovery factor (RF)] and economic (net present value (NPV)) parameters during a chemical EOR method [114].

9.8 Field applications

Some of the combined chemical EOR methods (e.g., WAG, SAG, PAG, and ASP) are known and most used in the oil industry (Table 9.1). This issue may be related to lower economic feasibility other methods, simplicity in operation, ease of simulation, and applicability for different reservoirs globally. These are the important reasons for implementing these known EOR methods. The execution of other methods may be applicable in the future after an adequate understanding about the underlying mechanisms, development of different simulators for better modeling of different phenomena and mechanisms, economic and operation justification of these methods when their ability for the higher oil recovery performance is proven, and maybe further progress in technical aspects to perform with lower costs. The WAG method is a widely used and known

Table 9.2 The screening criteria for the combined chemical EOR methods.

EOR methods	Gravity (API)	Viscosity (cp)	Composition	Oil saturation (%PV)	Formation type	Net thickness (ft)	Average permeability (md)	Depth (ft)	Temperature (°F)
WAG	33–39	0.3–0.9		NC	Sandstone preferred		130–1000	7545–8887	194–253
SAG	22 <	16 <		20–53	Sandstone preferred	NC	50–60	2500–5300	122–155
ASP	20 >	35 <	Light, intermediate, some organic acid for alkaline injection	35 >	Sandstone preferred	NC	10 >	9000 <	200 <
Heated CEWAG/ WAG	12–25	0.3–0.9		15–85	Sandstone preferred		130–1000	500–2950	75–135
PAG	15 <	150 <	High percent of light hydrocarbons	20 >	Sandstone preferred	Wide range	10 >	2500–9000	200 <

NC: not critical.

(Adapted from Gbadamosi AO, Kiwalabye J, Junin R, Augustine A. A review of gas enhanced oil recovery schemes used in the North Sea. J Pet Explor Prod Technol 2018;8:1373–87; Taber JJ, Martin FD, Seright RS, et al.. EOR screening criteria revisited-part 2: applications and impact of oil prices. SPE Reserv Eng 1997;12:199–206; Taber JJ, Martin FD, Seright RS. EOR screening criteria revisited-part 1: introduction to screening criteria and enhanced recovery field projects. SPE Reserv Eng 1997;12:189–98; Dandan H, Yuanbing W, Xiaocui G, Songhao H, Mingsheng F, et al.. Optimisation and design for WAG-CO2 combined with soft micro gel SMG in complicated carbonate reservoirs containing with high permeability streaks. SPE Middle East Oil Gas show conf, 2017; Al Adasani A, Bai B. Analysis of EOR projects and updated screening criteria. J Petrol Sci Eng 2011;79:10–24.)

process in the tertiary EOR methods. Several field studies around the world have been reported both in limestone and sandstone reservoirs [122–127]. All field studies showed that this technique has higher oil recovery than waterflooding or gas flooding. Christensen et al. reviewed 59 WAG projects from different locations. They showed that about 5%–10% of the original oil in place (OOIP) higher oil recovery is achieved in successful WAG projects after secondary EOR methods. They expressed that most of the WAG projects were performed as a miscible process in onshore sandstone reservoirs [19]. ASP method is the second grade after the WAG method from the field experiences. Several successful ASP projects have been implemented in different countries, mostly in the United States and China [52, 128–131]. The first ASP process in China was commenced in the Daqing oil field in 1994 [98]. The oil recovery was forecast to be increased by 20% OOIP over waterflooding from numerical simulations and core flooding [98]. The West Kiehl Field was subjected to ASP injection in 1987, and an oil recovery enhancement of 16% OOIP over waterflooding was calculated [99]. Li et al. simulated polymer alternating gas (PAG) injection in TR78 of the North Burbank Unit to overcome challenges of high heterogeneity and high permeability in the reservoir layers. They showed that 7% OOIP higher oil recovery could be achieved regarding WAG in this field [68]. There is no field experience from the Heated CEWAG/WAG injection. This method has been reported from laboratory experiments for semiheavy and heavy oil reservoirs in which higher oil recovery was achieved over the routine WAG process [45–47]. Several field studies have reported a higher performance for low salinity water methods [100, 132–135], although a few field studies are available for hybrid low salinity water methods [100, 136]. Ocampo et al. performed a pilot study of surfactant alternating gas (SAG) injection in the northern part of the Cusiana field in Colombia, South America. Increasing oil recovery, lower gas injectivity, and decreasing GOR were observed [96]. Several lab-scale studies have been reported for the ASG method [39].

9.9 Field challenges

Although the combined chemical methods provide a higher oil recovery, their implementation in oil fields encounters several challenges and difficulties. An efficient decision and strategy should be made about these challenges before initiating the injection process. Waste management of produced fluids, corrosion of surface and downhole facilities, scale formation, asphaltene precipitation, decreasing injectivity, chemical degradation and retention, reservoir heterogeneity, and hydrate formation are typical and crucial issues for field application of these methods [137–139]. These challenges are apart from the economic issues related to the cost of chemicals and their required operational facilities for injection.

Asphaltene precipitation is a common phenomenon during chemical EOR methods, especially in miscible floods or those that include light *n*-alkanes (especially C1–C4) and

CO_2 gas [140]. The dissolution of CO_2 or other injectants (e.g., surfactant) in reservoir oil causes a decrease in oil solubility regarding asphaltenes. Hence, these compounds destabilized and precipitated [141]. The clogging of reservoir pores, production facilities plugging, and wettability alteration are well-known problems of asphaltenes that impact oil recovery performance and efficiency [137, 142]. Using inhibitors is a common strategy to avoid asphaltene precipitation [143]. Several kinds of research have recently reported the effects of nanoparticles on asphaltene reduction [144–146]. The scale is an inorganic substance that is deposited in operational facilities and causes flow restriction. Increasing pH during flooding, especially in the ASP method, led to the deposition of calcium carbonate. This problem is severe when seawater or water with high divalent ions is used. The sequestering divalent ions and desalting injected water are common ways to inhibit scale formation [147]. Some inhibitors were also suggested in the literature to reduce scale formation problems [148]. The injection of chemical agents and CO_2 gas plays a key role in the occurrence of corrosion. This phenomenon imposes extra costs on the economy of EOR projects. CO_2 with water mixing is responsible for producing corrosive agents during EOR methods [149].

9.10 Economic and environmental feasibility

The primary environmental concern about the combined chemical EOR methods is that the produced water is oversalted and contains chemicals, radioactive substances, and toxic heavy metals. High caution should be exercised to avoid contaminating farmland and drinking water sources. The filtration of produced water and reinjection into the reservoir is a highly recommended strategy [137]. Generally, the CO_2-based EOR methods are environmentally friendly methods in which greenhouse gas (i.e., CO_2) is stored in the oil reservoir and additional oil is recovered [150].

The economic evaluation of the EOR project gives critical instructions that are the basis for deciding to embark on a project [151]. Manufacturing costs and product revenues should be considered cautiously to evaluate the viability of any project. Capital expenditure (CAPEX), operational expenditure (OPEX), chemical and labor costs are important factors in manufacturing costs evaluation [152]. Maximizing net present value (NPV) is a primary goal for each EOR project. NPV is defined as the difference between all positive cash flow and initial investments over a period of time [152]. Unfortunately, the published economic feasibility evaluation of the combined chemical EOR methods is scarce in the literature. Most of the CO_2 EOR projects are evaluated economically if the CO_2 sources are available. These methods are also successful in the storage of CO_2 [153]. The combined EOR methods' economic feasibility is highly related to the oil price, amount of incremental oil recovery over other methods, and chemicals' cost.

References

[1] Levitt D, Pope GA, et al. Selection and screening of polymers for enhanced-oil recovery. In: SPE symp. improv. oil recover; 2008.

[2] Sorbie KS. Polymer-improved oil recovery. Scotl Blackie Son; 2013.

[3] Olajire AA. Review of ASP EOR (alkaline surfactant polymer enhanced oil recovery) technology in the petroleum industry: prospects and challenges. Energy 2014;77:963–82.

[4] Green DW, Willhite GP. Enhanced oil recovery. vol. 6. Richardson, TX: Henry L. Doherty Memorial Fund of AIME, Society of Petroleum Engineers; 1998.

[5] Lyons WC, Plisga GJ. Standard handbook of petroleum and natural gas engineering. Elsevier; 2011.

[6] Aadland RC, Jakobsen TD, Heggset EB, Long-Sanouiller H, Simon S, Paso KG, et al. High-temperature core flood investigation of nanocellulose as a green additive for enhanced oil recovery. Nanomaterials 2019;9:665.

[7] Gbadamosi AO, Junin R, Manan MA, Agi A, Yusuff AS. An overview of chemical enhanced oil recovery: recent advances and prospects. Int Nano Lett 2019;9:171–202.

[8] Derikvand Z, Rezaei A, Parsaei R, Riazi M, Torabi F. A mechanistic experimental study on the combined effect of Mg2+, Ca2+, and SO42-ions and a cationic surfactant in improving the surface properties of oil/water/rock system. Colloids Surfaces A Physicochem Eng Asp 2019;587, 124327.

[9] Morrow NR, Tang G, Valat M, Xie X. Prospects of improved oil recovery related to wettability and brine composition. J Petrol Sci Eng 1998;20:267–76.

[10] Katende A, Sagala F. A critical review of low salinity water flooding: mechanism, laboratory and field application. J Mol Liq 2019;278:627–49.

[11] Olabode O, Alaigba D, Oramabo D, Bamigboye O. Modelling low-salinity water flooding as a tertiary oil recovery technique. Model Simul Eng 2020;, 6485826.

[12] Yildiz HO, Morrow NR. Effect of brine composition on recovery of Moutray crude oil by waterflooding. J Petrol Sci Eng 1996;14:159–68.

[13] Rezaei A, Abdollahi H, Derikvand Z, Hemmati-Sarapardeh A, Mosavi A, Nabipour N. Insights into the effects of pore size distribution on the flowing behavior of carbonate rocks: linking a nano-based enhanced oil recovery method to rock typing. Nanomaterials 2020;10:972.

[14] Rezaei A, Riazi M, Escrochi M, Elhaei R. Integrating surfactant, alkali and nano-fluid flooding for enhanced oil recovery: a mechanistic experimental study of novel chemical combinations. J Mol Liq 2020;308, 113106.

[15] Wang Z, Zhao X, Bai Y, Zhao M, Li Y, Cs W. Performance comparison of different alkaline/surfactant binary combination flooding system. Oilf Chem 2015;4:020.

[16] Orr Jr FM, Silva MK, Lien CL, Pelletier MT, et al. Laboratory experiments to evaluate field prospects for CO2 flooding. J Petrol Tech 1982;34:888–98.

[17] Jarrell PM, Fox CE, Stein MH, Webb SL. Practical aspects of CO2 flooding. vol. 22. Richardson, TX: Society of Petroleum Engineers; 2002.

[18] Van Gool FR, Currie PK, et al. An improved model for the liquid-loading process in gas wells. In: Prod. oper. symp; 2007.

[19] Christensen JR, Stenby EH, Skauge A. Review of WAG field experience. SPE Reserv Eval Eng 2001;4:97–106.

[20] Hustad OS, Holt T. Gravity stable displacement of oil by hydrocarbon gas after waterflooding. In: SPE/DOE enhanc. oil recover. symp. Society of Petroleum Engineers; 1992.

[21] Rogers JD, Grigg RB. A literature analysis of the WAG injectivity abnormalities in the CO2 process. In: SPE/DOE improv. oil recover. symp. Society of Petroleum Engineers; 2000.

[22] Pariani GJ, McColloch KA, Warden SL, Edens DR. An approach to optimize economics in a West Texas CO2 flood. J Petrol Tech 1992;44:981–4.

[23] Sanchez NL. Management of water alternating gas (WAG) injection projects. In: Lat. Am. Caribb. pet. eng. conf. Society of Petroleum Engineers; 1999.

[24] Awan AR, Teigland R, Kleppe J. A survey of North Sea enhanced-oil-recovery projects initiated during the years 1975 to 2005. SPE Reserv Eval Eng 2008;11:497–512.

[25] Panda M, Ambrose JG, Beuhler G, McGguire PL. Optimized eor design for the Eileen west end area, Greater Prudhoe bay. SPE Reserv Eval Eng 2009;12:25–32.
[26] Bahagio DNT. Ensemble optimization of CO2 WAG EOR; 2013.
[27] Chen S, Li H, Yang D, Tontiwachwuthikul P. Optimal parametric design for water-alternating-gas (WAG) process in a CO2-miscible flooding reservoir. J Can Pet Technol 2010;49:75–82.
[28] Shahverdi H. Characterization of three-phase flow and WAG injection in oil reservoirs; 2012.
[29] Christensen JR, Larsen M, Nicolaisen H, et al. Compositional simulation of water-alternating-gas processes. In: SPE annu. tech. conf. exhib; 2000.
[30] Juanes R, Spiteri EJ, Orr Jr FM, Blunt MJ. Impact of relative permeability hysteresis on geological CO2 storage. Water Resour Res 2006;42:1–13.
[31] Tabatabaei Nezhad SAR, Rahimzadeh Mojarad M, Paitakhti Oskouei SJ, Moghadas JS, Farahmand DR. Experimental study on applicability of water alternating CO2 injection in the secondary and tertiary recovery. In: Int. oil conf. exhib. Mex. Society of Petroleum Engineers; 2006.
[32] Afzali S, Rezaei N, Zendehboudi S. A comprehensive review on enhanced oil recovery by water alternating gas (WAG) injection. Fuel 2018;227:218–46.
[33] Rezaei A, Derikvand Z, Parsaei R, Imanivarnosfaderani M. Surfactant-silica nanoparticle stabilized N2-foam flooding: a mechanistic study on the effect of surfactant type and temperature. J Mol Liq 2021;325:115091.
[34] Renkema WJ, Rossen WR. Success of SAG foam processes in heterogeneous reservoirs. In: IPTC 2007 int. pet. technol. conf. European Association of Geoscientists & Engineers; 2007. p. cp-147.
[35] Salehi MM, Safarzadeh MA, Sahraei E, Nejad SAT. Comparison of oil removal in surfactant alternating gas with water alternating gas, water flooding and gas flooding in secondary oil recovery process. J Petrol Sci Eng 2014;120:86–93.
[36] Hadlow RE. Update of industry experience with CO2 injection. In: SPE annu. tech. conf. exhib. Society of Petroleum Engineers; 1992.
[37] Chakravarthy D, Muralidharan V, Putra E, Hidayati D, Schechter DS. Mitigating oil bypassed in fractured cores during CO2 flooding using WAG and polymer gel injections. In: SPE/DOE symp. improv. oil recover. Society of Petroleum Engineers; 2006.
[38] Shi JX, Rossen WR. Improved surfactant-alternating-gas foam process to control gravity override. In: SPE/DOE improv. oil recover. symp. Society of Petroleum Engineers; 1998.
[39] Phukan R, Gogoi SB, Tiwari P. Alkaline-surfactant-alternated-gas/CO2 flooding: effects of key parameters. J Petrol Sci Eng 2019;173:547–57.
[40] Cottin C, Morel DC, Levitt D, Cordelier PR, Pope GA. (Alkali) surfactant gas injection: attractive laboratory results under the harsh salinity and temperature conditions of Middle East carbonates. In: Abu Dhabi int. pet. conf. exhib. Society of Petroleum Engineers; 2012.
[41] Farad S, Nsamba HK, Al Hassan A, Joseph W, Kabenge I. Effect of pH and slug ratio of alkaline surfactant polymer alternating gas flooding on oil recovery. Am Assoc Sci Technol 2016;3:47–52.
[42] Tian H, Gao P, Chen Z, Li Y, Li Y, Wang Y, et al. Compositions and abundances of sulfate-reducing and sulfur-oxidizing microorganisms in water-flooded petroleum reservoirs with different temperatures in China. Front Microbiol 2017;8:143.
[43] Wei L, Ma F, Zhao G. Composition and dynamics of sulfate-reducing bacteria during the waterflooding process in the oil field application. Bioresour Technol 2010;101:2643–50.
[44] Fink J. Petroleum engineer's guide to oil field chemicals and fluids. Gulf Professional Publishing; 2015.
[45] Dorostkar MJ, Mohebbi A, Sarrafi A, Soltani A. A laboratory study of hot WAG injection into fractured and conventional sand packs. Pet Sci 2009;6:400–4.
[46] Ebadati A, Akbari E, Davarpanah A. An experimental study of alternative hot water alternating gas injection in a fractured model. Energy Explor Exploit 2019;37:945–59.
[47] Gholamzadeh MA, Hashemi P, Dorostkar MJ, et al. New improved oil recovery from heavy and semi-heavy oil reservoirs by implementing immiscible heated WAG injection. In: Trinidad Tobago energy resour. conf; 2010.
[48] Austine J, Van Batenburg DW, Southwick JG, Zarubinska MA, Paramanathan S, Bouwmeester RCM, et al. Laboratory Evaluation of Inter-Well Partitioning Tracers for the determination of remaining oil saturation after ASP flooding. In: SPE Asia Pacific enhanc. oil recover. conf; 2015.

[49] Shen P, Wang J, Yuan S, Zhong T, Jia X. Study of enhanced-oil-recovery mechanism of alkali/surfactant/polymer flooding in porous media from experiments. SPE J 2009;14:237–44.
[50] Gao S, Li H, Li H, et al. Laboratory investigation of combination of alkaline-surfactant-polymer for Daqing EOR. SPE Reserv Eng 1995;10:194–7.
[51] Al-Sahhaf T, Suttar Ahmed A, Elkamel A. Producing ultralow interfacial tension at the oil/water interface. Pet Sci Technol 2002;20:773–88.
[52] Sheng JJ. A comprehensive review of alkaline-surfactant-polymer (ASP) flooding. In: SPE West. Reg. AAPG pacific Sect. Meet. 2013 Jt. tech. conf; 2013.
[53] Krumrine PH, Falcone Jr JS, Campbell TC, et al. Surfactant flooding 1: the effect of alkaline additives on IFT, surfactant adsorption, and recovery efficiency. Soc Pet Eng J 1982;22:503–13.
[54] Walters K, Jones DM, et al. The extensional viscosity behavior of polymeric liquids of use in EOR. In: SPE int. symp. oilf. chem; 1989.
[55] Kumar S, Mandal A. Studies on interfacial behavior and wettability change phenomena by ionic and non-ionic surfactants in presence of alkalis and salt for enhanced oil recovery. Appl Surf Sci 2016;372:42–51.
[56] Guo H, Li Y, Wang F, Yu Z, Chen Z, Wang Y, et al. ASP flooding: theory and practice progress in China. J Chem 2017;2017:1–18.
[57] Jia ZW, Yang QY, Yuan M, Hou Z, Zhang L. Experimental study on influential factors of ASP displacement efficiency in Daqing oilfield. Acta Petrol Sin 2006;27:102–5.
[58] Cheng J, Zhou W, Wang Q, Cao G, Bai W, Zhao C, et al. Technical breakthrough in production engineering ensures economic development of ASP flooding in Daqing oilfield. In: SPE Asia Pacific Oil Gas conf. exhib; 2014.
[59] French TR, Burchfield TE, et al. Design and optimization of alkaline flooding formulations. In: SPE/DOE enhanc. oil recover. symp; 1990.
[60] Gregersen CS, Kazempour M, Alvarado V. ASP design for the Minnelusa formation under low-salinity conditions: impacts of anhydrite on ASP performance. Fuel 2013;105:368–82.
[61] Hirasaki G, Miller CA, Puerto M, et al. Recent advances in surfactant EOR. SPE J 2011;16:889–907.
[62] Soleimani Zohr Shiri M, Henderson W, Mucalo MR. A review of the lesser-studied microemulsion-based synthesis methodologies used for preparing nanoparticle systems of the noble metals, Os, Re, Ir and Rh. Materials (Basel) 2019;12:1896.
[63] Zulkifli NN, Mahmood SM, Akbari S, Manap AAA, Kechut NI, Elrais KA. Evaluation of new surfactants for enhanced oil recovery applications in high-temperature reservoirs. J Pet Explor Prod Technol 2020;10:283–96.
[64] Sorbie KS. Polymer-improved oil recovery. Springer Science & Business Media; 2013.
[65] Wang D, Cheng J, Xia H, Li Q, Shi J, et al. Viscous-elastic fluids can mobilize oil remaining after water-flood by force parallel to the oil-water interface. In: SPE Asia Pacific improv. oil recover. conf; 2001.
[66] Wang D, Cheng J, Yang Q, Wenchao G, Qun L, Chen F, et al. Viscous-elastic polymer can increase microscale displacement efficiency in cores. In: SPE annu. tech. conf. exhib; 2000.
[67] Jeong MS, Cho J, Choi J, Lee JH, Lee KS. Compositional simulation on the flow of polymeric solution alternating CO2 through heavy oil reservoir. Adv Mech Eng 2014;6:978465.
[68] Li W, Dong Z, Sun J, Schechter DS, et al. Polymer-alternating-gas simulation: a case study. In: SPE EOR conf. Oil Gas West Asia; 2014.
[69] Van Doren J, Douma SG, LBM W, Kraaijevanger J, De Zwart AH, et al. Adjoint-based optimization of polymer flooding. In: SPE enhanc. oil recover. conf; 2011.
[70] Zhang Y, Huang SS, Luo P. Coupling immiscible CO2 technology and polymer injection to maximize EOR performance for heavy oils. J Can Pet Technol 2010;49:25–33.
[71] Al-Abri A, Amin R. Phase behaviour, fluid properties and recovery efficiency of immiscible and miscible condensate displacements by SCCO 2 injection: experimental investigation. Transp Porous Media 2010;85:743–56.
[72] Majidaie S, Onur M, Tan IM. An experimental and numerical study of chemically enhanced water alternating gas injection. Pet Sci 2015;12:470–82.
[73] Mungan N. Carbon dioxide flooding-fundamentals. J Can Pet Technol 1981;20:87–92.

[74] Chen C, Balhoff M, Mohanty KK. Effect of reservoir heterogeneity on improved shale oil recovery by CO 2 huff-n-puff. In: SPE unconv. resour. conf; 2013.
[75] Jamal MS, Alnuaim S, Awotunde AA, Khan RA, et al. Optimal parameter selection in a polymer alternating gas PAG process. In: SPE Kingdom Saudi Arab. annu. tech. symp. exhib; 2016.
[76] Shalabi EW. Modeling the effect of injecting low salinity water on oil recovery. Austin, Texas, USA: The University of Texas at Austin; 2014. Ph. D. thesis.
[77] Dang CTQ, Nghiem LX, Chen Z, Nguyen QP, Nguyen NTB, et al. State-of-the art low salinity waterflooding for enhanced oil recovery. In: SPE Asia Pacific Oil Gas conf. exhib; 2013.
[78] Tang GQ, Morrow NR, et al. Salinity, temperature, oil composition, and oil recovery by waterflooding. SPE Reserv Eng 1997;12:269–76.
[79] Kumar M, Fogden A, Morrow NR, Buckley JS. Mechanisms of improved oil recovery from sandstone by low salinity flooding. Petrophysics 2011;52:428–36.
[80] Loahardjo N, Xie X, Morrow NR. Oil recovery by sequential waterflooding of mixed-wet sandstone and limestone. Energy Fuel 2010;24:5073–80.
[81] McGuire PL, Chatham JR, Paskvan FK, Sommer DM, Carini FH, et al. Low salinity oil recovery: an exciting new EOR opportunity for Alaska's North Slope. In: SPE West. reg. meet; 2005.
[82] Thyne GD, Siyambalagoda Gamage PH, et al. Evaluation of the effect of low salinity waterflooding for 26 fields in Wyoming. In: SPE annu. tech. conf. exhib; 2011.
[83] Skrettingland K, Holt T, Tweheyo MT, Skjevrak I. Snorre low salinity water injection-core flooding experiments and single well field pilot. Society of Petroleum Engineers; 2010.
[84] Dang TQC, Chen Z, Nguyen TBN, Bae W. Numerical simulation of surfactant adsorption in micellar polymer flooding. Energy Sources, Part A Recover Util Environ Eff 2014;36:1774–85.
[85] Alagic E, Skauge A. Combined low salinity brine injection and surfactant flooding in mixed− wet sandstone cores. Energy Fuel 2010;24:3551–9.
[86] Kozaki C. Efficiency of low salinity polymer flooding in sandstone cores; 2012.
[87] Mohammadi H, Jerauld G, et al. Mechanistic modeling of the benefit of combining polymer with low salinity water for enhanced oil recovery. In: SPE improv. oil recover. symp; 2012.
[88] Fjelde I, Asen SM, Omekeh AV, et al. Low salinity water flooding experiments and interpretation by simulations. In: SPE improv. oil recover. symp; 2012.
[89] Kulkarni MM, Rao DN. Experimental investigation of miscible and immiscible water-alternating-gas (WAG) process performance. J Petrol Sci Eng 2005;48:1–20.
[90] Webb KJ, Black CJJ, Al-Ajeel H, et al. Low salinity oil recovery-log-inject-log. In: SPE/DOE symp. improv. oil recover; 2004.
[91] Pourafshary P, Moradpour N. Hybrid EOR methods utilizing low-salinity water. In: Enhanc. oil recover. process. technol. IntechOpen; 2019.
[92] Hill LB, Li X, Wei N. CO2-EOR in China: a comparative review. Int J Greenhouse Gas Control 2020;103:103173.
[93] Belazreg L, Mahmood SM. Water alternating gas incremental recovery factor prediction and WAG pilot lessons learned. J Pet Explor Prod Technol 2020;10:249–69.
[94] Vishnyakov V, Suleimanov B, Salmanov A, Zeynalov E. 11—Water altering gas injection. In: Primer on enhanced oil recovery. Gulf Professional Publishing; 2020. p. 127–39.
[95] Xu Q, Rossen WR, et al. Experimental study of gas injection in surfactant-alternating-gas foam process. In: SPE annu. tech. conf. exhib; 2003.
[96] Ocampo A, Restrepo A, Cifuentes H, Hester J, Orozco N, Gil C, et al. Successful foam EOR pilot in a mature volatile oil reservoir under miscible gas injection. In: IPTC 2013 int. pet. technol. conf; 2013. p. cp- -350.
[97] Jain AK, Dhawan AK, Misra TR. ASP flood pilot in Jhalora (K-IV)—a case study. SPE Oil and Gas India Conference and Exhibition; 2012. p. 153667.
[98] Demin W, Zhenhua Z, Jiecheng C, Jingchun Y, Shutang G, Li L, et al. Pilot test of alkaline surfactant polymer flooding in Daqing oil field. SPE Reserv Eng 1997;12:229–33.
[99] Meyers JJ, Pitts MJ, Wyatt K, et al. Alkaline-surfactant-polymer flood of the West Kiehl, Minnelusa Unit. In: SPE/DOE enhanc. oil recover. symp; 1992.

[100] Rotondi M, Callegaro C, Masserano F, Bartosek M, et al. Low salinity water injection: eni' s experience. In: Abu Dhabi int. pet. exhib. conf; 2014.
[101] Li W, Schechter DS, et al. Using polymer alternating gas to maximize CO 2 flooding performance. In: SPE energy resour. conf; 2014.
[102] Abbas AH, Abdullah DS, Jaafar MZ, Sulaiman WRW, Agi A. Comparative numerical study for polymer alternating gas (PAG) flooding in high permeability condition. SN Appl Sci 2020;2:1–12.
[103] Parada Minakowski CH. An artificial neural network based tool-box for screening and designing improved oil recovery methods; 2007.
[104] Brown JS. Compositional simulation model for carbon dioxide flooding with improved fluid trapping, A. Colorado School of Mines; 2014.
[105] Wang J, Liu H. A novel model and sensitivity analysis for viscoelastic polymer flooding in offshore oilfield. J Ind Eng Chem 2014;20:656–67.
[106] Zhang Z, Li J, Zhou J. Microscopic roles of "viscoelasticity" in HPMA polymer flooding for EOR. Transp Porous Media 2011;86:199–214.
[107] Manzoor AA. Modeling and simulation of polymer flooding with time-varying injection pressure. ACS Omega 2020;5:5258–69.
[108] Delshad M, Pope GA, Sepehrnoori K. Volume II: Technical documentation for UTCHEM 2011–7, A three-dimensional chemical flood simulator. Technical documentation. Austin, Texas: Cent Pet Geosystems Eng Univ Texas Austin; 2011.
[109] Jafari A, Hasani M, Hosseini M, Gharibshahi R. Application of CFD technique to simulate enhanced oil recovery processes: current status and future opportunities. Pet Sci 2020;17:434–56.
[110] Afsharpoor A, Balhoff MT, Bonnecaze R, Huh C. CFD modeling of the effect of polymer elasticity on residual oil saturation at the pore-scale. J Petrol Sci Eng 2012;94:79–88.
[111] Jackson GT, Balhoff MT, Huh C, Delshad M. CFD-based representation of non-Newtonian polymer injectivity for a horizontal well with coupled formation-wellbore hydraulics. J Petrol Sci Eng 2011;78:86–95.
[112] Ersahin A. An artificial neural network approach for evaluating the performance of cyclic steam injection in naturally fractured heavy oil reservoirs; 2016.
[113] Alvarado V, Ranson A, Hernandez K, Manrique E, Matheus J, Liscano T, et al. Selection of EOR/IOR opportunities based on machine learning. In: Eur. pet. conf; 2002.
[114] Ahmadi MA, Pournik M. A predictive model of chemical flooding for enhanced oil recovery purposes: application of least square support vector machine. Petroleum 2016;2:177–82.
[115] Shalev-Shwartz S, Ben-David S. Understanding machine learning: from theory to algorithms. Cambridge university press; 2014.
[116] Mohri M, Rostamizadeh A, Talwalkar A. Foundations of machine learning. MIT press; 2018.
[117] Thanh HV, Sugai Y, Sasaki K. Application of artificial neural network for predicting the performance of CO 2 enhanced oil recovery and storage in residual oil zones. Sci Rep 2020;10:1–16.
[118] Ahmadi MA, Bahadori A. A simple approach for screening enhanced oil recovery methods: application of artificial intelligence. Pet Sci Technol 2016;34:1887–93.
[119] Alqahtani AS. New polymer rheology models based on machine learning; 2019.
[120] Ramos GAR, Akanji L. Application of artificial intelligence for technical screening of enhanced oil recovery methods. J Oil Gas Petrochem Sci 2017;1–12.
[121] Rashid Z, Wilfred CD, Gnanasundaram N, Arunagiri A, Murugesan T. A comprehensive review on the recent advances on the petroleum asphaltene aggregation. J Petrol Sci Eng 2019;176:249–68.
[122] Kane AV, et al. Performance review of a large-scale CO2-WAG enhanced recovery project, SACROC Unit Kelly-Snyder Field. J Petrol Tech 1979;31:217–31.
[123] Holm LW, et al. Propane-gas-water miscible floods in watered-out areas of the adena field, Colorado. J Petrol Tech 1972;24:1–264.
[124] Blanton JR, McCaskill N, Herbeck EF, et al. Performance of a propane slug pilot in a watered-out SandSouth Ward field. J Petrol Tech 1970;22:1–209.
[125] Burleigh LH, et al. ESPs successfully meet challenges of CO 2 use in tertiary oil recovery. J Petrol Tech 2010;62:24–7.

[126] Ghahfarokhi RB, Pennell S, Matson M, Linroth M, et al. Overview of CO2 injection and WAG sensitivity in SACROC. In: SPE improv. oil recover. conf; 2016.
[127] Jaturakhanawanit S, Wannakomol A, et al. Water alternating gas injection for enhanced oil recovery in the Phitsanulok Basin. Suranaree J Sci Technol 2011;18:267.
[128] McInnis LE, Hunter KD, Ellis-Toddington TT, Grawbarger DJ. Case study of the Mannville B ASP flood. In: SPE enhanc. oil recover. conf; 2013.
[129] Pitts MJ, Dowling P, Wyatt K, Surkalo H, Adams KC, et al. Alkaline-surfactant-polymer flood of the Tanner Field. In: SPE/DOE symp. improv. oil recover; 2006.
[130] Zhang Z-L. Asp Flooding Pilot test in Honggang oilfield. J Southwest Pet Inst 2001;23:47–9.
[131] Wan X, Huang M, Yu H, Zhang Y, Chen W, Gao E, et al. Pilot test of weak alkaline system ASP flooding in secondary layer with small well spacing. In: Int. oil gas conf. exhib. China; 2006.
[132] Suijkerbuijk B, Sorop TG, Parker AR, Masalmeh SK, Chmuzh IV, Karpan VM, et al. Low salinity waterflooding at west-salym: laboratory experiments and field forecasts. In: SPE improv. oil recover. symp; 2014.
[133] Okasha TM, Alshiwaish A, et al. Effect of brine salinity on interfacial tension in Arab-D carbonate reservoir, Saudi Arabia. In: SPE Middle East oil gas show conf; 2009.
[134] Robbana E, Buikema TA, Mair C, Williams D, Mercer DJ, Webb KJ, et al. Low salinity enhanced oil recovery-laboratory to day one field implementation-LoSal EOR into the Clair Ridge Project. In: Abu Dhabi int. pet. conf. exhib; 2012.
[135] Seccombe J, Lager A, Jerauld G, Jhaveri B, Buikema T, Bassler S, et al. Demonstration of low-salinity EOR at interwell scale, Endicott field, Alaska. In: SPE improv. oil recover. symp; 2010.
[136] Al-Qattan A, Sanaseeri A, Al-Saleh Z, Singh BB, Al-Kaaoud H, Delshad M, et al. Low salinity waterflood and low salinity polymer injection in the Wara Reservoir of the Greater Burgan Field. In: SPE EOR conf. oil gas West Asia; 2018.
[137] Kumar S, Mandal A. A comprehensive review on chemically enhanced water alternating gas/CO2 (CEWAG) injection for enhanced oil recovery. J Petrol Sci Eng 2017;157:696–715.
[138] Levitt D, Klimenko A, Jouenne S, Passade-Boupat N, Cordelier P, Morel D, et al. Designing and injecting a chemical formulation for a successful offshore chemical EOR pilot in a high-temperature, high-salinity, low-permeability carbonate field. In: SPE improv. oil recover. conf; 2016.
[139] Adepoju OO, Hussein H, Chawathe A, et al. Assessment of chemical performance uncertainty in chemical EOR simulations. In: SPE reserv. simul. conf; 2017.
[140] Ghedan S. Global laboratory experience of CO2-EOR flooding. In: SPE/EAGE reserv. charact. simul. conf; 2009. p. cp- -170.
[141] Fakher SM. Asphaltene stability in crude oil during carbon dioxide injection and its impact on oil recovery: a review, data analysis, and experimental study; 2019.
[142] Zanganeh P, Ayatollahi S, Alamdari A, Zolghadr A, Dashti H, Kord S. Asphaltene deposition during CO2 injection and pressure depletion: a visual study. Energy Fuel 2012;26:1412–9.
[143] Vargas FM, Creek JL, Chapman WG. On the development of an asphaltene deposition simulator. Energy Fuel 2010;24:2294–9.
[144] Franco CA, Nassar NN, Montoya T, Ruíz MA, Cortés FB. Influence of asphaltene aggregation on the adsorption and catalytic behavior of nanoparticles. Energy Fuel 2015;29:1610–21.
[145] Nassar NN, Betancur S, Acevedo S, Franco CA, Cortés FB. Development of a population balance model to describe the influence of shear and nanoparticles on the aggregation and fragmentation of asphaltene aggregates. Ind Eng Chem Res 2015;54:8201–11.
[146] Bagherpour S, Riazi M, Riazi M, Cortés FB, Mousavi SH. Investigating the performance of carboxylate-alumoxane nanoparticles as a novel chemically functionalized inhibitor on Asphaltene precipitation. ACS Omega 2020. https://doi.org/10.1021/acsomega.0c01732.
[147] Karazincir O, Thach S, Wei W, Prukop G, Malik T, Dwarakanath V. Scale formation prevention during ASP flooding. In: Spe int. symp. oilf. chem. Society of Petroleum Engineers; 2011.
[148] Wang Q, Liang F, Al-Nasser W, Al-Dawood F, Al-Shafai T, Al-Badairy H, et al. Laboratory study on efficiency of three calcium carbonate scale inhibitors in the presence of EOR chemicals. Petroleum 2018;4:375–84.

[149] Feng W, Deping Z, Guojun Y, Ruosheng P, Shuai X, et al. CO2 flooding WAG safety control technology. In: SPE Asia Pacific Oil Gas conf. exhib; 2013.
[150] Kokal S, Al-Kaabi A. Enhanced oil recovery: challenges & opportunities. World Pet Counc Off Publ 2010;64:64–9.
[151] AlGhnemi MF. Chemical EOR economics of Sabriyah Lower Burgan reservoir: a reservoir sector model study. Colorado School of Mines; 2018.
[152] Shadiya OO. Social, economic and environmental metrics for the sustainable optimization of chemical and petroleum processes. Oklahoma State University; 2010.
[153] Gozalpour F, Ren SR, Tohidi B. CO2 EOR and storage in oil reservoir. Oil Gas Sci Technol 2005;60:537–46.

CHAPTER 10

Chemical enhanced oil recovery in unconventional reservoirs

Ali Mirzaalian Dastjerdi[a] and Sefatallah Ashoorian[b]
[a]Chemical and Petroleum Engineering Department, Sharif University of Technology, Tehran, Iran
[b]Institute of Petroleum Engineering, School of Chemical Engineering, College of Engineering, University of Tehran, Tehran, Iran

10.1 Introduction

Unconventional reservoirs typically refer to any underground hydrocarbon resources which require specific operation processes for their recovery. They usually contain an enormous amount of oil and gas, which makes them attractive as a proper energy supply source despite difficulties associated with their recovery operation. The distribution of unconventional reservoirs around the world is shown in Fig. 10.1. As can be seen, a considerable portion of these resources is within some developed countries which have soaring energy demands, such as China, the United States, and Canada. This has brought much attention to unconventional reservoirs during the past decade.

Unconventional reservoirs are typically defined as low-quality reservoirs that need stimulation to produce at commercial flow rates and recovery factors [2]. Tight/shale gas or oil, heavy oil and tar sand, and coalbed methane (CBM) are examples of such reservoirs. Unconventional reservoirs require nontraditional as well as innovative technologies for development, production, and management. Although they hold worldwide attention due to the increasing energy demand, their operations pose technological challenges. For instance, a huge portion of natural gas in the United States is currently supplied by unconventional shale gas reservoirs [3]. However, the primary recovery factor from these reservoirs is between 2% and 8% [4], which is far less than their conventional counterparts. To compensate for such low recovery factors, different approaches have been proposed to aid hydrocarbon recovery. Enhanced oil recovery (EOR) techniques are among the most important methods, which are offered to increase recovery from such reservoirs.

The past few decades have witnessed many studies in the investigation of chemical injection to improve oil recovery from conventional reservoirs. This technique also gains attention for boosting recovery from unconventional reservoirs. In this chapter, chemical EOR (CEOR) for hydrocarbon recovery from unconventional reservoirs will be discussed.

Chemical Methods
https://doi.org/10.1016/B978-0-12-821931-7.00004-3

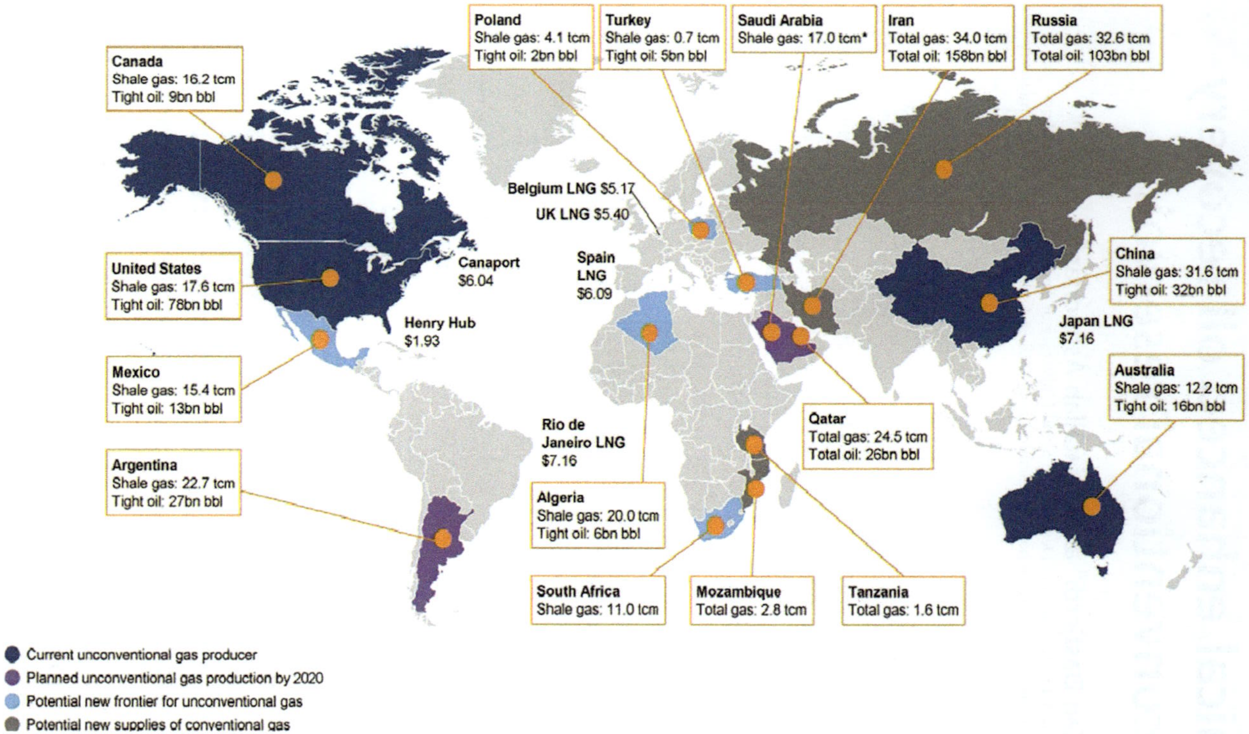

Fig. 10.1 An overview of unconventional reservoirs distribution around the world [1].

10.2 Unconventional reservoirs description

Conventional hydrocarbon resources commonly accumulate in convenient structural traps, i.e., porous and permeable formations capped by a sealed layer that prevents hydrocarbon from escaping. Owing to their favorable geological structures, these reservoirs typically do not require large-scale stimulation for hydrocarbon production. On the other hand, unconventional reservoirs typically exist in tight formations, making them hard to extract. Moreover, becuase of their formation conditions, they have a lower quality which also intensifies their recovery challenges. However, unconventional reservoirs contain a larger amount of hydrocarbon compared to conventional ones and are also more abundant [5]. The hydrocarbon resource triangle is shown in Fig. 10.2. It can be seen that the majority of hydrocarbon resources exist in unconventional reservoirs. This enormous abundance is motivating countries to develop them to fulfill their energy demands.

According to Fig. 10.2, a wide variety of geological formations, including tight gas sands, gas shales, heavy oil sands, coalbed methane reservoirs, oil shales, and gas hydrates, are categorized under unconventional reservoirs. Generally, an unconventional reservoir is characterized by low to ultralow permeability and low to moderate porosity. Typically, reservoirs with permeability lower than 0.1 mD are categorized as unconventional

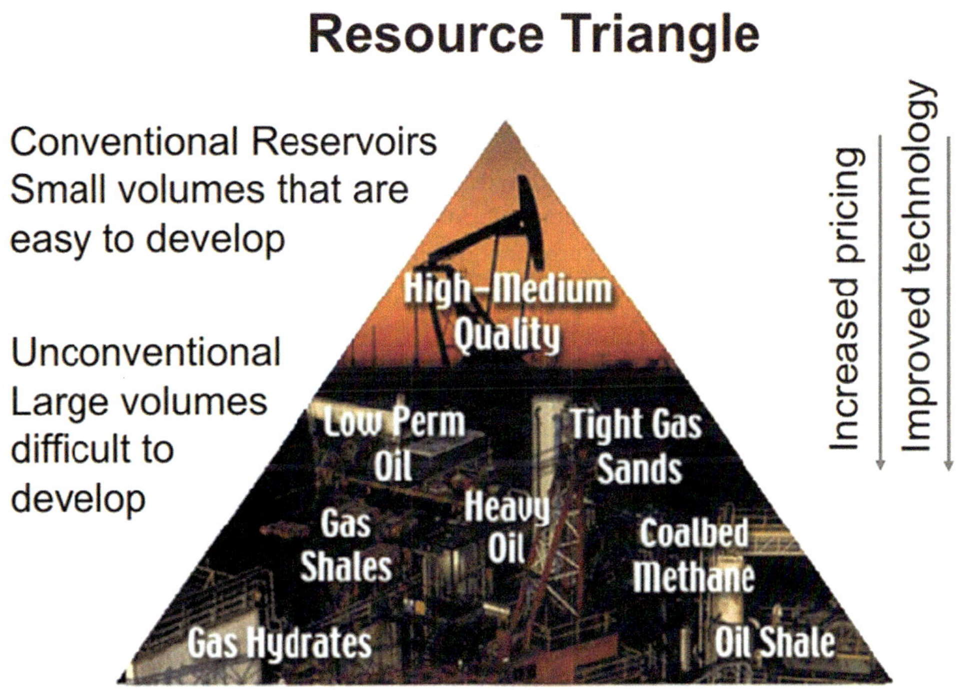

Fig. 10.2 Resource triangle [2].

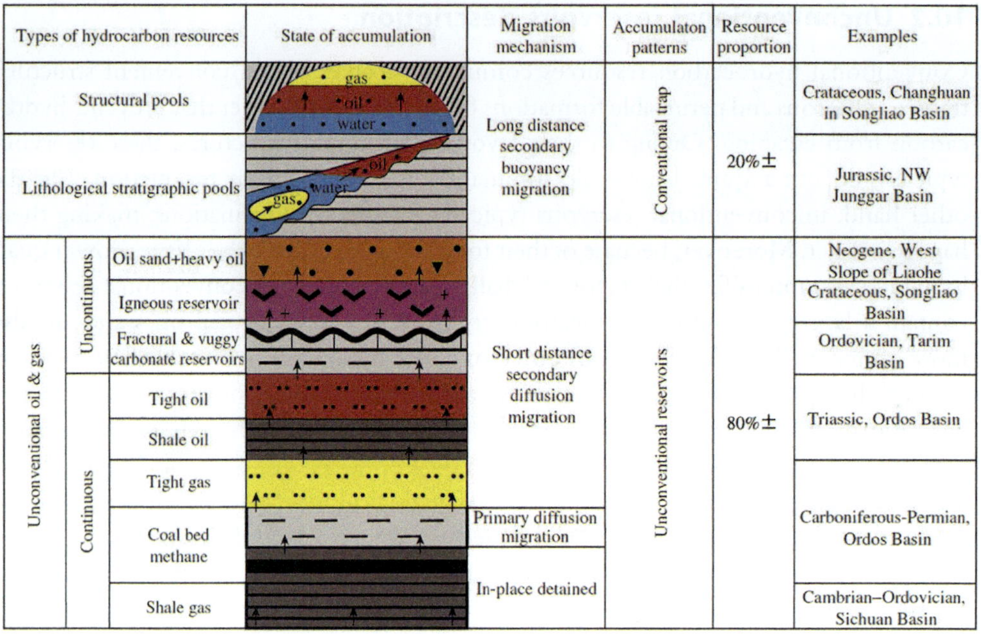

Fig. 10.3 Different hydrocarbon reservoir types [9].

reservoirs [2, 5], but this is not clear-cut in practice because of the heterogeneous nature of the hydrocarbon reservoirs. In addition to their ultralow permeability, the smaller pore throat size (<100nm) [6, 7], as well as their pore network wettability, which compromises hydrophilic and hydrophobic pores [8], are two other unique features that differentiate unconventional reservoirs from conventional ones. Due to these structural challenges, the development of unconventional reservoirs requires special focus on the mechanical and geotechnical rock properties [3]. Different reservoir types and hydrocarbon accumulation patterns are shown in Fig. 10.3 [9].

10.2.1 Rock mineralogy

The recovery performance of any hydrocarbon reservoir is highly dependent on its rock properties. Reservoir rock properties also play a vital role in reservoir modeling as well as interpretation of data in the development stages. The primary information classified in rock properties is rock mineralogy, grain size distribution, absolute and relative permeabilities, formation factor, and residual oil/water content. Typically, a detailed description of rock mineralogy in unconventional reservoirs is more important than that in conventional ones since rock/fluid interactions have a more profound impact on fluid flow in micro/nanopores than conventional reservoirs. Therefore, a comprehensive understanding of unconventional reservoirs mineralogy is required.

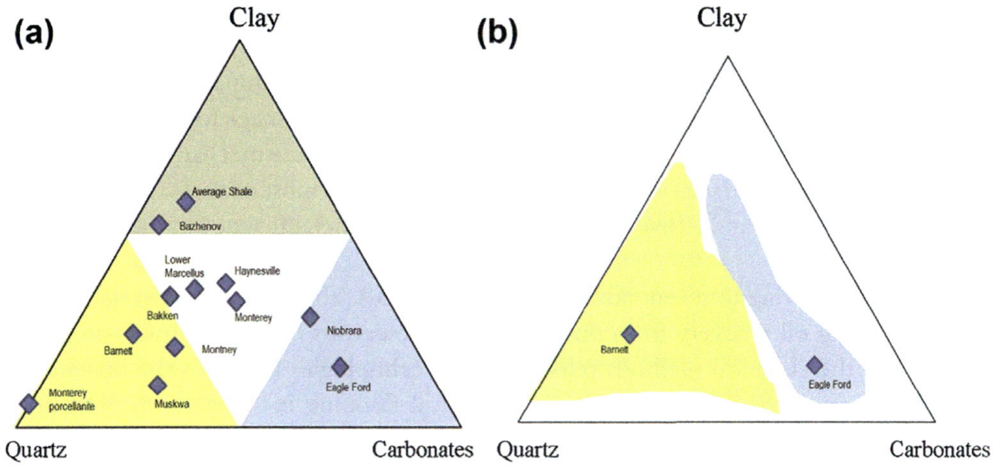

Fig. 10.4 The mineral composition of several unconventional fields: (A) ternary plot of mineral compositions of some known shale reservoirs and (B) heterogeneity in mineral composition of two shale-gas reservoirs [5].

From a mineralogical perspective, regarding the fact that a significant number of unconventional reservoirs are shale formations, geologists have mineralogically described shales with three dominant components—silica, clay, and carbonates—using a ternary diagram shown in Fig. 10.4. Accordingly, mineralogy is significantly important to analyze reservoir quality. For example, the clay content of good organic shale is lower than that of conventional shales imposing less clay-bound water, resulting in more water-wet rock. As shown in Fig. 10.4, some famous unconventional reservoirs, such as the Bazhenov shale reservoir, contain relatively high clay content that is a negative factor for production; therefore, they need other favorable factors like an intensive fracture network to make them desirable for commercial oil production.

10.2.2 Proposed mechanisms

The economic evaluation of any reservoir depends on the ultimate recovery factor that can be achieved in any field development stage. This parameter is highly affected by reservoir porosity and permeability. In conventional reservoirs, the recovery factor can generally reach 30%–40% after waterflooding, while it is relatively less than 10% in unconventional reservoirs. Hence, secondary actions are required to improve the oil recovery from these reservoirs, namely hydraulic fracturing, huff-n-puff gas injection, thermal recovery, and chemical flooding. Although most of these methods can be applied to heavy oil and bitumen reservoirs [3, 10] in most cases, for other unconventional reservoir types, e.g., shale oil reservoirs or CBMs, the decision would be narrowed down to hydraulic fracturing and huff-n-puff techniques [11–16]. Nevertheless, the use of chemical additives, mostly surfactants in conjunction with such techniques, has also been suggested.

There are several recovery mechanisms in tight oil reservoirs impacting fluid displacement, including imbibition for the water-wet state, gravity drainage for the intermediate to oil-wet state, diffusion, and solubilization [3, 4]. Among these mechanisms, wettability modification and interfacial tension (IFT) reduction are the most critical factors in recovery from tight/shale and CBM reservoirs and provide higher efficiency in these reservoirs [4, 5, 17].

Recently, owing to recent advancements in CEOR, the potential of these methods for improving oil recovery from unconventional reservoirs has gained much attention. Typically, the chemical methods refer to EOR technologies based on the injection of chemical components [5]. Accordingly, chemical flooding includes surfactants, polymers, alkali, alkali/polymer, and alkali/surfactant/polymer (ASP) flooding. Although such methods have been successfully employed in conventional reservoirs, their performance in unconventional reservoirs is not fully established yet. Therefore, many simulations and laboratory studies have been carried out on the topic. Surfactant flooding is one of the most important techniques in this regard for boosting oil recovery from shale and tight reservoirs, which has gained increasing attention over the past years [6–12].

Laboratory studies note that surfactant addition to the aqueous phase can improve the aqueous phase imbibition into the oil-saturated matrix, leading to countercurrent oil production [13, 14]. Surfactants are amphiphilic compounds with a hydrophobic tail and hydrophilic head [8], which reduce IFT and alter the surface wettability, resulting in improving water imbibition. Based on their head charge, surfactants are classified into these four types: nonionic (without any charge), anionic (carrying a negative charge), cationic (carrying a positive charge), and zwitterionic (carrying both negative and positive charges) [15]. The mechanism of wettability alteration by cationic surfactants is demonstrated in Fig. 10.5.

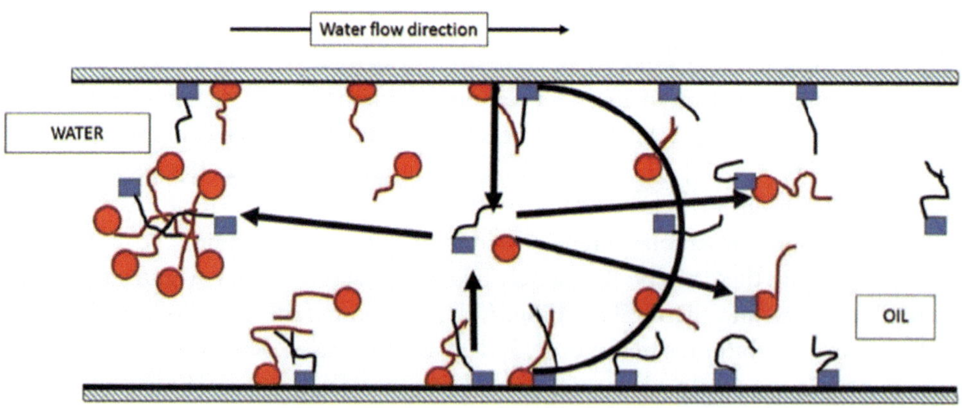

Fig. 10.5 The wettability alteration by a cationic surfactant. *Circles* represent the cationic surfactant molecules and *squares* represent the anionic surface-active material present in the oil [18, 19].

Generally, changing capillary pressure occurs by reducing IFT and altering wettability due to the flow and/or diffusion of surfactants into the matrix. Consequently, the water imbibes spontaneously into the rock and expels the oil from the pores. However, owing to the characteristic features of tight rocks, the precise mechanism by which the spontaneous imbibition of surfactant solutions improves oil recovery remains controversial [16].

The effect of IFT reduction on oil recovery can be expressed by the definition of the capillary number. The capillary number, N_C, which is a ratio of viscous forces to capillary forces, is defined as:

$$N_C = \frac{u\mu}{\sigma \cos\theta} \tag{10.1}$$

where μ is the viscosity of the displacing fluid, u is the displacing Darcy velocity, θ is the contact angle, and σ is the IFT between the displacing fluid and the displaced fluid (oil). Regarding the experimental data, residual saturations are reduced by increasing the capillary number. As seen in Fig. 10.6, the relation between residual saturation and the capillary number is expressed by the capillary desaturation curve (CDC) [17, 18]. Note that the properties and arrangements of the pores within the media and the distribution of fluids within the pores can be reflected by CDC [20].

Surfactants can significantly affect the capillary number by either altering the wettability or reducing the IFT value between the injected and reservoir fluids. The proper usage of surfactants can lead to ultralow IFT values, increasing the capillary number by 2–3 orders of magnitude, which results in a higher recovery efficiency. This will lead

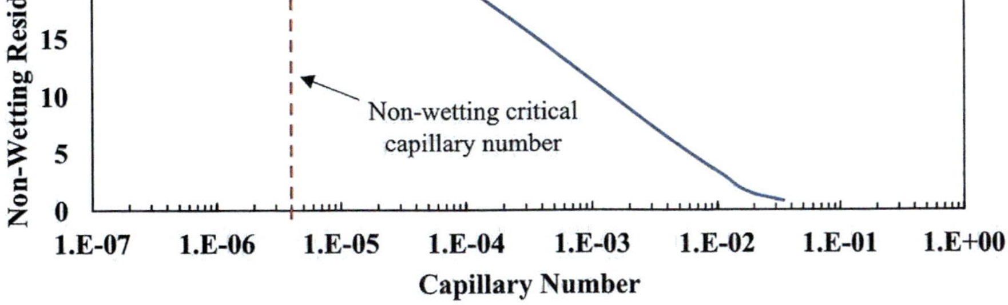

Fig. 10.6 The schematic of the capillary desaturation curve (CDC) [20].

to the reduction of the capillary trapping, which helps oil droplets to flow easily through pore throats and form oil banks [19]. It is noteworthy to mention that surfactant flooding is mainly applied to enhance oil recovery in tight reservoirs, particularly shale oil and coalbed methane reservoirs.

In addition to surfactants, polymers have also been utilized to aid oil recovery from unconventional reservoirs. Polymers have mainly been applied to heavy oil and bitumen reservoirs. Polymers can improve oil recovery by modifying the displacing fluid viscosity. Typically, when waterflooding is applied to heavy oil reservoirs, due to the remarkable difference between the viscosity of the injected fluid and the reservoir fluid, fingering takes place. The simulated fingering phenomena in a heavy oil reservoir can be seen in Fig. 10.7 [21]. Polymers can increase the displacing fluid viscosity, which results in lower mobility ratios. This prevents fingering that finally leads to a higher sweep efficiency and more oil recovery [22, 23]. Meanwhile, the fractional flow curve shifts right

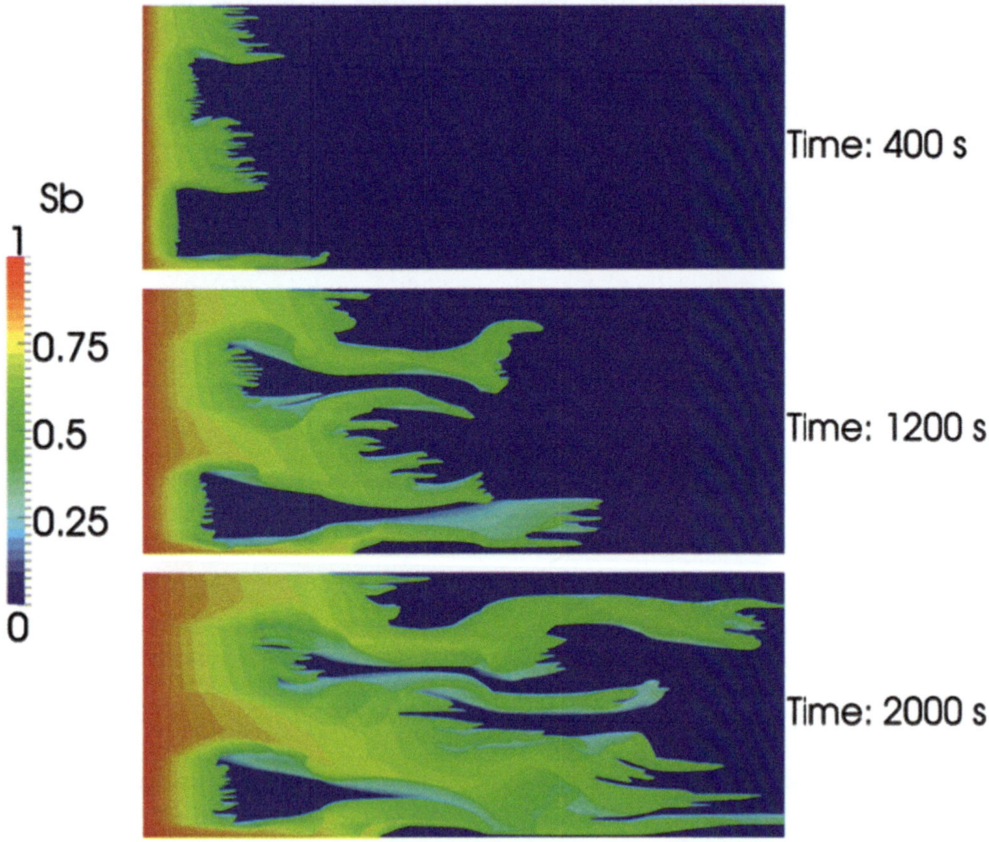

Fig. 10.7 Viscous fingering in a heavy oil reservoir simulated by computational fluid dynamic (CFD) (note that the water saturation field is Sb) [21].

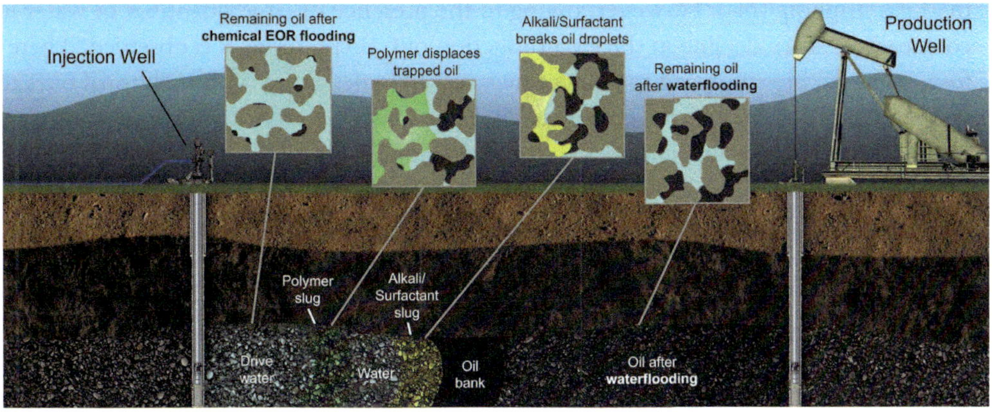

Fig. 10.8 Different mechanisms of chemical flooding [25].

forward due to the decreasing water to oil mobility ratio, which diverts the injected water from swept zones to unswept ones.

Although polymers have shown great potential for EOR, there are some challenges that need to be addressed. One of the significant challenges regarding polymer injection is the injectivity issue. In recent years, owing to the advancement in horizontal well drillings, this issue has been dramatically compensated; however, care should still be taken to ensure a successful polymer flooding [24].

Finally, it is noteworthy to mention that due to the complexities associated with unconventional reservoirs, the field application of CEOR is still at its early stages and requires more lab and pilot studies. An overview of the different mechanisms of chemical flooding for EOR is shown in Fig. 10.8.

10.3 Coalbed methane reservoirs

Natural gas, mainly methane, which is accumulated in the cleats and micropores of coal, is called coalbed methane (CBM) [26]. CBM reservoirs are classified as unconventional reservoirs since the gas resides in low permeable coal streams. The CBM reserve has been estimated at 18,000 TCF only in the top 20 countries [3]. Also, the CBM resources provide 7% of the total natural gas consumption in the United States [3]. Along with the United States, Russia, China, Canada, and Australia are the following top countries that have the most CBM reserves and production [27].

The microporosity, low permeability, and high adsorption characteristics of CBMs make gas extraction from such reservoirs a challenging task [28]. Typically, hydraulic fracturing and gas injection have been utilized in CBMs for gas recovery. Hydraulic fracturing is applied to create and connect fractures in coal formation. Gas injection also helps methane recovery by creating competitive adsorption of a more reactive gas (CO_2) or inert gas stripping (N_2). However, owing to the presence of large quantities of water in coal seams, a sharp decline in the gas production of CBM reservoirs is observed. Therefore, there is a pressing need to explore new technologies along with traditional ones to boost gas extraction from these challenging reservoirs [11].

CEOR is one possible choice in this regard. For example, a few researchers [12, 29] proposed that surfactants have a positive effect on acid-treated coal in addition to acidizing hydraulic fracturing. Guanhua et al. [12, 29] conducted studies to find an optimal fluid performance test using anionic, nonionic, and cationic surfactants, whose procedure is shown in Fig. 10.9. The results revealed that sodium dodecyl sulfate (SDS) was the best surface-active additive for cleaning fracturing fluids among the other tested surfactants. Accordingly, they concluded that the efficiency of anionic and nonionic surfactants in reducing the surface tension of the solution is higher than that of the cationic surfactants. The key conclusion of their experimental tests on coal samples are: (1) SDS promotes the pressure relief effect of the acidified coal body, (2) SDS reduces the strength of the treated coal sample, and (3) SDS increases the porosity of coal.

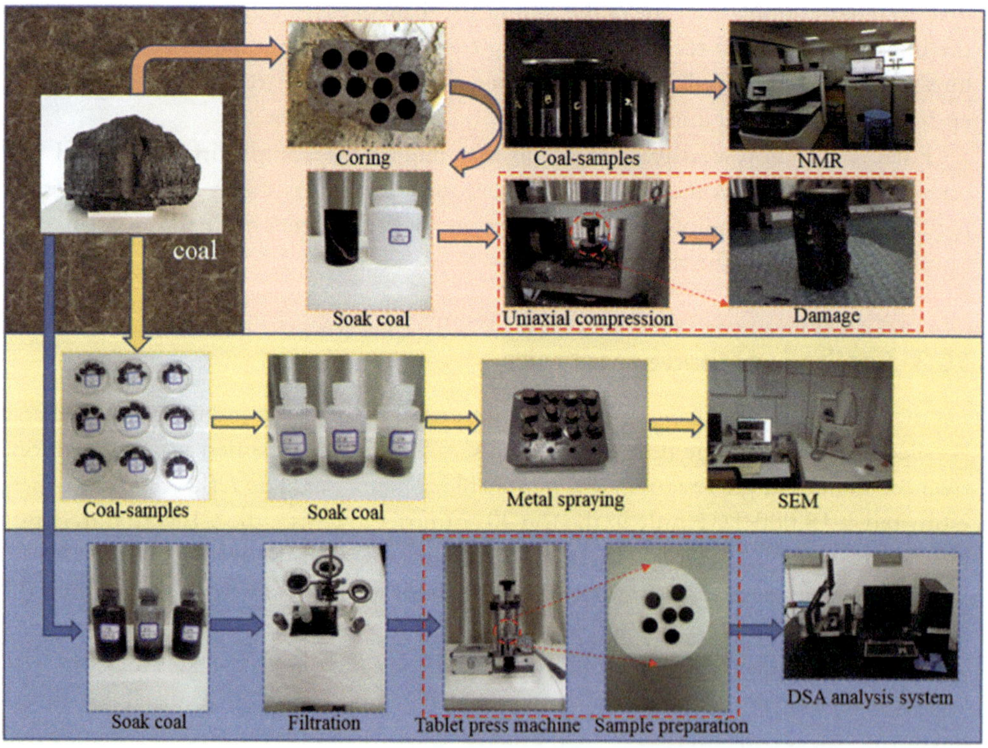

Fig. 10.9 The experimental equipment and procedures which were used to investigate the effect of SDS on CBMs [12].

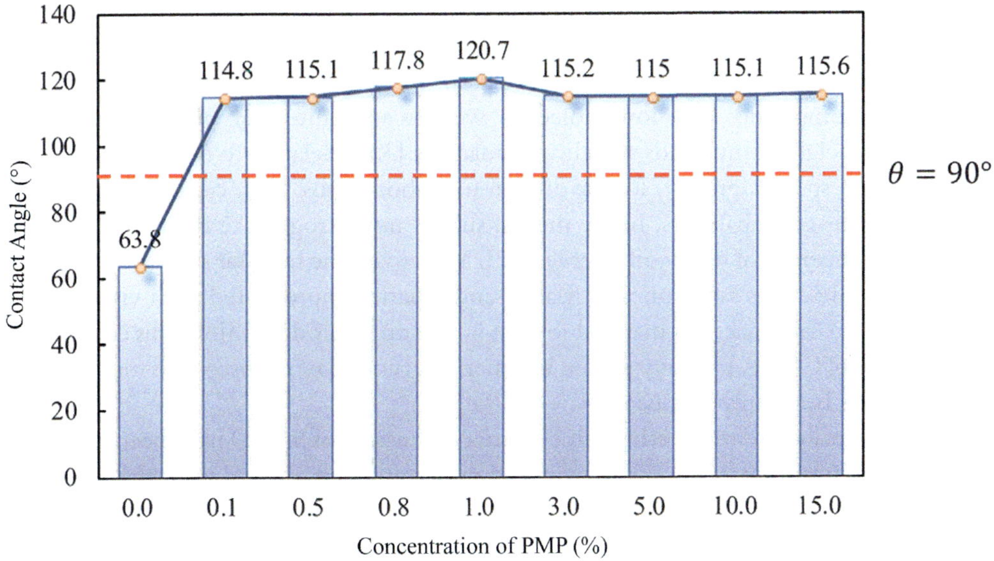

Fig. 10.10 Change in the contact angle of CBM samples treated by different PMP concentrations [11].

Furthermore, Cai et al. [30] investigated the impact of SDS, sodium dodecylbenzene sulfonate (SDBS) within tetrahydrofuran (THF) solution from coalbed methane gas mixture via hydrate-based methane at 279.15 K and 1.50–4.50 MPa. They discovered that a mixture of SDS within the THF solution (THF-SDS) is an appropriately integrated additive for recovering CH_4 from coalbed methane gas mixture. As the presence of a large quantity of water reduces the gas production from CBMs, injection of fluorocarbon surfactant perfluorooctyl methacrylate monomer-containing polymethacrylate (PMP) also has shown a tremendous effect on wettability by the gas wettability alteration (GWA) approach [11]. As shown in Fig. 10.10, the contact angle of all treated samples is more than 90 degrees, which clearly indicates the profound impact of chemical treatment of the coalbed samples, which increases the water-wetness of coalbed samples. These results demonstrated that by selecting a proper chemical agent, a critical challenge regarding water production in CBMs could be diminished [11].

According to the above discussions, it can be concluded that a suitable choice of surfactants as a chemical additive can improve gas recovery from CBM reservoirs.

10.4 Heavy oil and tar sands

Recently, in light of energy demand and the depletion of conventional hydrocarbon resources, the extraction of oil from heavy oil and bitumen reservoirs has gained much attention. These resources are approximated to be about 10 trillion barrels globally, almost three times the conventional oil-in-place (OOIP) [23, 31]. The majority of the heavy oil reservoirs are spread in Canada, Latin America, China, and the United States. Typically, heavy oil is defined as crude oil having a viscosity above 100 or 1000 cP and a gravity of less than 20 API [32]. It contains relatively low proportions of volatile compounds with low molecular weights and quite high proportions of high molecular weight compounds with lower volatility [33]. Relative to conventional crude oil, the high specific gravity, low hydrogen to carbon ratios, high carbon residues, and high contents of asphaltenes, heavy metals, sulfur, and nitrogen are the common characteristic properties of heavy oil reserves [33]. Moreover, the term tar sand or bituminous sands often describes sandstones or friable sand (quartz) impregnated with viscous bitumen (a hydro carbonaceous material soluble in carbon disulfide) [33]. In this chapter, the application of CEOR in heavy oil and bitumen reservoirs having viscosities mainly above 1000 cP will be mainly discussed.

Traditionally, thermal methods like in situ combustion or SAGD have been utilized to produce oil from the heavy oil and tar sand reservoirs. However, during the past years, becuase of high oil prices and technological advancement in chemical additives, polymer flooding has also gained a reputation for use in such reservoirs. The applicability of polymer flooding for heavy oil reservoirs has recently been the subject of many laboratory and simulation investigations. It is shown that there is a viscosity-sensitive region (the so-called S-shaped curve) within which the oil recovery will significantly increase by polymer flooding through an increase in the polymer solution viscosity [31, 34]. Wang et al. carried on 28 sand packs polymer flooding for five different heavy oils with viscosities ranging from 430 to 5500 mPa s to investigate the incremental oil recovery and the effective viscosity of the polymer solution [34]. Their findings revealed that a minimum polymer concentration is required to achieve efficient polymer flooding in extra heavy oil reservoirs. In addition, it is shown that the onset point and the sensitive range in polymer concentration are dependent on oil viscosity [35]. The S-shape curve showing incremental oil recovery versus the viscosity of the polymer solution is illustrated in Fig. 10.11 [34, 36].

Recently, the low salinity polymer core flooding as well as the configuration of a pilot test was utilized for enhancing oil recovery from an extra heavy oil reservoir in Venezuela. The experimental results show that low salinity polymer flooding requires a lower polymer consumption for a specific viscosity target, and it brings a higher incremental recovery than conventional polymer flooding [37]. The high performance of low salinity polymer flooding in this case is attributed to wettability alteration and ion exchange between the rock surface and the low salinity polymer solution [37].

More recently, Zhou et al. conducted a comprehensive study on the polymer-enhanced foam (PEF) flooding in extra heavy oil samples from thin oil reservoirs with a viscosity of more than 1000 cP. They employed micromodel, core flooding, and simulation to investigate the PEF behavior. They concluded that PEF flooding for heavy oil recovery in thin reservoirs excels over surfactant-polymer (SP) flooding and water flooding by 11.7% and 26.7%, respectively [38].

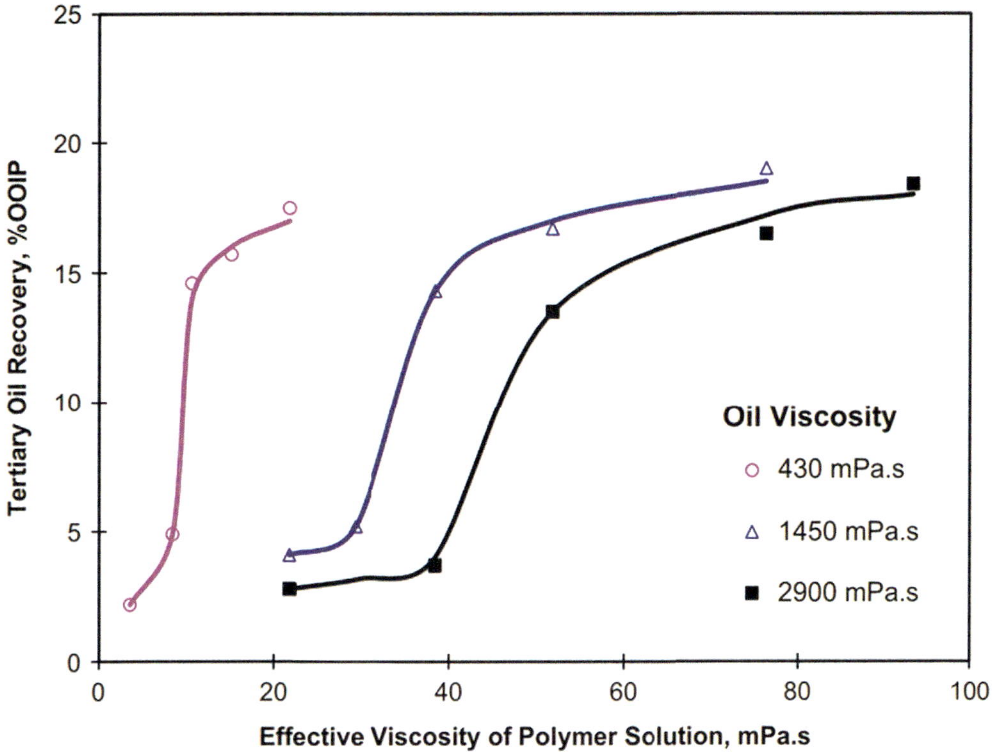

Fig. 10.11 S-shaped relationship between oil recovery and polymer viscosity [36].

In addition to polymers, other chemical additives such as alkali and surfactant and their combination with polymer were also investigated to measure their performance for oil recovery in heavy oil reservoirs. For example, Dong et al. carried out a series of flood tests to screen the chemical formulas of two alkalis (alkyl ether sulfates, sodium dodecyl sulfate) and surfactant to increase oil recovery from the Brintnell reservoir. They stated that the synergic effect of the alkali and surfactant would result in easy emulsification of the Brintnell oil in brine, which could have a promising impact on recovery [39]. Different chemical flooding systems (alkaline-only, polymer-only, alkaline-surfactant (AS), alkaline-polymer (AP), and ASP systems) have also been exerted on Court oil by sand pack flooding. The results revealed that the oil recovery was correlated with pressure drops rather than IFT, which denotes that the recovery enhancement was mainly due to the plugging of water channels rather than the IFT reduction [40]. Moreover, the comparison of chemical slugs for enhancement of heavy oil recovery demonstrated that the combination of polymer, alkali, and surfactant yielded the highest oil recovery, mainly due to the plugging of water channels after the first water flooding in heavy oil reservoirs [41]. The above discussions show that although adding a surfactant and an alkali along with the polymers might yield better recovery in some cases, the central role in these types of unconventional reservoirs is played by polymers.

Finally, it should be noted that the promising results obtained by new methods like PEF or low salinity polymer flooding along with conventional methods like ASP are the harbinger of new opportunities to enhance heavy oil production from unconventional reservoirs via CEOR.

10.5 Shale oil and gas reservoirs

The total gas and oil in shale resources across the globe are estimated at around 10,000 TCF and 6.05 trillion barrels, respectively [42, 43]. Fig. 10.12 indicates the top 10 countries with the most shale gas reserves. The production from shale gas resources was initiated successfully in the United States and extended to Canada and subsequently to other countries to fulfill their energy demands [42]. In recent years, more than half of the total US oil and gas production was extracted from shale and tight formations. Owing to the increasing trend in hydrocarbon recovery from low permeable reservoirs, it is expected that the share of such sources in the energy market increases in the upcoming years [45].

This clearly illustrates the importance of shale oil and gas reservoirs in today's market compared to the past. However, like other unconventional resources, the initial recovery from these reservoirs is less than 10% [45], which forces engineers to find a solution to tackle challenges. In this regard, chemical flooding, particularly surfactant flooding, is investigating a novel technique for enhancing recovery from these reservoirs owing to the fact that chemicals have revealed a profound impact on oil recovery from conventional reservoirs.

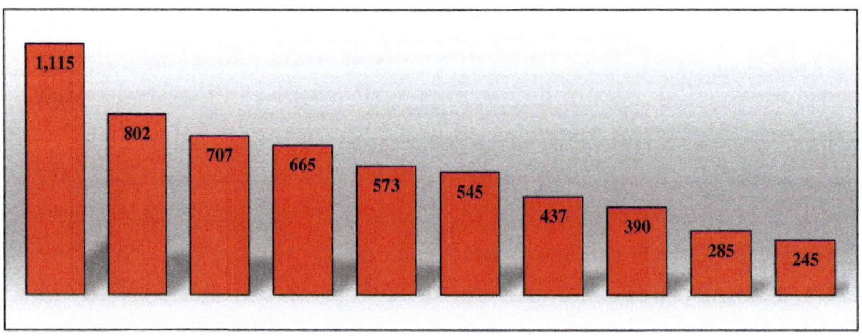

Fig. 10.12 Top 10 countries in shale gas reserves (in trillion cubic feet (TCF)) [44].

10.5.1 Experimental studies

The major productivity of tight reservoirs is by means of horizontal wells with transverse artificial fractures that connect complex natural fractures [46]. With regard to this fact, surfactant injection is employed for these reservoirs, which leads to taking advantage of both fractures and surfactants for improving hydrocarbon recovery. In this regard, injected surfactants make contact with the formation matrix provided by complex natural fractures [47]. As explained earlier, surfactants help oil recovery by wettability alteration and IFT reduction. Also, the capillary force is more important in the shale micro- and nanopores than gravity force. Accordingly, wettability alteration from oil-wet to water-wet can improve oil recovery as the capillary pressure shifts from a negative to a positive value. Fig. 10.13 demonstrates how surfactants adsorb on the rock surface, resulting in wettability alteration from oil-wet to water-wet. The figure clearly shows that the surfactants adsorb to the positive charge of a surface by their hydroxyl groups which form an attraction force. Further, the capillary pressure is related to IFT values; hence, capillary forces will decrease by IFT reduction as well [48].

The selection of a proper chemical is an essential step to achieve the most favorable performance for any CEOR process. Designing an optimized surfactant flood initially requires developing a fully qualified chemical system, i.e., it should be compatible with the fluid system, well, and formation fluids. To this end, several laboratory studies have been conducted to evaluate the efficiency of various surfactants in recent years. Different types of surfactants, including anionic, nonionic, and mixtures, have been utilized to study the oil recovery from shale reservoirs through fracturing fluids [49]. The initial

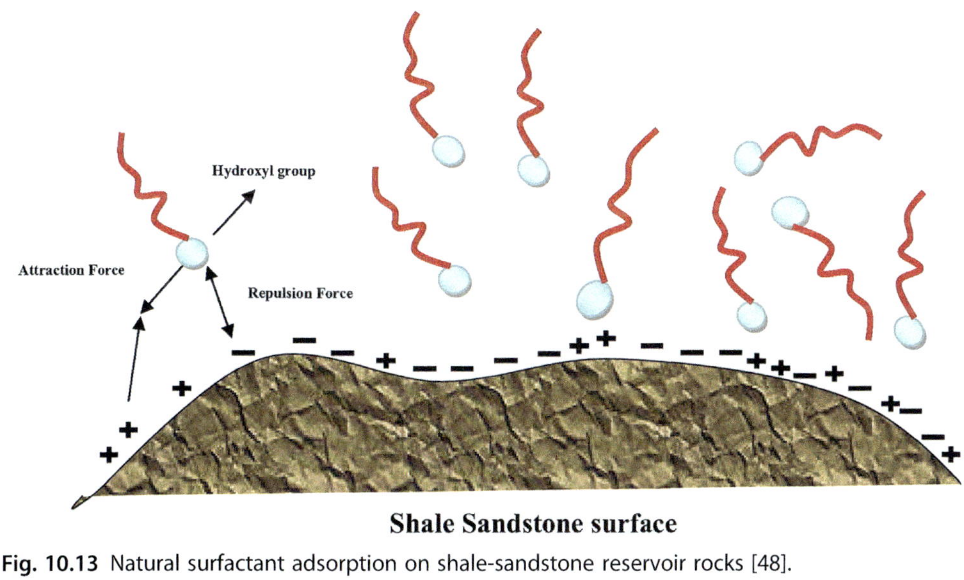

Fig. 10.13 Natural surfactant adsorption on shale-sandstone reservoir rocks [48].

studies by Wang et al. [50], Chen et al. [51], and Alvarez et al. [14–16] claimed that the anionic surfactants had boosted more oil recovery than other types of surfactants. More oil recovery is believed to be the result of a change in the rock surface wettability to a water-wet state, enhancing the oil relative permeability and water imbibition. Therefore, the anionic surfactants are known as effective chemicals that can alter the rock wettability to a more water-wet surface through adsorption. Fig. 10.14 shows the changes in wettability and zeta potential during adsorption of anionic surfactants on the Eagle Ford samples. In this figure, it is clear that these three surfactants alter the rock surface toward the water-wet and impose a more negative charge on the surface, which was measured by the zeta potential [52]. The cumulative oil recovery by three various anionic surfactants from the Eagle Ford samples, which researchers reported, is shown in Fig. 10.15.

Besides, Zeng et al. [49] carried out a large number of experimental tests to evaluate the efficiency of various kinds of surfactants regarding their aqueous stability, contact angle measurement, and spontaneous imbibition. They reported that blended surfactants in low salinity brine alter the wettability from oil-wet to water-wet. The oil recovery from these samples was varied from 17% to 21% for all cases, significantly higher than other types of surfactants. Note that the efficiency of hybrid EOR (a mixture of anionic and nonionic surfactants with water) depends on the compatibility of surfactants with brine, their stability under reservoir conditions, and their ability to reduce IFT at the oil/brine interface [53]. Moreover, the results of experimental studies in the core samples of some well-known shale fields illustrate that cationic surfactants have a higher adsorption rate and oil recovery rate in core samples from Wolfcamp, Eagle Ford, and part of Bakken, whose calcite and dolomite compositions are dominant (above 50%) [41, 42], while anionic surfactants have a higher adsorption rate and oil recovery rate in core samples from Permian, Sichuan, and part of Bakken, whose quartz and clay compositions are above 50% [18, 33, 43]. However, these results have not been verified by field scale or pilot tests yet.

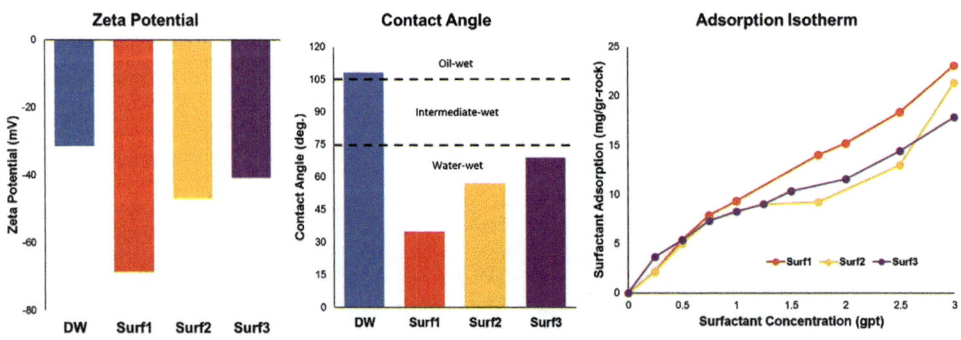

Fig. 10.14 Zeta potential (left), contact angle (middle), and adsorption isotherm (right) measurement results of the Eagle Ford sample treated by three different anionic surfactants [52].

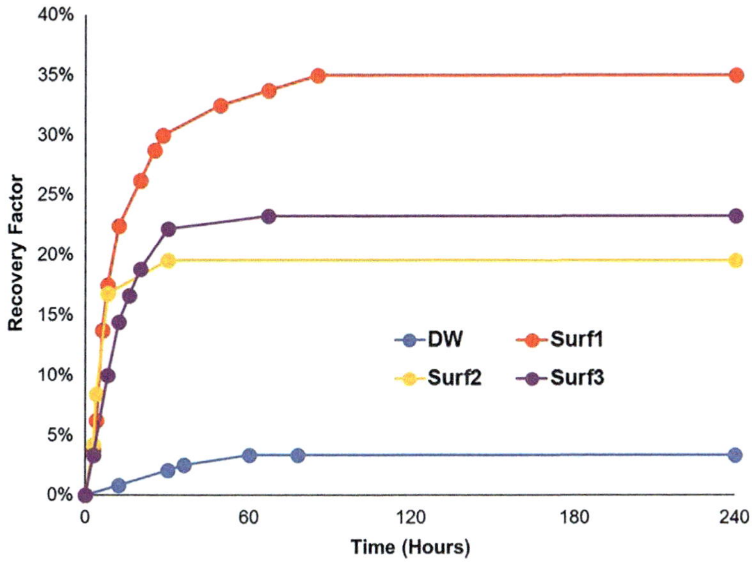

Fig. 10.15 Oil cumulative recovery from the spontaneous imbibition by three different anionic surfactants from the Eagle Ford samples [52].

As a result, there is a consensus on the profound impact of blended surfactants in improving the oil recovery from shale samples, which is a step forward in the application of surfactants in shale reservoirs and paves the way for new researches for a better understanding of surfactants mechanisms in this type of reservoirs at micro/nanoscales.

As reported, CO_2 huff-n-puff is an effective EOR method in shale reservoirs. In this method, gas enters the matrix from fractures through convection and diffusion, which leads to an oil swelling. As a result, the gas can come out of the oil and push the oil into the fracture during the production phase. Accordingly, reports show that adding chemical blends such as anionic surfactants to CO_2 huff-n-puff has a more profound impact than pure CO_2 huff-n-puff on oil recovery (64% recovery against 40% for pure huff-n-puff). These procedures activate beneficial mechanisms: (1) The surfactant can lower the interfacial tension and alter wettability, and (2) CO_2 dissolves in oil and swells the oil [54]. The developed instrument for chemical blend plus CO_2 huff-n-puff experiments is shown in Fig. 10.16.

Despite many promising works conducted on the surfactant application as a CEOR method in shale reservoirs, selecting the most suitable surfactant for shale reservoirs is still under debate. Also, the enhancement mechanism(s) of surfactant in this reservoir type is not clear yet.

Taking all the aforementioned studies into account, there are some hints that can be concluded by reviewing the past researches. First, surfactants that create unstable macroemulsions in tight reservoirs should be avoided since they increase the oil

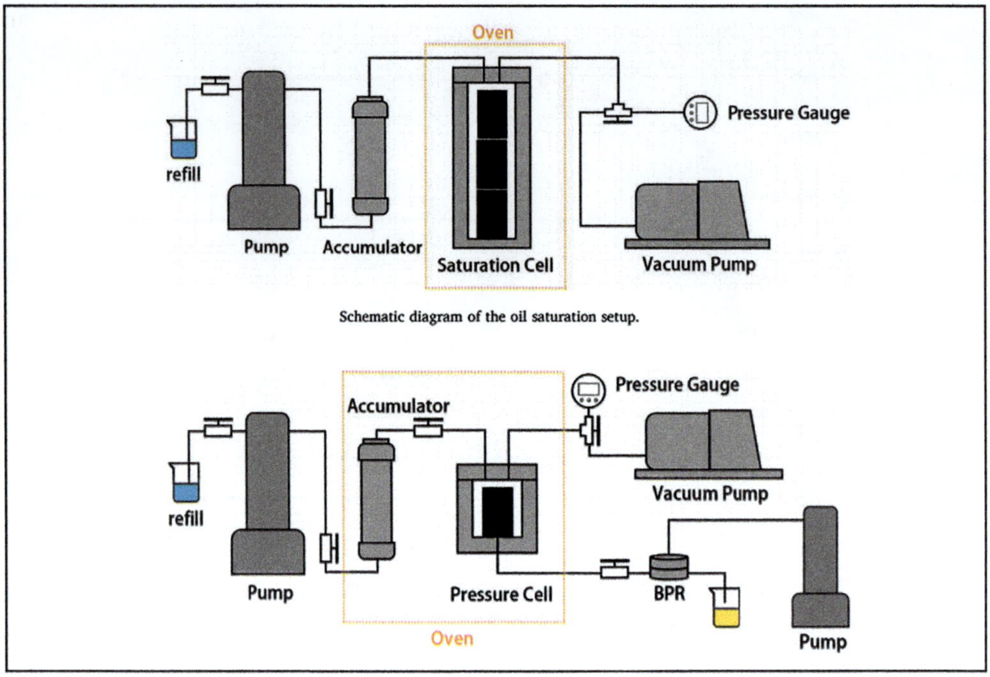

Fig. 10.16 Schematic diagram of a huff-n-puff experimental setup developed by Zeng et al. [54].

viscosity and reduce the oil recovery rate; moreover, the ultralow IFT value that is created under surfactant flooding is not preferable because the capillary force is the driving force of water imbibition in these types of reservoirs [55–57]. Further, it is noteworthy that most of the wettability-alteration surfactants chosen for changing wettability in shales and tight rocks are in the same categories as the ones selected for conventional reservoirs, which include anionic surfactants like sulfates or sulfonates, nonionic surfactants like ethoxylated alcohols, and cationic surfactants like quaternary ammonium salts [14, 15, 46, 58].

Finally, it should be noted that besides technical criteria, economic and environmental considerations should also be addressed before selecting the desired surfactant (or any chemical) for any CEOR application [59].

10.5.2 Modeling and simulation

In the previous sections, numerous studies regarding the EOR potential of chemical flooding in unconventional reservoirs were given. However, like conventional reservoirs, these methods need to be modeled and simulated before any actual operation on the field scale. In this section, we will review some of the latest studies performed on the application of CEOR in shale reservoirs.

Several studies [60–63] have surveyed huff-n-puff surfactant flooding and surfactant addition to fracture fluid to aid incremental oil recovery in hydraulically fractured wells in liquid-rich shale reservoirs by various modeling ways. Indeed, in most studies, the chemical components, particularly surfactants, are used as an aided additive to other enhancement tools, although they have not been employed as a separate procedure.

Although most of these studies emphasize that wettability alteration and IFT reduction are the main mechanisms to improve oil recovery in such reservoirs, each mechanism's dominance is still controversial. Based on numerical modeling, surfactants that alter the wettability toward water-wet or intermediate water-wet are more favorable, due to the more desirable capillary pressure and water relative permeability, than those surfactants which lower the IFT [60].

Further, models have illustrated that huff-n-puff surfactant flooding can double the recovery from primary production ranging from 1.1% to 25.2%. The sensitivity analysis on parameters like capillary pressure, surfactant adsorption, surfactant injection and soaking times, nonwetting (oil) phase hysteresis, matrix permeability, fracture length, and fracture conductivity demonstrates the number of huff-puff cycles and surfactant adsorptions that are the key elements in oil recovery.

As discussed earlier, the CDCs can clearly demonstrate how the injected fluid aids residual oil recovery and relative permeability improvement. Typically, it is a challenging task to measure the CDC uncertainty in the lab. Accordingly, Monte Carlo modeling has been employed to assess the effect of CDC uncertainties on the feasibility of surfactant EOR once it adds to the fractured fluid in shale reservoirs. As indicated in Fig. 10.17, water-wetting and higher-IFT surfactants are the optimum surfactant criteria for shale reservoirs because of the maximum incremental oil recovery and the minimum risks associated with the CDC uncertainty [63]. The benefit of high IFT surfactants is due to the fact that the primary recovery mechanism in shale reservoirs is spontaneous imbibition; then, because ultralow IFT causes a significant capillary reduction, these types of IFT decrease the water imbibition process.

Comprehensive modeling on the performance of surfactants in tight reservoirs was done by Sheng [64]. He used a numerical simulation approach to examine the surfactant and alkali performance in fractured shale and tight rocks. The results revealed the importance of the initial wettability state of rock since it can affect the chemicals diffusion rate. It is shown that the chemicals diffuse very slowly in the oil-wet shale rocks or tight matrix, which hinders the wettability alteration. Therefore, he concluded that the change in relative permeability curves is much more effective than altering the capillary pressure. According to their results, the surfactant solution with ultralow IFT cannot enter the matrix in shale and tight reservoirs; hence, using these surfactants to achieve ultralow IFT may not be effective. To sum up, it was concluded that the surfactant must be able to change wettability as well as maintain high IFT to enhance the imbibition in a shale or tight reservoir [47, 64]. The results of these two subsequent modeling studies are shown in Fig. 10.18. This figure demonstrates that not only is reaching a water-wet status crucial

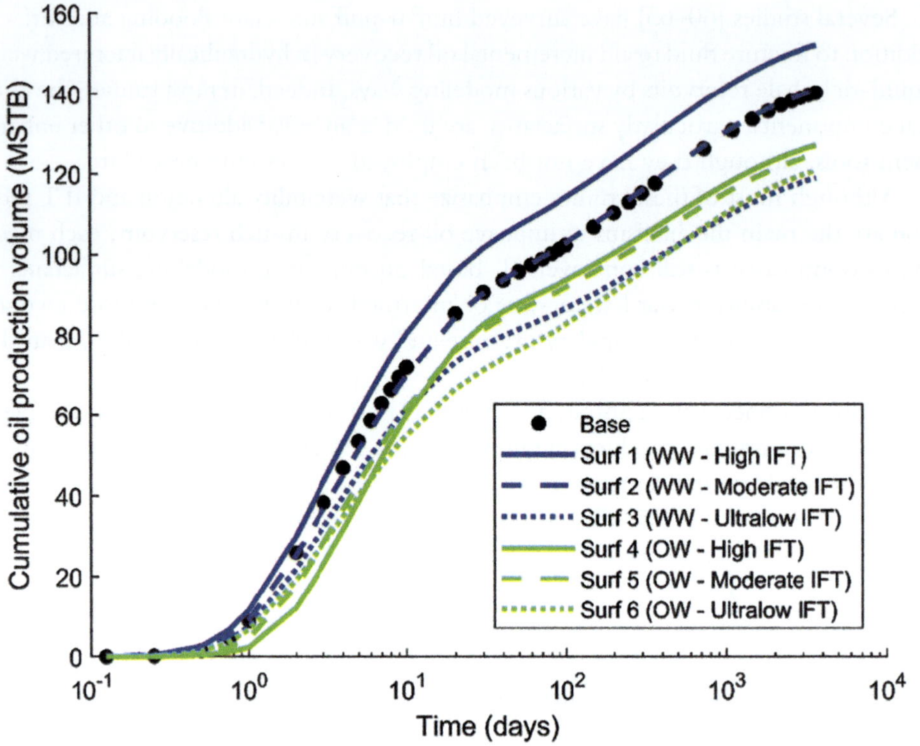

Fig. 10.17 Effect of different surfactants on the preliminary oil production forecasts [63].

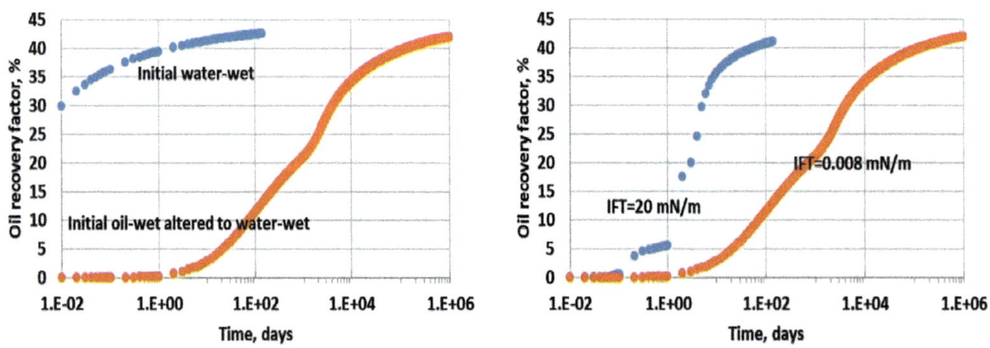

Fig. 10.18 The oil recovery factor versus time: (A) comparison of oil recovery curves of the initial water-wet case with the case of initially oil-wet altered to water-wet by surfactant; (B) comparison of oil recovery curves of different IFT values when the core is initially oil-wet but altered to water-wet by surfactant [47].

for higher oil recovery in tight shale reservoirs, but also the effects of IFT reduction could be detrimental to the recovery once the wettability is changed to water-wet due to a lower capillary driven force [47]. These results are supported by experimental studies that have been done by Liang et al. [55], Mirzaei et al. [56], and Li et al. [57].

Some numerical analyses were conducted on the dispersion coefficients of some surfactants in tight core reservoirs with different permeabilities, such as Bakkan rock. They concluded that the dispersion coefficient of surfactants is around 10^{-5} to 10^{-7} cm^2/s [65, 66]. These results are pretty important as the dispersion coefficient affects the shut-in time needed after the fracturing process or the frequency of surfactant huff-n-puff for maximizing the oil recovery rate [55].

10.6 Field examples

Despite numerous experimental and simulation studies on the application of CEOR for unconventional reservoirs, a few attempts have been made to apply CEOR methods on a field scale so far. However, chemical additives, especially surfactants with other EOR and stimulation techniques, have shown great potential for hydrocarbon recovery in such reservoirs. In this section, some of the applications of chemical additives in the field scale for unconventional reservoirs will be presented.

10.6.1 Surfactant flooding

The only field investigation of CEOR in shaly and tight reservoirs reported so far has been done in two pairs of horizontal wells on one pad site of the Mahu tight oil reservoir (Ma-141 play). To minimize the adsorption rate of surfactants, a new form of surfactant, a diluted suspension of oil droplets (i.e., O/W microemulsions), which are stabilized by the surfactants called diluted microemulsion (DME), has been utilized. After screening laboratory tests on different surfactants/DMEs, the chosen DME-aided slick water is compared with the conventional slick water in these four wells. Both wells where DME has been used have higher friction reduction rates, higher oil production rates, and smaller water flowback rates than control wells. The detailed analysis illustrated that the DME could enhance oil production and reduce the decline rate of the reservoir pressure by propagating within the tight rock, making the surface of the rock more water-wet, and thus enhancing the fracturing fluid imbibition [55].

In this field, researchers used the slick water with the conventional friction reducer in wells A1 and B1, while they used the slick water with DME in wells A2 and B2. For example, as reported in Fig. 10.19, the cumulative oil production in well B2 has increased by about 18.6%. Besides, as DME has increased the imbibition of fracture fluid, it decreases the decline rate of reservoir pressure. It has thus enhanced the oil recovery, which reported that these results are entirely in agreement with the experimental results.

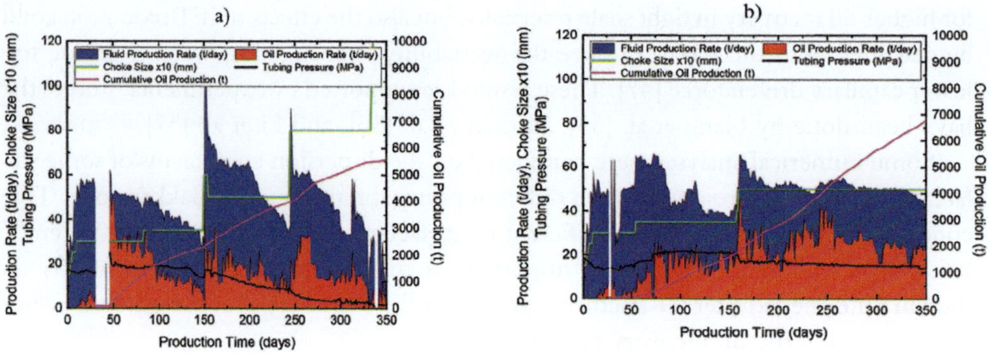

Fig. 10.19 The production history of (A) well B1 and (B) well B2 [55].

10.6.2 Polymer flooding

Despite a rare field example for CEOR in shale and tight reservoirs, there are some pilot and field examples in extra heavy oil fields. Most of these pilot and field applications were made in heavy oil fields in Canada, but there are a few cases in the United States, China, and Oman [24]. Further, although there were some polymer injections before 2010, owing to the low oil price, they were not commercial and economical. The reservoir properties and the properties of the injected polymer in some extra heavy oil reservoirs are reported in Tables 10.1 and 10.2, respectively.

Table 10.1 The properties of extra heavy oil reservoirs flooded by polymer [24].

Field	Country	Porosity (%)	Permeability (mD)	T (°F)	API gravity	Formation water salinity (ppm)	Viscosity at T_r (cP)
Pelican Lake	Canada	28–32	300–3000	53.6–62.6	11.5–16.5	6000–10,000	600–10,000
East Bodo	Canada	30	1000	68	14	25,000–29,000	600–2000
Seal	Canada	27–33	300–5800	68	10–12	22,000	3000–7000

Table 10.2 The properties of injected polymer in extra heavy oil reservoirs [24].

Field	Formation	Polymer	Polymer solution viscosity (cP)	Polymer concentration (ppm)	Injection water salinity (ppm)
Pelican Lake	Wabiskaw "A"	HPAM	13–25	600	8000–26,000
East Bodo	Lloydminster	HPAM	50–60	1500	3700
Seal	Bluesky	HPAM	25–45	1000–1500	2500–11,000

The most well-known cases are the Pelican Lake oil field and Murphy oil field. The oil viscosities in these fields were 600–10,000 cP and 5000–12,000 cP, respectively. The polymer flooding results reveal that the oil recovery can increment 20%–30% of OOIP for these fields [24, 67]. Also, polymer flooding can be employed successfully for extra heavy oil reservoirs with a viscosity of up to 12,000 cP. The Mooney case experience shows that the success of polymer flooding in heavy oil reservoirs depends on heterogeneities and the presence of initial mobile water, which leads to an early breakthrough [67].

10.7 Screening criteria/challenging

Surfactant/chemical selection is the first step that needs to be addressed in any CEOR application. The selection of a proper chemical is usually divided into different stages. The different stages of a surfactant selection for CEOR applications are shown in Fig. 10.20. It is seen that the surfactant compatibility and thermal stability at reservoir condition are the primary factors which should be considered. Then the IFT and adsorption rate should be taken into account for the core flood stage. After laboratory investigations, the modeling stage should also be performed to compare the laboratory results with the results of the upscaling model. It is evident that performing any field development plan, from research to pilot tests, is a time-consuming task. Typically, all parts of EOR research for any field take more than a decade to be potentially ready to run on a full field scale. This fact is mainly related to the challenges scientists face through their researches. As mentioned before, the challenges in studying unconventional reservoirs are much more than those in other reservoir types due to the complexities associated with them. Besides, the typical screening criteria used in conventional reservoirs cannot directly be applied to those in unconventional ones. For instance, in surfactant-aided EOR, ultralow IFT values can be achieved, which leads to in situ microemulsion formation. This can rapidly mobilize the invaded water in the matrix adjacent to hydraulic fractures. However, it cannot activate the capillary desaturation of the water phase under production conditions of tight oil reservoirs, which restrains the degree of matrix permeability regaining [55, 68]. As a result, the screening of surfactants for tight oil reservoirs is different from that for conventional ones.

Although there are several studies on surfactant application in unconventional reservoirs for enhancing oil recovery at a laboratory scale, there is a long way from laboratory tests to use chemicals in field development. However, it is worthwhile to say that compared to the gas-EOR, surfactant injection for EOR purposes can be economically viable under the current condition of low oil prices as the surfactant-EOR neither alters the current fracturing procedure nor needs high-tech injection facilities [44]. The ultralow permeability of shale samples is the main issue facing scientists while trying to analyze them. For example, performing a core flooding test is tedious work compared to

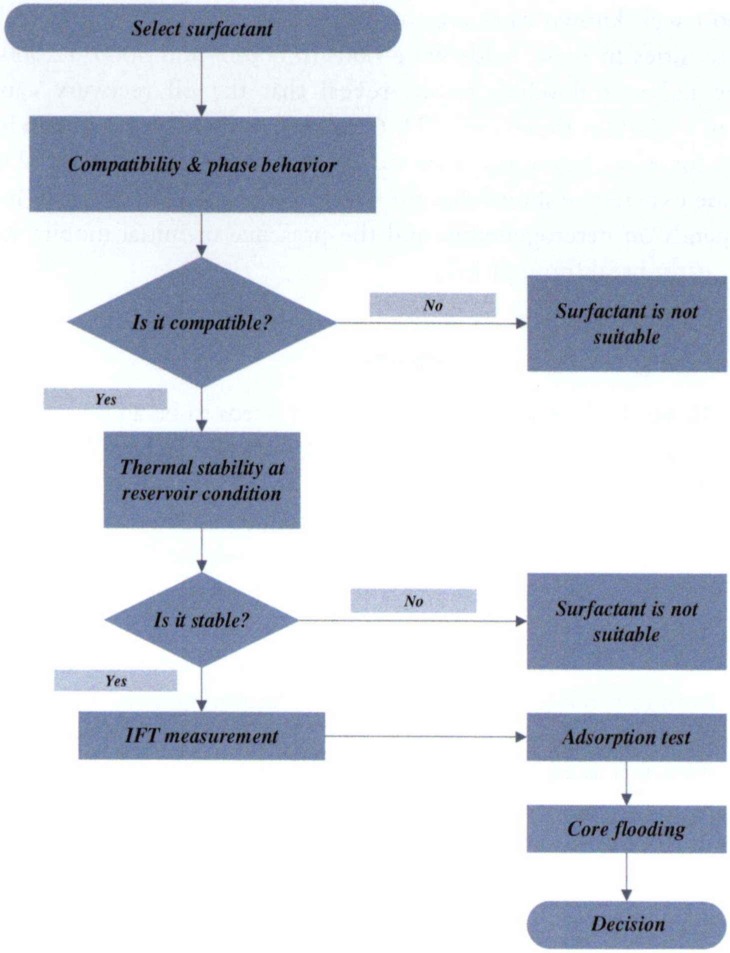

Fig. 10.20 Surfactant selection flowchart [18].

conventional core plugs. This is also intensified during field applications. There are also some other problems that make their study a challenging task and subsequently hinder their development. Due to the unconventional reservoirs' properties, such as small pores and large specific surface areas, the surfactant adsorption is pretty significant, which causes less surfactant penetration. The surfactant adsorption is measured only with the crushed rock samples, which can be underestimated compared to that in the intact rock [55]. Besides, as the temperature of shale reservoirs is high, the stability of surfactants at the reservoir condition is a key factor. In addition, experts must create a balancing condition for the temperature, pH, water salinity, and divalent-cation content of aqueous fluids.

In terms of polymer flooding in extra heavy oil reservoirs, the main challenges for the technical success of the polymer flood include the quality of the injection water, ability to achieve the desired viscosity of the polymer solution, reduced injectivity/productivity issues. Furthermore, polymer solution degradation and retention, insufficient polymer slug size, clay swelling and migration, incompatibility and scale formation, fracture induction and/or propagation, channeling problems, solution quality assurance, downhole and surface facility issues, and environmental limitations are among the other issues which need to be taken into account before any field application [24]. Finally, some economic factors, such as the oil price, capital and operating costs (e.g., polymer and source water consumption), and a discount rate restrict the economic success of polymer flooding in heavy oil reservoirs.

References

[1] World Energy Resources. Unconventional gas, a global phenomenon, www.worldenergy.org; 2016.
[2] Holditch SA. Unconventional oil and gas resource development—let's do it right. J Unconv Oil Gas Resour 2013;1:2–8.
[3] Satter A, Iqbal GM. Reservoir engineering: the fundamentals, simulation, and management of conventional and unconventional recoveries. Gulf Professional Publishing; 2015.
[4] Balasubramanian S, Chen P, Bose S, Alzahabi A, Thakur G. Recent advances in enhanced oil recovery technologies for unconventional oil reservoirs. In: Offshore technology conference; 2018.
[5] Ma YZ, Holditch S. Unconventional oil and gas resources handbook: evaluation and development. Gulf Professional Publishing; 2015.
[6] Hou Y, He S, Harris NB, Yi J, Wang Y, Zhang J, et al. The effects of shale composition and pore structure on gas adsorption potential in highly mature marine shales, Lower Paleozoic, central Yangtze, China. Can J Earth Sci 2017;54:1033–48.
[7] Liu K, Ostadhassan M, Zhou J, Gentzis T, Rezaee R. Nanoscale pore structure characterization of the Bakken shale in the USA. Fuel 2017;209:567–78.
[8] Javaheri A, Dehghanpour H, Wood J. Imbibition oil recovery from tight rocks with dual-wettability pore-network a Montney case study. In: SPE unconventional resources conference; 2017.
[9] Zou C. Introduction. In: Zou C, editor. Unconventional petroleum geology. Boston: Elsevier; 2013. p. 1–26 [chapter 1].
[10] Liu P, Shi L, Liu P, Li L, Hua D. Experimental study of high-temperature CO_2 foam flooding after hot-water injection in developing heavy oil reservoirs. J Pet Sci Eng 2020;185:106597.
[11] Jia L, Li K, Zhou J, Yan Z, Wang Y, Mahlalela BM. Experimental study on enhancing coal-bed methane production by wettability alteration to gas wetness. Fuel 2019;255:115860.
[12] Guanhua N, Hongchao X, Shang L, Qian S, Dongmei H, Yanying C, et al. The effect of anionic surfactant (SDS) on pore-fracture evolution of acidified coal and its significance for coalbed methane extraction. Adv Powder Technol 2019;30:940–51.
[13] Alvarez J, Schechter D. Altering wettability in Bakken shale by surfactant additives and potential of improving oil recovery during injection of completion fluids. In: SPE improved oil recovery conference; 2016.
[14] Alvarez JO, Saputra IWR, Schechter DS. Potential of improving oil recovery with surfactant additives to completion fluids for the Bakken. Energy Fuel 2017;31:5982–94.
[15] Alvarez JO, Schechter DS. Improving oil recovery in the Wolfcamp unconventional liquid reservoir using surfactants in completion fluids. J Pet Sci Eng 2017;157:806–15.
[16] Alvarez JO, Schechter DS. Wettability alteration and spontaneous imbibition in unconventional liquid reservoirs by surfactant additives. SPE Reserv Eval Eng 2017;20:107–17.
[17] Elturki M, Imqam A. Application of enhanced oil recovery methods in unconventional reservoirs: a review and data analysis. In: 54th U.S. rock mechanics/geomechanics symposium; 2020.
[18] Kamal MS, Hussein IA, Sultan AS. Review on surfactant flooding: phase behavior, retention, IFT, and field applications. Energy Fuel 2017;31:7701–20.
[19] Standnes DC, Austad T. Wettability alteration in chalk: 2. Mechanism for wettability alteration from oil-wet to water-wet using surfactants. J Pet Sci Eng 2000;28:123–43.

[20] Zivar D, Pourafshary P, Moradpour N. Capillary desaturation curve: does low salinity surfactant flooding significantly reduce the residual oil saturation? J Pet Explor Prod Technol 2021;11:783–94.
[21] Horgue P, Soulaine C, Franc J, Guibert R, Debenest G. An open-source toolbox for multiphase flow in porous media. Comput Phys Commun 2015;187:217–26.
[22] Scott AJ, Romero-Zerón L, Penlidis A. Evaluation of polymeric materials for chemical enhanced oil recovery. Processes 2020;8:361.
[23] Santos R, Loh W, Bannwart A, Trevisan O. An overview of heavy oil properties and its recovery and transportation methods. Braz J Chem Eng 2014;31:571–90.
[24] Saboorian-Jooybari H, Dejam M, Chen Z. Heavy oil polymer flooding from laboratory core floods to pilot tests and field applications: half-century studies. J Pet Sci Eng 2016;142:85–100.
[25] Druetta P, Raffa P, Picchioni F. Chemical enhanced oil recovery and the role of chemical product design. Appl Energy 2019;252:113480.
[26] Rogers RE. Coalbed methane: principles and practice. Prentice Hall; 1994.
[27] Thakur P. Global reserves of coal bed methane and prominent coal basins. In: Advanced reservoir and production engineering for coal bed methane. Gulf Professional Publishing; 2017. p. 1–15 [chapter 1].
[28] Hu Z-X, Hu X-M, Cheng W-M, Lu W. Influence of synthetic conditions on the performance of melamine–phenol–formaldehyde resin microcapsules. High Perform Polym 2019;31:197–210.
[29] Ni G, Li Z, Xie H. The mechanism and relief method of the coal seam water blocking effect (WBE) based on the surfactants. Powder Technol 2018;323:60–8.
[30] Cai J, Xu C-G, Chen Z-Y, Li X-S. Recovery of methane from coal-bed methane gas mixture via hydrate-based methane separation method by adding anionic surfactants. Energy Sources A: Recovery Util Environ Eff 2018;40:1019–26.
[31] Wang J, Dong M. A laboratory study of polymer flooding for improving heavy oil recovery. In: Canadian international petroleum conference; 2007.
[32] Dusseault M. Comparing Venezuelan and Canadian heavy oil and tar sands. In: Canadian international petroleum conference; 2001.
[33] Speight JG. Introduction to enhanced recovery methods for heavy oil and tar sands. Gulf Professional Publishing; 2016.
[34] Wang J, Dong M. Optimum effective viscosity of polymer solution for improving heavy oil recovery. J Pet Sci Eng 2009;67:155–8.
[35] Asghari K, Nakutnyy P. Experimental results of polymer flooding of heavy oil reservoirs. In: Canadian international petroleum conference; 2008.
[36] Guo Z, Dong M, Chen Z, Yao J. A fast and effective method to evaluate the polymer flooding potential for heavy oil reservoirs in Western Canada. J Pet Sci Eng 2013;112:335–40.
[37] Hernández E, Valero E, Rodríguez I, Guerra E, Espinoza J, Briceño R, et al. Enhancing the recovery of an extra heavy oil reservoir by using low salinity polymer flooding. In: SPE Latin American and Caribbean petroleum engineering conference; 2020.
[38] Zhou W, Xin C, Chen S, Yu Q, Wang K. Polymer-enhanced foam flooding for improving heavy oil recovery in thin reservoirs. Energy Fuel 2020;34:4116–28.
[39] Dong M, Ma S, Liu Q. Enhanced heavy oil recovery through interfacial instability: a study of chemical flooding for Brintnell heavy oil. Fuel 2009;88:1049–56.
[40] Zhang H, Dong M, Zhao S. Which one is more important in chemical flooding for enhanced court heavy oil recovery, lowering interfacial tension or reducing water mobility? Energy Fuel 2010;24:1829–36.
[41] Zhou X, Dong M, Maini B. The dominant mechanism of enhanced heavy oil recovery by chemical flooding in a two-dimensional physical model. Fuel 2013;108:261–8.
[42] Aguilera R. Shale gas reservoirs: theoretical, practical and research issues. Pet Res 2016;1:10–26. 2016/09/01.
[43] Anon. World Energy resources unconventional gas, a global phenomenon. World Energy Council; 2016. https://www.worldenergy.org/.
[44] Anon. A look at the most abundant shale gas reserves around the world. Oil & Gas Middle east; February 27, 2014. Available from: https://www.oilandgasmiddleeast.com/article-12003-top-10-top-10-shale-gas-reserves-in-the-world.
[45] Sheng JJ. Critical review of field EOR projects in shale and tight reservoirs. J Pet Sci Eng 2017;159:654–65.

[46] Liu J, Sheng JJ. Experimental investigation of surfactant enhanced spontaneous imbibition in Chinese shale oil reservoirs using NMR tests. J Ind Eng Chem 2019;72:414–22.
[47] Sheng JJ. What type of surfactants should be used to enhance spontaneous imbibition in shale and tight reservoirs? J Pet Sci Eng 2017;159:635–43.
[48] Ahmadi MA, Shadizadeh SR. Experimental investigation of a natural surfactant adsorption on shale-sandstone reservoir rocks: static and dynamic conditions. Fuel 2015;159:15–26.
[49] Zeng T, Miller CS, Mohanty K. Application of surfactants in shale chemical EOR at high temperatures. In: SPE improved oil recovery conference; 2018.
[50] Wang D, Butler R, Zhang J, Seright R. Wettability survey in Bakken shale with surfactant-formulation imbibition. SPE Reserv Eval Eng 2012;15:695–705.
[51] Chen P, Mohanty K. Surfactant-mediated spontaneous imbibition in carbonate rocks at harsh reservoir conditions. SPE J 2013;18:124–33.
[52] Saputra IWR, Park KH, Zhang F, Adel IA, Schechter DS. Surfactant-assisted spontaneous imbibition to improve oil recovery on the Eagle Ford and Wolfcamp shale oil reservoir: laboratory to field analysis. Energy Fuel 2019;33:6904–20.
[53] Afanasev P, Scerbacova A, Tsyshkova A, Mukhina E, Grishin P, Grishaev V, et al. Compositions of anionic and non-ionic surfactants within a hybrid EOR technology for unconventional hydrocarbon reservoirs. In: SPE Russian petroleum technology conference; 2019.
[54] Zeng T, Miller CS, Mohanty KK. Combination of a chemical blend with CO_2 huff-n-puff for enhanced oil recovery in oil shales. J Pet Sci Eng 2020;194:107546.
[55] Liang T, Zhao X, Yuan S, Zhu J, Liang X, Li X, et al. Surfactant-EOR in tight oil reservoirs: current status and a systematic surfactant screening method with field experiments. J Pet Sci Eng 2021;196:108097.
[56] Mirzaei M, DiCarlo DA, Pope GA. Visualization and analysis of surfactant imbibition into oil-wet fractured cores. SPE J 2016;21:101–11.
[57] Li Y, Pope GA, Lu J, Churchwell L, Tagavifar M, Weerasooriya UP. Scaling of low-interfacial-tension imbibition in oil-wet carbonates. SPE J 2017;22:1349–61.
[58] Huang H, Babadagli T, Chen X, Li H. Performance comparison of novel chemical agents in improving oil recovery from tight sands through spontaneous imbibition. In: SPE international conference on oilfield chemistry; 2019.
[59] Soleimani Zohr Shiri M, Henderson W, Mucalo MR. A review of the lesser-studied microemulsion-based synthesis methodologies used for preparing nanoparticle systems of the noble metals, Os, Re, Ir and Rh. Materials 2019;12:1896.
[60] Lotfollahi M, Beygi MR, Abouie A, Sepehrnoori K, Wheeler MF, DiCarlo DA. Optimization of surfactant flooding in tight oil reservoirs. In: Unconventional resources technology conference, Austin, Texas, 24–26 July 2017; 2017. p. 3156–68.
[61] Sun R, Pu H, Yu W, Miao J, Zhao JX. Simulation-based enhanced oil recovery predictions from wettability alteration in the Middle Bakken tight reservoir with hydraulic fractures. Fuel 2019;253:229–37.
[62] Weirong L, Zhenzhen D, Gang L, Cai W, Huijie W. The study of surfactant huff and puff of horizontal well with SRV in tight oil reservoirs. In: International petroleum technology conference; 2019.
[63] Wijaya N, Sheng JJ. Uncertainty quantification of shale capillary desaturation curves for surfactant EOR in shale through fracturing fluids using predictive modeling. Fuel 2021;283:118857.
[64] Sheng JJ. Performance analysis of chemical flooding in fractured shale and tight reservoirs. Asia Pac J Chem Eng 2018;13:e2147.
[65] Wang D, Zhang J, Butler R, Olatunji K. Scaling laboratory-data surfactant-imbibition rates to the field in fractured-shale formations. SPE Reserv Eval Eng 2016;19:440–9.
[66] Chevalier T, Labaume J, Gautier S, Chevallier E, Chabert M. A novel experimental approach for accurate evaluation of chemical EOR processes in tight reservoir rocks. In: SPE improved oil recovery conference; 2018.
[67] Delamaide E, Bazin B, Rousseau D, Degre G. Chemical EOR for heavy oil: the Canadian experience. In: SPE EOR conference at oil and gas West Asia; 2014.
[68] Liang T, Achour SH, Longoria RA, DiCarlo DA, Nguyen QP. Flow physics of how surfactants can reduce water blocking caused by hydraulic fracturing in low permeability reservoirs. J Pet Sci Eng 2017;157:631–42.

CHAPTER 11

Formation damage during chemical flooding

Amin Rezaei[a] and Saman Bagherpour[b]
[a]Department of Petroleum Engineering, School of Chemical and Petroleum Engineering, Shiraz University, Shiraz, Iran
[b]Separation Processes & Nanotechnology Lab, Faculty of Caspian, College of Engineering, University of Tehran, Tehran, Iran

11.1 Introduction

The term improved oil recovery (IOR) refers to a pool of methodologies applied to enhance the efficiency at which hydrocarbons are withdrawn from a reservoir. Among others, IOR techniques may refer to what is known as enhanced oil recovery (EOR, e.g., CO_2 injection, thermal recovery, and chemical flooding), secondary recovery approaches (e.g., gas/water injection), hydraulic fracturing, horizontal drilling, and multilateral well facilities [1]. Focusing on the EOR, chemical enhanced oil recovery (CEOR) has been largely regarded by industrial practitioners as well as researchers during the recent past. In a CEOR plan, one or more chemicals are injected into a hydrocarbon reservoir either alone or in combination with other chemicals/additives. This can be done by flooding the reservoir with either a single chemical agent [e.g., alkaline (A), surfactant (S), or polymer (P)], a binary pair of chemicals [e.g., alkaline-polymer (AP), alkaline-surfactant (AS), or surfactant-polymer (SP)] or a tertiary combination of agents [e.g., alkaline-surfactant-polymer (ASP)]. Advantages of the CEOR techniques come with adverse side effects, however, in terms of formation damage (FD) to the reservoir—a well-known problem in the reservoir engineering that not only affects the performance of oil recovery negatively but also leads to significant challenges that are much costly to address [2].

In the past, relatively high oil prices have made the practitioners largely ignore the contributions of the IOR into FD despite the well-established fact that the EOR has much to do with FD. Reviewing the causes and effects of FD, Civan (2015) published a study where the focus was on solutions for assessing, controlling, and even remediating the FD [3]. FD has been defined as the adverse alteration of various properties (i.e., chemical, physical, or mechanical) of the reservoir formation. In many cases, this comes in the form of impaired permeability upon implementing different EOR processes. Indeed, EOR processes are known to modify physical, chemical, and/or thermodynamic conditions to which the formation is exposed, thereby inducing FD in the form of water/gas bubble blockage, fluid-rock incompatibility, fine/sand migration,

Chemical Methods
https://doi.org/10.1016/B978-0-12-821931-7.00006-7

precipitation of organic/inorganic matter, and changes in pore surface properties, structure, and mechanical features.

Although there are cases where the FD boosts the oil recovery by, for example, increasing the sweep efficiency upon fines migration and the resultant blockage of highly permeable regions [4–7], the much more common effects of FD include degrading the performance of hydrocarbon recovery and lowering the productivity and/or injectivity of the wells drilled into the reservoir dramatically [8, 9].

According to Porter [10], as far as the FD is concerned, similar to many other cases, prevention is much easier and better off than a post-occurrence remedy, if any. The type of damage induced into the formation is known to be dependent on the type of EOR technique. That is, one may end up with unrealistic results if he/she fails to take into consideration the engineering activities performed on a reservoir when investigating the degree to which the reservoir is susceptible to a particular type of FD. Put differently, oil companies must comprehend the cause-effect relationships between different EOR techniques and the FD to optimize the EOR plan based on reservoir characteristics in such a way to achieve maximum possible technical and economic advantages.

By definition, FD refers to a wide spectrum of mechanisms that influence not only the reservoir (i.e., matrix and fractures) properties but also the efficiency at which multiphase fluids (e.g., hydrocarbons, water, fine particles, EOR agents, etc.) can flow within the reservoir formation. The influence can take various forms, usually including alterations to injectivity/productivity of the well and hydrocarbon recovery [11]. FDs have been classified under four major classes, namely mechanical, chemical, biological, and thermal FDs [12]. Focusing on the chemical FD mechanisms [13], one can identify two subcategories: (i) fluid-fluid incompatibilities (e.g., inorganic scale and organic asphaltene deposition, foam/emulsion blockage, and hydrate formation) and (ii) rock-fluid incompatibilities (e.g., swelling/de-flocculation of clay, changed wettability, and adsorption of ionic/surfactant/polymer species on rock surface). Mechanical FDs are most commonly caused by migration of fines/sands or other particles through the reservoir [8], phase trapping due to excessive capillary force in multiphase fluid flow [14], and pressure-induced compaction/dilatation of reservoir rock [15]. Temperature variations can contribute to dissolving and transforming minerals and trigger thermally induced wettability changes [16]. Last but not least, bacterial activities in reservoir formation can accelerate crude oil souring, mineral erosion, and porosity reduction by blockage of pores, all of whom may bring about positive and/or negative effects [17]. Briefly, Fig. 11.1 shows the possible mechanisms for FDs during CEOR processes [2, 18].

Various IOR technologies may end up with different impacts on the medium, changing the medium physically, chemically, thermoelectrically, mechanically, and/or biologically. Following this chapter, we discuss the association of different types of FD with CEOR techniques in common sandstone/carbonate reservoirs. In this respect, we focus on the range of different FD problems that may arise upon implementing different

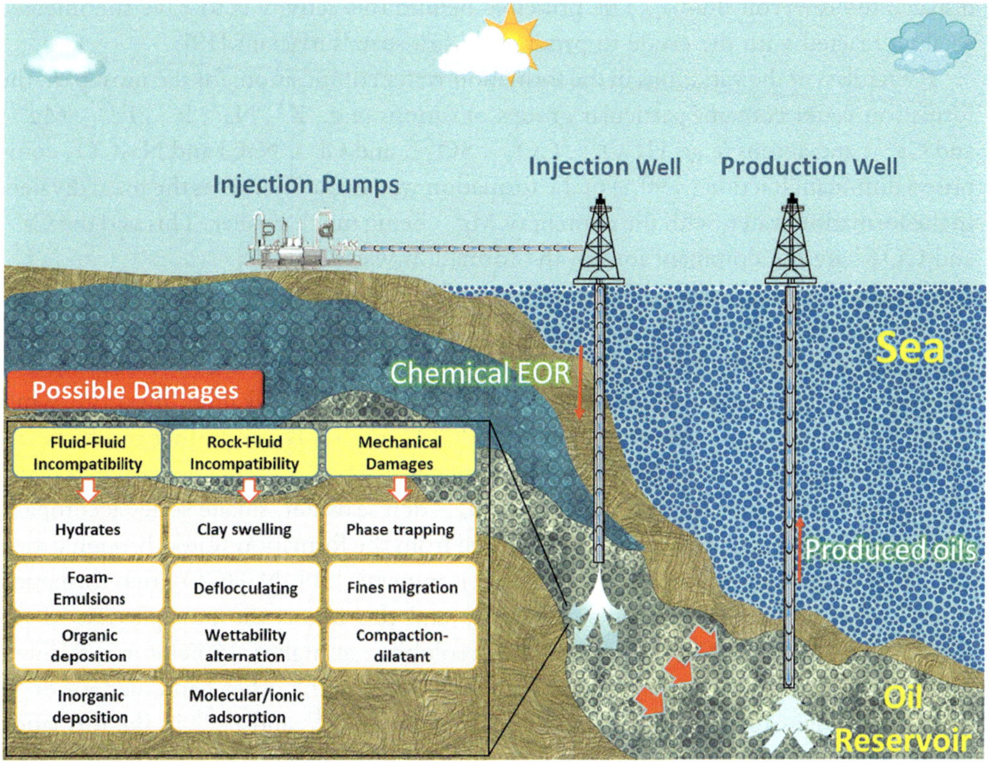

Fig. 11.1 The possible mechanisms of formation damage during chemical injection for enhanced oil recovery purpose.

CEOR methods in various hydrocarbon-bearing reservoir types, rather than barely repeating fundamental theories for each EOR technique. This shall include both conventional and unconventional reservoirs. For that, we present a comprehensive review of the literature published so far. Based on this review, we then proceed to identify the associated risks and opportunities of FD, with a summary of the newest developments on FD control in the course of EOR processes further laid out. Acknowledging the highly uncertain nature of the EOR-induced FDs, an integrated framework is further developed for risk assessment, opportunity evaluation, and FD management to help optimize IOR plans in relation to FD.

11.2 Problems associated with different chemical methods

11.2.1 Alkaline flooding

Alkaline flooding is an EOR technique where alkaline solutions (e.g., sodium orthosilicate, sodium carbonate, and sodium hydroxide) are injected into a reservoir to help displace the reservoir fluid(s). The principle behind this activity is to have the injected alkaline reacted with the crude to produce in-situ soap/surfactant [19].

Regardless of the variations in the formation water composition, for the most part, the formation water contains particular groups of cations (e.g., K^+, Na^+, F^{3+}, Fe^{2+}, Mg^{2+}, and Ca^{2+}) and anions (e.g., HCO^{3-}, CO_3^{2-}, SO_4^{2-}, and Cl^-). NaCl and Na_2CO_3 comprise a dominant fraction (>90%) of the formation water. Ca^{2+} serves as the main divalent in the formation water, with the content of Mg^{2+} being much smaller. This said the Ca^{2+} and CO_3^{2-} are the dominant ions in the formation water [20, 21].

Alkaline flooding-induced FDs are most commonly sourced from the incompatibility of the alkaline-formation water mixture with the reservoir rocks. As an example, increased dosages of CO_3^{2-}, OH^-, and SiO_3^{2-} in the formation after it is subjected to alkaline injection. Here the OH^- sources directly from the alkaline injection, while CO_3^{2-} is produced from the conversion of HCO_3 into CO_3^{2-} and other products and SiO_3^{2-} is formed as the alkaline reacts with minerals. Particular scales are known to occur with metal ions. For example, with some SO_4^{2-}-rich seawater, sulfate scales accompany Ca^{2+}, Mg^{2+}, Ba^{2+}, and Sr^{2+} as they are formed from the formation water through cation exchange interaction of the alkaline with rock minerals [20]. $NaHCO_3$-rich formation water contains excessive amounts of HCO_3^- and this boosts the precipitation of the $CaCO_3$. However, $CaCO_3$ exhibits reduced solubility at higher temperatures, facilitating its precipitation at higher reservoir temperatures. On the other hand, deposition of $FeCO_3$ is anticipated only at higher concentrations of Fe^{2+} and Fe^{3+} in the formation water. As depicted in Fig. 11.2, the interactions resulting in scale formation in the presence of various ions are presented.

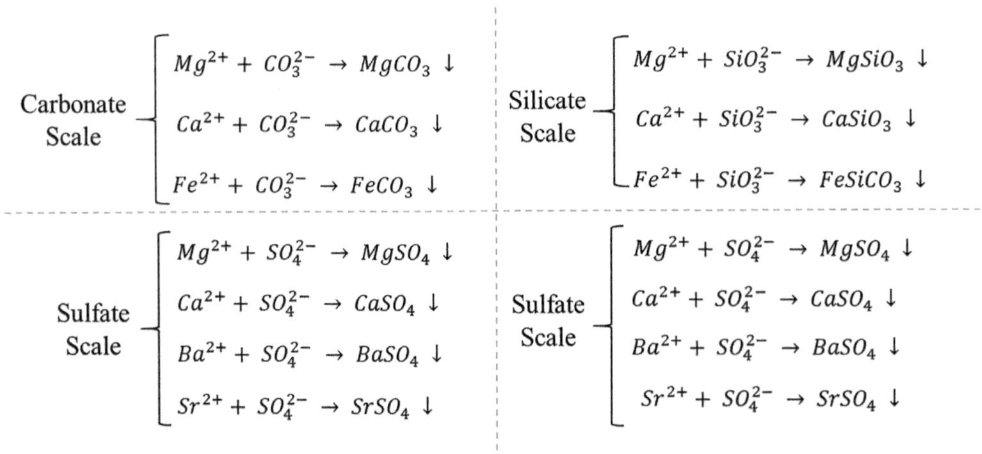

Fig. 11.2 The process to formation of inorganic scales during the injection of alkaline solutions. *(Modified after Yuan B, Wood DA. A comprehensive review of formation damage during enhanced oil recovery. J Petrol Sci Eng 2018;167:287–99.)*

As these scales precipitate, the formation has its permeability decreased remarkably, thereby lowering the effectiveness of alkaline flooding. Such precipitation can further contribute to a wide variety of problems in the flow line, safety valves and choke, pumping systems, and corrosion of subscale equipment in both surface facilities and pipework. As an example of such challenges, the alkaline flooding plan in the Wilming Field (California, USA) came to early termination due to nothing but the precipitation of Mg_2SiO_4 and $CaCO_3$ scales, even after an acidization stage which was, later on, proved ineffective for removing the precipitates [18]. Even worse, a SO_4^{2-}-rich formation water increases the chances of precipitation of sulfate scale, a very complicated and cost-intensive germ to remove due to its acid-insoluble nature.

At the other end of the spectrum, an alkaline agent can facilitate the dissolution of clay and some other minerals, thereby opening new channels for the hydrocarbon fluid to flow through the reservoir (i.e., enhancing the permeability of reservoir rock). This is however, a rare outcome and the common effect of alkaline flooding is the adverse migration of fines leading to permeability blockage, i.e., a well-known type of FD. Moreover, clay-mineral interactions tend to trigger migration of fines and/or swelling of clays by changing the configuration of electrostatic attractions/repulsions between the particles [22]. Alternatively, oil emulsions may develop with water droplets of various sizes depending on the concentration of the alkaline; these droplets can then block the pore throats, degrading the reservoir permeability. Indeed, the higher the concentration of the alkaline, the further the conversion of the initially oil-in-water (OIW) emulsion to a water-in-oil (WIO) emulsion (Fig. 11.3), and this lowers the ability of the reservoir fluid to flow, i.e., providing higher oil phase viscosity [23].

According to an empirical work by Cheng, as far as alkaline flooding is concerned, one can reduce the dissolution of minerals and ion exchange by using Na_2CO_3, as the alkaline agent, rather than either silicate or sodium hydroxide. Performing flooding tests on sands of Ottawa and Wilmington and some dolomite samples, Cheng concluded that

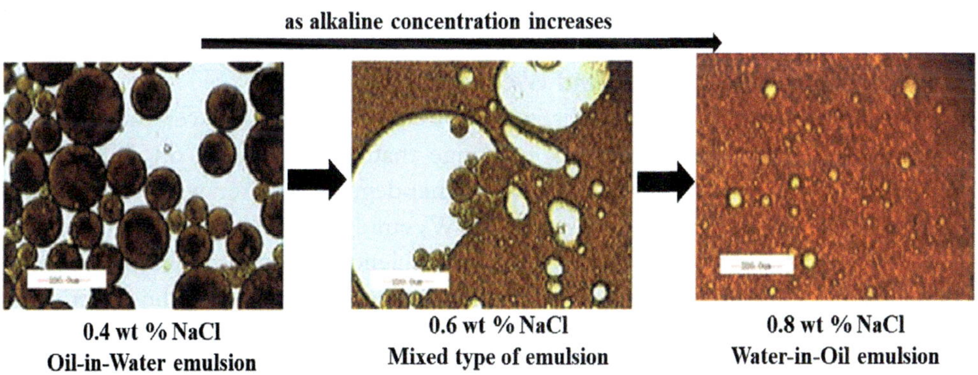

Fig. 11.3 Emulsion type conversion (from OIW to WIO) upon changes in the alkaline concentration [18].

the FD-inducing precipitation can be attenuated by avoiding Na_4SiO_4, with the efficiency of such EOR plan being superior with Na_2CO_3 rather than NaOH [24].

Aimed to address the issues related to the use of strong alkalis, Flateen et al. proposed the use of a weak alkali as an alternative to strong alkalis: sodium metaborate [25]. In another piece of research, Berger and Lee evaluated an organic alkali for replacing inorganic alkalis [26].

11.2.2 Surfactant flooding

In many CEOR plans, surfactants are used to facilitate the process. In such cases, one should ensure that a single-phase flow is established to avoid the so-called phase trapping due to precipitation, liquid crystallization, or formation of any second liquid phase. Surfactant injection may succeed to achieve the anticipated results depending on many factors including the crude-brine interaction efficiency, reservoir setting, and some other conditions that are specific to the surfactant used, e.g., the surfactant retention time and thermal and aqueous compatibility and stability of the surfactant. Most dominantly, the following FD mechanisms may be triggered by a surfactant flooding process:

11.2.2.1 Excessive adsorption and retention of surfactants

Once injected as an EOR method, it is a primary objective to reduce surfactant adsorption on the rock surfaces to an appropriate degree. Nevertheless, too much surfactant adsorption is undesirable as it can lead to FD in the form of permeability reduction in sandstone reservoirs [27]. Indeed, too much adsorption of surfactants or retention of them due to phase trapping [28, 29] is known to negatively affect the reservoir rock permeability and porosity, thereby decreasing the effectiveness of the entire EOR plan as the surfactant can no more treat the oil-water interfaces properly. This problem can be addressed by providing for lower salinity of the injected fluid and adding alcohol-derived methanol to water-based surfactant mixes, thereby inhibiting too much adsorption, agglomeration, and deposition of surfactants.

11.2.2.2 Wettability alteration and W/O emulsion blockage

Extensive emulsification of a surfactant in oil/water tends to lower the interfacial tension (IFT) of the resultant fluid, a change that contributes to oil scale off and dispersion out of rock surfaces, establishing higher degrees of water-wettability on rock surfaces. Development of an oil-in-water (OIW) emulsion can help control the mobility of the reservoir fluid and enhance the efficiency of oil production thanks to increased viscosity exhibited by the aqueous phase. This can, on the other hand, severely block the pore throats, degrading the reservoir rock permeability—an issue to be considered in especially low-permeability formations [30]. At the other end of the spectrum, a surfactant-assisted development of a WIO at, for example, higher salinities make the emulsion even more viscous than the pure crude, complicating the motion of the crude towards the wellbore [31].

11.2.2.3 *Surfactant precipitation and phase trapping*

As an anionic surfactant comes in contact with a cationic fluid/mineral in the formation, the surfactant is likely to be lost in the form of precipitate [32]. The incompatibility of crude with the multivalent cations found in formation water can lead to the production of sulfonate sediments (e.g., calcium and magnesium sulfonates) that tend to clog the pore throats and hence lower the success rate of surfactant flooding. In addition, instability of the aqueous and oleic phases and the resultant separation may end up setting the scene for phase trapping due to multiphase flows under elevated capillary forces [33]. Among others, mineralization state, temperature, ionic strength, and pH value are the main parameters controlling the sedimentation of surfactants [34]. In this respect, one can opt for elevated fluid temperatures and/or pH values and attenuated extent of mineralization. For example, one can effectively lower the sedimentation of calcium and magnesium sulfonates by limiting the presence of Ca^{2+} and Mg^{2+} in the reservoir fluid.

11.2.3 Polymer flooding

Polymer flooding works primarily by making the displacing phase more viscous, thereby elevating the mobility control and hence oil recovery. In the meantime, FD is an anticipated outcome when the reservoir is suffering from harsh conditions in the form of, for example, high salinity levels and exposure to calcium, elevated temperatures, and long injection periods. The mentioned issue is particularly observed with polymers that are soluble in water and takes the form of decreased thickening capability and flocculation of the polymer [35]. Strangely, there are cases where even larger pore throats than the moving particles cannot save the formation against FD. Emphasizing such a setting, Veizen and Leeriooljer [36] demonstrated that even having sevenfold larger pore throats than the particles cannot save a porous medium from being clogged. This emphasizes the importance of identifying potential FD mechanisms that can occur in a specific reservoir prior to practicing the polymer flooding. The followings are some of the main reasons why the reservoir formation is hit by FD in the course of polymer flooding [37].

11.2.3.1 *Polymer retention in the porous medium*

Optimal polymer retention time contributes to lower permeability of the aqueous phase, thereby lowering the mobility ratio and hence enhancing the sweep efficacy. However, as mentioned earlier, similar to the case with surfactants, too much adsorption of polymer on the rock surface can bring the pore throats to closure, i.e., damaging the formation in terms of permeability and porosity reduction. So far, researchers have elaborated on three different polymer retention mechanisms, namely physical adsorption [38], mechanical trapping [39], and hydraulic retention [40].

As the name suggests, physical adsorption points out to the accumulation of polymer molecules on the opening of pores on the rock surface by means of physical interactions like Van der Waals force and hydrogen bonding with OH groups on the rock face. Indeed, the physical nature of this mechanism roots in the fact that it involves no chemical adsorption due to lack of any chemical bonding during this process. Under full dispersion of the polymer, the primary retention mechanism for the polymer is expectedly the physical adsorption. In their work, Tang et al. investigated the way polymers were adsorbed across a reservoir of sandstone in the Daqing Oilfield (China) [41]. Based on their findings, they figured out that the flow of the polymer in the pores within the sandstones leads to easy retention of the polymer in the pore space. For the most part, clay minerals exhibit larger capacities for adsorbing the polymers, as compared to the matrix of the rock. Lanlei et al. published a research paper in 2011 where they worked on polymer aggregation morphology following a polymer flooding into the formation; they based their work on high-magnification SEM analyses [42]. The results indicated the retention of the polyacrylamide (PAM) in the reservoir formation following the polymer flooding. This was made possible through the aggregation and adsorption of polymer molecules on the surface of the rock.

Mechanical trapping simply defines a situation where polymer molecules are trapped in smaller pore throats. Ultraheavy-molecule polymers suffer from more serious degrees of mechanical trapping. As an example, intensive agglomeration of polymer molecules into micro-gel structures in the vicinity of the well through which the polymer injection is being performed characterizes the mechanical trapping.

As another mechanism of polymer retention, one may rely on changing the flow direction or velocity, i.e., hydraulic retention. Herein, despite the blockage of tiny pores by polymers (e.g., because of the mechanical trapping), a forced modification to the streamlines can extend the polymer retention. The resultant ascendance in velocity would keep the detained polymer inside the rock matrix while worsening the retention phenomenon. Hydraulic retention becomes increasingly more significant at higher polymer molecular weights and lower core permeabilities. Among other advantages, hydraulic retention helps extend the retention time of the polymer in the cores and indirectly elevates the injection pressure. Published in 2009, the work of Fu et al. suggested that the polymer retention and adsorption capacities change with the injection rate. That said, low-permeability cores were found to exhibit much more significant capacities for polymer retention and adsorption, as compared to high-permeability cores [43]. As shown in Fig. 11.4, high-permeability cores tended to have the polymer molecules detached from their primary retention sites as a result of increased injection rates, thereby lowering the polymer retention. At lower permeabilities, a boost of injection rate sets the scene for the polymer molecules to go through smaller pore throats after undergoing a tensile deformation, making the problems even worse.

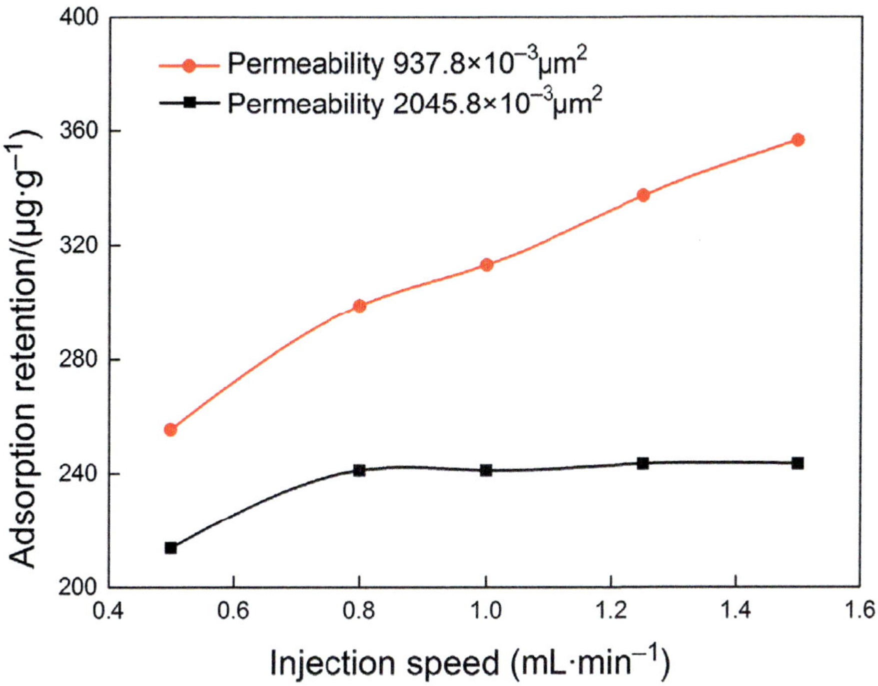

Fig. 11.4 Impact of injection rate on polymer retention for various rock permeabilities [43].

In an attempt to optimize polymer retention especially in terms of avoiding excessive polymer retention, Dai [2] presented several strategies, as follows: limiting the extent of mineralization, optimizing the reaction temperature and the molecular weight, concentration, and injection rate of the injection fluid.

11.2.3.2 Incompatibility of polymer solutions with formation fluids and minerals

Strongly electrophilic cations derived from inorganic salts may couple with carboxyl groups on polymers to form ionic pairs; this decreases the force at which polymer particles repulse one another electrostatically, ending up with increased chances of polymer coil contraction and precipitation of inorganic species [44]. Possible cross-linking of ferric cations (Fe^{3+}) existing in the formation water with polymers can give rise to hydrogels that partly plug the permeability of the reservoir rock [45]. Polymer injection may lower the formation temperature as the injected fluid is much cooler than the reservoir temperature. This facilitates the precipitation of inorganic species that block the pore space. In order to address the mentioned issues, one may use such deoxidants such as Na_2SO_3 or $NaHSO_3$ to trigger a reaction with cation contents of the formation water and produce inorganic precipitates. The paraffin content of the crude expels out in the form of wax upon a drop of the formation temperature by polymer injection. Water-wet rocks promote the precipitation and adsorption of polymers. Such polymer adsorption, however, tends to limit the flowability of the formation fluid when occurred in tiny pores. On the other hand, polymer adsorption on clay grains tends to mitigate, partially, the problem of fine/sand migration into the formation fluid. Some of the workarounds proposed for controlling the polymer flooding-induced FD are as follows: keeping the injected polymer temperature close enough to the reservoir temperature, preinjection filtering of polymer solution to remove suspended particles, and adopting optimal additives to make the injected fluid adequately compatible with the existing formation fluids. Indeed, prior to polymer injection for IOR, the to-be-injected mixture must undergo various laboratory screening assessments to verify the polymer injectivity, adsorption capacity, and brine compatibility, and characterize its degradation microbial, chemically and mechanically.

11.3 Combined CEOR processes

Combining different CEOR processes may bring about particular advantages should the combined chemicals be compatible with one another and remain stable in the mix. Otherwise, extensive FD is anticipated.

11.3.1 Alkaline-surfactant (AS) flooding

A way of enhancing the performance of alkaline flooding is to combine it with some surfactant to come with the so-called alkaline-surfactant (AS) flooding. According to Rudin et al. (1994), such combinations can weaken the IFT in equilibrium state while adding to the interfacial resistance of the resultant emulsion [46]. This, in turn, improves the capacity of the emulsion to withhold oil within the stream of water although it can also plug the pores upon an agglomeration of the emulsified species. One may end up with lower adsorption of surfactants and sulfonate on the surface of kaolinite minerals by introducing an alkali to make the mineral surface even more negatively charged [2]. This adsorption control measure can then promote the crude sweep efficiency though the sulfonate is prone to detachment, causing some FD.

11.3.2 Surfactant-polymer (SP) flooding

Developed by combining some surfactant with some polymer, the flooding of a binary surfactant-polymer (SP) fluid provides for lower IFT. Synergizing the features of both the polymer and the surfactant, such a combination contributes to EOR significantly. At surfactant concentrations beyond the critical concentration of micelle, flooding of the miscible combination of a micro-emulsion with a polymer can greatly enhance the sweep efficiency [30]. When it comes to SP flooding, the surfactant part of the injected fluid can drastically weaken the IFT, thereby helping to achieve improved crude displacement in the reservoir.

Chelation can be generated among structures and polymers in specific conditions. This phenomenon is highly effective in augmenting the compound system viscosity. The displaced phase viscosity can also be raised, resulting from the surfactant emulsification impact. By adding polymers, the displacing phase viscosity is enormously increased, leading to improving the performance of mobility control. Consequently, the residual oil is washed resulting from the surfactant swept yield. Besides, in order to provide reaction with the divalent cations and anions, polymer molecules operate as sacrificial agents. The performance of this polymer causes the reduction of surfactant consumptions resulting in the emulsion stability improvement. Furthermore, the synergistic impact of polymer and surfactant associated with combining polymers and surfactants provides privileges. In the polymer presence, the precipitation of surfactants will not occur while injecting the polymer-surfactant combination. In addition, the deactivation of surfactants will not happen due to the metal ions impact. Moreover, the high reduction of IFT resulting from surfactant indicates the polymers viscoelasticity properties [47].

Surfactants utilized in the displacement of oil commonly should provide some properties. High solubility, cost-effectiveness, and an existing wide range of resources for providing these materials are considered as the most important properties of surfactants. Additionally, they must be consistent with the fluids of formation. After injecting, surfactants should also exhibit high stability in the formation even in low concentration conditions. They should also provide sufficient IFT reduction and maintain their stability in various salt concentrations and temperatures, as well as assisting the small adsorption retention at the surface of reservoir rocks. Recently, nonionic-anionic surfactants, anionic surfactants, and nonionic surfactants are assumed as the most important surfactants employed in oil recovery applications. Lignosulfonate, petroleum sulfonate, as well as carboxylate types, apply as anionic surfactants in the oil recovery territory. Tween-type surfactants are also categorized in the nonionic surfactant groups used in oil recovery applications.

Furthermore, separation of phases, surfactant precipitation, and adsorption of emulsification result from the formation damage occurred by polymers, as well as flooding the combination of surfactant and polymer, in the oil recovery area.

11.3.3 Alkaline-polymer (AP) flooding

The absence of mobility control of alkaline is the one important obstacle related to alkaline flooding. This phenomenon is attributed to the great ratio of mobility related to the phase which is displacing, as well as the phase which is displaced. For solving this problem, adding appropriate polymers is proposed. One disadvantage of this solution is that by increasing the alkaline dosage, the hydrolysis reaction of polymers occurs, leading to the reduction of polymer viscosity. Therefore, the sodium ions that existed in the solution neutralize the groups of carboxyl related to polymer structures, and it causes the coil up of polymers [48]. While using alkaline, increasing the mineral surface negative charge decreases the polymer adsorption [49].

For polymer and alkaline combination flooding, the hydrolysis of polymer and viscosity reduction of the polymer solution is the main interaction that happened between alkaline and polymer. It also causes the in situ surfactant generation containing a salinity lower than the salinity of alkaline solution injected during the alkaline-polymer flooding process. To solve this problem, adding a specific synthetic surfactant is proposed to augment the generated soap-surfactant (in situ surfactant) salinity, which increases the impact of alkaline-polymer flooding in real reservoirs.

11.3.4 Alkaline-surfactant-polymer (ASP) flooding

ASP flooding provides some privileges, including considerable IFT reduction, reduction in the adsorption of chemical agents, and appreciable mobility control. However, ASP flooding may cause intense formation damage resulting in drawbacks in real reservoirs. In the presence of formation, the equilibrium state related to the reservoir rocks and formation fluids is disrupted while injecting the mixture of ASP fluids. However, ASP flooding constructs an intricate system that contains formation rock, displacing fluid (formation/synthetic brine), and crude oil consisting of soluble gas. ASP complex incorporates formation fluids for physical and chemical reactions. This system also provides reactions between ASP and minerals present in the reservoir rocks. Some results such as minerals dissolution, minerals oxidation, oxidation of different hydrocarbons, the ion exchange reactions between formation waters and minerals of reservoir rocks, as well as pH and temperature variations related to formation can be attributed to applying ASP systems. Scale formation leads to chemical precipitation because the ions reach an oversaturation state under determined pressure, temperature, and PH conditions [50]. The scale formation and corrosion results from different properties of chemical agents existing in the ASP system. Among various chemical agents, the most important compound is assumed as alkaline, which leads to a scaling factor. The more the potent alkali exists in ASP complexes, the more corrosion phenomenon occurs, and consequently, the more scale formation happens. Alkaline flooding has been employed by the USA since 1980. Close to the injection well related to alkaline, the efficiency of fluid attributed to the wells of production has dramatically declined after a few months. As alkali reacts with minerals of reservoir rocks, as well as reacts with fluids that exist in the reservoir, partially blocking occurred in the formation. In other words, formation blocking was originated through the generation of alkali scale along with silica aluminate scale. Raimondi et al. reported the increment of gypsum scale amount associated with the well production when the alkaline flooding experiments were conducted in the oilfield of North Ward-Estes. The silica scale deposition and calcium carbonate scale sedimentation were investigated by Chenget al. [51]. They consider some factors, including ion strength of materials in the solution, temperature parameter, pH value, and pressure in their study. In this research, they introduced an equation that assisted in predicting the sedimentation of a mixture of silicon and calcium [51]. In order to test this prediction, it was applied in the oilfield of Daqing. After ASP flooding, this equation illustrated a more than 90% accuracy rate.

On the other hand, IFT decreases thanks to the presence of surfactant while injecting the mixture of ASP. Moreover, surfactant assists the reduction velocity attributed to the scale deposition. However, a potent preference for adsorption with $CaCO_3$ is exhibited by surfactants. The inorganic salt sediments adsorb most of the surfactants applied in the process of production and injection. Therefore, the environmental impacts merely deteriorate a little surfactant amount utilized in the ASP flooding process. In addition, the inorganic scale is adsorbed by polymer and their product after the destruction of polymers, resulting in the transformation of the inorganic scale to organic scale. The calcium carboxylate sedimentation can also occur through the reaction between the carboxylic groups produced from the PAM hydrolysis reaction and calcium. Consequently, the presence of polymer in the ASP flooding process assists the occurrence of scaling [52].

Due to complicated environmental conditions related to different oilfields, various parameters can influence the scaling formation while ASP injections. Some of these factors are the temperature and pressure of the underground environment, the composition of the minerals related to formation, ionic concentration, reservoir temperature and pressure, pH value, and colloidal properties [53].

11.4 Techniques to mitigate FD during EOR processes

Various antiscaling techniques were designed considering the composition of scales, their formation processes, and controlling parameters, such as chemical/physical procedures to decrease the possibility of scale formation. A list of the applicable strategies to prevent scaling problems in the vicinity of the well, as well as in the reservoir formation zones was presented by Wu et al. [54]. They proposed the following techniques to prevent scaling during ASP flooding: (i) using scale inhibitors dissolved in the fluid used for fracturing operation, (ii) application of a mixture of proppant and compact scale inhibitors, (iii) employing fracture proppants, which are coated with scale inhibitors, (iv) cyclic squeeze treatments using chemical inhibitors; this method is applicable for the near-wellbore damages, which not only prevent damages but also due to the ease of detaching from the rock surfaces, can be expelled by the flowing fluids [55].

11.5 Identification of risk and opportunity of EOR projects

Due to the dynamic aspects, multifacet, and complexity of the FD objects, the need for comprehensive analysis as well as holistic management to prevent/mitigate FD during EOR projects is of utmost importance. However, in some cases, it is a common issue that some FDs may emerge while implementing a mitigation operation to remedy a specific FD. Thus, the risk of imposing severe damages and worsening the FD mechanisms becomes highlighted. Accordingly, the need for a comprehensive workflow, as well as an integrated risk assessment and opportunity management to represent a detailed EOR plan along with thorough predictions of the possible FDs is clear and urgent [5]. Therefore, several stages with systematic procedures should be taken to achieve an integrated risk management/opportunity assessment plan. To this end, it is crucial to identify, assess, and rank the operations in regard to the FD problems and the corresponding uncertainties. Fig. 11.5 illustrates a detailed procedure to have a better understanding of the causes/actions to prevent FD issues. Utilizing a broad range of facilities, an integrated workflow with a qualitative/quantitative risk assessment to monitor and manage the opportunities/risks associated with the EOR projects is presented.

In order to have a better understanding and a thorough prediction program of the possible damages while implementation of different EOR processes, it is crucial to obtain reliable datasets, including laboratory and field data. Also, performing pilot tests coupled with simulation studies can improve the prediction of possible FDs [56]. A full set of the required data to accurately predict FD issues associated with various EOR plans is as follows: (i) laboratory measurements, such as porosity and permeability of rock samples and the mechanical properties of reservoir rocks, (ii) field-scale data, such as well production/injection rates and surface/subsurface pressures, well test and well logging data (formation permeability, porosity, skin, phase saturations, productivity indices, and mechanical failure) [18], that covers quantitative, semiquantitative, and qualitative methods. To put it in other words, by moving from the technical aspects of the issue to the risk management aspects, the quantitative, semiquantitative, and qualitative assessment approaches become important as they refer to the numerical relationships for improving resource recovery, prioritization of specific risks and opportunities, and recognizing the scope of risks and opportunities, and their cause(s) and consequence(s), respectively. Wood and Yuan focused on the association between types of FDs and their corresponding influence diagrams to better identify the causes and impacts of damage [57]. They introduced the bow tie and butterfly influence diagrams to show the complexity of the relationships. So, they classified the FD projects into two groups: (a) simple projects that merely focus on the causes and impacts of the damages associated with the applied EOR method, such as fines migration, and (b) complex projects, such as carbon capture and storage (CCS), which require not only underground risk assessment, but also need information about surface risks.

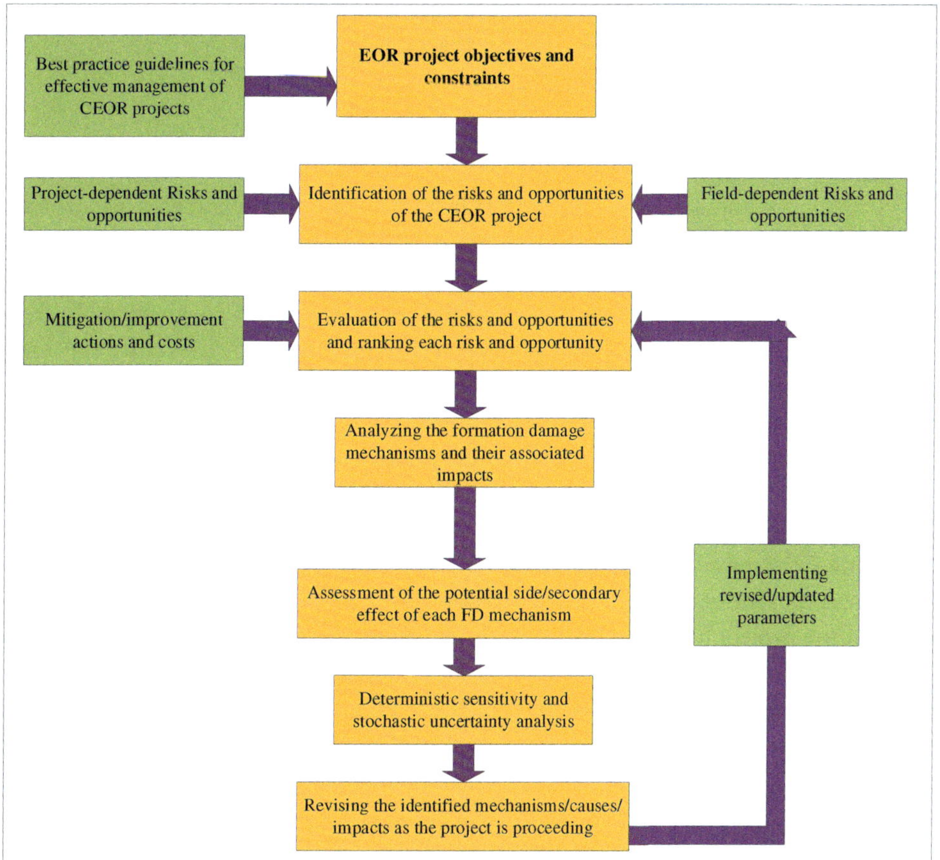

Fig. 11.5 A comprehensive workflow for risk/opportunity evaluation, classification, and management of FD subjects influencing EOR designs. *(Modified after Yuan B, Wood DA. A comprehensive review of formation damage during enhanced oil recovery. J Petrol Sci Eng 2018;167:287–99.)*

Each FD issue can be classified by employing qualitative methods and recognizing the relationships between the possibility and impacts of FD problems. By mastering these relationships and knowing the technical points related to each EOR method, a petroleum engineer can precisely prioritize the problems associated with each FD issue, and the required corresponding action (exploitation or mitigation). Using such analysis, one can evaluate how to improve the level of an identified risk from "critical to mitigation risks" to a higher level, for instance, "manageable risks." The diagram showed in Fig. 11.5 proposes an identification process for risk/opportunity analysis and management concerning the EOR plans.

Finally, it is noteworthy that the evaluation and supervision of the risk/opportunity of the EOR projects will be doubly important when not only considering the technical and regulatory subjects of the project but also having a look at the market uncertainties and financial problems throughout the project life cycle.

11.6 Summary and conclusions

This chapter aimed at elaborating on the FD in relation to CEOR practices, to formulate a framework for optimization of EOR plans by controlling or even leveraging the FD. So far, some scholars have pointed out the importance of FD studies during CEOR processes, but there are still rooms for comprehensive and integrated studies to cover not only the FD mechanisms associated with various CEOR techniques but also investigate the mitigation/prevention methods along with the risk/opportunity analysis and management. In the at-hand chapter, the potential FDs in different CEOR methods were discussed, and some recommendations regarding mitigation/prevention approaches were explained to help reservoir engineers in addressing the complicated problems linked to FD. Furthermore, a procedure/workflow to identify and control risks and opportunities concerning FDs during EOR projects was presented. This workflow blends the quantitative, semiqualitative, and qualitative techniques considering all the technical, regulatory, and financial aspects of the project to deliver an accurate and rigorous estimation of the potential risks. So, the importance of FD analysis was mentioned as a critical parameter that should be considered to ensure the success of a CEOR plan. With this in view, conducting appropriate experiments along with simulation studies to evaluate possible damages during the pointed EOR plan will improve the chance of success in an EOR project.

References

[1] Walsh M, Lake LW. A generalized approach to primary hydrocarbon recovery. U.S. Department of Energy, Office of Scientific and Technical Information (OSTI); 2003. https://www.osti.gov/etdeweb/biblio/20371600.

[2] Dai C. Formation damage during chemical flooding. In: Form. Damage Dur. Improv. Oil Recover. Elsevier; 2018. p. 275–304.

[3] Civan F. Reservoir formation damage. Gulf Professional Publishing; 2015.

[4] Yuan B, Moghanloo RG, Zheng D. Analytical evaluation of nanoparticle application to mitigate fines migration in porous media. SPE J 2016;21:2317–32.

[5] Frequin D, Bedrikovetsky P, Zitha PLJ. CT scan study of the leak-off of oil-based drilling fluids into saturated media. Society of Petroleum Engineers: SPE Eur. Form. Damage Conf. Exhib; 2013.

[6] Moghadasi R, Rostami A, Hemmati-Sarapardeh A, Motie M. Application of Nanosilica for inhibition of fines migration during low salinity water injection: experimental study, mechanistic understanding, and model development. Fuel 2019;242:846–62.

[7] Bagherpour S, Rashidi A, Mousavi SH, Izadi N, Hamidpour E. Experimental investigation of carboxylate-alumoxane nanoparticles for the enhanced oil recovery performance. Colloid Surf A Physicochem Eng Asp 2019;563:37–49.

[8] Bedrikovetsky PG, Siqueira FD, CJA F, Serra de Souza AL. Quantitative theory for fines migration and formation damage. In: SPE Int. Symp. Exhib. Form. Damage Control, Society of Petroleum Engineers; 2010.

[9] Rezaei A, Abdollahi H, Derikvand Z, Hemmati-Sarapardeh A, Mosavi A, Nabipour N. Insights into the effects of pore size distribution on the flowing behavior of carbonate rocks: linking a Nano-based enhanced oil recovery method to rock typing. Nanomaterials 2020;10:972.
[10] Porter KE. An overview of formation damage (includes associated paper 20014). J Petrol Tech 1989;41:780–6.
[11] Harper TR, Buller DC. Formation damage and remedial stimulation. Clay Miner 1986;21:735–51.
[12] Faergestad I. Formation damage. Schlumberger Oilf Rev 2016.
[13] Bennion DB. An overview of formation damage mechanisms causing a reduction in the productivity and injectivity of oil and gas producing formations. J Can Pet Technol 2002;41.
[14] Derikvand Z, Rezaei A, Parsaei R, Riazi M, Torabi F. A mechanistic experimental study on the combined effect of Mg^{2+}, Ca^{2+}, and SO42-ions and a cationic surfactant in improving the surface properties of oil/water/rock system. Colloid Surf A Physicochem Eng Asp 2019;124327.
[15] Razavifar M, Mukhametdinova A, Nikooee E, Burukhin A, Rezaei A, Cheremisin A, et al. Rock porous structure characterization: a critical assessment of various state-of-the-art techniques. Transp Porous Media 2021;136:431–56.
[16] Romanova UG, Ma T, Piwowar M, Strom R, Stepic J. Thermal formation damage and relative permeability of oil sands of the lower cretaceous formations in Western Canada. Society of Petroleum Engineers: SPE Canada Heavy Oil Tech. Conf; 2015.
[17] Ezeuko CC, Sen A, Gates ID. Modelling biofilm-induced formation damage and biocide treatment in subsurface geosystems. J Microbial Biotechnol 2013;6:53–66.
[18] Yuan B, Wood DA. A comprehensive review of formation damage during enhanced oil recovery. J Petrol Sci Eng 2018;167:287–99.
[19] Liu Q. Interfacial phenomena in enhanced heavy oil recovery by alkaline flood. ProQuest; 2007.
[20] Sheng JJ. Formation damage in chemical enhanced oil recovery processes. Asia-Pacific J Chem Eng 2016;11:826–35.
[21] Moghadasi R, Rostami A, Tatar A, Hemmati-Sarapardeh A. An experimental study of Nanosilica application in reducing calcium sulfate scale at high temperatures during high and low salinity water injection. J Petrol Sci Eng 2019;179:7–18.
[22] Assef Y, Arab D, Pourafshary P. Application of nanofluid to control fines migration to improve the performance of low salinity water flooding and alkaline flooding. J Petrol Sci Eng 2014;124:331–40.
[23] Ge J, Feng A, Zhang G, Jiang P, Pei H, Li R, et al. Study of the factors influencing alkaline flooding in heavy-oil reservoirs. Energy Fuel 2012;26:2875–82.
[24] Cheng KH. In: Chemical consumption during alkaline flooding: a comparative evaluation. SPE Enhanc. oil Recover. Symp., Society of Petroleum Engineers; 1986.
[25] Flaaten A, Nguyen QP, Pope GA, Zhang J. In: A systematic laboratory approach to low-cost, high-performance chemical flooding. SPE Symp. Improv. Oil Recover., Society of Petroleum Engineers; 2008.
[26] Berger PD, Lee CH. In: Improved ASP process using organic alkali. SPE/DOE Symp. Improv. oil Recover., Society of Petroleum Engineers; 2006.
[27] Panga MKR, Ooi YS, Koh PL, Chan KS, Enkababian PG, Cheneviere P, et al. Wettability alteration for water block prevention in high temperature gas wells. In: SPE Eur. Annu. Conf. Exhib., Society of Petroleum Engineers; 2006.
[28] Rezaei A, Riazi M, Escrochi M, Elhaei R. Integrating surfactant, alkali and nano-fluid flooding for enhanced oil recovery: a mechanistic experimental study of novel chemical combinations. J Mol Liq 2020;113106.
[29] Rezaei A, Khodabakhshi A, Esmaeili A, Razavifar M. Effects of initial wettability and different surfactant-silica nanoparticles flooding scenarios on oil-recovery from carbonate rocks. In: Petroleum; 2021.
[30] Rezaei A, Derikvand Z, Parsaei R, Imanivarnosfaderani M. Surfactant-silica nanoparticle stabilized N2-foam flooding: a mechanistic study on the effect of surfactant type and temperature. J Mol Liq 2021;325:115091.
[31] Xu F, Guo X, Wang W, Zhang N, Jia S, Wang X. Case study: numerical simulation of surfactant flooding in low permeability oil filed. In: SPE Enhanc. Oil Recover. Conf., Society of Petroleum Engineers; 2011.

[32] Stellner KL, Scamehorn JF. Surfactant precipitation in aqueous solutions containing mixtures of anionic and nonionic surfactants. J Am Oil Chem Soc 1986;63:566–74.
[33] Nelson RC, Pope GA. Phase relationships in chemical flooding. Soc Pet Eng J 1978;18:325–38.
[34] Martin MM, Rockholm DC, Martin JS. Effects of surfactants, pH, and certain cations on precipitation of proteins by tannins. J Chem Ecol 1985;11:485–94.
[35] Conti J, Holtberg P, Diefenderfer J, LaRose A, Turnure JT, Westfall L. International energy outlook 2016 with projections to 2040. Washington, DC United States: USDOE Energy Information Administration (EIA); 2016.
[36] Van Velzen JFG, Leerlooijer K. Impairment of a water injection well by suspended solids: testing and prediction. In: Int. Symp. Form. damage Control; 1992. p. 477–90.
[37] Vela S, Peaceman DW, Sandvik EI. Evaluation of polymer flooding in a layered reservoir with crossflow, retention, and degradation. Soc Pet Eng J 1976;16:82–96.
[38] Elhaei R, Kharrat R, Madani M. Stability, flocculation, and rheological behavior of silica suspension-augmented polyacrylamide and the possibility to improve polymer flooding functionality. J Mol Liq 2021;322:114572.
[39] Garti N, Zour H. The effect of surfactants on the crystallization and polymorphic transformation of glutamic acid. J Cryst Growth 1997;172:486–98.
[40] Bhardwaj AK, Shainberg I, Goldstein D, Warrington DN, Levy G. Water retention and hydraulic conductivity of cross-linked polyacrylamides in sandy soils. Soil Sci Soc Am J 2007;71:406–12.
[41] Tang H, Meng Y, Yang X. A study of adsorption consumption of polyacrylamide on reservoir minerals. Oilf Chem 2001;18:343–6.
[42] Lanlei G. Study of Polymer's adsorption and retention law and performance change. Chem Eng Oil Gas/Shi You Yu Tian Ran Qi Hua Gong 2011;40.
[43] Fu M, Xiong F. Study on the plugging mechanism of polymer flooding in Henan oilfield. Drill Prod Technol 2009;32(4):77–9.
[44] Yan H, Blanford CF, Holland BT, Smyrl WH, Stein A. General synthesis of periodic macroporous solids by templated salt precipitation and chemical conversion. Chem Mater 2000;12:1134–41.
[45] Fletcher AJP, Lamb SP, Clifford PJ. Formation damage from polymer solutions: factors governing injectivity. SPE Reserv Eng 1992;7:237–46.
[46] Rudin J, Bernard C, Wasan DT. Effect of added surfactant on interfacial tension and spontaneous emulsification in alkali/acidic oil systems. Ind Eng Chem Res 1994;33:1150–8.
[47] Cui Z, Song H, Yu J, Jiang J, Wang F. Synthesis of N-(3-Oxapropanoxyl) dodecanamide and its application in surfactant-polymer flooding. J Surfactant Deterg 2011;14:317–24.
[48] Green DW, Willhite GP. Enhanced oil recovery. vol. 6. Richardson, TX: Henry L. Doherty Memorial Fund of AIME, Society of Petroleum Engineers; 1998.
[49] Krumrine PH, Falcone Jr JS. In: Surfactant, polymer, and alkali interactions in chemical flooding processes. SPE Oilf. Geotherm. Chem. Symp., Society of Petroleum Engineers; 1983.
[50] Karazincir O, Thach S, Wei W, Prukop G, Malik T, Dwarakanath V. In: Scale formation prevention during ASP flooding. SPE Int. Symp. Oilf. Chem., Society of Petroleum Engineers; 2011.
[51] Cheng J, Wang Q, Wang J, Li C, Shi W. Ca/Si scale sedimentary model in the strong-base ASP flooding block and scaling prediction. Acta Petrol Sin 2016;37(5):653–9.
[52] Wang Y, Cheng JC. The scaling characteristics and adaptability of mechanical recovery during ASP flooding. Journal-Daqing Pet Inst 2003;27:20–2.
[53] Li J, Li T, Yan J, Zuo X, Zheng Y, Yang F. In: Silicon containing scale forming characteristics and how scaling impacts sucker rod pump in ASP flooding. Asia Pacific Oil Gas Conf. Exhib., Society of Petroleum Engineers; 2009.
[54] Wu X. Formation damage by inorganic deposition. In: Form. Damage Dur. Improv. Oil Recover. Elsevier; 2018. p. 217–42.
[55] Graham AL, Boak LS, Neville A, Sorbie KS. In: How minimum inhibitor concentration (MIC) and sub-MIC concentrations affect bulk precipitation and surface scaling rates. SPE Int. Symp. Oilf. Chem., Society of Petroleum Engineers; 2005.
[56] Haimes YY. Risk modeling, assessment, and management. vol. 40. John Wiley & Sons; 2005.
[57] Wood DA, Yuan B. Integrated risks assessment and management of IOR/EOR projects: a formation damage view. In: Form. Damage Dur. Improv. Oil Recover. Elsevier; 2018. p. 587–631.

Index

Note: Page numbers followed by *f* indicate figures, and *t* indicate tables.

B

C

D

T

U

9780128219317